Mein Kampf

Mein Kampf

Adolf Hitler

Translated by
Ralph Manheim

A Mariner Book
HOUGHTON MIFFLIN COMPANY
BOSTON · NEW YORK

First Mariner Books edition 1999

For information about permission to reproduce selections
from this book, write to Permissions,
Houghton Mifflin Company, 215 Park Avenue South,
New York, New York 10003.

ISBN 0-395-95105-4
ISBN 0-395-92503-7 (pbk.)

Printed in the United States of America

QUM 10 9 8 7 6 5 4 3 2 1

DEDICATION

On November 9, 1923, at 12.30 in the afternoon, in front of the Feldherrnhalle as well as in the courtyard of the former War Ministry, the following men fell, with loyal faith in the resurrection of their people:

ALFARTH, FELIX, *businessman, b. July 5, 1901*
BAURIEDL, ANDREAS, *hatter, b. May 4, 1879*
CASELLA, THEODOR, *bank clerk, b. August 8, 1900*
EHRLICH, WILHELM, *bank clerk, b. August 19, 1894*
FAUST, MARTIN, *lank clerk, b. January 27, 1901*
HECHENBERGER, ANTON, *locksmith, b. September 28, 1902*
KÖRNER, OSKAR, *businessman, b. January 4, 1875*
KUHN, KARL, *headwaiter, b. July 26, 1897*
LAFORCE, KARL, *student of engineering, b. October 28, 1904*
NEUBAUER, KURT, *valet, b. March 27, 1899*
PAPE, CLAUS VON, *businessman, b. August 16, 1904*
PFORDTEN, THEODOR VON DER, *County Court Councillor, b. May 14, 1873*
RICKMERS, JOHANN, *retired Cavalry Captain, b. May 7, 1881*
SCHEUBNER–RICHTER, MAX ERWIN VON, *Doctor of Engineering, b. January 9, 1884*
STRANSKY, LORENZ, RITTER VON, *engineer, b. March 14, 1889*
WOLF, WILHELM, *businessman, b. October 19, 1898*

So-called national authorities denied these dead heroes a common grave.

Therefore I dedicate to them, for common memory, the first volume of this work. As its blood witnesses, may they shine forever, a glowing example to the followers of our movement.

Adolf Hitler

LANDSBERG AM LECH
FORTRESS PRISON
October 16, 1924

PREFACE

On April 1, 1924, I entered upon my prison term in the fortress of Landsberg am Lech, as sentenced by the People's Court in Munich on that day.

Thus, after years of uninterrupted work, an opportunity was for the first time offered me to embark upon a task which many had demanded and which I myself felt to be worth while for the movement. I decided to set forth, in two volumes, the aims of our movement, and also to draw a picture of its development. From this it will be possible to learn more than from any purely doctrinaire treatise.

At the same time I have had occasion to give an account of my own development, in so far as this is necessary for the understanding of the first as well as the second volume, and in so far as it may serve to destroy the foul legends about my person dished up in the Jewish press.

I do not address this work to strangers, but to those adherents of the movement who belong to it with their hearts, and whose intelligence is eager for a more penetrating enlightenment. I know that men are won over less by the written than by the spoken word, that every great movement on this earth owes its growth to great orators and not to great writers.

Nevertheless, for a doctrine to be disseminated uniformly and coherently, its basic elements must be set down for all time. To this end I wish to contribute these two volumes as foundation stones in our common edifice.

The Author

LANDSBERG AM LECH
FORTRESS PRISON

CONTENTS

Translator's Note

MEIN KAMPF is written in the style of a self-educated modern South German with a gift for oratory. Of course this picture does not begin to characterize Hitler the man, but it does, I think, account for the elements of his style.

Beginning in his Vienna period, Hitler was a voracious newspaper reader. The style of the Austrian press, as Karl Kraus never wearied of pointing out, was slovenly, illogical, pretentious. Even the grammar, doubtless because of the large number of Czechs, Hungarians, and other foreigners in the trade, was uncommonly bad. Hitler inveighed against the Viennese melting pot, but was unconsciously influenced by its literary style.

He must also have read popular pamphlets on history, psychology, racist biology, and political subjects. He never attempted to systematize his knowledge; he retained, for the most part, disjointed facts that met some personal need, and phrases that appealed to his oratorical sense. But the main source of his pet phrases was the theater and the opera. He is full of popular quotations from Goethe and Schiller, and largely unintelligible flights of Wagnerian terminology. There is no indication that he ever read any of the German, let alone foreign classics, from which he might have gathered some feeling for stylistic principles.

Hitler has been called a paranoiac; at all events, his view of the world is highly personal. Even where he is discussing theoretical matters like 'the state,' 'race,' etc., he seldom pursues any logic inherent in the subject matter. He makes the most extraordinary allegations without so much as an attempt to prove them. Often there is no visible connection between one paragraph and the next. The logic is purely psychological: Hitler is fighting his persecutors, magnifying his person, creating a dream-world in which he can be an important figure. In more concrete passages he is combating political adversaries in his own movement, but even here the continuity is mystifying, because he never tells us whom he is arguing against, but sets up every political expedient as a universal principle.

This personalism makes Hitler a poor observer. His style is without color and movement. Images are rare, and when they do appear, they tend to be purely verbal and impossible to visualize, like the 'cornerstone for the end of German domination in the monarchy,' or forcing 'the less strong and less healthy back into the womb of the eternal unknown.' The mixed metaphor is almost a specialty of modern German journalism, but Hitler, with his eyes closed to the visual world, was an expert in his own right. Pöhner, for example, was 'a thorn in the eyes of venal officials.'

A non-German of Hitler's intellectual level would in some ways write quite differently. Germany was a land of high general culture, with the largest reading public of any country in the world. In the lower middle class, there was a tremendous educational urge. People who in other countries would read light novels and popular magazines devoured works on art, science, history, and above all philosophy. Certain philosophical phrases became journalistic clichés. Hitler is forever speaking of 'concepts,' of things 'as such' (*an sich*). Moreover, he is constantly at pains to show that he, too, is cultured. Hence the long, intricate sentences in which he frequently gets lost; hence such sententious bombast as the opening lines of Chapter Ten.

The absence of movement and development in *Mein Kampf* is surely connected with Hitler's lack of concern for the objective world. But its stylistic expression, the preponderance of substantives over verbs, again shows the influence of German journalism. Many German writers, including some academicians, seem to feel that the substantive is the strongest and most impressive part of speech. This tendency is found even in German police reports. Instead of saying that a man was arrested, they will say that his arrest took place. This predilection for substantives is a salient feature of Hitler's style.

Here and there, amid his ponderous reflections, Hitler is suddenly shaken with rage. He casts off his intellectual baggage and writes a speech, eloquent and vulgar.

*   *   *

Most of Hitler's stylistic peculiarities represent no problem for the translator. The mixed metaphors are just as mixed in one language as in the other. A lapse of grammatical logic can occur in any language. An English-language Hitler might be just as redundant as the German one; a half-educated writer, without clear ideas, generally feels that to say a thing only once is rather slight.

There are, however, certain traits of Hitler's style that are peculiarly German and do present a problem in translation. Chief among these are the length of the sentences, the substantives, and the German particles.

A translation must not necessarily be good English, but it must be English such as some sort of English author — in this case, let us say, a poor one — might write. On the other hand, it would be wrong to make Hitler an English-speaking rabble-rouser, because his very style is necessarily German.

No non-German would write such labyrinthine sentences. The translator's task — often a feat of tightrope-walking — is to render the ponderousness and even convey a German flavor, without writing German-American. In general I have cut down the sentences only when the length made them unintelligible in English. (The German language with its cases and genders does enable the reader to find his way through tangles which in a non-inflected language would be inextricable.) Contrary to the general opinion, the German text contains only one or two sentences that make no sense at first reading.

The substantives are a different matter. Here it has been

necessary to make greater changes, because in many cases the use of verbal nouns is simply incompatible with the English language. No pedant, no demagogue, no police clerk writes that way. I have used the construction where it seemed conceivable in English, elsewhere reluctantly abandoned it. German stylists may say that Hitler's piling up of substantives is bad German, but the fact remains that numerous German writers do the same thing, while this failing is almost non-existent in English.

In approaching Hitler's use of particles, it must be remembered that he was at home in the Lower Bavarian dialect. Even without the dialect, much German prose, some not of the worst quality, abounds in these useless little words: *wohl, ja, denn, schon, noch, eigentlich,* etc. The South Germans are especially addicted to them, and half of Hitler's sentences are positively clogged with particles, not to mention such private favorites as *besonders* and *damals* which he strews about quite needlessly. His particles even have a certain political significance, for in the petit bourgeois mind they are, like carved furniture, an embodiment of the home-grown German virtues, while their avoidance is viewed with suspicion as foreign and modernistic. Unfortunately, they must largely be sacrificed in translation. There are no English equivalents, and an attempt to translate them results in something like the language of the Katzenjammer Kids. Sometimes, however, it is possible to give a similar impression of wordiness by other means.

*   *   *

The translation follows the first edition. The more interesting changes made in later German editions have been indicated in the notes. Where Hitler's formulations challenge the reader's credulity, I have quoted the German original in the notes. Seeing is believing.

INTRODUCTION

FOR YEARS *Mein Kampf* stood as proof of the blindness and complacency of the world. For in its pages Hitler announced — long before he came to power — a program of blood and terror in a self-revelation of such overwhelming frankness that few among its readers had the courage to believe it. Once again it was demonstrated that there was no more effective method of concealment than the broadest publicity.

Mein Kampf was written in white-hot hatred. But this hatred was aimed, not at the author's enemies, but at his supposed friends. The book is an oratorical denunciation of collaborators who largely shared the political aim of the author, but refused to accept his methods.

As the discussion goes on, the tone of the debate grows louder and more impassioned. For to Hitler the only principle which really counts is his claim to supreme leadership, and when this claim is disputed by people whom, in his loveless soul, he secretly despises, he is driven half mad with hatred.

It was in 1922 that Hitler first planned a small pamphlet the title of which was to be *Eine Abrechnung* ('Settling Accounts'). He wanted to prove that the leaders of the German right wing parties were wrong in their methods, lacked strength, imagination, and political judgment, and were bound to fail. He discussed the booklet at length with his close friends, Rudolf Hess and Dietrich Eckart. Eckart himself had compiled a little book from his talks with Hitler. Its title was *Bolshevism from Moses to Lenin*, and it attempted to prove that Judaism was the great destructive force which had ruined Western civilization. The book was brilliantly written, and, although it claimed to express Hitler's ideas, these were expounded in a way which was more Eckart's than Hitler's.

Hitler must have realized the growing necessity of stating his own principles in his own language, in a more definite, more authoritative form. He probably had an uneasy feeling that some of his followers, although admiring him as an orator, did not believe him to be a clear or profound thinker. People whom he himself regarded as his inferiors endeavored to outline what they called the fundamental principles of his own party, and quarreled about these principles over the head of their leader.

Gottfried Feder, who had collaborated in drafting the party's program, and who considered himself the master-mind of the whole movement, wrote a booklet about the aims of Naziism. In it he insisted strongly on his own particular ideas, and laid stress on economic principles that were distinctly socialistic, such as abolition of capital interest. At the same time, Alfred Rosenberg, Hitler's mentor, published an equally 'authentic' comment on the party aims differing sharply from Feder's. Rosenberg insisted on certain racial and sociological aspects of the issue, such as the inequality of men, and the superiority of the Nordic or Aryan type.

True, Feder and Rosenberg had one thing in common: both attacked the Jews. But their points of view were entirely different. To Feder, the Jews were profiteers in a vicious economic system. To Rosenberg, they were the creators of the modern world, the standard-bearers of the twentieth century and of all that he and his fellows hated: freedom of speech and press, egalitarian justice, representative government, free access of every citizen to all professions and public services without discrimination of birth, wealth, or religion. The decisive factor in human society was, to Rosenberg, the outstanding personality, the great lone wolf, the creative superman. This idea was not new, but it well fitted Hitler's purpose, for, at that time, democratic and dictatorial tendencies still struggled side by side in the Nazi movement, and men like Feder claimed to be Hitler's equals.

To settle these disturbing quarrels between his followers, to strengthen his own authority as leader and idea-man of his movement, Hitler decided to broaden the scope of his projected book, and to add the story of his life, brushing up to heroic size the picture of his inglorious youth, and picturing himself as one whose boyhood had been marred by poverty, but who had risen by his own strength and intelligence to be the leader of a great political party, thus making himself a symbol of the humiliated but unconquerable German people and a victim of the same social forces which he was then attacking.

But the main aim of his book was to show that he was the only founder and builder of the National Socialist movement. He never mentions the name of Captain Ernst Roehm, who contributed at least as much as himself to the rise of the party. He says little about Anton Drexler, the real founder of the movement. He makes no mention of Hermann Esser and gives no idea of the indispensable rôle played by Julius Streicher. His only reference to Gregor Strasser, his most powerful rival for the party leadership, is an obscure joke about his predilection for card-playing. Strasser is never mentioned by name.

After the unsuccessful *putsch* of November, 1923, Hitler spent about thirteen months in a so-called prison. There he wrote most of what later became the first volume of his famous book. He intended to call it, 'A Four and One-Half Year Struggle against Lies, Stupidity, and Cowardice: Settling Accounts with the Destroyers of the National Socialist Movement.' His idea was to tell the inside story of his *putsch* and its collapse. But then he dropped this plan and the title lost its meaning. It seems that one of his associates, Max Amann, had the idea of simply calling the whole *Mein Kampf* ('My Struggle'). The subtitle of the first volume remained *Eine Abrechnung* ('Settling Accounts'), although it no longer made sense.

The first volume was dictated by the author to two men who were in prison with him. One was Emil Maurice, a strong-arm man not too well fitted for the job. Then came Rudolf Hess, who had escaped to Austria, returned, was arrested, and went to the same prison as his master. Together they finished the first volume. Hess has often been called a disciple of Karl Haushofer, the military geographer. No doubt Haushofer influenced some parts of *Mein Kampf*, but it is misleading to represent him as Hitler's guiding genius. The essential parts of the book do not concern questions of foreign policy or military geography, but of race, propaganda, and political education.

The first volume was published in the fall of 1925. It disappointed many readers, who had expected a revelation or at least a dramatic story, not a philosophy; ill-founded, undocumented, and badly written. 'Many' readers, incidentally, should not be taken too literally. It is true that Amann claimed at that time to have sold twenty-three thousand copies during the first year, but this is a very doubtful boast. The average party member did not read the book, and among the leaders it was a common saying that Hitler was an extraordinary speaker, a great leader, a political genius, but 'it's too bad he had to write that silly book.'

The second edition of the first volume, and the second volume from the beginning, were supervised by one Josef Cerny, at that time staff member of the *Völkischer Beobachter*, the party paper. The text became somewhat more civilized. The difference between the first edition and the later ones is not as great as is often believed. The most important of these differences concerns the structure of the Nazi Party. When Hitler published the first edition of the first volume, the Nazi Party was still a democratic organization which elected officials. Two years later the 'leader-principle,' or dictatorship within the party, was firmly established. Hitler had reached the goal for which the whole book was written. The 'unalterable' program of the party, the so-called twenty-five points, called for a 'central parliament' as the highest authority of the future Nazi state. The first edition of *Mein Kampf* had to accept this principle. The second edition changed all that. From now on the party aim was definitely set toward dictatorship.

The second volume came out in December, 1926, and was read by even fewer people than the first; certainly by none outside the still small National Socialist Party. For some years to come the book remained a financial burden on the shoulders of Max Amann, Hitler's business manager. Amann had been appointed, in 1925, manager of the Franz Eher Verlag, the party's publishing house virtually owned by Hitler, and to this day he has remained at the head of this source of Hitler's private fortune. He knew that Hitler needed financial resources in order to control the party effectively; he saw it more clearly than Hitler himself, who has no sense for business. Amann hoped that the book would become a source of private revenue. Two or three years before Hitler came to power, the sales of *Mein Kampf* began to rise, and later they rose astronomically. Amann went on printing the book from the same poor type, on the same cheap paper as before. Everybody was forced to buy it. It was presented as a gift to newly-wed couples, but the license fee was doubled. *Mein Kampf* made Hitler rich. It became a best-seller second only to the Bible.

The book may well be called a kind of satanic Bible. To the author — although he was shrewd enough not to state it explicitly himself, but to have it said by his spiritual adviser, Alfred Rosenberg — the belief in human equality is a kind of hypnotic spell exercised by world-conquering Judaism with the help of the Christian Churches. Later the Jews invented the mass-seduction of liberal democracy; in the last stage Marxism was their tool. By preaching the principle of human equality, Judaism has attempted to extirpate the feeling of pride from the soul of the Aryan race, to rob them of their leadership. To give back to these noble races their former consciousness of superiority by inculcating the principle that men are *not* equal is the theoretical purpose of *Mein Kampf*.

Compared to the spirit of the book, the pragmatical parts are of less importance. Nevertheless, they have aroused more interest, for they are easier to understand. Foreign policies, blueprints for world conquests, prescriptions for alliances or wars — most readers flung themselves upon these passages after the author had become a leading actor on the world stage. But a little reflection would show that in the pathless field of politics you cannot follow a schedule. When Hitler set down his plans for a future foreign policy, he was a mere schemer. Adolf Hitler, the author of *Mein Kampf*, was not yet Adolf Hitler, the Führer of Germany; he lacked the experience, the responsibilities, the knowledge. In fact, he was nothing more than an organizer of street fights, an impresario of mass meetings, the leader of a virtually non-existent party, who shrewdly gambled with rival leaders of similar unimportance.

The soundest appraisal of these utterances on foreign policy was made by Adolf Hitler himself, in the course of a libel suit in Munich. To a fat and bored judge, who, staring through gold-rimmed spectacles, meditated whether to fine both parties five hundred or a thousand marks, Adolf Hitler explained: 'In political life there is no such thing as principles of foreign policy. The programmatic principles of my party are its doctrine on the racial problem and its fight against pacifism and international-ism. But foreign policy in itself is merely a means to an end. In questions of foreign policy I shall never admit that I am tied by anything.' Nevertheless, *Mein Kampf* is packed with prin-ciples of foreign policy, all of which have been taken more seri-ously by others than by the author himself.

What gives *Mein Kampf* its terrific import is not the aims but the methods. Whether Hitler proclaims war against Russia or friendship with Britain, a crusade against the United States or a plot with Japan, conquest by land or by sea, revolt against the rich or the poor — all these plans and schemes mean nothing. He has changed them again and again, because 'in questions of foreign policy I never shall admit that I am tied by anything.' But the fact that all of his schemes, even his friendships, mean bloodshed ('an alliance which does not imply the intention of going to war would be meaningless') — that is what gives this foreign policy its sinister significance. Whether he speaks of art, of education, of economics, he always sees blood. He does not like a certain kind of artist or educator, and that will be reason enough to kill them. 'We shall do away with them radi-cally,' this is his typical slogan. The light-heartedness with which he threatens murder at the slightest provocation is perhaps even more frightful than the threats themselves.

That such a man could go so far toward realizing his ambitions, and — above all — could find millions of willing tools and helpers; that is a phenomenon the world will ponder for centuries to come.

KONRAD HEIDEN

MEIN KAMPF

VOLUME ONE

A RECKONING

CHAPTER
I

In the House of My Parents

Today it seems to me providential that Fate should have chosen Braunau on the Inn as my birthplace. For this little town lies on the boundary between two German states which we of the younger generation at least have made it our life work to reunite by every means at our disposal.

German-Austria must return to the great German mother country, and not because of any economic considerations. No, and again no: even if such a union were unimportant from an economic point of view; yes, even if it were harmful, it must nevertheless take place. One blood demands one Reich. Never will the German nation possess the moral right to engage in colonial politics until, at least, it embraces its own sons within a single state. Only when the Reich borders include the very last German, but can no longer guarantee his daily bread, will the moral right to acquire foreign soil arise from the distress of our own people. Their sword will become our plow, and from the tears of war the daily bread of future generations will grow. And so this little city on the border seems to me the symbol of a great mission. And in another respect as well, it looms as an admonition to the present day. More than a hundred years ago, this insignificant place had the distinction of being immortalized in the annals at least of German history, for it was the scene of a tragic catastrophe which gripped the entire German nation. At the time of our fatherland's deepest humiliation, Johannes Palm [1] of Nuremberg, burgher, bookseller, uncompromising nationalist and French hater, died there for the Germany which he loved so passionately even in her misfortune. He had stubbornly refused to denounce his accomplices, who were in fact his superiors. In this he resembled Leo Schlageter.[2] And like him, he was denounced to the French by a representative of his government. An Augsburg police chief won this unenviable fame, thus furnishing an example for our modern German officials in Herr Severing's [3] Reich.

In this little town on the Inn, gilded by the rays of German martyrdom, Bavarian by blood, technically Austrian,[4] lived my parents in the late eighties of the past century; my father a dutiful civil servant, my mother giving all her being to the household, and devoted above all to us children in eternal, loving care. Little remains in my memory of this period, for after a few years my father had to leave the little border city he had learned to love, moving down the Inn to take a new position in Passau, that is, in Germany proper.

In those days constant moving was the lot of an Austrian customs official. A short time later, my father was sent to Linz, and there he was finally pensioned. Yet, indeed, this was not to mean 'rest' for the old gentleman. In his younger days, as the son of a poor cottager, he couldn't bear to stay at home. Before

---

[1] Johann Philipp Palm was executed in 1806 by the French garrison of Nuremberg for publishing a pamphlet attacking Napoleon.

[2] A free corps leader who performed acts of sabotage against the French occupation authorities in the Ruhr. In the summer of 1923 he was captured by the French authorities, court-martialed and shot.

[3] With brief interruptions the Social Democrat Carl Severing was Prussian Minister of the Interior from 1920 until 1932, when Chancellor Von Papen dissolved the Prussian government. As Minister of the Interior he was in charge of the Prussian police. This, coupled with the fact that he successfully combated the influence of the Rightist secret leagues in the Reichswehr, earned him the special hatred of the Nazis.

[4] Braunau on the Inn, in Upper Austria, directly across from the German (Bavarian) border.

he was even thirteen, the little boy laced his tiny knapsack and ran away from his home in the Waldviertel.[1] Despite the attempts of 'experienced' villagers to dissuade him, he made his way to Vienna, there to learn a trade. This was in the fifties of the past century. A desperate decision, to take to the road with only three gulden for travel money, and plunge into the unknown. By the time the thirteen-year-old grew to be seventeen, he had passed his apprentice's examination, but he was not yet content. On the contrary. The long period of hardship, endless misery, and suffering he had gone through strengthened his determination to give up his trade and become 'something better.' Formerly the poor boy had regarded the priest as the embodiment of all humanly attainable heights; now in the big city, which had so greatly widened his perspective, it was the rank of civil servant. With all the tenacity of a young man whom suffering and care had made 'old' while still half a child, the seventeen-year-old clung to his new decision — he did enter the civil service. And after nearly twenty-three years, I believe, he reached his goal. Thus he seemed to have fulfilled a vow which he had made as a poor boy: that he would not return to his beloved native village until he had made something of himself.

His goal was achieved; but no one in the village could remember the little boy of former days, and to him the village had grown strange.

When finally, at the age of fifty-six, he went into retirement, he could not bear to spend a single day of his leisure in idleness. Near the Upper Austrian market village of Lambach he bought a farm, which he worked himself, and thus, in the circuit of a long and industrious life, returned to the origins of his forefathers.

It was at this time that the first ideals took shape in my breast. All my playing about in the open, the long walk to school, and particularly my association with extremely 'husky' boys, which sometimes caused my mother bitter anguish, made me the very opposite of a stay-at-home. And though at that time I scarcely

---

[1] Waldviertel, the mountainous section at the extreme west of Lower Austria, north of the Danube.

had any serious ideas as to the profession I should one day pursue, my sympathies were in any case not in the direction of my father's career. I believe that even then my oratorical talent was being developed in the form of more or less violent arguments with my schoolmates. I had become a little ringleader; at school I learned easily and at that time very well, but was otherwise rather hard to handle. Since in my free time I received singing lessons in the cloister at Lambach, I had excellent opportunity to intoxicate myself with the solemn splendor of the brilliant church festivals. As was only natural, the abbot seemed to me, as the village priest had once seemed to my father, the highest and most desirable ideal. For a time, at least, this was the case. But since my father, for understandable reasons, proved unable to appreciate the oratorical talents of his pugnacious boy, or to draw from them any favorable conclusions regarding the future of his offspring, he could, it goes without saying, achieve no understanding for such youthful ideas. With concern he observed this conflict of nature.

As it happened, my temporary aspiration for this profession was in any case soon to vanish, making place for hopes more suited to my temperament. Rummaging through my father's library, I had come across various books of a military nature, among them a popular edition of the Franco-German War of 1870-71.[1] It consisted of two issues of an illustrated periodical from those years, which now became my favorite reading matter. It was not long before the great heroic struggle had become my greatest inner experience. From then on I became more and more enthusiastic about everything that was in any way connected with war or, for that matter, with soldiering.

But in another respect as well, this was to assume importance for me. For the first time, though as yet in a confused form, the question was forced upon my consciousness: Was there a difference — and if so what difference — between the Germans who fought these battles and other Germans? Why hadn't Austria taken part in this war; why hadn't my father and all the others fought?

[1] *Eine Volksausgabe des deutsch-französischen Krieges von 1870-71.*

Are we not the same as all other Germans?

Do we not all belong together? This problem began to gnaw at my little brain for the first time. I asked cautious questions and with secret envy received the answer that not every German was fortunate enough to belong to Bismarck's Reich.

This was more than I could understand.

* * *

It was decided that I should go to high school.

From my whole nature, and to an even greater degree from my temperament, my father believed he could draw the inference that the humanistic *Gymnasium* [1] would represent a conflict with my talents. A *Realschule* seemed to him more suitable. In this opinion he was especially strengthened by my obvious aptitude for drawing; a subject which in his opinion was neglected in the Austrian *Gymnasiums*. Another factor may have been his own laborious career which made humanistic study seem impractical in his eyes, and therefore less desirable. It was his basic opinion and intention [2] that, like himself, his son would and must become a civil servant. It was only natural that the hardships of his youth should enhance his subsequent achievement in his eyes, particularly since it resulted exclusively from his own energy and iron diligence. It was the pride of the self-made man which made him want his son to rise to the same position in life, or, of course, even higher if possible, especially since, by his own industrious life, he thought he would be able to facilitate his child's development so greatly.

It was simply inconceivable to him that I might reject what

[1] The *Gymnasium* more or less corresponds to our Latin school, the *Realschule* to our technical high school. The former, with its emphasis on the liberal arts, enjoyed greater social prestige than the *Realschule* before the War of 1914.

[2] The German is '*Willensmeinung*,' an extraordinary word, perhaps Hitler's own invention. Literal translation would be 'opinion of the will.'

had become the content of his whole life. Consequently, my father's decision was simple, definite, and clear; in his own eyes I mean, of course. Finally, a whole lifetime spent in the bitter struggle for existence had given him a domineering nature, and it would have seemed intolerable to him to leave the final decision in such matters to an inexperienced boy, having as yet no sense of responsibility. Moreover, this would have seemed a sinful and reprehensible weakness in the exercise of his proper parental authority and responsibility for the future life of his child, and, as such, absolutely incompatible with his concept of duty.

And yet things were to turn out differently.

Then barely eleven years old, I was forced into opposition for the first time in my life. Hard and determined as my father might be in putting through plans and purposes once conceived, his son was just as persistent and recalcitrant in rejecting an idea which appealed to him not at all, or in any case very little.

I did not want to become a civil servant.

Neither persuasion nor 'serious' arguments made any impression on my resistance. I did not want to be a civil servant, no, and again no. All attempts on my father's part to inspire me with love or pleasure in this profession by stories from his own life accomplished the exact opposite. I yawned and grew sick to my stomach at the thought of sitting in an office, deprived of my liberty; ceasing to be master of my own time and being compelled to force the content of a whole life into blanks that had to be filled out.[1]

And what thoughts could this prospect arouse in a boy who in reality was really anything but 'good' in the usual sense of the word? School work was ridiculously easy, leaving me so much free time that the sun saw more of me than my room. When today my political opponents direct their loving attention to the examination of my life, following it back to those childhood days, and discover at last to their relief what intolerable pranks this 'Hitler' played even in his youth, I thank Heaven that a portion

[1] *'sondern in auszufüllende Formulare den Inhalt eines ganzen Lebens zwängen zu müssen.'*

of the memories of those happy days still remains with me. Woods and meadows were then the battlefields on which the 'conflicts' which exist everywhere in life were decided.

In this respect my attendance at the *Realschule*, which now commenced, made little difference.

But now, to be sure, there was a new conflict to be fought out.

As long as my father's intention of making me a civil servant encountered only my theoretical distaste for the profession, the conflict was bearable. Thus far, I had to some extent been able to keep my private opinions to myself; I did not always have to contradict him immediately. My own firm determination never to become a civil servant sufficed to give me complete inner peace. And this decision in me was immutable. The problem became more difficult when I developed a plan of my own in opposition to my father's. And this occurred at the early age of twelve. How it happened, I myself do not know, but one day it became clear to me that I would become a painter, an artist. There was no doubt as to my talent for drawing; it had been one of my father's reasons for sending me to the *Realschule*, but never in all the world would it have occurred to him to give me professional training in this direction. On the contrary. When for the first time, after once again rejecting my father's favorite notion, I was asked what I myself wanted to be, and I rather abruptly blurted out the decision I had meanwhile made, my father for the moment was struck speechless.

'Painter? Artist?'

He doubted my sanity, or perhaps he thought he had heard wrong or misunderstood me. But when he was clear on the subject, and particularly after he felt the seriousness of my intention, he opposed it with all the determination of his nature. His decision was extremely simple, for any consideration of what abilities I might really have was simply out of the question.

'Artist, no, never as long as I live!' But since his son, among various other qualities, had apparently inherited his father's stubbornness, the same answer came back at him. Except, of course, that it was in the opposite sense.

And thus the situation remained on both sides. My father did not depart from his 'Never!' And I intensified my 'Oh, yes!'

The consequences, indeed, were none too pleasant. The old man grew embittered, and, much as I loved him, so did I. My father forbade me to nourish the slightest hope of ever being allowed to study art. I went one step further and declared that if that was the case I would stop studying altogether. As a result of such 'pronouncements,' of course, I drew the short end; the old man began the relentless enforcement of his authority. In the future, therefore, I was silent, but transformed my threat into reality. I thought that once my father saw how little progress I was making at the *Realschule*, he would let me devote myself to my dream, whether he liked it or not.

I do not know whether this calculation was correct. For the moment only one thing was certain: my obvious lack of success at school. What gave me pleasure I learned, especially everything which, in my opinion, I should later need as a painter. What seemed to me unimportant in this respect or was otherwise unattractive to me, I sabotaged completely. My report cards at this time, depending on the subject and my estimation of it, showed nothing but extremes. Side by side with 'laudable' and 'excellent,' stood 'adequate' or even 'inadequate.' By far my best accomplishments were in geography and even more so in history. These were my favorite subjects, in which I led the class.

If now, after so many years, I examine the results of this period, I regard two outstanding facts as particularly significant:

First: *I became a nationalist.*[1]

Second: *I learned to understand and grasp the meaning of history.* Old Austria was a '*state of nationalities.*'

[1] Hitler's early nationalism had, of course, nothing to do with Austrian patriotism, but was the Pan-Germanism of the *Los-von-Rom* (Away-from-Rome) movement founded by Ritter Georg von Schönerer. It stood for union of Germany with the German parts of Austria and must be distinguished from the Pan-German movement of Germany, which was an out-and-out conspiracy for German world domination. Schönerer's movement was strongly anti-Semitic.

By and large, a subject of the German Reich, at that time at least, was absolutely unable to grasp the significance of this fact for the life of the individual in such a state. After the great victorious campaign of the heroic armies in the Franco-German War, people had gradually lost interest in the Germans living abroad; some could not, while others were unable to appreciate their importance.[1] Especially with regard to the German-Austrians, the degenerate dynasty was only too frequently confused with the people, which at the core was robust and healthy.

What they failed to appreciate was that, unless the German in Austria had really been of the best blood, he would never have had the power to set his stamp on a nation of fifty-two million souls to such a degree that, even in Germany, the erroneous opinion could arise that Austria was a German state. This was an absurdity fraught with the direst consequences, and yet a glowing testimonial to the ten million Germans in the *Ostmark*.[2] Only a handful of Germans in the Reich had the slightest conception of the eternal and merciless struggle for the German language, German schools, and a German way of life. Only today, when the same deplorable misery is forced on many millions of Germans from the Reich, who under foreign rule dream of their common fatherland and strive, amid their longing, at least to preserve their holy right to their mother tongue, do wider circles understand what it means to be forced to fight for one's nationality. Today perhaps some can appreciate the greatness of the Germans in the Reich's old *Ostmark*, who, with no one but themselves to depend on, for centuries protected the Reich against incursions from the East, and finally carried on an exhausting guerrilla warfare to maintain the German language frontier, at a

[1] '*zum Teil dieses auch gar nicht mehr zu würdigen vermocht oder wohl auch nicht gekonnt.*'

[2] Eastern Mark, established by Charlemagne as a buffer against the Avars. Later called Austria. The name was revived by nineteenth-century nationalists who looked back with longing at the unity and strength of the old empire.

time when the Reich was highly interested in colonies, but not
in its own flesh and blood at its very doorstep.

As everywhere and always, in every struggle, there were, in
this fight for the language in old Austria, three strata:

The fighters, the lukewarm, and the traitors.

This sifting process began at school. For the remarkable fact
about the language struggle is that its waves strike hardest perhaps in the school, since it is the seed-bed of the coming generation. It is a struggle for the soul of the child, and to the child its
first appeal is addressed:

'German boy, do not forget you are a German,' and, 'Little
girl, remember that you are to become a German mother.'

Anyone who knows the soul of youth will be able to understand
that it is they who lend ear most joyfully to such a battle-cry.
They carry on this struggle in hundreds of forms, in their own
way and with their own weapons. They refuse to sing un-
German songs. The more anyone tries to alienate them from
German heroic grandeur, the wilder becomes their enthusiasm:
they go hungry to save pennies for the grown-ups' battle fund;
their ears are amazingly sensitive to un-German teachers, and at
the same time they are incredibly resistant; they wear the forbidden insignia of their own nationality and are happy to be
punished or even beaten for it. Thus, on a small scale they are a
faithful reflection of the adults, except that often their convictions are better and more honest.

I, too, while still comparatively young, had an opportunity to
take part in the struggle of nationalities in old Austria. Collections were taken for the *Südmark* [1] and the school association;
we emphasized our convictions by wearing corn-flowers [2] and red,
black, and gold colors; 'Heil' was our greeting, and instead of the

[1] *Südmark.* Another term for Austria. Apparently devised in imitation
of the old imperial Marks by the *Verein für Deutschtum im Ausland*, founded
in 1881 to defend the endangered nationality of Germans in the border
territories.

[2] The corn-flower was the emblem of Germans loyal to the imperial House
of Hohenzollern and of the Austrian Pan-Germans.

imperial anthem we sang '*Deutschland über Alles*,' despite warnings and punishments. In this way the child received political
training in a period when as a rule the subject of a so-called national state knew little more of his nationality than its language.
It goes without saying that even then I was not among the lukewarm. In a short time I had become a fanatical 'German
Nationalist,' though the term was not identical with our present
party concept.

This development in me made rapid progress; by the time I
was fifteen I understood the difference between dynastic '*patriotism*' and folkish [1] '*nationalism*'; and even then I was interested
only in the latter.

For anyone who has never taken the trouble to study the inner
conditions of the Habsburg monarchy, such a process may not
be entirely understandable. In this country the instruction in
world history had to provide the germ for this development,
since to all intents and purposes there is no such thing as a
specifically Austrian history. The destiny of this state is so much
bound up with the life and development of all the Germans that
a separation of history into German and Austrian does not seem
conceivable. Indeed, when at length Germany began to divide
into two spheres of power, this division itself became German
history.

The insignia of former imperial glory, preserved in Vienna, still
seem to cast a magic spell; they stand as a pledge that these twofold destinies are eternally one.

The elemental cry of the German-Austrian people for union
with the German mother country, that arose in the days when the
Habsburg state was collapsing, was the result of a longing that
slumbered in the heart of the entire people — a longing to return
to the never-forgotten ancestral home. But this would be in-

---

[1] *Völkisch.* The word first appeared in nationalist literature about 1875
and is merely a Germanization of 'nationalist.' Where possible the nationalists avoided the use of foreign words. Since 1900 it has been in wide use
among nationalist circles, from the Pan-German League down to the National Socialists.

explicable if the historical education of the individual German-Austrian had not given rise to so general a longing. In it lies a well which never grows dry; which, especially in times of forgetfulness, transcends all momentary prosperity and by constant reminders of the past whispers softly of a new future.

Instruction in world history in the so-called high schools is even today in a very sorry condition. Few teachers understand that the aim of studying history can never be to learn historical dates and events by heart and recite them by rote; that what matters is not whether the child knows exactly when this or that battle was fought, when a general was born, or even when a monarch (usually a very insignificant one) came into the crown of his forefathers. No, by the living God, this is very unimportant.

To 'learn' history means to seek and find the forces which are the causes leading to those effects which we subsequently perceive as historical events.

The art of reading as of learning is this: *to retain the essential, to forget the non-essential.*

Perhaps it affected my whole later life that good fortune sent me a history teacher who was one of the few to observe this principle in teaching and examining. Dr. Leopold Pötsch, my professor at the *Realschule* in Linz, embodied this requirement to an ideal degree. This old gentleman's manner was as kind as it was determined, his dazzling eloquence not only held us spellbound but actually carried us away. Even today I think back with gentle emotion on this gray-haired man who, by the fire of his narratives, sometimes made us forget the present; who, as if by enchantment, carried us into past times and, out of the millennial veils of mist, molded dry historical memories into living reality. On such occasions we sat there, often aflame with enthusiasm, and sometimes even moved to tears.

What made our good fortune all the greater was that this teacher knew how to illuminate the past by examples from the present, and how from the past to draw inferences for the present. As a result he had more understanding than anyone else for all the daily problems which then held us breathless. He used our budding nationalistic fanaticism as a means of educating us, frequently appealing to our sense of national honor. By this alone he was able to discipline us little ruffians more easily than would have been possible by any other means.

This teacher made history my favorite subject.

And indeed, though he had no such intention, it was then that I became a little revolutionary.

For who could have studied German history under such a teacher without becoming an enemy of the state which, through its ruling house, exerted so disastrous an influence on the destinies of the nation?

And who could retain his loyalty to a dynasty which in past and present betrayed the needs of the German people again and again for shameless private advantage?

Did we not know, even as little boys, that this Austrian state had and could have no love for us Germans?

Our historical knowledge of the works of the House of Habsburg was reinforced by our daily experience. In the north and south the poison of foreign nations gnawed at the body of our nationality, and even Vienna was visibly becoming more and more of an un-German city. The Royal House Czechized wherever possible, and it was the hand of the goddess of eternal justice and inexorable retribution which caused Archduke Francis Ferdinand, the most mortal enemy of Austrian-Germanism, to fall by the bullets which he himself had helped to mold. For had he not been the patron of Austria's Slavization from above!

Immense were the burdens which the German people were expected to bear, inconceivable their sacrifices in taxes and blood, and yet anyone who was not totally blind was bound to recognize that all this would be in vain. What pained us most was the fact that this entire system was morally whitewashed by the alliance with Germany, with the result that the slow extermination of Germanism in the old monarchy was in a certain sense sanctioned by Germany itself. The Habsburg hypocrisy, which enabled the Austrian rulers to create the outward appearance that Austria was a German state, raised the hatred toward this house to flaming indignation and at the same time — contempt.

Only in the Reich itself, the men who even then were called to power saw nothing of all this. As though stricken with blindness, they lived by the side of a corpse, and in the symptoms of rottenness saw only the signs of 'new' life.

The unholy alliance of the young Reich and the Austrian sham state contained the germ of the subsequent World War and of the collapse as well.

In the course of this book I shall have occasion to take up this problem at length. Here it suffices to state that even in my earliest youth I came to the basic insight which never left me, but only became more profound:

That Germanism could be safeguarded only by the destruction of Austria, and, furthermore, that national sentiment is in no sense identical with dynastic patriotism; that above all the House of Habsburg was destined to be the misfortune of the German nation.

Even then I had drawn the consequences from this realization: ardent love for my German-Austrian homeland, deep hatred for the Austrian state.

*   *   *

The habit of historical thinking which I thus learned in school has never left me in the intervening years. To an ever-increasing extent world history became for me an inexhaustible source of understanding for the historical events of the present; in other words, for politics. I do not want to 'learn' it, I want it to instruct me.

Thus, at an early age, I had become a political 'revolutionary,' and I became an artistic revolutionary at an equally early age.

The provincial capital of Upper Austria had at that time a theater which was, relatively speaking, not bad. Pretty much of everything was produced. At the age of twelve I saw *Wilhelm Tell* for the first time, and a few months later my first opera,

Lohengrin. I was captivated at once. My youthful enthusiasm for the master of Bayreuth knew no bounds. Again and again I was drawn to his works, and it still seems to me especially fortunate that the modest provincial performance left me open to an intensified experience later on.

All this, particularly after I had outgrown my adolescence (which in my case was an especially painful process), reinforced my profound distaste for the profession which my father had chosen for me. My conviction grew stronger and stronger that I would never be happy as a civil servant. The fact that by this time my gift for drawing had been recognized at the *Realschule* made my determination all the firmer.

Neither pleas nor threats could change it one bit.

I wanted to become a painter and no power in the world could make me a civil servant.

Yet, strange as it may seem, with the passing years I became more and more interested in architecture.

At that time I regarded this as a natural complement to my gift as a painter, and only rejoiced inwardly at the extension of my artistic scope.

I did not suspect that things would turn out differently.

* * *

The question of my profession was to be decided more quickly than I had previously expected.

In my thirteenth year I suddenly lost my father. A stroke of apoplexy felled the old gentleman who was otherwise so hale, thus painlessly ending his earthly pilgrimage, plunging us all into the depths of grief. His most ardent desire had been to help his son forge his career, thus preserving him from his own bitter experience. In this, to all appearances, he had not succeeded. But, though unwittingly, he had sown the seed for a future which at that time neither he nor I would have comprehended.

For the moment there was no outward change.

My mother, to be sure, felt obliged to continue my education in accordance with my father's wish; in other words, to have me study for the civil servant's career. I, for my part, was more than ever determined absolutely not to undertake this career. In proportion as my schooling departed from my ideal in subject matter and curriculum, I became more indifferent at heart. Then suddenly an illness came to my help and in a few weeks decided my future and the eternal domestic quarrel. As a result of my serious lung ailment, a physician advised my mother in most urgent terms never to send me into an office. My attendance at the *Realschule* had furthermore to be interrupted for at least a year. The goal for which I had so long silently yearned, for which I had always fought, had through this event suddenly become reality almost of its own accord.

Concerned over my illness, my mother finally consented to take me out of the *Realschule* and let me attend the Academy.

These were the happiest days of my life and seemed to me almost a dream; and a mere dream it was to remain. Two years later, the death of my mother put a sudden end to all my high-flown plans.

It was the conclusion of a long and painful illness which from the beginning left little hope of recovery. Yet it was a dreadful blow, particularly for me. I had honored my father, but my mother I had loved.

Poverty and hard reality now compelled me to take a quick decision. What little my father had left had been largely exhausted by my mother's grave illness; the orphan's pension to which I was entitled was not enough for me even to live on, and so I was faced with the problem of somehow making my own living.

In my hand a suitcase full of clothes and underwear; in my heart an indomitable will, I journeyed to Vienna. I, too, hoped to wrest from Fate what my father had accomplished fifty years before; I, too, wanted to become 'something' — but on no account a civil servant.

CHAPTER
II

Years of Study and Suffering in Vienna

W<small>HEN</small> my mother died, Fate, at least in one respect, had made its decisions.

In the last months of her sickness, I had gone to Vienna to take the entrance examination for the Academy. I had set out with a pile of drawings, convinced that it would be child's play to pass the examination. At the *Realschule* I had been by far the best in my class at drawing, and since then my ability had developed amazingly; my own satisfaction caused me to take a joyful pride in hoping for the best.

Yet sometimes a drop of bitterness put in its appearance: my talent for painting seemed to be excelled by my talent for drawing, especially in almost all fields of architecture. At the same time my interest in architecture as such increased steadily, and this development was accelerated after a two weeks' trip to Vienna which I took when not yet sixteen. The purpose of my trip was to study the picture gallery in the Court Museum, but I had eyes for scarcely anything but the Museum itself. From morning until late at night, I ran from one object of interest to another, but it was always the buildings which held my primary interest. For hours I could stand in front of the Opera, for hours I could gaze at the Parliament; the whole Ring Boulevard seemed to me like an enchantment out of *The Thousand-and-One-Nights*.

Now I was in the fair city for the second time, waiting with burning impatience, but also with confident self-assurance, for the result of my entrance examination. I was so convinced that I would be successful that when I received my rejection, it struck me as a bolt from the blue. Yet that is what happened. When I presented myself to the rector, requesting an explanation for my non-acceptance at the Academy's school of painting, that gentleman assured me that the drawings I had submitted incontrovertibly showed my unfitness for painting, and that my ability obviously lay in the field of architecture; for me, he said, the Academy's school of painting was out of the question, the place for me was the School of Architecture. It was incomprehensible to him that I had never attended an architectural school or received any other training in architecture. Downcast, I left von Hansen's [1] magnificent building on the Schillerplatz, for the first time in my young life at odds with myself. For what I had just heard about my abilities seemed like a lightning flash, suddenly revealing a conflict with which I had long been afflicted, although until then I had no clear conception of its why and wherefore.

In a few days I myself knew that I should some day become an architect.

To be sure, it was an incredibly hard road; for the studies I had neglected out of spite at the *Realschule* were sorely needed. One could not attend the Academy's architectural school without having attended the building school at the *Technik*, and the latter required a high-school degree. I had none of all this. The fulfillment of my artistic dream seemed physically impossible.

When after the death of my mother I went to Vienna for the third time, to remain for many years, the time which had meanwhile elapsed had restored my calm and determination. My old defiance had come back to me and my goal was now clear and definite before my eyes. I wanted to become an architect, and obstacles do not exist to be surrendered to, but only to be broken. I was determined to overcome these obstacles, keeping before my

[1] Theophil von Hansen (1813–1891), Danish architect, built many of Vienna's best-known buildings, including the Academy of Art, the Stock Exchange, and the Reichsrat (parliament). His work was in the historical, archaeological style of the period.

eyes the image of my father, who had started out as the child of a village shoemaker, and risen by his own efforts to be a government official. I had a better foundation to build on, and hence my possibilities in the struggle were easier, and what then seemed to be the harshness of Fate, I praise today as wisdom and Providence. While the Goddess of Suffering took me in her arms, often threatening to crush me, my will to resistance grew, and in the end this will was victorious.

I owe it to that period that I grew hard and am still capable of being hard. And even more, I exalt it for tearing me away from the hollowness of comfortable life; for drawing the mother's darling out of his soft downy bed and giving him 'Dame Care' for a new mother; for hurling me, despite all resistance, into a world of misery and poverty, thus making me acquainted with those for whom I was later to fight.

*　　　*　　　*

In this period my eyes were opened to two menaces of which I had previously scarcely known the names, and whose terrible importance for the existence of the German people I certainly did not understand: Marxism and Jewry.

To me Vienna, the city which, to so many, is the epitome of innocent pleasure, a festive playground for merrymakers, represents, I am sorry to say, merely the living memory of the saddest period of my life.

Even today this city can arouse in me nothing but the most dismal thoughts. For me the name of this Phaeacian city [1] represents five years of hardship and misery. Five years in which I was forced to earn a living, first as a day laborer, then as a small painter; a truly meager living which never sufficed to appease even my daily hunger. Hunger was then my faithful bodyguard;

---

[1] Phaeacian city. The allusion to the happy isle of the Phaeacians is more popular in Germany than in English-speaking countries. Hitler's use of it does not mean that he has read the *Odyssey*.

he never left me for a moment and partook of all I had, share and share alike. Every book I acquired aroused his interest; a visit to the Opera prompted his attentions for days at a time; my life was a continuous struggle with this pitiless friend. And yet during this time I studied as never before. Aside from my architecture and my rare visits to the Opera, paid for in hunger, I had but one pleasure: my books.

At that time I read enormously and thoroughly. All the free time my work left me was employed in my studies. In this way I forged in a few years' time the foundations of a knowledge from which I still draw nourishment today.

And even more than this:

In this period there took shape within me a world picture and a philosophy which became the granite foundation of all my acts. In addition to what I then created, I have had to learn little; and I have had to alter nothing.

On the contrary.

Today I am firmly convinced that basically and on the whole all creative ideas appear in our youth, in so far as any such are present. I distinguish between the wisdom of age, consisting solely in greater thoroughness and caution due to the experience of a long life, and the genius of youth, which pours out thoughts and ideas with inexhaustible fertility, but cannot for the moment develop them because of their very abundance. It is this youthful genius which provides the building materials and plans for the future, from which a wiser age takes the stones, carves them and completes the edifice, in so far as the so-called wisdom of age has not stifled the genius of youth.

*　　　*　　　*

The life which I had hitherto led at home differed little or not at all from the life of other people. Carefree, I could await the new day, and there was no social problem for me. The environment of my youth consisted of petty-bourgeois circles, hence of a world having very little relation to the purely manual worker. For, strange as it may seem at first glance, the cleft between this class, which in an economic sense is by no means so brilliantly situated, and the manual worker is often deeper than we imagine. The reason for this hostility, as we might almost call it, lies in the fear of a social group, which has but recently raised itself above the level of the manual worker, that it will sink back into the old despised class, or at least become identified with it. To this, in many cases, we must add the repugnant memory of the cultural poverty of this lower class, the frequent vulgarity of its social intercourse; the petty bourgeois' own position in society, however insignificant it may be, makes any contact with this outgrown stage of life and culture intolerable.

Consequently, the higher classes feel less constraint in their dealings with the lowest of their fellow men than seems possible to the 'upstart.'

For anyone is an upstart who rises by his own efforts from his previous position in life to a higher one.

Ultimately this struggle, which is often so hard, kills all pity. Our own painful struggle for existence destroys our feeling for the misery of those who have remained behind.

In this respect Fate was kind to me. By forcing me to return to this world of poverty and insecurity, from which my father had risen in the course of his life, it removed the blinders of a narrow petty-bourgeois upbringing from my eyes. Only now did I learn to know humanity, learning to distinguish between empty appearances or brutal externals and the inner being.

*   *   *

After the turn of the century, Vienna was, socially speaking, one of the most backward cities in Europe.

Dazzling riches and loathsome poverty alternated sharply. In the center and in the inner districts you could really feel the pulse of this realm of fifty-two millions, with all the dubious magic of the national melting pot. The Court with its dazzling glamour attracted wealth and intelligence from the rest of the country like a magnet. Added to this was the strong centralization of the Habsburg monarchy in itself.

It offered the sole possibility of holding this medley of nations together in any set form. But the consequence was an extraordinary concentration of high authorities in the imperial capital.

Yet not only in the political and intellectual sense was Vienna the center of the old Danube monarchy, but economically as well. The host of high officers, government officials, artists, and scholars was confronted by an even greater army of workers, and side by side with aristocratic and commercial wealth dwelt dire poverty. Outside the palaces on the Ring loitered thousands of unemployed, and beneath this *Via Triumphalis* of old Austria dwelt the homeless in the gloom and mud of the canals.

In hardly any German city could the social question have been studied better than in Vienna. But make no mistake. This 'studying' cannot be done from lofty heights. No one who has not been seized in the jaws of this murderous viper can know its poison fangs. Otherwise nothing results but superficial chatter and false sentimentality. Both are harmful. The former because it can never penetrate to the core of the problem, the latter because it passes it by. I do not know which is more terrible: inattention to social misery such as we see every day among the majority of those who have been favored by fortune or who have risen by their own efforts, or else the snobbish, or at times tactless and obtrusive, condescension of certain women of fashion in skirts or in trousers, who 'feel for the people.' In any event, these gentry sin far more than their minds, devoid of all instinct, are capable of realizing. Consequently, and much to their own amazement, the result of their social 'efforts' is always nil, frequently, in fact, an indignant rebuff; though this, of course, is passed off as a proof of the people's ingratitude.

Such minds are most reluctant to realize that social endeavor has nothing in common with this sort of thing; that above all it can raise no claim to gratitude, since its function is not to distribute favors but to restore rights.

I was preserved from studying the social question in such a way. By drawing me within its sphere of suffering, it did not seem to invite me to 'study,' but to experience it in my own skin. It was none of its doing that the guinea pig came through the operation safe and sound.

* * *

An attempt to enumerate the sentiments I experienced in that period could never be even approximately complete; I shall describe here only the most essential impressions, those which often moved me most deeply, and the few lessons which I derived from them at the time.

* * *

The actual business of finding work was, as a rule, not hard for me, since I was not a skilled craftsman, but was obliged to seek my daily bread as a so-called helper and sometimes as a casual laborer.

I adopted the attitude of all those who shake the dust of Europe from their feet with the irrevocable intention of founding a new existence in the New World and conquering a new home. Released from all the old, paralyzing ideas of profession and position, environment and tradition, they snatch at every livelihood that offers itself, grasp at every sort of work, progressing step by step to the realization that honest labor, no matter of what sort, disgraces no one. I, too, was determined to leap into this new world, with both feet, and fight my way through.

I soon learned that there was always some kind of work to be had, but equally soon I found out how easy it was to lose it.

The uncertainty of earning my daily bread soon seemed to me one of the darkest sides of my new life.

The 'skilled' worker does not find himself out on the street as
frequently as the unskilled; but he is not entirely immune to this
fate either. And in his case the loss of livelihood owing to lack of
work is replaced by the lock-out, or by going on strike himself.

In this respect the entire economy suffers bitterly from the
individual's insecurity in earning his daily bread.

The peasant boy who goes to the big city, attracted by the
easier nature of the work (real or imaginary), by shorter hours,
but most of all by the dazzling light emanating from the metropolis, is accustomed to a certain security in the matter of livelihood. He leaves his old job only when there is at least some
prospect of a new one. For there is a great lack of agricultural
workers, hence the probability of any long period of unemployment is in itself small. It is a mistake to believe that the young
fellow who goes to the big city is made of poorer stuff than his
brother who continues to make an honest living from the peasant
sod. No, on the contrary: experience shows that all those elements which emigrate consist of the healthiest and most energetic
natures, rather than conversely. Yet among these 'emigrants'
we must count, not only those who go to America, but to an equal
degree the young farmhand who resolves to leave his native village for the strange city. He, too, is prepared to face an uncertain
fate. As a rule he arrives in the big city with a certain amount of
money; he has no need to lose heart on the very first day if he has
the ill fortune to find no work for any length of time. But it is
worse if, after finding a job, he soon loses it. To find a new one,
especially in winter, is often difficult if not impossible. Even so,
the first weeks are tolerable. He receives an unemployment benefit from his union funds and manages as well as possible. But
when his last cent is gone and the union, due to the long duration
of his unemployment, discontinues its payments, great hardships
begin. Now he walks the streets, hungry; often he pawns and
sells his last possessions; his clothing becomes more and more
wretched; and thus he sinks into external surroundings which, on
top of his physical misfortune, also poison his soul. If he is
evicted and if (as is so often the case) this occurs in winter, his

misery is very great. At length he finds some sort of job again. But the old story is repeated. The same thing happens a second time, the third time perhaps it is even worse, and little by little he learns to bear the eternal insecurity with greater and greater indifference. At last the repetition becomes a habit.

And so this man, who was formerly so hard-working, grows lax in his whole view of life and gradually becomes the instrument of those who use him only for their own base advantage. He has so often been unemployed through no fault of his own that one time more or less ceases to matter, even when the aim is no longer to fight for economic rights, but to destroy political, social, or cultural values in general. He may not be exactly enthusiastic about strikes, but at any rate he has become indifferent.

With open eyes I was able to follow this process in a thousand examples. The more I witnessed it, the greater grew my revulsion for the big city which first avidly sucked men in and then so cruelly crushed them.

When they arrived, they belonged to their people; after remaining for a few years, they were lost to it.

I, too, had been tossed around by life in the metropolis; in my own skin I could feel the effects of this fate and taste them with my soul. One more thing I saw: the rapid change from work to unemployment and vice versa, plus the resultant fluctuation of income, end by destroying in many all feeling for thrift, or any understanding for a prudent ordering of their lives. It would seem that the body gradually becomes accustomed to living on the fat of the land in good times and going hungry in bad times. Indeed, hunger destroys any resolution for reasonable budgeting in better times to come by holding up to the eyes of its tormented victim an eternal mirage of good living and raising this dream to such a pitch of longing that a pathological desire puts an end to all restraint as soon as wages and earnings make it at all possible. The consequence is that once the man obtains work he irresponsibly forgets all ideas of order and discipline, and begins to live luxuriously for the pleasures of the moment. This upsets even the small weekly budget, as even here any intelligent apportionment is lacking; in the beginning it suffices for five days instead of seven, later only for three, finally scarcely for one day, and in the end it is drunk up in the very first night.

Often he has a wife and children at home. Sometimes they, too, are infected by this life, especially when the man is good to them on the whole and actually loves them in his own way. Then the weekly wage is used up by the whole family in two or three days; they eat and drink as long as the money holds out and the last days they go hungry. Then the wife drags herself out into the neighborhood, borrows a little, runs up little debts at the food store, and in this way strives to get through the hard last days of the week. At noon they all sit together before their meager and sometimes empty bowls, waiting for the next payday, speaking of it, making plans, and, in their hunger, dreaming of the happiness to come.

And so the little children, in their earliest beginnings, are made familiar with this misery.

It ends badly if the man goes his own way from the very beginning and the woman, for the children's sake, opposes him. Then there is fighting and quarreling, and, as the man grows estranged from his wife, he becomes more intimate with alcohol. He is drunk every Saturday, and, with her instinct of self-preservation for herself and her children, the woman has to fight to get even a few pennies out of him; and, to make matters worse, this usually occurs on his way from the factory to the barroom. When at length he comes home on Sunday or even Monday night, drunk and brutal, but always parted from his last cent, such scenes often occur that God have mercy!

I have seen this in hundreds of instances. At first I was repelled or even outraged, but later I understood the whole tragedy of this misery and its deeper causes. These people are the unfortunate victims of bad conditions!

Even more dismal in those days were the housing conditions. The misery in which the Viennese day laborer lived was frightful to behold. Even today it fills me with horror when I think of these wretched caverns, the lodging houses and tenements, sordid scenes of garbage, repulsive filth, and worse.

What was — and still is — bound to happen some day, when the stream of unleashed slaves pours forth from these miserable dens to avenge themselves on their thoughtless fellow men!

For thoughtless they are!

Thoughtlessly they let things slide along, and with their utter lack of intuition fail even to suspect that sooner or later Fate must bring retribution, unless men conciliate Fate while there is still time.

How thankful I am today to the Providence which sent me to that school! In it I could no longer sabotage the subjects I did not like. It educated me quickly and thoroughly.

If I did not wish to despair of the men who constituted my environment at that time, I had to learn to distinguish between their external characters and lives and the foundations of their development. Only then could all this be borne without losing heart. Then, from all the misery and despair, from all the filth and outward degeneration, it was no longer human beings that emerged, but the deplorable results of deplorable laws; and the hardship of my own life, no easier than the others, preserved me from capitulating in tearful sentimentality to the degenerate products of this process of development.

No, this is not the way to understand all these things!

Even then I saw that only a twofold road could lead to the goal of improving these conditions:

The deepest sense of social responsibility for the creation of better foundations for our development, coupled with brutal determination in breaking down incurable tumors.

Just as Nature does not concentrate her greatest attention in preserving what exists, but in breeding offspring to carry on the species, likewise, in human life, it is less important artificially to alleviate existing evil, which, in view of human nature, is ninety-nine per cent impossible, than to ensure from the start healthier channels for a future development.

During my struggle for existence in Vienna, it had become clear to me that

Social activity must never and on no account be directed toward

philanthropic flim-flam, but rather toward the elimination of the basic deficiencies in the organization of our economic and cultural life that must — or at all events can — lead to the degeneration of the individual.

The difficulty of applying the most extreme and brutal methods against the criminals who endanger the state lies not least in the uncertainty of our judgment of the inner motives or causes of such contemporary phenomena.

This uncertainty is only too well founded in our own sense of guilt regarding such tragedies of degeneration; be that as it may, it paralyzes any serious and firm decision and is thus partly responsible for the weak and half-hearted, because hesitant, execution of even the most necessary measures of self-preservation.

Only when an epoch ceases to be haunted by the shadow of its own consciousness of guilt will it achieve the inner calm and outward strength brutally and ruthlessly to prune off the wild shoots and tear out the weeds.

Since the Austrian state had practically no social legislation or jurisprudence, its weakness in combating even malignant tumors was glaring.

*   *   *

I do not know what horrified me most at that time: the economic misery of my companions, their moral and ethical coarseness, or the low level of their intellectual development.

How often does our bourgeoisie rise in high moral indignation when they hear some miserable tramp declare that it is all the same to him whether he is a German or not, that he feels equally happy wherever he is, as long as he has enough to live on!

This lack of 'national pride' is most profoundly deplored, and horror at such an attitude is expressed in no uncertain terms.

How many people have asked themselves what was the real reason for the superiority of their own sentiments?

How many are aware of the infinite number of separate memories of the greatness of our national fatherland in all the fields of cultural and artistic life, whose total result is to inspire them with just pride at being members of a nation so blessed?

How many suspect to how great an extent pride in the fatherland depends on knowledge of its greatness in all these fields?

Do our bourgeois circles ever stop to consider to what an absurdly small extent this prerequisite of pride in the fatherland is transmitted to the 'people'?

Let us not try to condone this by saying that 'it is no better in other countries,' and that in those countries the worker avows his nationality 'notwithstanding.' Even if this were so, it could serve as no excuse for our own omissions. But it is not so; for the thing that we constantly designate as 'chauvinistic' education, for example among the French people, is nothing other than extreme emphasis on the greatness of France in all the fields of culture, or, as the Frenchman puts it, of 'civilization.' The fact is that the young Frenchman is not brought up to be objective, but is instilled with the most subjective conceivable view, in so far as the importance of the political or cultural greatness of his fatherland is concerned.

This education will always have to be limited to general and extremely broad values which, if necessary, must be engraved in the memory and feeling of the people by eternal repetition.

But to the negative sin of omission is added in our country the positive destruction of the little which the individual has the good fortune to learn in school. The rats that politically poison our nation gnaw even this little from the heart and memory of the broad masses, in so far as this has not been previously accomplished by poverty and suffering.

Imagine, for instance, the following scene:

In a basement apartment, consisting of two stuffy rooms, dwells a worker's family of seven. Among the five children there is a boy of, let us assume, three years. This is the age in which the first impressions are made on the consciousness of the child. Talented persons retain traces of memory from this period down to advanced old age. The very narrowness and overcrowding of the room does not lead to favorable conditions. Quarreling and wrangling will very frequently arise as a result. In these circumstances, people do not live with one another, they press against one another. Every argument, even the most trifling, which in a spacious apartment can be reconciled by a mild segregation, thus solving itself, here leads to loathsome wrangling without end. Among the children, of course, this is still bearable; they always fight under such circumstances, and among themselves they quickly and thoroughly forget about it. But if this battle is carried on between the parents themselves, and almost every day in forms which for vulgarity often leave nothing to be desired, then, if only very gradually, the results of such visual instruction must ultimately become apparent in the children. The character they will inevitably assume if this mutual quarrel takes the form of brutal attacks of the father against the mother, of drunken beatings, is hard for anyone who does not know this milieu to imagine. At the age of six the pitiable little boy suspects the existence of things which can inspire even an adult with nothing but horror. Morally poisoned, physically undernourished, his poor little head full of lice, the young 'citizen' goes off to public school. After a great struggle he may learn to read and write, but that is about all. His doing any homework is out of the question. On the contrary, the very mother and father, even in the presence of the children, talk about his teacher and school in terms which are not fit to be repeated, and are more inclined to curse the latter to their face than to take their little offspring across their knees and teach them some sense. All the other things that the little fellow hears at home do not tend to increase his respect for his dear fellow men. Nothing good remains of humanity, no institution remains unassailed; beginning with his teacher and up to the head of the government, whether it is a question of religion or of morality as such, of the state or society, it is all the same, everything is reviled in the most obscene terms and dragged into the filth of the basest possible outlook. When at the age of fourteen the young man is discharged from school, it is hard to decide what is stronger in him: his incredible stupidity as far as any real knowledge and ability are concerned, or the corrosive insolence of his behavior, combined with an immorality, even at this age, which would make your hair stand on end.

What position can this man — to whom even now hardly anything is holy, who, just as he has encountered no greatness, conversely suspects and knows all the sordidness of life — occupy in the life into which he is now preparing to emerge?

The three-year-old child has become a fifteen-year-old despiser of all authority. Thus far, aside from dirt and filth, this young man has seen nothing which might inspire him to any higher enthusiasm.

But only now does he enter the real university of this existence.

Now he begins the same life which all along his childhood years [1] he has seen his father living. He hangs around the street corners and bars, coming home God knows when; and for a change now and then he beats the broken-down being which was once his mother, curses God and the world, and at length is convicted of some particular offense and sent to a house of correction.

There he receives his last polish.

And his dear bourgeois fellow men are utterly amazed at the lack of 'national enthusiasm' in this young 'citizen.'

Day by day, in the theater and in the movies, in backstairs literature and the yellow press, they see the poison poured into the people by bucketfuls, and then they are amazed at the low 'moral content,' the 'national indifference,' of the masses of the people.

As though trashy films, yellow press, and such-like dung [2] could furnish the foundations of a knowledge of the greatness of our fatherland! — quite aside from the early education of the individual.

What I had never suspected before, I quickly and thoroughly learned in those years:

The question of the 'nationalization' of a people is, among other

---

[1] '*die Jahre der Kindheit entlang.*'

[2] '*ähnliche Jauche.*' In the second edition this is toned down to '*ähnliches*' (the like).

*things, primarily a question of creating healthy social conditions as a
foundation for the possibility of educating the individual. For only
those who through school and upbringing learn to know the cultural,
economic, but above all the political, greatness of their own fatherland
can and will achieve the inner pride in the privilege of being a member of such a people. And I can fight only for something that I love,
love only what I respect, and respect only what I at least know.*

*   *   *

Once my interest in the social question was aroused, I began
to study it with all thoroughness. It was a new and hitherto unknown world which opened before me.

In the years 1909 and 1910, my own situation had changed
somewhat in so far as I no longer had to earn my daily bread as
a common laborer. By this time I was working independently
as a small draftsman and painter of watercolors. Hard as this
was with regard to earnings — it was barely enough to live on —
it was good for my chosen profession. Now I was no longer dead
tired in the evening when I came home from work, unable to
look at a book without soon dozing off. My present work ran
parallel to my future profession. Moreover, I was master of my
own time and could apportion it better than had previously been
possible.

I painted to make a living and studied for pleasure.

Thus I was able to supplement my visual instruction in the
social problem by theoretical study. I studied more or less all of
the books I was able to obtain regarding this whole field, and for
the rest immersed myself in my own thoughts.

I believe that those who knew me in those days took me for an
eccentric.

Amid all this, as was only natural, I served my love of architecture with ardent zeal. Along with music, it seemed to me the
queen of the arts: under such circumstances my concern with it
was not 'work,' but the greatest pleasure. I could read and draw

until late into the night, and never grow tired. Thus my faith grew that my beautiful dream for the future would become reality after all, even though this might require long years. I was firmly convinced that I should some day make a name for myself as an architect.

In addition, I had the greatest interest in everything connected with politics, but this did not seem to me very significant. On the contrary: in my eyes this was the self-evident duty of every thinking man. Anyone who failed to understand this lost the right to any criticism or complaint.

In this field, too, I read and studied much.

By 'reading,' to be sure, I mean perhaps something different than the average member of our so-called 'intelligentsia.'

I know people who 'read' enormously, book for book, letter for letter, yet whom I would not describe as 'well-read.' True, they possess a mass of 'knowledge,' but their brain is unable to organize and register the material they have taken in. They lack the art of sifting what is valuable for them in a book from that which is without value, of retaining the one forever, and, if possible, not even seeing the rest, but in any case not dragging it around with them as useless ballast. For reading is no end in itself, but a means to an end. It should primarily help to fill the framework constituted by every man's talents and abilities; in addition, it should provide the tools and building materials which the individual needs for his life's work, regardless whether this consists in a primitive struggle for sustenance or the satisfaction of a high calling; secondly, it should transmit a general world view. In both cases, however, it is essential that the content of what one reads at any time should not be transmitted to the memory in the sequence of the book or books, but like the stone of a mosaic should fit into the general world picture in its proper place, and thus help to form this picture in the mind of the reader. Otherwise there arises a confused muddle of memorized facts which not only are worthless, but also make their unfortunate possessor conceited. For such a reader now believes himself in all seriousness to be 'educated,' to understand something of life, to have knowledge, while in reality, with every new acquisition of this kind of 'education,' he is growing more and more removed from the world until, not infrequently, he ends up in a sanitarium or in parliament.

Never will such a mind succeed in culling from the confusion of his 'knowledge' anything that suits the demands of the hour, for his intellectual ballast is not organized along the lines of life, but in the sequence of the books as he read them and as their content has piled up in his brain. If Fate, in the requirements of his daily life, desired to remind him to make a correct application of what he had read, it would have to indicate title and page number, since the poor fool would otherwise never in all his life find the correct place. But since Fate does not do this, these bright boys in any critical situation come into the most terrible embarrassment, cast about convulsively for analogous cases, and with mortal certainty naturally find the wrong formulas.

If this were not true, it would be impossible for us to understand the political behavior of our learned and highly placed government heroes, unless we decided to assume outright villainy instead of pathological propensities.

On the other hand, a man who possesses the art of correct reading will, in studying any book, magazine, or pamphlet, instinctively and immediately perceive everything which in his opinion is worth permanently remembering, either because it is suited to his purpose or generally worth knowing. Once the knowledge he has achieved in this fashion is correctly coordinated within the somehow existing picture of this or that subject created by the imagination, it will function either as a corrective or a complement, thus enhancing either the correctness or the clarity of the picture. Then, if life suddenly sets some question before us for examination or answer, the memory, if this method of reading is observed, will immediately take the existing picture as a norm, and from it will derive all the individual items regarding these questions, assembled in the course of decades, submit them to the mind for examination and reconsideration, until the question is clarified or answered.

Only this kind of reading has meaning and purpose.

An orator, for example, who does not thus provide his intelligence with the necessary foundation will never be in a position cogently to defend his view in the face of opposition, though it may be a thousand times true or real. In every discussion his memory will treacherously leave him in the lurch; he will find neither grounds for reinforcing his own contentions nor any for confuting those of his adversary. If, as in the case of a speaker, it is only a question of making a fool of himself personally, it may not be so bad, but not so when Fate predestines such a know-it-all incompetent to be the leader of a state.

Since my earliest youth I have endeavored to read in the correct way, and in this endeavor I have been most happily supported by my memory and intelligence. Viewed in this light, my Vienna period was especially fertile and valuable. The experiences of daily life provided stimulation for a constantly renewed study of the most varied problems. Thus at last I was in a position to bolster up reality by theory and test theory by reality, and was preserved from being stifled by theory or growing banal through reality.

In this period the experience of daily life directed and stimulated me to the most thorough theoretical study of two questions in addition to the social question.

Who knows when I would have immersed myself in the doctrines and essence of Marxism if that period had not literally thrust my nose into the problem!

* * *

What I knew of Social Democracy in my youth was exceedingly little and very inaccurate.

I was profoundly pleased that it should carry on the struggle for universal suffrage and the secret ballot. For even then my intelligence told me that this must help to weaken the Habsburg régime which I so hated. In the conviction that the Austrian

Empire could never be preserved except by victimizing its Germans, but that even the price of a gradual Slavization of the German element by no means provided a guaranty of an empire really capable of survival, since the power of the Slavs to uphold the state must be estimated as exceedingly dubious, I welcomed every development which in my opinion would inevitably lead to the collapse of this impossible state which condemned ten million Germans to death. The more the linguistic Babel corroded and disorganized parliament, the closer drew the inevitable hour of the disintegration of this Babylonian Empire, and with it the hour of freedom for my German-Austrian people. Only in this way could the *Anschluss* with the old mother country be restored.

Consequently, this activity of the Social Democracy was not displeasing to me. And the fact that it strove to improve the living conditions of the worker, as, in my innocence, I was still stupid enough to believe, likewise seemed to speak rather for it than against it. What most repelled me was its hostile attitude toward the struggle for the preservation of Germanism, its disgraceful courting of the Slavic 'comrade,' who accepted this declaration of love in so far as it was bound up with practical concessions, but otherwise maintained a lofty and arrogant reserve, thus giving the obtrusive beggars their deserved reward.

Thus, at the age of seventeen the word 'Marxism' was as yet little known to me, while 'Social Democracy' and socialism seemed to me identical concepts. Here again it required the fist of Fate to open my eyes to this unprecedented betrayal of the peoples.

Up to that time I had known the Social Democratic Party only as an onlooker at a few mass demonstrations, without possessing even the slightest insight into the mentality of its adherents or the nature of its doctrine; but now, at one stroke, I came into contact with the products of its education and 'philosophy.' And in a few months I obtained what might otherwise have required decades: an understanding of a pestilential whore,[1] cloak-

---

[1] '*Pesthure.*'   Second edition has '*Pestilenz.*'

ing herself as social virtue and brotherly love, from which I hope humanity will rid this earth with the greatest dispatch, since otherwise the earth might well become rid of humanity.

My first encounter with the Social Democrats occurred during my employment as a building worker.

From the very beginning it was none too pleasant. My clothing was still more or less in order, my speech cultivated, and my manner reserved. I was still so busy with my own destiny that I could not concern myself much with the people around me. I looked for work only to avoid starvation, only to obtain an opportunity of continuing my education, though ever so slowly. Perhaps I would not have concerned myself at all with my new environment if on the third or fourth day an event had not taken place which forced me at once to take a position. I was asked to join the organization.

My knowledge of trade-union organization was at that time practically non-existent. I could not have proved that its existence was either beneficial or harmful. When I was told that I had to join, I refused. The reason I gave was that I did not understand the matter, but that I would not let myself be forced into anything. Perhaps my first reason accounts for my not being thrown out at once. They may perhaps have hoped to convert me or break down my resistance in a few days. In any event, they had made a big mistake. At the end of two weeks I could no longer have joined, even if I had wanted to. In these two weeks I came to know the men around me more closely, and no power in the world could have moved me to join an organization whose members had meanwhile come to appear to me in so unfavorable a light.

During the first days I was irritable.

At noon some of the workers went to the near-by taverns while others remained at the building site and ate a lunch which, as a rule, was quite wretched. These were the married men whose wives brought them their noonday soup in pathetic bowls. Toward the end of the week their number always increased, why I did not understand until later. On these occasions politics was discussed.

I drank my bottle of milk and ate my piece of bread somewhere off to one side, and cautiously studied my new associates or reflected on my miserable lot. Nevertheless, I heard more than enough; and often it seemed to me that they purposely moved closer to me, perhaps in order to make me take a position. In any case, what I heard was of such a nature as to infuriate me in the extreme. These men rejected everything: the nation as an invention of the 'capitalistic' (how often was I forced to hear this single word!) classes; the fatherland as an instrument of the bourgeoisie for the exploitation of the working class; the authority of law as a means for oppressing the proletariat; the school as an institution for breeding slaves and slaveholders; religion as a means for stultifying the people and making them easier to exploit; morality as a symptom of stupid, sheeplike patience, etc. There was absolutely nothing which was not drawn through the mud of a terrifying depth.[1]

At first I tried to keep silent. But at length it became impossible. I began to take a position and to oppose them. But I was forced to recognize that this was utterly hopeless until I possessed certain definite knowledge of the controversial points. And so I began to examine the sources from which they drew this supposed wisdom. I studied book after book, pamphlet after pamphlet.

From then on our discussions at work were often very heated. I argued back, from day to day better informed than my antagonists concerning their own knowledge, until one day they made use of the weapon which most readily conquers reason: terror and violence. A few of the spokesmen on the opposing side forced me either to leave the building at once or be thrown off the scaffolding. Since I was alone and resistance seemed hopeless, I preferred, richer by one experience, to follow the former counsel.

I went away filled with disgust, but at the same time so agitated that it would have been utterly impossible for me to turn my back on the whole business. No, after the first surge of indignation, my stubbornness regained the upper hand. I was

[1] *'In den Kot einer entsetzlichen Tiefe.'*

determined to go to work on another building in spite of my experience. In this decision I was reinforced by Poverty which, a few weeks later, after I had spent what little I had saved from my wages, enfolded me in her heartless arms. I had to go back whether I wanted to or not. The same old story began anew and ended very much the same as the first time.

I wrestled with my innermost soul: are these people human, worthy to belong to a great nation?

A painful question; for if it is answered in the affirmative, the struggle for my nationality really ceases to be worth the hardships and sacrifices which the best of us have to make for the sake of such scum; and if it is answered in the negative, our nation is pitifully poor in *human beings*.

On such days of reflection and cogitation, I pondered with anxious concern on the masses of those no longer belonging to their people and saw them swelling to the proportions of a menacing army.

With what changed feeling I now gazed at the endless columns of a mass demonstration of Viennese workers that took place one day as they marched past four abreast! For nearly two hours I stood there watching with bated breath the gigantic human dragon slowly winding by. In oppressed anxiety, I finally left the place and sauntered homeward. In a tobacco shop on the way I saw the *Arbeiter-Zeitung*, the central organ of the old Austrian Social Democracy. It was available in a cheap people's café, to which I often went to read newspapers; but up to that time I had not been able to bring myself to spend more than two minutes on the miserable sheet, whose whole tone affected me like moral vitriol. Depressed by the demonstration, I was driven on by an inner voice to buy the sheet and read it carefully. That evening I did so, fighting down the fury that rose up in me from time to time at this concentrated solution of lies.

More than any theoretical literature, my daily reading of the Social Democratic press enabled me to study the inner nature of these thought-processes.

For what a difference between the glittering phrases about freedom, beauty, and dignity in the theoretical literature, the delusive welter of words seemingly expressing the most profound and laborious wisdom, the loathsome humanitarian morality — all this written with the incredible gall that comes with prophetic certainty — and the brutal daily press, shunning no villainy, employing every means of slander, lying with a virtuosity that would bend iron beams, all in the name of this gospel of a new humanity. The one is addressed to the simpletons of the middle, not to mention the upper, educated, 'classes,' the other to the masses.

For me immersion in the literature and press of this doctrine and organization meant finding my way back to my own people.

What had seemed to me an unbridgable gulf became the source of a greater love than ever before.

Only a fool can behold the work of this villainous poisoner and still condemn the victim. The more independent I made myself in the next few years, the clearer grew my perspective, hence my insight into the inner causes of the Social Democratic successes. I now understood the significance of the brutal demand that I read only Red papers, attend only Red meetings, read only Red books, etc. With plastic clarity I saw before my eyes the inevitable result of this doctrine of intolerance.

The psyche of the great masses is not receptive to anything that is half-hearted and weak.

Like the woman, whose psychic state is determined less by grounds of abstract reason than by an indefinable emotional longing for a force which will complement her nature, and who, consequently, would rather bow to a strong man than dominate a weakling, likewise the masses love a commander more than a petitioner and feel inwardly more satisfied by a doctrine, tolerating no other beside itself, than by the granting of liberalistic freedom with which, as a rule, they can do little, and are prone to feel that they have been abandoned. They are equally unaware of their shameless spiritual terrorization and the hideous abuse of their human freedom, for they absolutely fail to suspect the inner insanity of the whole doctrine. All they see is the ruthless force and brutality of its calculated manifestations, to which they always submit in the end.

If Social Democracy is opposed by a doctrine of greater truth, but equal brutality of methods, the latter will conquer, though this may require the bitterest struggle.

Before two years had passed, the theory as well as the technical methods of Social Democracy were clear to me.

I understood the infamous spiritual terror which this movement exerts, particularly on the bourgeoisie, which is neither morally nor mentally equal to such attacks; at a given sign it unleashes a veritable barrage of lies and slanders against whatever adversary seems most dangerous, until the nerves of the attacked persons break down and, just to have peace again, they sacrifice the hated individual.

However, the fools obtain no peace.

The game begins again and is repeated over and over until fear of the mad dog results in suggestive paralysis.

Since the Social Democrats best know the value of force from their own experience, they most violently attack those in whose nature they detect any of this substance which is so rare. Conversely, they praise every weakling on the opposing side, sometimes cautiously, sometimes loudly, depending on the real or supposed quality of his intelligence.

They fear an impotent, spineless genius less than a forceful nature of moderate intelligence.

But with the greatest enthusiasm they commend weaklings in both mind and force.

They know how to create the illusion that this is the only way of preserving the peace, and at the same time, stealthily but steadily, they conquer one position after another, sometimes by silent blackmail, sometimes by actual theft, at moments when the general attention is directed toward other matters, and either does not want to be disturbed or considers the matter too small to raise a stir about, thus again irritating the vicious antagonist.

This is a tactic based on precise calculation of all human weaknesses, and its result will lead to success with almost mathematical certainty unless the opposing side learns to combat poison gas with poison gas.

It is our duty to inform all weaklings that this is a question of to be or not to be.

I achieved an equal understanding of the importance of physical terror toward the individual and the masses.

Here, too, the psychological effect can be calculated with precision.

Terror at the place of employment, in the factory, in the meeting hall, and on the occasion of mass demonstrations will always be successful unless opposed by equal terror.

In this case, to be sure, the party will cry bloody murder; though it has long despised all state authority, it will set up a howling cry for that same authority and in most cases will actually attain its goal amid the general confusion: it will find some idiot of a higher official who, in the imbecilic hope of propitiating the feared adversary for later eventualities, will help this world plague to break its opponent.

The impression made by such a success on the minds of the great masses of supporters as well as opponents can only be measured by those who know the soul of a people, not from books, but from life. For while in the ranks of their supporters the victory achieved seems a triumph of the justice of their own cause, the defeated adversary in most cases despairs of the success of any further resistance.

The more familiar I became, principally with the methods of physical terror, the more indulgent I grew toward all the hundreds of thousands who succumbed to it.

What makes me most indebted to that period of suffering is that it alone gave back to me my people, taught me to distinguish the victims from their seducers.

The results of this seduction can be designated only as victims. For if I attempted to draw a few pictures from life, depicting the essence of these 'lowest' classes, my picture would not be complete without the assurance that in these depths I also found bright spots in the form of a rare willingness to make sacrifices, of loyal comradeship, astonishing frugality, and modest reserve, especially among the older workers. Even though these virtues were steadily vanishing in the younger generation, if only through the general effects of the big city, there were many, even among the young men, whose healthy blood managed to dominate the foul tricks of life. If in their political activity, these good, often kind-hearted people nevertheless joined the mortal enemies of our nationality, thus helping to cement their ranks, the reason was that they neither understood nor could understand the baseness of the new doctrine, and that no one else took the trouble to bother about them, and finally that the social conditions were stronger than any will to the contrary that may have been present. The poverty to which they sooner or later succumbed drove them into the camp of the Social Democracy.

Since on innumerable occasions the bourgeoisie has in the clumsiest and most immoral way opposed demands which were justified from the universal human point of view, often without obtaining or even justifiably expecting any profit from such an attitude, even the most self-respecting worker was driven out of the trade-union organization into political activity.

Millions of workers, I am sure, started out as enemies of the Social Democratic Party in their innermost soul, but their resistance was overcome in a way which was sometimes utterly insane; that is, when the bourgeois parties adopted a hostile attitude toward every demand of a social character. Their simple, narrow-minded rejection of all attempts to better working conditions, to introduce safety devices on machines, to prohibit child labor and protect the woman, at least in the months when she was bearing the future national comrade under her heart, contributed to drive the masses into the net of Social Democracy which gratefully snatched at every case of such a disgraceful attitude. Never can our political bourgeoisie make good its sins in this direction, for by resisting all attempts to do away with social abuses, they sowed hatred and seemed to justify even the assertions of the mortal enemies of the entire nation, to the effect that only the Social Democratic Party represented the interests of the working people

Thus, to begin with, they created the moral basis for the actual existence of the trade unions, the organization which has always been the most effective pander to the political party.

In my Viennese years I was forced, whether I liked it or not, to take a position on the trade unions.

Since I regarded them as an inseparable ingredient of the Social Democratic Party as such, my decision was instantaneous and — mistaken.

I flatly rejected them without thinking.

And in this infinitely important question, as in so many others, Fate itself became my instructor.

The result was a reversal of my first judgment.

By my twentieth year I had learned to distinguish between a union as a means of defending the general social rights of the wage-earner, and obtaining better living conditions for him as an individual, and the trade union as an instrument of the party in the political class struggle.

The fact that Social Democracy understood the enormous importance of the trade-union movement assured it of this instrument and hence of success; the fact that the bourgeoisie were not aware of this cost them their political position. They thought they could stop a logical development by means of an impertinent 'rejection,' but in reality they only forced it into illogical channels. For to call the trade-union movement in itself unpatriotic is nonsense and untrue to boot. Rather the contrary is true. If trade-union activity strives and succeeds in bettering the lot of a class which is one of the basic supports of the nation, its work is not only not anti-patriotic or seditious, but 'national' in the truest sense of the word. For in this way it helps to create the social premises without which a general national education is unthinkable. It wins the highest merit by eliminating social cankers, attacking intellectual as well as physical infections, and thus helping to contribute to the general health of the body politic.

Consequently, the question of their necessity is really superfluous.

As long as there are employers with little social understanding or a deficient sense of justice and propriety, it is not only the right but the duty of their employees, who certainly constitute a part of our nationality, to protect the interests of the general public against the greed and unreason of the individual; for the preservation of loyalty and faith in a social group is just as much to the interest of a nation as the preservation of the people's health.

Both of these are seriously menaced by unworthy employers who do not feel themselves to be members of the national community as a whole. From the disastrous effects of their greed or ruthlessness grow profound evils for the future.

To eliminate the causes of such a development is to do a service to the nation and in no sense the opposite.

Let no one say that every individual is free to draw the consequences from an actual or supposed injustice; in other words, to leave his job. No! This is shadow-boxing and must be regarded as an attempt to divert attention. Either the elimination of bad, unsocial conditions serves the interest of the nation or it does not. If it does, the struggle against them must be carried on with weapons which offer the hope of success. The individual worker, however, is never in a position to defend himself against the power of the great industrialist, for in such matters it cannot be superior justice that conquers (if that were recognized, the whole struggle would stop from lack of cause) — no, what matters here is superior power. Otherwise the sense of justice alone would bring the struggle to a fair conclusion, or, more accurately speaking, the struggle could never arise.

No, if the unsocial or unworthy treatment of men calls for resistance, this struggle, as long as no legal judicial authorities have been created for the elimination of these evils, can only be decided by superior power. And this makes it obvious that the power of the employer concentrated in a single person can only be countered by the mass of employees banded into a single person, if the possibility of a victory is not to be renounced in advance.

Thus, trade-union organization can lead to a strengthening of the social idea in its practical effects on daily life, and thereby to an elimination of irritants which are constantly giving cause for dissatisfaction and complaints.

If this is not the case, it is to a great extent the fault of those who have been able to place obstacles in the path of any legal regulation of social evils or thwart them by means of their political influence.

Proportionately as the political bourgeoisie did not understand, or rather did not want to understand, the importance of trade-union organization, and resisted it, the Social Democrats took possession of the contested movement. Thus, far-sightedly it created a firm foundation which on several critical occasions has stood up when all other supports failed. In this way the intrinsic purpose was gradually submerged, making place for new aims.

It never occurred to the Social Democrats to limit the movement they had thus captured to its original task.

No, that was far from their intention.

In a few decades the weapon for defending the social rights of man had, in their experienced hands, become an instrument for the destruction of the national economy. And they did not let themselves be hindered in the least by the interests of the workers. For in politics, as in other fields, the use of economic pressure always permits blackmail, as long as the necessary unscrupulousness is present on the one side, and sufficient sheeplike patience on the other.

Something which in this case was true of both sides.

*   *   *

By the turn of the century, the trade-union movement had ceased to serve its former function. From year to year it had entered more and more into the sphere of Social Democratic politics and finally had no use except as a battering-ram in the class struggle. Its purpose was to cause the collapse of the whole arduously constructed economic edifice by persistent blows, thus,

the more easily, after removing its economic foundations, to prepare the same lot for the edifice of state. Less and less attention
was paid to defending the real needs of the working class, and
finally political expediency made it seem undesirable to relieve
the social or cultural miseries of the broad masses at all, for otherwise there was a risk that these masses, satisfied in their desires,
could no longer be used forever as docile shock troops.

The leaders of the class struggle looked on this development
with such dark foreboding and dread that in the end they rejected any really beneficial social betterment out of hand, and
actually attacked it with the greatest determination.

And they were never at a loss for an explanation of a line of
behavior which seemed so inexplicable.

By screwing the demands higher and higher, they made their
possible fulfillment seem so trivial and unimportant that they
were able at all times to tell the masses that they were dealing
with nothing but a diabolical attempt to weaken, if possible in
fact to paralyze, the offensive power of the working class in the
cheapest way, by such a ridiculous satisfaction of the most elementary rights. In view of the great masses' small capacity for
thought, we need not be surprised at the success of these
methods.

The bourgeois camp was indignant at this obvious insincerity
of Social Democratic tactics, but did not draw from it the slightest inference with regard to their own conduct. The Social
Democrats' fear of really raising the working class out of the
depths of their cultural and social misery should have inspired
the greatest exertions in this very direction, thus gradually
wresting the weapon from the hands of the advocates of the
class struggle.

This, however, was not done.

Instead of attacking and seizing the enemy's position, the
bourgeoisie preferred to let themselves be pressed to the wall and
finally had recourse to utterly inadequate makeshifts, which remained ineffectual because they came too late, and, moreover,
were easy to reject because they were too insignificant. Thus,

in reality, everything remained as before, except that the discontent was greater.

Like a menacing storm-cloud, the 'free trade union' hung, even then, over the political horizon and the existence of the individual.

It was one of the most frightful instruments of terror against the security and independence of the national economy, the solidity of the state, and personal freedom.

And chiefly this was what made the concept of democracy a sordid and ridiculous phrase, and held up brotherhood to everlasting scorn in the words: 'And if our comrade you won't be, we'll bash your head in — one, two, three!'

And that was how I became acquainted with this friend of humanity. In the course of the years my view was broadened and deepened, but I have had no need to change it.

*   *   *

The greater insight I gathered into the external character of Social Democracy, the greater became my longing to comprehend the inner core of this doctrine.

The official party literature was not much use for this purpose. In so far as it deals with economic questions, its assertions and proofs are false; in so far as it treats of political aims, it lies. Moreover, I was inwardly repelled by the new-fangled pettifogging phraseology and the style in which it was written. With an enormous expenditure of words, unclear in content or incomprehensible as to meaning, they stammer an endless hodgepodge of phrases purportedly as witty as in reality they are meaningless. Only our decadent metropolitan bohemians can feel at home in this maze of reasoning and cull an 'inner experience' from this dung-heap of literary dadaism, supported by the proverbial modesty of a section of our people who always detect profound wisdom in what is most incomprehensible to them personally. However, by balancing the theoretical untruth and nonsense of this doctrine with the reality of the phenomenon, I gradually obtained a clear picture of its intrinsic will.

At such times I was overcome by gloomy foreboding and malignant fear. Then I saw before me a doctrine, comprised of egotism and hate, which can lead to victory pursuant to mathematical laws, but in so doing must put an end to humanity.

Meanwhile, I had learned to understand the connection between this doctrine of destruction and the nature of a people of which, up to that time, I had known next to nothing.

Only a knowledge of the Jews provides the key with which to comprehend the inner, and consequently real, aims of Social Democracy.

The erroneous conceptions of the aim and meaning of this party fall from our eyes like veils, once we come to know this people, and from the fog and mist of social phrases rises the leering grimace of Marxism.

*   *   *

Today it is difficult, if not impossible, for me to say when the word 'Jew' first gave me ground for special thoughts. At home I do not remember having heard the word during my father's lifetime. I believe that the old gentleman would have regarded any special emphasis on this term as cultural backwardness. In the course of his life he had arrived at more or less cosmopolitan views which, despite his pronounced national sentiments, not only remained intact, but also affected me to some extent.

Likewise at school I found no occasion which could have led me to change this inherited picture.

At the *Realschule*, to be sure, I did meet one Jewish boy who was treated by all of us with caution, but only because various experiences had led us to doubt his discretion and we did not particularly trust him; but neither I nor the others had any thoughts on the matter.

Not until my fourteenth or fifteenth year did I begin to come across the word 'Jew,' with any frequency, partly in connection with political discussions. This filled me with a mild distaste, and I could not rid myself of an unpleasant feeling that always came over me whenever religious quarrels occurred in my presence.

At that time I did not think anything else of the question.

There were few Jews in Linz. In the course of the centuries their outward appearance had become Europeanized and had taken on a human look; in fact, I even took them for Germans. The absurdity of this idea did not dawn on me because I saw no distinguishing feature but the strange religion. The fact that they had, as I believed, been persecuted on this account sometimes almost turned my distaste at unfavorable remarks about them into horror.

Thus far I did not so much as suspect the existence of an organized opposition to the Jews.

Then I came to Vienna.

Preoccupied by the abundance of my impressions in the architectural field, oppressed by the hardship of my own lot, I gained at first no insight into the inner stratification of the people in this gigantic city. Notwithstanding that Vienna in those days counted nearly two hundred thousand Jews among its two million inhabitants, I did not see them. In the first few weeks my eyes and my senses were not equal to the flood of values and ideas. Not until calm gradually returned and the agitated picture began to clear did I look around me more carefully in my new world, and then among other things I encountered the Jewish question.

I cannot maintain that the way in which I became acquainted with them struck me as particularly pleasant. For the Jew was still characterized for me by nothing but his religion, and therefore, on grounds of human tolerance, I maintained my rejection of religious attacks in this case as in others. Consequently, the tone, particularly that of the Viennese anti-Semitic press, seemed to me unworthy of the cultural tradition of a great nation. I was oppressed by the memory of certain occurrences in the Middle Ages, which I should not have liked to see repeated. Since the newspapers in question did not enjoy an outstanding reputation (the reason for this, at that time, I myself did not precisely know), I regarded them more as the products of anger and envy than the results of a principled, though perhaps mistaken, point of view.

I was reinforced in this opinion by what seemed to me the far more dignified form in which the really big papers answered all these attacks, or, what seemed to me even more praiseworthy, failed to mention them; in other words, simply killed them with silence.

I zealously read the so-called world press (*Neue Freie Presse, Wiener Tageblatt*, etc.) and was amazed at the scope of what they offered their readers and the objectivity of individual articles. I respected the exalted tone, though the flamboyance of the style sometimes caused me inner dissatisfaction, or even struck me unpleasantly. Yet this may have been due to the rhythm of life in the whole metropolis.

Since in those days I saw Vienna in that light, I thought myself justified in accepting this explanation of mine as a valid excuse.

But what sometimes repelled me was the undignified fashion in which this press curried favor with the Court. There was scarcely an event in the Hofburg which was not imparted to the readers either with raptures of enthusiasm or plaintive emotion, and all this to-do, particularly when it dealt with the 'wisest monarch' of all time, almost reminded me of the mating cry of a mountain cock.

To me the whole thing seemed artificial.

In my eyes it was a blemish upon liberal democracy.

To curry favor with this Court and in such indecent forms was to sacrifice the dignity of the nation.

This was the first shadow to darken my intellectual relationship with the 'big' Viennese press.

As I had always done before, I continued in Vienna to follow events in Germany with ardent zeal, quite regardless whether they were political or cultural. With pride and admiration, I compared the rise of the Reich with the wasting away of the Austrian state. If events in the field of foreign politics filled me, by and large, with undivided joy, the less gratifying aspects of internal life often aroused anxiety and gloom. The struggle which at that time was being carried on against William II did not meet with my approval. I regarded him not only as the German Emperor, but first and foremost as the creator of a German fleet. The restrictions of speech imposed on the Kaiser by the Reichstag angered me greatly because they emanated from a source which in my opinion really hadn't a leg to stand on, since in a single session these parliamentarian imbeciles gabbled more nonsense than a whole dynasty of emperors, including its very weakest numbers, could ever have done in centuries.

I was outraged that in a state where every idiot not only claimed the right to criticize, but was given a seat in the Reichstag and let loose upon the nation as a 'lawgiver,' the man who bore the imperial crown had to take 'reprimands' from the greatest babblers' club of all time.

But I was even more indignant that the same Viennese press which made the most obsequious bows to every rickety horse in the Court, and flew into convulsions of joy if he accidentally swished his tail, should, with supposed concern, yet, as it seemed to me, ill-concealed malice, express its criticisms of the German Kaiser. Of course it had no intention of interfering with conditions within the German Reich — oh, no, God forbid — but by placing its finger on these wounds in the friendliest way, it was fulfilling the duty imposed by the spirit of the mutual alliance, and, conversely, fulfilling the requirements of journalistic truth, etc. And now it was poking this finger around in the wound to its heart's content.

In such cases the blood rose to my head.

It was this which caused me little by little to view the big papers with greater caution.

And on one such occasion I was forced to recognize that one of the anti-Semitic papers, the *Deutsches Volksblatt*, behaved more decently.

Another thing that got on my nerves was the loathsome cult for France which the big press, even then, carried on. A man couldn't help feeling ashamed to be a German when he saw these saccharine hymns of praise to the 'great cultural nation.' This wretched licking of France's boots more than once made me throw down one of these 'world newspapers.' And on such occasions I sometimes picked up the *Volksblatt*, which, to be sure, seemed to me much smaller, but in these matters somewhat more appetizing. I was not in agreement with the sharp anti-Semitic tone, but from time to time I read arguments which gave me some food for thought.

At all events, these occasions slowly made me acquainted with the man and the movement, which in those days guided Vienna's destinies: Dr. Karl Lueger [1] and the Christian Social Party.

When I arrived in Vienna, I was hostile to both of them.

The man and the movement seemed 'reactionary' in my eyes.

My common sense of justice, however, forced me to change this judgment in proportion as I had occasion to become acquainted with the man and his work; and slowly my fair judgment turned to unconcealed admiration. Today, more than ever, I regard this man as the greatest German mayor of all times.

How many of my basic principles were upset by this change in my attitude toward the Christian Social movement!

My views with regard to anti-Semitism thus succumbed to the passage of time, and this was my greatest transformation of all.

It cost me the greatest inner soul struggles, and only after months of battle between my reason and my sentiments did my reason begin to emerge victorious. Two years later, my sentiment had followed my reason, and from then on became its most loyal guardian and sentinel.

At the time of this bitter struggle between spiritual education and cold reason, the visual instruction of the Vienna streets had performed invaluable services. There came a time when I no longer, as in the first days, wandered blindly through the mighty city; now with open eyes I saw not only the buildings but also the people.

[1] Karl Lueger (1844–1910). In 1897, as a member of the anti-Semitic Christian Social Party, he became mayor of Vienna and kept the post until his death. At first opposed by the Court for his radical nationalism and anti-Semitism, toward the end of his career he became more moderate and was reconciled with the Emperor.

Once, as I was strolling through the Inner City, I suddenly encountered an apparition in a black caftan and black hair locks. Is this a Jew? was my first thought.

For, to be sure, they had not looked like that in Linz. I observed the man furtively and cautiously, but the longer I stared at this foreign face, scrutinizing feature for feature, the more my first question assumed a new form:

Is this a German?

As always in such cases, I now began to try to relieve my doubts by books. For a few hellers I bought the first anti-Semitic pamphlets of my life. Unfortunately, they all proceeded from the supposition that in principle the reader knew or even understood the Jewish question to a certain degree. Besides, the tone for the most part was such that doubts again arose in me, due in part to the dull and amazingly unscientific arguments favoring the thesis.

I relapsed for weeks at a time, once even for months.

The whole thing seemed to me so monstrous, the accusations so boundless, that, tormented by the fear of doing injustice, I again became anxious and uncertain.

Yet I could no longer very well doubt that the objects of my study were not Germans of a special religion, but a people in themselves; for since I had begun to concern myself with this question and to take cognizance of the Jews, Vienna appeared to me in a different light than before. Wherever I went, I began to see Jews, and the more I saw, the more sharply they became distinguished in my eyes from the rest of humanity. Particularly the Inner City and the districts north of the Danube Canal swarmed with a people which even outwardly had lost all resemblance to Germans.

And whatever doubts I may still have nourished were finally dispelled by the attitude of a portion of the Jews themselves.

Among them there was a great movement, quite extensive in Vienna, which came out sharply in confirmation of the national character of the Jews: this was the *Zionists*.

It looked, to be sure, as though only a part of the Jews approved this viewpoint, while the great majority condemned and inwardly rejected such a formulation. But when examined more closely, this appearance dissolved itself into an unsavory vapor of pretexts advanced for mere reasons of expedience, not to say lies. For the so-called liberal Jews did not reject the Zionists as non-Jews, but only as Jews with an impractical, perhaps even dangerous, way of publicly avowing their Jewishness.

Intrinsically they remained unalterably of one piece.

In a short time this apparent struggle between Zionistic and liberal Jews disgusted me; for it was false through and through, founded on lies and scarcely in keeping with the moral elevation and purity always claimed by this people.

The cleanliness of this people, moral and otherwise, I must say, is a point in itself. By their very exterior you could tell that these were no lovers of water, and, to your distress, you often knew it with your eyes closed. Later I often grew sick to my stomach from the smell of these caftan-wearers. Added to this, there was their unclean dress and their generally unheroic appearance.

All this could scarcely be called very attractive; but it became positively repulsive when, in addition to their physical uncleanliness, you discovered the moral stains on this 'chosen people.'

In a short time I was made more thoughtful than ever by my slowly rising insight into the type of activity carried on by the Jews in certain fields.

Was there any form of filth or profligacy, particularly in cultural life, without at least one Jew involved in it?

If you cut even cautiously into such an abscess, you found, like a maggot in a rotting body, often dazzled by the sudden light — a kike![1]

What had to be reckoned heavily against the Jews in my eyes was when I became acquainted with their activity in the press, art, literature, and the theater. All the unctuous reassurances helped little or nothing. It sufficed to look at a billboard, to study

---

[1] *Sowie man nur vorsichtig in eine solche Geschwulst hineinschnitt, fand man, wie die Made im faulenden Leibe, oft ganz geblendet vom plötzlichen Lichte, ein Jüdlein.*

the names of the men behind the horrible trash they advertised, to make you hard for a long time to come. This was pestilence, spiritual pestilence, worse than the Black Death of olden times, and the people was being infected with it! It goes without saying that the lower the intellectual level of one of these art manufacturers, the more unlimited his fertility will be, and the scoundrel ends up like a garbage separator, splashing his filth in the face of humanity. And bear in mind that there is no limit to their number; bear in mind that for one Goethe Nature easily can foist on the world ten thousand of these scribblers who poison men's souls like germ-carriers of the worse sort, on their fellow men.

It was terrible, but not to be overlooked, that precisely the Jew, in tremendous numbers, seemed chosen by Nature for this shameful calling.

Is this why the Jews are called the 'chosen people'?

I now began to examine carefully the names of all the creators of unclean products in public artistic life. The result was less and less favorable for my previous attitude toward the Jews. Regardless how my sentiment might resist, my reason was forced to draw its conclusions.

The fact that nine tenths of all literary filth, artistic trash, and theatrical idiocy can be set to the account of a people, constituting hardly one hundredth of all the country's inhabitants, could simply not be talked away; it was the plain truth.

And I now began to examine my beloved 'world press' from this point of view.

And the deeper I probed, the more the object of my former admiration shriveled. The style became more and more unbearable; I could not help rejecting the content as inwardly shallow and banal; the objectivity of exposition now seemed to me more akin to lies than honest truth; and the writers were — Jews.

A thousand things which I had hardly seen before now struck my notice, and others, which had previously given me food for thought, I now learned to grasp and understand.

I now saw the liberal attitude of this press in a different light;

the lofty tone in which it answered attacks and its method of killing them with silence now revealed itself to me as a trick as clever as it was treacherous; the transfigured raptures of their theatrical critics were always directed at Jewish writers, and their disapproval never struck anyone but Germans. The gentle pinpricks against William II revealed its methods by their persistency, and so did its commendation of French culture and civilization. The trashy content of the short story now appeared to me as outright indecency, and in the language I detected the accents of a foreign people; the sense of the whole thing was so obviously hostile to Germanism that this could only have been intentional.

But who had an interest in this?

Was all this a mere accident?

Gradually I became uncertain.

The development was accelerated by insights which I gained into a number of other matters. I am referring to the general view of ethics and morals which was quite openly exhibited by a large part of the Jews, and the practical application of which could be seen.

Here again the streets provided an object lesson of a sort which was sometimes positively evil.

The relation of the Jews to prostitution and, even more, to the white-slave traffic, could be studied in Vienna as perhaps in no other city of Western Europe, with the possible exception of the southern French ports. If you walked at night through the streets and alleys of Leopoldstadt,[1] at every step you witnessed proceedings which remained concealed from the majority of the German people until the War gave the soldiers on the eastern front occasion to see similar things, or, better expressed, forced them to see them.

When thus for the first time I recognized the Jew as the cold-hearted, shameless, and calculating director of this revolting vice traffic in the scum of the big city, a cold shudder ran down my back.

---

[1] Second District of Vienna, separated from the main part of the city by the Danube Canal. Formerly the ghetto, it still has a predominantly Jewish population.

But then a flame flared up within me. I no longer avoided discussion of the Jewish question; no, now I sought it. And when I learned to look for the Jew in all branches of cultural and artistic life and its various manifestations, I suddenly encountered him in a place where I would least have expected to find him.

When I recognized the Jew as the leader of the Social Democracy, the scales dropped from my eyes. A long soul struggle had reached its conclusion.

Even in my daily relations with my fellow workers, I observed the amazing adaptability with which they adopted different positions on the same question, sometimes within an interval of a few days, sometimes in only a few hours. It was hard for me to understand how people who, when spoken to alone, possessed some sensible opinions, suddenly lost them as soon as they came under the influence of the masses. It was often enough to make one despair. When, after hours of argument, I was convinced that now at last I had broken the ice or cleared up some absurdity, and was beginning to rejoice at my success, on the next day to my disgust I had to begin all over again; it had all been in vain. Like an eternal pendulum their opinions seemed to swing back again and again to the old madness.

All this I could understand: that they were dissatisfied with their lot and cursed the Fate which often struck them so harshly; that they hated the employers who seemed to them the heartless bailiffs of Fate; that they cursed the authorities who in their eyes were without feeling for their situation; that they demonstrated against food prices and carried their demands into the streets: this much could be understood without recourse to reason. But what inevitably remained incomprehensible was the boundless hatred they heaped upon their own nationality, despising its greatness, besmirching its history, and dragging its great men into the gutter.

This struggle against their own species, their own clan, their own homeland, was as senseless as it was incomprehensible. It was unnatural.

It was possible to cure them temporarily of this vice, but only for days or at most weeks. If later you met the man you thought you had converted, he was just the same as before.

His old unnatural state had regained full possession of him.

*   *   *

I gradually became aware that the Social Democratic press was directed predominantly by Jews; yet I did not attribute any special significance to this circumstance, since conditions were exactly the same in the other papers. Yet one fact seemed conspicuous: there was not one paper with Jews working on it which could have been regarded as truly national, according to my education and way of thinking.

I swallowed my disgust and tried to read this type of Marxist press production, but my revulsion became so unlimited in so doing that I endeavored to become more closely acquainted with the men who manufactured these compendiums of knavery.

From the publisher down, they were all Jews.

I took all the Social Democratic pamphlets I could lay hands on and sought the names of their authors: Jews. I noted the names of the leaders; by far the greatest part were likewise members of the 'chosen people,' whether they were representatives in the Reichsrat or trade-union secretaries, the heads of organizations or street agitators. It was always the same gruesome picture. The names of the Austerlitzes, Davids, Adlers, Ellenbogens, etc., will remain forever graven in my memory. One thing had grown clear to me: the party with whose petty representatives I had been carrying on the most violent struggle for months was, as to leadership, almost exclusively in the hands of a foreign people; for, to my deep and joyful satisfaction, I had at last come to the conclusion that the Jew was no German.

Only now did I become thoroughly acquainted with the seducer of our people.

A single year of my sojourn in Vienna had sufficed to imbue me with the conviction that no worker could be so stubborn that he would not in the end succumb to better knowledge and better explanations. Slowly I had become an expert in their own doctrine and used it as a weapon in the struggle for my own profound conviction.

Success almost always favored my side.

The great masses could be saved, if only with the gravest sacrifice in time and patience.

But a Jew could never be parted from his opinions.

At that time I was still childish enough to try to make the madness of their doctrine clear to them; in my little circle I talked my tongue sore and my throat hoarse, thinking I would inevitably succeed in convincing them how ruinous their Marxist madness was; but what I accomplished was often the opposite. It seemed as though their increased understanding of the destructive effects of Social Democratic theories and their results only reinforced their determination.

The more I argued with them, the better I came to know their dialectic. First they counted on the stupidity of their adversary, and then, when there was no other way out, they themselves simply played stupid. If all this didn't help, they pretended not to understand, or, if challenged, they changed the subject in a hurry, quoted platitudes which, if you accepted them, they immediately related to entirely different matters, and then, if again attacked, gave ground and pretended not to know exactly what you were talking about. Whenever you tried to attack one of these apostles, your hand closed on a jelly-like slime which divided up and poured through your fingers, but in the next moment collected again. But if you really struck one of these fellows so telling a blow that, observed by the audience, he couldn't help but agree, and if you believed that this had taken you at least one step forward, your amazement was great the next day. The Jew had not the slightest recollection of the day before, he rattled off his same old nonsense as though nothing at all had happened, and, if indignantly challenged, affected amazement; he couldn't remember a thing, except that he had proved the correctness of his assertions the previous day.

Sometimes I stood there thunderstruck.

I didn't know what to be more amazed at: the agility of their tongues or their virtuosity at lying.

Gradually I began to hate them.

All this had but one good side: that in proportion as the real leaders or at least the disseminators of Social Democracy came within my vision, my love for my people inevitably grew. For who, in view of the diabolical craftiness of these seducers, could damn the luckless victims? How hard it was, even for me, to get the better of this race of dialectical liars! And how futile was such success in dealing with people who twist the truth in your mouth, who without so much as a blush disavow the word they have just spoken, and in the very next minute take credit for it after all.

No. The better acquainted I became with the Jew, the more forgiving I inevitably became toward the worker.

In my eyes the gravest fault was no longer with him, but with all those who did not regard it as worth the trouble to have mercy on him, with iron righteousness giving the son of the people his just deserts, and standing the seducer and corrupter up against the wall.

Inspired by the experience of daily life, I now began to track down the sources of the Marxist doctrine. Its effects had become clear to me in individual cases; each day its success was apparent to my attentive eyes, and, with some exercise of my imagination, I was able to picture the consequences. The only remaining question was whether the result of their action in its ultimate form had existed in the mind's eye of the creators, or whether they themselves were the victims of an error.

I felt that both were possible.

In the one case it was the duty of every thinking man to force himself to the forefront of the ill-starred movement, thus perhaps averting catastrophe; in the other, however, the original founders of this plague of the nations must have been veritable devils; for only in the brain of a monster — not that of a man — could the plan of an organization assume form and meaning, whose activity must ultimately result in the collapse of human civilization and the consequent devastation of the world.

In this case the only remaining hope was struggle, struggle with all the weapons which the human spirit, reason, and will can devise, regardless on which side of the scale Fate should lay its blessing.

Thus I began to make myself familiar with the founders of this doctrine, in order to study the foundations of the movement. If I reached my goal more quickly than at first I had perhaps ventured to believe, it was thanks to my newly acquired, though at that time not very profound, knowledge of the Jewish question. This alone enabled me to draw a practical comparison between the reality and the theoretical flim-flam of the founding fathers of Social Democracy, since it taught me to understand the language of the Jewish people, who speak in order to conceal or at least to veil their thoughts; their real aim is not therefore to be found in the lines themselves, but slumbers well concealed between them.

For me this was the time of the greatest spiritual upheaval I have ever had to go through.

I had ceased to be a weak-kneed cosmopolitan and become an anti-Semite.

Just once more — and this was the last time — fearful, oppressive thoughts came to me in profound anguish.

When over long periods of human history I scrutinized the activity of the Jewish people, suddenly there rose up in me the fearful question whether inscrutable Destiny, perhaps for reasons unknown to us poor mortals, did not with eternal and immutable resolve, desire the final victory of this little nation.

Was it possible that the earth had been promised as a reward to this people which lives only for this earth?

Have we an objective right to struggle for our self-preservation, or is this justified only subjectively within ourselves?

As I delved more deeply into the teachings of Marxism and thus in tranquil clarity submitted the deeds of the Jewish people to contemplation, Fate itself gave me its answer.

The Jewish doctrine of Marxism rejects the aristocratic principle of Nature and replaces the eternal privilege of power and strength by the mass of numbers and their dead weight. Thus it denies the value of personality in man, contests the significance of nationality and race, and thereby withdraws from humanity the premise of its existence and its culture. As a foundation of the universe, this doctrine would bring about the end of any order intellectually conceivable to man. And as, in this greatest of all recognizable organisms, the result of an application of such a law could only be chaos, on earth it could only be destruction for the inhabitants of this planet.

If, with the help of his Marxist creed, the Jew is victorious over the other peoples of the world, his crown will be the funeral wreath of humanity and this planet will, as it did thousands [1] of years ago, move through the ether devoid of men.

Eternal Nature inexorably avenges the infringement of her commands.

Hence today I believe that I am acting in accordance with the will of the Almighty Creator: *by defending myself against the Jew, I am fighting for the work of the Lord.*

[1] Changed to 'millions' in second edition.

General Political Considerations Based on My Vienna Period

TODAY it is my conviction that in general, aside from cases of unusual talent, a man should not engage in public political activity before his thirtieth year. He should not do so, because up to this time, as a rule, he is engaged in molding a general platform, on the basis of which he proceeds to examine the various political problems and finally establishes his own position on them. Only after he has acquired such a basic philosophy, and the resultant firmness of outlook on the special problems of the day, is he, inwardly at least, mature enough to be justified in partaking in the political leadership of the general public.

Otherwise he runs the risk of either having to change his former position on essential questions, or, contrary to his better knowledge and understanding, of clinging to a view which reason and conviction have long since discarded. In the former case this is most embarrassing to him personally, since, what with his own vacillations, he cannot justifiably expect the faith of his adherents to follow him with the same unswerving firmness as before; for those led by him, on the other hand, such a reversal on the part of the leader means perplexity and not rarely a certain feeling of shame toward those whom they hitherto opposed. In the second case, there occurs a thing which, particularly today, often confronts us: in the same measure as the leader ceases to believe in what he says, his arguments become shallow and flat, but he tries to make up for it by vileness in his choice of means. While he himself has given up all idea of fighting seriously for his political revelations (a man does not die for something which he himself does not believe in), his demands on his supporters become correspondingly greater and more shameless until he ends up by sacrificing the last shred of leadership and turning into a 'politician'; in other words, the kind of man whose only real conviction is lack of conviction, combined with offensive impertinence and an art of lying, often developed to the point of complete shamelessness.

If to the misfortune of decent people such a character gets into a parliament, we may as well realize at once that the essence of his politics will from now on consist in nothing but an heroic struggle for the permanent possession of his feeding-bottle for himself and his family. The more his wife and children depend on it, the more tenaciously he will fight for his mandate. This alone will make every other man with political instincts his personal enemy; in every new movement he will scent the possible beginning of his end, and in every man of any greatness the danger which menaces him through that man.

I shall have more to say about this type of parliamentary bedbug.

Even a man of thirty will have much to learn in the course of his life, but this will only be to supplement and fill in the framework provided him by the philosophy he has basically adopted. When he learns, his learning will not have to be a revision of principle, but a supplementary study, and his supporters will not have to choke down the oppressive feeling that they have hitherto been falsely instructed by him. On the contrary: the visible, organic growth of the leader will give them satisfaction, for when he learns, he will only be deepening their own philosophy. And this in their eyes will be a proof for the correctness of the views they have hitherto held.

A leader who must depart from the platform of his general philosophy as such, because he recognizes it to be false, behaves with decency only if, in recognizing the error of his previous insight, he is prepared to draw the ultimate consequence. In such

a case he must, at the very least, renounce the public exercise of
any further political activity. For since in matters of basic knowledge he has once succumbed to an error, there is a possibility
that this will happen a second time. And in no event does he retain the right to continue claiming, not to mention demanding,
the confidence of his fellow citizens.

How little regard is taken of such decency today is attested by
the general degeneracy of the rabble which contemporaneously
feel justified in 'going into' politics.

Hardly a one of them is fit for it.

I had carefully avoided any public appearance, though I think
that I studied politics more closely than many other men. Only
in the smallest groups did I speak of the things which inwardly
moved or attracted me. This speaking in the narrowest circles
had many good points: I learned to orate less, but to know people
with their opinions and objections that were often so boundlessly
primitive. And I trained myself, without losing the time and
occasion for the continuance of my own education. It is certain
that nowhere else in Germany was the opportunity for this so
favorable as in Vienna.

                    *   *   *

General political thinking in the old Danubian monarchy was
just then broader and more comprehensive in scope than in old
Germany, excluding parts of Prussia, Hamburg, and the North
Sea coast, at the same period. In this case, to be sure, I understand, under the designation of 'Austria,' that section of the
great Habsburg Empire which, in consequence of its German settlement, not only was the historic cause of the very formation of
this state, but whose population, moreover, exclusively demonstrated that power which for so many centuries was able to give
this structure, so artificial in the political sense, its inner cultural
life. As time progressed, the existence and future of this state
came to depend more and more on the preservation of this nuclear
cell of the Empire.

If the old hereditary territories were the heart of the Empire, continually driving fresh blood into the circulatory stream of political and cultural life, Vienna was the brain and will in one.

Its mere outward appearance justified one in attributing to this city the power to reign as a unifying queen amid such a conglomeration of peoples, thus by the radiance of her own beauty causing us to forget the ugly symptoms of old age in the structure as a whole.

The Empire might quiver and quake beneath the bloody battles of the different nationalities, yet foreigners, and especially Germans, saw only the charming countenance of this city. What made the deception all the greater was that Vienna at that time seemed engaged in what was perhaps its last and greatest visible revival. Under the rule of a truly gifted mayor, the venerable residence of the Emperors of the old régime awoke once more to a miraculous youth. The last great German to be born in the ranks of the people who had colonized the Ostmark was not officially numbered among so-called 'statesmen'; but as mayor of Vienna, this 'capital and imperial residence,' Dr. Lueger conjured up one amazing achievement after another in, we may say, every field of economic and cultural municipal politics, thereby strengthening the heart of the whole Empire, and indirectly becoming a statesman greater than all the so-called 'diplomats' of the time.

If the conglomeration of nations called 'Austria' nevertheless perished in the end, this does not detract in the least from the political ability of the Germans in the old Ostmark, but was the necessary result of the impossibility of permanently maintaining a state of fifty million people of different nationalities by means of ten million people, unless certain definite prerequisites were established in time.

The ideas of the German-Austrian were more than grandiose.

He had always been accustomed to living in a great empire and had never lost his feeling for the tasks bound up with it. He was the only one in this state who, beyond the narrow boundaries of the crown lands, still saw the boundaries of the Reich; indeed, when Fate finally parted him from the common fatherland, he kept on striving to master the gigantic task and preserve for the German people what his fathers had once wrested from the East in endless struggles. In this connection it should be borne in mind that this had to be done with divided energy; for the heart and memory of the best never ceased to feel for the common mother country, and only a remnant was left for the homeland.

The general horizon of the German-Austrian was in itself comparatively broad. His economic connections frequently embraced almost the entire multiform Empire. Nearly all the big business enterprises were in his hands; the directing personnel, both technicians and officials, were in large part provided by him. He was also in charge of foreign trade in so far as the Jews had not laid their hands on this domain, which they have always seized for their own. Politically, he alone held the state together. Military service alone cast him far beyond the narrow boundaries of his homeland. The German-Austrian recruit might join a German regiment, but the regiment itself might equally well be in Herzegovina, Vienna, or Galicia. The officers' corps was still German, the higher officials predominantly so. Finally, art and science were German. Aside from the trash of the more modern artistic development, which a nation of Negroes might just as well have produced, the German alone possessed and disseminated a truly artistic attitude. In music, architecture, sculpture, and painting, Vienna was the source supplying the entire dual monarchy in inexhaustible abundance, without ever seeming to go dry itself.

Finally, the Germans directed the entire foreign policy if we disregard a small number of Hungarians.

And yet any attempt to preserve this Empire was in vain, for the most essential premise was lacking.

For the Austrian state of nationalities there was only one possibility of overcoming the centrifugal forces of the individual nations. Either the state was centrally governed, hence internally organized along the same lines, or it was altogether inconceivable.

At various lucid moments this insight dawned on the 'supreme' authority. But as a rule it was soon forgotten or shelved as difficult of execution. Any thought of a more federative organization of the Empire was doomed to failure owing to the lack of a strong political germ-cell of outstanding power. Added to this were the internal conditions of the Austrian state which differed essentially from the German Empire of Bismarck. In Germany it was only a question of overcoming political conditions, since there was always a common cultural foundation. Above all, the Reich, aside from little foreign splinters, embraced members of only one people.

In Austria the opposite was the case.

Here the individual provinces, aside from Hungary, lacked any political memory of their own greatness, or it had been erased by the sponge of time, or at least blurred and obscured. In the period when the principle of nationalities was developing, however, national forces rose up in the various provinces, and to counteract them was all the more difficult as on the rim of the monarchy national states began to form whose populations, racially equivalent or related to the Austrian national splinters, were now able to exert a greater power of attraction than, conversely, remained possible for the German-Austrian.

Even Vienna could not forever endure this struggle.

With the development of Budapest into a big city, she had for the first time a rival whose task was no longer to hold the entire monarchy together, but rather to strengthen a part of it. In a short time Prague was to follow her example, then Lemberg, Laibach, etc. With the rise of these former provincial cities to national capitals of individual provinces, centers formed for more or less independent cultural life in these provinces. And only then did the politico-national instincts obtain their spiritual foundation and depth. The time inevitably approached when these dynamic forces of the individual peoples would grow stronger than the force of common interests, and that would be the end of Austria.

Since the death of Joseph II [1] the course of this development

---

[1] Joseph II, Holy Roman Emperor (1765-1790). Until 1780 he shared the throne with his mother, Maria Theresa. From 1780 till his death he ruled alone. Hostile to the clergy and nobility, he made a great effort to unify and

was clearly discernible. Its rapidity depended on a series of factors which in part lay in the monarchy itself and in part were the result of the Empire's momentary position on foreign policy.

If the fight for the preservation of this state was to be taken up and carried on in earnest, only a ruthless and persistent policy of centralization could lead to the goal. First of all, the purely formal cohesion had to be emphasized by the establishment in principle of a uniform official language, and the administration had to be given the technical implement without which a unified state simply cannot exist. Likewise a unified state-consciousness could only be bred for any length of time by schools and education. This was not feasible in ten or twenty years; it was inevitably a matter of centuries; for in all questions of colonization, persistence assumes greater importance than the energy of the moment.

It goes without saying that the administration as well as the political direction must be conducted with strict uniformity. To me it was infinitely instructive to ascertain why this did not occur, or rather, why it was not done.[1] He who was guilty of this omission was alone to blame for the collapse of the Empire.

Old Austria more than any other state depended on the greatness of her leaders. The foundation was lacking for a national state, which in its national basis always possesses the power of survival, regardless how deficient the leadership as such may be. A homogeneous national state can, by virtue of the natural inertia of its inhabitants, and the resulting power of resistance, sometimes withstand astonishingly long periods of the worst administration or leadership without inwardly disintegrating. At such times it often seems as though there were no more life in such a body, as though it were dead and done for, but one fine day the supposed corpse suddenly rises and gives the rest of

modernize the Empire in accordance with the principles of the Enlightenment. For a time Joseph removed the Hungarian crown to Vienna, but restored it after bitter opposition from the Hungarians. He had a plan to join Bavaria to Austria by giving Charles Theodore, Elector of the Palatinate (to which Bavaria then belonged), the Austrian Netherlands in return.

[1] *'Warum dies nicht geschah, oder besser, warum dies nicht getan.'*

humanity astonishing indications of its unquenchable vital force.

It is different, however, with an empire not consisting of similar peoples, which is held together, not by common blood but by a common fist. In this case the weakness of leadership will not cause a hibernation of the state, but an awakening of all the individual instincts which are present in the blood, but cannot develop in times when there is a dominant will. Only by a common education extending over centuries, by common tradition, common interests, etc., can this danger be attenuated. Hence the younger such state formations are, the more they depend on the greatness of leadership, and if they are the work of outstanding soldiers and spiritual heroes, they often crumble immediately after the death of the great solitary founder. But even after centuries these dangers cannot be regarded as overcome; they only lie dormant, often suddenly to awaken as soon as the weakness of the common leadership and the force of education and all the sublime traditions can no longer overcome the impetus of the vital urge of the individual tribes.

Not to have understood this is perhaps the tragic guilt of the House of Habsburg.

For only a single one of them did Fate once again raise high the torch over the future of his country, then it was extinguished forever.

Joseph II, Roman Emperor of the German nation, saw with fear and trepidation how his House, forced to the outermost corner of the Empire, would one day inevitably vanish in the maelstrom of a Babylon of nations unless at the eleventh hour the omissions of his forefathers were made good. With superhuman power this 'friend of man' braced himself against the negligence of his ancestors and endeavored to retrieve in one decade what centuries had failed to do. If he had been granted only forty years for his work, and if after him even two generations had continued his work as he began it, the miracle would probably have been achieved. But when, after scarcely ten years on the throne, worn in body and soul, he died, his work sank with him into the grave, to awaken no more and sleep forever in the

Capuchin crypt. His successors were equal to the task neither in mind nor in will.

When the first revolutionary lightnings of a new era flashed through Europe, Austria, too, slowly began to catch fire, little by little. But when the fire at length broke out, the flame was fanned less by social or general political causes than by dynamic forces of national origin.

The revolution of 1848 may have been a class struggle everywhere, but in Austria it was the beginning of a new racial war. By forgetting or not recognizing this origin and putting themselves in the service of the revolutionary uprising, the Germans sealed their own fate. They helped to arouse the spirit of 'Western democracy,' which in a short time removed the foundations of their own existence.

With the formation of a parliamentary representative body without the previous establishment and crystallization of a common state language, the cornerstone had been laid for the end of German domination of the monarchy.[1] From this moment on the state itself was lost. All that followed was merely the historic liquidation of an empire.

To follow this process of dissolution was as heartrending as it was instructive. This execution of an historical sentence was carried out in detail in thousands and thousands of forms. The fact that a large part of the people moved blindly through the manifestations of decay showed only that the gods had willed Austria's destruction.

I shall not lose myself in details on this point, for that is not the function of this book. I shall only submit to a more thorough-going observation those events which are the ever-unchanging causes of the decline of nations and states, thus possessing significance for our time as well, and which ultimately contributed to securing the foundations of my own political thinking.

*   *   *

[1] ' ... war er Grundstein zum Ende Vorherrschaft des Deutschtums in der Monarchie gelegt worden.'

At the head of those institutions which could most clearly have revealed the erosion of the Austrian monarchy, even to a shopkeeper not otherwise gifted with sharp eyes, was one which ought to have had the greatest strength — parliament, or, as it was called in Austria, the Reichsrat.

Obviously the example of this body had been taken from England, the land of classical 'democracy.' From there the whole blissful institution was taken and transferred as unchanged as possible to Vienna.

The English two-chamber system was solemnly resurrected in the *Abgeordnetenhaus* and the *Herrenhaus*. Except that the 'houses' themselves were somewhat different. When Barry raised his parliament buildings from the waters of the Thames, he thrust into the history of the British Empire and from it took the decorations for the twelve hundred niches, consoles, and pillars of his magnificent edifice. Thus, in their sculpture and painting, the House of Lords and the House of Commons became the nation's Hall of Fame.

This was where the first difficulty came in for Vienna. For when Hansen, the Danish builder, had completed the last pinnacle on the marble building of the new parliament, there was nothing he could use as decoration except borrowings from antiquity. Roman and Greek statesmen and philosophers now embellish this opera house of Western democracy, and in symbolic irony the *quadrigae* fly from one another in all four directions above the two houses, in this way giving the best external expression of the activities that went on inside the building.

The 'nationalities' had vetoed the glorification of Austrian history in this work as an insult and provocation, just as in the Reich itself it was only beneath the thunder of World War battles that they dared to dedicate Wallot's Reichstag Building to the German people by an inscription.

When, not yet twenty years old, I set foot for the first time in the magnificent building on the Franzensring to attend a session of the House of Deputies as a spectator and listener, I was seized with the most conflicting sentiments.

I had always hated parliament, but not as an institution in itself. On the contrary, as a freedom-loving man I could not even conceive of any other possibility of government, for the idea of any sort of dictatorship would, in view of my attitude toward the House of Habsburg, have seemed to me a crime against freedom and all reason.

What contributed no little to this was that as a young man, in consequence of my extensive newspaper reading, I had, without myself realizing it, been inoculated with a certain admiration for the British Parliament, of which I was not easily able to rid myself. The dignity with which the Lower House there fulfilled its tasks (as was so touchingly described in our press) impressed me immensely. Could a people have any more exalted form of self-government?

But for this very reason I was an enemy of the Austrian parliament. I considered its whole mode of conduct unworthy of the great example. To this the following was now added:

The fate of the Germans in the Austrian state was dependent on their position in the Reichsrat. Up to the introduction of universal and secret suffrage, the Germans had had a majority, though an insignificant one, in parliament. Even this condition was precarious, for the Social Democrats, with their unreliable attitude in national questions, always turned against German interests in critical matters affecting the Germans — in order not to alienate the members of the various foreign nationalities. Even in those days the Social Democracy could not be regarded as a German party. And with the introduction of universal suffrage the German superiority ceased even in a purely numerical sense. There was no longer any obstacle in the path of the further de-Germanization of the state.

For this reason my instinct of national self-preservation caused me even in those days to have little love for a representative body in which the Germans were always misrepresented rather than represented. Yet these were deficiencies which, like so many others, were attributable, not to the thing in itself, but to the Austrian state. I still believed that if a German majority were restored in the representative bodies, there would no longer be any reason for a principled opposition to them, that is, as long as the old state continued to exist at all.

These were my inner sentiments when for the first time I set foot in these halls as hallowed as they were disputed. For me, to be sure, they were hallowed only by the lofty beauty of the magnificent building. A Hellenic miracle on German soil!

How soon was I to grow indignant when I saw the lamentable comedy that unfolded beneath my eyes!

Present were a few hundred of these popular representatives who had to take a position on a question of most vital economic importance.

The very first day was enough to stimulate me to thought for weeks on end.

The intellectual content of what these men said was on a really depressing level, in so far as you could understand their babbling at all; for several of the gentlemen did not speak German, but their native Slavic languages or rather dialects. I now had occasion to hear with my own ears what previously I had known only from reading the newspapers. A wild gesticulating mass screaming all at once in every different key, presided over by a good-natured old uncle who was striving in the sweat of his brow to revive the dignity of the House by violently ringing his bell and alternating gentle reproofs with grave admonitions.

I couldn't help laughing.

A few weeks later I was in the House again. The picture was changed beyond recognition. The hall was absolutely empty. Down below everybody was asleep. A few deputies were in their places, yawning at one another; one was 'speaking.' A vice-president of the House was present, looking into the hall with obvious boredom.

The first misgivings arose in me. From now on, whenever time offered me the slightest opportunity, I went back and, with silence and attention, viewed whatever picture presented itself, listened to the speeches in so far as they were intelligible, studied the more or less intelligent faces of the elect of the peoples of this woe-begone state — and little by little formed my own ideas.

A year of this tranquil observation sufficed totally to change or eliminate my former view of the nature of this institution. My innermost position was no longer against the misshapen form which this idea assumed in Austria; no, by now I could no longer accept the parliament as such. Up till then I had seen the misfortune of the Austrian parliament in the absence of a German majority; now I saw that its ruination lay in the whole nature and essence of the institution as such.

A whole series of questions rose up in me.

I began to make myself familiar with the democratic principle of majority rule as the foundation of this whole institution, but devoted no less attention to the intellectual and moral values of these gentlemen, supposedly the elect of the nations, who were expected to serve this purpose.

Thus I came to know the institution and its representatives at once.

In the course of a few years, my knowledge and insight shaped a plastic model of that most dignified phenomenon of modern times: the parliamentarian. He began to impress himself upon me in a form which has never since been subjected to any essential change.

Here again the visual instruction of practical reality had prevented me from being stifled by a theory which at first sight seemed seductive to so many, but which none the less must be counted among the symptoms of human degeneration.

The Western democracy of today is the forerunner of Marxism which without it would not be thinkable. It provides this world plague with the culture in which its germs can spread. In its most extreme form, parliamentarianism created a 'monstrosity of excrement and fire,'[1] in which, however, sad to say, the 'fire' seems to me at the moment to be burned out.

I must be more than thankful to Fate for laying this question

---

[1] '*Spottgeburt aus Dreck und Feuer.*' Should be '*von Dreck und Feuer.*' Goethe's *Faust*, Part 1, 5356: Faust to Mephistopheles.

before me while I was in Vienna, for I fear that in Germany at that time I would have found the answer too easily. For if I had first encountered this absurd institution known as 'parliament' in Berlin, I might have fallen into the opposite fallacy, and not without seemingly good cause have sided with those who saw the salvation of the people and the Reich exclusively in furthering the power of the imperial idea, and who nevertheless were alien and blind at once to the times and the people involved.

In Austria this was impossible.

Here it was not so easy to go from one mistake to the other. If parliament was worthless, the Habsburgs were even more worthless — in no event, less so. To reject 'parliamentarianism' was not enough, for the question still remained open: what then? The rejection and abolition of the Reichsrat would have left the House of Habsburg the sole governing force, a thought which, especially for me, was utterly intolerable.

The difficulty of this special case led me to a more thorough contemplation of the problem as such than would otherwise have been likely at such tender years.

What gave me most food for thought was the obvious absence of any responsibility in a single person.

The parliament arrives at some decision whose consequences may be ever so ruinous — nobody bears any responsibility for this, no one can be taken to account. For can it be called an acceptance of responsibility if, after an unparalleled catastrophe, the guilty government resigns? Or if the coalition changes, or even if parliament is itself dissolved?

Can a fluctuating majority of people ever be made responsible in any case?

Isn't the very idea of responsibility bound up with the individual?

But can an individual directing a government be made practically responsible for actions whose preparation and execution must be set exclusively to the account of the will and inclination of a multitude of men?

Or will not the task of a leading statesman be seen, not in the birth of a creative idea or plan as such, but rather in the art of making the brilliance of his projects intelligible to a herd of sheep and blockheads, and subsequently begging for their kind approval?

Is it the criterion of the statesman that he should possess the art of persuasion in as high degree as that of political intelligence in formulating great policies or decisions? Is the incapacity of a leader shown by the fact that he does not succeed in winning for a certain idea the majority of a mob thrown together by more or less savory accidents?

Indeed, has this mob ever understood an idea before success proclaimed its greatness?

Isn't every deed of genius in this world a visible protest of genius against the inertia of the mass?

And what should the statesman do, who does not succeed in gaining the favor of this mob for his plans by flattery?

Should he buy it?

Or, in view of the stupidity of his fellow citizens, should he renounce the execution of the tasks which he has recognized to be vital necessities? Should he resign or should he remain at his post?

In such a case, doesn't a man of true character find himself in a hopeless conflict between knowledge and decency, or rather honest conviction?

Where is the dividing line between his duty toward the general public and his duty toward his personal honor?

Mustn't every true leader refuse to be thus degraded to the level of a political gangster?

And, conversely, mustn't every gangster feel that he is cut out for politics, since it is never he, but some intangible mob, which has to bear the ultimate responsibility?

Mustn't our principle of parliamentary majorities lead to the demolition of any idea of leadership?

Does anyone believe that the progress of this world springs from the mind of majorities and not from the brains of individuals?

Or does anyone expect that the future will be able to dispense with this premise of human culture?

Does it not, on the contrary, today seem more indispensable than ever?

By rejecting the authority of the individual and replacing it by the numbers of some momentary mob, the parliamentary principle of majority rule sins against the basic aristocratic principle of Nature, though it must be said that this view is not necessarily embodied in the present-day decadence of our upper ten thousand.

The devastation caused by this institution of modern parliamentary rule is hard for the reader of Jewish newspapers to imagine, unless he has learned to think and examine independently. It is, first and foremost, the cause of the incredible inundation of all political life with the most inferior, and I mean the most inferior, characters of our time. Just as the true leader will withdraw from all political activity which does not consist primarily in creative achievement and work, but in bargaining and haggling for the favor of the majority, in the same measure this activity will suit the small mind and consequently attract it.

The more dwarfish one of these present-day leather-merchants is in spirit and ability, the more clearly his own insight makes him aware of the lamentable figure he actually cuts — that much more will he sing the praises of a system which does not demand of him the power and genius of a giant, but is satisfied with the craftiness of a village mayor, preferring in fact this kind of wisdom to that of a Pericles. And this kind doesn't have to torment himself with responsibility for his actions. He is entirely removed from such worry, for he well knows that, regardless what the result of his 'statesmanlike' bungling may be, his end has long been written in the stars: one day he will have to cede his place to another equally great mind, for it is one of the characteristics of this decadent system that the number of great statesmen increases in proportion as the stature of the individual decreases. With increasing dependence on parliamentary majorities it will inevitably continue to shrink, since on the one hand great minds will refuse to be the stooges of idiotic incompetents and bigmouths, and on the other, conversely, the representatives of the

majority, hence of stupidity, hate nothing more passionately
than a superior mind.

For such an assembly of wise men of Gotham, it is always a
consolation to know that they are headed by a leader whose intelligence is at the level of those present: this will give each one
the pleasure of shining from time to time — and, above all, if
Tom can be master, what is to prevent Dick and Harry from
having their turn too?

This invention of democracy is most intimately related to a
quality which in recent times has grown to be a real disgrace, to
wit, the cowardice of a great part of our so-called 'leadership.'
What luck to be able to hide behind the skirts of a so-called majority in all decisions of any real importance!

Take a look at one of these political bandits. How anxiously
he begs the approval of the majority for every measure, to assure
himself of the necessary accomplices, so he can unload the responsibility at any time. And this is one of the main reasons why
this type of political activity is always repulsive and hateful to
any man who is decent at heart and hence courageous, while it
attracts all low characters — and anyone who is unwilling to
take personal responsibility for his acts, but seeks a shield, is a
cowardly scoundrel. When the leaders of a nation consist of such
vile creatures, the results will soon be deplorable. Such a nation
will be unable to muster the courage for any determined act; it
will prefer to accept any dishonor, even the most shameful, rather
than rise to a decision; for there is no one who is prepared of his
own accord to pledge his person and his head for the execution of
a dauntless resolve.

For there is one thing which we must never forget: in this, too,
the majority can never replace the man. It is not only a representative of stupidity, but of cowardice as well. And no more
than a hundred empty heads make one wise man will an heroic
decision arise from a hundred cowards.

The less the responsibility of the individual leader, the more
numerous will be those who, despite their most insignificant
stature, feel called upon to put their immortal forces in the

service of the nation. Indeed, they will be unable to await their turn; they stand in a long line, and with pain and regret count the number of those waiting ahead of them, calculating almost the precise hour at which, in all probability, their turn will come. Consequently, they long for any change in the office hovering before their eyes, and are thankful for any scandal which thins out the ranks ahead of them. And if some man is unwilling to move from the post he holds, this in their eyes is practically a breach of a holy pact of solidarity. They grow vindictive, and they do not rest until the impudent fellow is at last overthrown, thus turning his warm place back to the public. And, rest assured, he won't recover the position so easily. For as soon as one of these creatures is forced to give up a position, he will try at once to wedge his way into the 'waiting-line' unless the hue and cry raised by the others prevents him.

The consequence of all this is a terrifying turn-over in the most important offices and positions of such a state, a result which is always harmful, but sometimes positively catastrophic. For it is not only the simpleton and incompetent who will fall victim to this custom, but to an even greater extent the real leader, if Fate somehow manages to put one in this place. As soon as this fact has been recognized, a solid front will form against him, especially if such a mind has not arisen from their own ranks, but none the less dares to enter into this exalted society. For on principle these gentry like to be among themselves and they hate as a common enemy any brain which stands even slightly above the zeros. And in this respect their instinct is as much sharper as it is deficient in everything else.

The result will be a steadily expanding intellectual impoverishment of the leading circles. The result for the nation and the state, everyone can judge for himself, excepting in so far as he himself is one of these kind of 'leaders.'

Old Austria possessed the parliamentary régime in its purest form.

To be sure, the prime ministers were always appointed by the Emperor and King, but this very appointment was nothing but

the execution of the parliamentary will. The haggling and bargaining for the individual portfolios represented Western democracy of the first water. And the results corresponded to the
principles applied. Particularly the change of individual personalities occurred in shorter and shorter terms, ultimately becoming a
veritable chase. In the same measure, the stature of the 'statesmen' steadily diminished until finally no one remained but that
type of parliamentary gangster whose statesmanship could only
be measured and recognized by their ability in pasting together
the coalitions of the moment; in other words, concluding those
pettiest of political bargains which alone demonstrate the fitness
of these representatives of the people for practical work.

Thus the Viennese school transmitted the best impressions in
this field.

But what attracted me no less was to compare the ability and
knowledge of these representatives of the people and the tasks
which awaited them. In this case, whether I liked it or not, I was
impelled to examine more closely the intellectual horizon of these
elect of the nations themselves, and in so doing, I could not avoid
giving the necessary attention to the processes which lead to the
discovery of these ornaments of our public life.

The way in which the real ability of these gentlemen was applied and placed in the service of the fatherland — in other words,
the technical process of their activity — was also worthy of
thorough study and investigation.

The more determined I was to penetrate these inner conditions,
to study the personalities and material foundations with dauntless and penetrating objectivity, the more deplorable became my
total picture of parliamentary life. Indeed, this is an advisable
procedure in dealing with an institution which, in the person of
its representatives, feels obliged to bring up 'objectivity' in every
second sentence as the only proper basis for every investigation
and opinion. Investigate these gentlemen themselves and the laws
of their sordid existence, and you will be amazed at the result.

There is no principle which, objectively considered, is as false
as that of parliamentarianism.

Here we may totally disregard the manner in which our fine representatives of the people are chosen, how they arrive at their office and their new dignity. That only the tiniest fraction of them rise in fulfillment of a general desire, let alone a need, will at once be apparent to anyone who realizes that the political understanding of the broad masses is far from being highly enough developed to arrive at definite general political views of their own accord and seek out the suitable personalities.

The thing we designate by the word 'public opinion' rests only in the smallest part on experience or knowledge which the individual has acquired by himself, but rather on an idea which is inspired by so-called 'enlightenment,' often of a highly persistent and obtrusive type.

Just as a man's denominational orientation is the result of upbringing, and only the religious need as such slumbers in his soul, the political opinion of the masses represents nothing but the final result of an incredibly tenacious and thorough manipulation of their mind and soul.

By far the greatest share in their political 'education,' which in this case is most aptly designated by the word 'propaganda,' falls to the account of the press. It is foremost in performing this 'work of enlightenment' and thus represents a sort of school for grown-ups. This instruction, however, is not in the hands of the state, but in the claws of forces which are in part very inferior. In Vienna as a very young man I had the best opportunity to become acquainted with the owners and spiritual manufacturers of this machine for educating the masses. At first I could not help but be amazed at how short a time it took this great evil power within the state to create a certain opinion even where it meant totally falsifying profound desires and views which surely existed among the public. In a few days a ridiculous episode had become a significant state action, while, conversely, at the same time, vital problems fell a prey to public oblivion, or rather were simply filched from the memory and consciousness of the masses.

Thus, in the course of a few weeks it was possible to conjure up names out of the void, to associate them with incredible hopes on the part of the broad public, even to give them a popularity which the really great man often does not obtain his whole life long; names which a month before no one had even seen or heard of, while at the same time old and proved figures of political or other public life, though in the best of health, simply died as far as their fellow men were concerned, or were heaped with such vile insults that their names soon threatened to become the symbol of some definite act of infamy or villainy. We must study this vile Jewish technique of emptying garbage pails full of the vilest slanders and defamations from hundreds and hundreds of sources at once, suddenly and as if by magic, on the clean garments of honorable men, if we are fully to appreciate the entire menace represented by these scoundrels of the press.

There is absolutely nothing one of these spiritual robber-barons will not do to achieve his savory aims.

He will poke into the most secret family affairs and not rest until his truffle-searching instinct digs up some miserable incident which is calculated to finish off the unfortunate victim. But if, after the most careful sniffing, absolutely nothing is found, either in the man's public or private life, one of these scoundrels simply seizes on slander, in the firm conviction that despite a thousand refutations something always sticks and, moreover, through the immediate and hundredfold repetition of his defamations by all his accomplices, any resistance on the part of the victim is in most cases utterly impossible; and it must be borne in mind that this rabble never acts out of motives which might seem credible or even understandable to the rest of humanity. God forbid! While one of these scum is attacking his beloved fellow men in the most contemptible fashion, the octopus covers himself with a veritable cloud of respectability and unctuous phrases, prates about 'journalistic duty' and such-like lies, and even goes so far as to shoot off his mouth at committee meetings and congresses — that is, occasions where these pests are present in large numbers — about a very special variety of 'honor,' to wit, the journalistic variety, which the assembled rabble gravely and mutually confirm.

These scum manufacture more than three quarters of the so-called 'public opinion,' from whose foam the parliamentarian Aphrodite arises. To give an accurate description of this process and depict it in all its falsehood and improbability, one would have to write volumes. But even if we disregard all this and examine only the given product along with its activity, this seems to me enough to make the objective lunacy of this institution dawn on even the naïvest mind.

This human error, as senseless as it is dangerous, will most readily be understood as soon as we compare democratic parliamentarianism with a truly Germanic democracy.

The distinguishing feature of the former is that a body of, let us say five hundred men, or in recent times even women, is chosen and entrusted with making the ultimate decision in any and all matters. And so for practical purposes they alone are the government; for even if they do choose a cabinet which undertakes the external direction of the affairs of state, this is a mere sham. In reality this so-called government cannot take a step without first obtaining the approval of the general assembly. Consequently, it cannot be made responsible for anything, since the ultimate decision never lies with it, but with the majority of parliament. In every case it does nothing but carry out the momentary will of the majority. Its political ability can only be judged according to the skill with which it understands how either to adapt itself to the will of the majority or to pull the majority over to its side. Thereby it sinks from the heights of real government to the level of a beggar confronting the momentary majority. Indeed, its most urgent task becomes nothing more than either to secure the favor of the existing majority, as the need arises, or to form a majority with more friendly inclinations. If this succeeds, it may 'govern' a little while longer; if it doesn't succeed, it can resign. The soundness of its purposes as such is beside the point.

For practical purposes, this excludes all responsibility.

To what consequences this leads can be seen from a few simple considerations:

The internal composition of the five hundred chosen repre-

sentatives of the people, with regard to profession or even individual abilities, gives a picture as incoherent as it is usually
deplorable. For no one can believe that these men elected by the
nation are elect of spirit or even of intelligence! It is to be hoped
that no one will suppose that the ballots of an electorate which is
anything else than brilliant will give rise to statesmen by the
hundreds. Altogether we cannot be too sharp in condemning the
absurd notion that geniuses can be born from general elections.
In the first place, a nation only produces a real statesman once
in a blue moon and not a hundred or more at once; and in the
second place, the revulsion of the masses for every outstanding
genius is positively instinctive. Sooner will a camel pass through
a needle's eye than a great man be 'discovered' by an election.

In world history the man who really rises above the norm of
the broad average usually announces himself personally.

As it is, however, five hundred men, whose stature is to say the
least modest, vote on the most important affairs of the nation,
appoint governments which in every single case and in every
special question have to get the approval of the exalted assembly,
so that policy is really made by five hundred.

And that is just what it usually looks like.

But even leaving the genius of these representatives of the
people aside, bear in mind how varied are the problems awaiting
attention, in what widely removed fields solutions and decisions
must be made, and you will realize how inadequate a governing
institution must be which transfers the ultimate right of decision
to a mass assembly of people, only a tiny fraction of which possess
knowledge and experience of the matter to be treated. The most
important economic measures are thus submitted to a forum,
only a tenth of whose members have any economic education to
show. This is nothing more nor less than placing the ultimate
decision in a matter in the hands of men totally lacking in every
prerequisite for the task.

The same is true of every other question. The decision is always
made by a majority of ignoramuses and incompetents, since the
composition of this institution remains unchanged while the

problems under treatment extend to nearly every province of public life and would thereby presuppose a constant turn-over in the deputies who are to judge and decide on them, since it is impossible to let the same persons decide matters of transportation as, let us say, a question of high foɪ eign policy. Otherwise these men would all have to be universal geniuses such as we actually seldom encounter once in centuries. Unfortunately we are here confronted, for the most part, not with 'thinkers,' but with dilettantes as limited as they are conceited and inflated, intellectual *demi-monde* of the worst sort. And this is the source of the often incomprehensible frivolity with which these gentry speak and decide on things which would require careful meditation even in the greatest minds. Measures of the gravest significance for the future of a whole state, yes, of a nation, are passed as though a game of *schafkopf* or *tarock*,[1] which would certainly be better suited to their abilities, lay on the table before them and not the fate of a race.

Yet it would surely be unjust to believe that all of the deputies in such a parliament were personally endowed with so little sense of responsibility.

No, by no means.

But by forcing the individual to take a position on such questions completely ill-suited to him, this system gradually ruins his character. No one will summon up the courage to declare: 'Gentlemen, I believe we understand nothing about this matter. I personally certainly do not.' (Besides, this would change matters little, for surely this kind of honesty would remain totally unappreciated, and what is more, our friends would scarcely allow one honorable jackass to spoil their whole game.) Anyone with a knowledge of people will realize that in such an illustrious company no one is eager to be the stupidest, and in certain circles honesty is almost synonymous with stupidity.

Thus, even the representative who at first was honest is thrown

[1] *Schafkopf* is a four-handed card-game widely played in Germany. *Tarock.* Three-handed card-game of Italian origin (*tarocco*), popular in Austria and southern Germany.

into this track of general falsehood and deceit. The very conviction that the non-participation of an individual in the business would in itself change nothing kills every honorable impulse which may rise up in this or that deputy. And finally, moreover, he may tell himself that he personally is far from being the worst among the others,[1] and that the sole effect of his collaboration is perhaps to prevent worse things from happening.

It will be objected, to be sure, that. though the individual deputy possesses no special understanding in this or that matter, his position has been discussed by the fraction which directs the policy of the gentleman in question, and that the fraction has its special committees which are more than adequately enlightened by experts anyway.

At first glance this seems to be true. But then the question arises: Why are five hundred chosen when only a few possess the necessary wisdom to take a position in the most important matters?

And this is the worm in the apple!

It is not the aim of our present-day parliamentarianism to constitute an assembly of wise men, but rather to compose a band of mentally dependent nonentities who are the more easily led in certain directions, the greater is the personal limitation of the individual. That is the only way of carrying on party politics in the malodorous present-day sense. And only in this way is it possible for the real wirepuller to remain carefully in the background and never personally be called to responsibility. For then every decision, regardless how harmful to the nation, will not be set to the account of a scoundrel visible to all, but will be unloaded on the shoulders of a whole fraction.

And thereby every practical responsibility vanishes. For responsibility can lie only in the obligation of an individual and not in a parliamentary bull session.

Such an institution can only please the biggest liars and sneaks of the sort that shun the light of day, because it is inevitably hateful to an honorable, straightforward man who welcomes personal responsibility.

[1] *'der Schlechteste unter den Anderen.'*

And that is why this type of democracy has become the instrument of that race which in its inner goals must shun the light of day, now and in all ages of the future. Only the Jew can praise an institution which is as dirty and false as he himself.

* * *

Juxtaposed to this is the truly Germanic democracy characterized by the free election of a leader and his obligation fully to assume all responsibility for his actions and omissions. In it there is no majority vote on individual questions, but only the decision of an individual who must answer with his fortune and his life for his choice.

If it be objected that under such conditions scarcely anyone would be prepared to dedicate his person to so risky a task, there is but one possible answer:

Thank the Lord, Germanic democracy means just this: that any old climber or moral slacker cannot rise by devious paths to govern his national comrades,[1] but that, by the very greatness of the responsibility to be assumed, incompetents and weaklings are frightened off.

But if, nevertheless, one of these scoundrels should attempt to sneak in, we can find him more easily, and mercilessly challenge him: Out, cowardly scoundrel! Remove your foot, you are besmirching the steps; the front steps of the Pantheon of history are not for sneak-thieves, but for heroes!

* * *

[1] '*Volksgenosse.*' Brockhaus defines: In contrast to the concept of citizen which is based on the idea of legal equality in the state, the designation for all members of the same national community (*Volksgemeinschaft*), especially those who form a working association in the service of the nation as a whole. As used by the National Socialists, it might be translated as 'racial comrades.' I have chosen the more neutral term 'national comrades' because the National Socialists did not coin the term and it occurs frequently in the speeches of parliamentarians who were not even noted for their anti-Semitism.

I had fought my way to this conclusion after two years attendance at the Vienna parliament.

After that I never went back.

The parliamentary régime shared the chief blame for the weakness, constantly increasing in the past few years, of the Habsburg
state. The more its activities broke the predominance of the
Germans, the more the country succumbed to a system of playing off the nationalities against one another. In the Reichsrat
itself this was always done at the expense of the Germans and
thereby, in the last analysis, at the expense of the Empire; for by
the turn of the century it must have been apparent even to the
simplest that the monarchy's force of attraction would no longer
be able to withstand the separatist tendencies of the provinces.

On the contrary.

The more pathetic became the means which the state had to
employ for its preservation, the more the general contempt for it
increased. Not only in Hungary, but also in the separate Slavic
provinces, people began to identify themselves so little with the
common monarchy that they did not regard its weakness as their
own disgrace. On the contrary, they rejoiced at such symptoms
of old age; for they hoped more for the Empire's death than for
its recovery.

In parliament, for the moment, total collapse was averted by
undignified submissiveness and acquiescence at every extortion,
for which the German had to pay in the end; and in the country,
by most skillfully playing off the different peoples against each
other. But the general line of development was nevertheless directed against the Germans. Especially since Archduke Francis
Ferdinand [1] became heir apparent and began to enjoy a certain

---

[1] Archduke Francis Ferdinand (1863–1914), brother of Francis Joseph,
became heir apparent to the Austrian throne in 1889 when Archduke Rudolf,
only son of Francis Joseph, committed suicide. Francis Ferdinand, who
favored a reorganization of the monarchy giving the Slavic population
equality with the Germans and Hungarians, was assassinated at Sarajevo
on June 28, 1914. Hitler's talk of Czechization is an exaggeration, to say
the least. The most that can be said is that some of the leaders of the
monarchy favored collaboration with the Slavic upper classes and some relief
for the oppressed Slavic populations.

influence, there began to be some plan and order in the policy of
Czechization from above. With all possible means, this future
ruler of the dual monarchy tried to encourage a policy of de-
Germanization, to advance it himself or at least to sanction it.
Purely German towns, indirectly through government officialdom, were slowly but steadily pushed into the mixed-language
danger zones. Even in Lower Austria this process began to make
increasingly rapid progress, and many Czechs considered Vienna
their largest city.

The central idea of this new Habsburg, whose family had
ceased to speak anything but Czech (the Archduke's wife, a
former Czech countess, had been morganatically married to the
Prince — she came from circles whose anti-German attitude
was traditional), was gradually to establish a Slavic state in
Central Europe which for defense against Orthodox Russia
should be placed on a strictly Catholic basis. Thus, as the
Habsburgs had so often done before, religion was once again
put into the service of a purely political idea, and what was
worse — at least from the German viewpoint — of a catastrophic
idea.

The result was more than dismal in many respects.

Neither the House of Habsburg nor the Catholic Church received the expected reward.

Habsburg lost the throne, Rome a great state.

For by employing religious forces in the service of its political
considerations, the crown aroused a spirit which at the outset it
had not considered possible.

In answer to the attempt to exterminate the Germans in the
old monarchy by every possible means, there arose the Pan-
German movement in Austria.

By the eighties the basic Jewish tendency of Manchester
liberalism had reached, if not passed, its high point in the monarchy. The reaction to it, however, as with everything in old
Austria, arose primarily from a social, not from a national standpoint. The instinct of self-preservation forced the Germans to
adopt the sharpest measures of defense. Only secondarily did

economic considerations begin to assume a decisive influence.
And so, two party formations grew out of the general political
confusion, the one with the more national, the other with the more
social, attitude, but both highly interesting and instructive for
the future.

After the depressing end of the War of 1866, the House of
Habsburg harbored the idea of revenge on the battlefield. Only
the death of Emperor Max of Mexico, whose unfortunate ex-
pedition was blamed primarily on Napoleon III and whose
abandonment by the French aroused general indignation, prevented a closer collaboration with France. Habsburg nevertheless lurked in wait. If the War of 1870–71 had not been so unique
a triumph, the Vienna Court would probably have risked a
bloody venture to avenge Sadowa. But when the first amazing
and scarcely credible, but none the less true, tales of heroism
arrived from the battlefields, the 'wisest' of all monarchs recognized that the hour was not propitious and put the best possible
face on a bad business.

But the heroic struggle of these years had accomplished an
even mightier miracle; for with the Habsburgs a change of position never arose from the urge of the innermost heart, but from
the compulsion of circumstances. However, the German people
of the old Ostmark were swept along by the Reich's frenzy of
victory, and looked on with deep emotion as the dream of their
fathers was resurrected to glorious reality.

For make no mistake: the truly German-minded Austrian had,
even at Königgrätz,[1] and from this time on, recognized the tragic
but necessary prerequisite for the resurrection of a Reich which
would no longer be — and actually was not — afflicted with the

---

[1] The Battle of Königgrätz (Sadowa), on July 3, 1866, was a decisive victory of the Prussians over the Austrians. By the ensuing treaties of Nikols-
burg and Prague, the Seven Weeks' War was ended; the North-German
states formed a North-German Confederation under Prussian leadership,
and the South-German states were forced into a defensive alliance with
Prussia (against France). This was an important step toward the founding
of the German Empire in 1871.

foul morass[1] of the old Union.[2] Above all, he had come to understand thoroughly, by his own suffering, that the House of Habsburg had at last concluded its historical mission and that the new
Reich could choose as Emperor only him whose heroic convictions
made him worthy to bear the 'Crown of the Rhine.' But how
much more was Fate to be praised for accomplishing this in-
vestiture in the scion of a house which in Frederick the Great had
given the nation a gleaming and eternal symbol of its resurrection.

But when after the great war the House of Habsburg began
with desperate determination slowly but inexorably to exterminate the dangerous German element in the dual monarchy (the
inner convictions of this element could not be held in doubt), for
such would be the inevitable result of the Slavization policy —
the doomed people rose to a resistance such as modern German
history had never seen.

For the first time, men of national and patriotic mind became
rebels.

Rebels, not against the nation and not against the state as such,
but rebels against a kind of government which in their conviction
would inevitably lead to the destruction of their own nationality.

For the first time in modern German history, traditional dynastic patriotism parted ways with the national love of fatherland
and people.

The Pan-German movement in German-Austria in the nineties
is to be praised for demonstrating in clear, unmistakable terms
that a state authority is entitled to demand respect and protection
only when it meets the interests of a people, or at least does not
harm them.

There can be no such thing as state authority as an end in itself,
for, if there were, every tyranny in this world would be unassailable and sacred.

---

[1] '... mit dem fauligen Marasmus des alten Bundes behaftet sein sollte.'

[2] Customs Union (*Zollverein*). The German Customs Union (*Zollverein*)
between the various large and petty independent states that then made up
Germany developed between 1815 and 1830 under the leadership of Prussia.
It formed the basis for the subsequent unification of the German states.

If, by the instrument of governmental power, a nationality is led toward its destruction, then rebellion is not only the right of every member of such a people — it is his duty.

And the question — when is this the case? — is decided not by theoretical dissertations, but by force and — results.

Since, as a matter of course, all governmental power claims the duty of preserving state authority — regardless how vicious it is, betraying the interests of a people a thousandfold — the national instinct of self-preservation, in overthrowing such a power and achieving freedom or independence, will have to employ the same weapons by means of which the enemy tries to maintain his power. Consequently, the struggle will be carried on with 'legal' means as long as the power to be overthrown employs such means; but it will not shun illegal means if the oppressor uses them.

In general it should not be forgotten that the highest aim of human existence is not the preservation of a state, let alone a government, but the preservation of the species.

And if the species itself is in danger of being oppressed or utterly eliminated, the question of legality is reduced to a subordinate rôle. Then, even if the methods of the ruling power are alleged to be legal a thousand times over, nonetheless the oppressed people's instinct of self-preservation remains the loftiest justification of their struggle with every weapon.

Only through recognition of this principle have wars of liberation against internal and external enslavement of nations on this earth come down to us in such majestic historical examples.

Human law cancels out state law.

And if a people is defeated in its struggle for human rights, this merely means that it has been found too light in the scale of destiny for the happiness of survival on this earth. For when a people is not willing or able to fight for its existence — Providence in its eternal justice has decreed that people's end.

The world is not for cowardly peoples.

*   *   *

How easy it is for a tyranny to cover itself with the cloak of so-called 'legality' is shown most clearly and penetratingly by the example of Austria.

The legal state power in those days was rooted in the anti-German soil of parliament with its non-German majorities — and in the equally anti-German ruling house. In these two factors the entire state authority was embodied. Any attempt to change the destinies of the German-Austrian people from this position was absurd. Hence, in the opinions of our friends the worshipers of state authority as such and of the 'legal' way, all resistance would have had to be shunned, as incompatible with legal methods. But this, with compelling necessity, would have meant the end of the German people in the monarchy — and in a very short time. And, as a matter of fact, the Germans were saved from this fate only by the collapse of this state.

The bespectacled theoretician, it is true, would still prefer to die for his doctrine than for his people.

Since it is men who make the laws, he believes that they live for the sake of these laws.

The Pan-German movement in Austria had the merit of completely doing away with this nonsense, to the horror of all theoretical pedants and other fetish-worshiping isolationists in the government.

Since the Habsburgs attempted to attack Germanism with all possible means, this party attacked the 'exalted' ruling house itself, and without mercy. For the first time it probed into this rotten state and opened the eyes of hundreds of thousands. To its credit be it said that it released the glorious concept of love of fatherland from the embrace of this sorry dynasty.

In the early days of its appearance, its following was extremely great, threatening to become a veritable avalanche. But the success did not last. When I came to Vienna, the movement had long been overshadowed by the Christian Social Party which had meanwhile attained power — and had indeed been reduced to almost complete insignificance.

This whole process of the growth and passing of the Pan-

German movement on the one hand, and the unprecedented rise of the Christian Social Party on the other, was to assume the deepest significance for me as a classical object of study.

When I came to Vienna, my sympathies were fully and wholly on the side of the Pan-German tendency.

That they mustered the courage to cry '*Hoch Hohenzollern*' impressed me as much as it pleased me; that they still regarded themselves as an only temporarily severed part of the German Reich, and never let a moment pass without openly attesting this fact, inspired me with joyful confidence; that in all questions regarding Germanism they showed their colors without reserve, and never descended to compromises, seemed to me the one still passable road to the salvation of our people; and I could not understand how after its so magnificent rise the movement should have taken such a sharp decline. Even less could I understand how the Christian Social Party at this same period could achieve such immense power. At that time it had just reached the apogee of its glory.

As I set about comparing these movements, Fate, accelerated by my otherwise sad situation, gave me the best instruction for an understanding of the causes of this riddle.

I shall begin my comparisons with the two men who may be regarded as the leaders and founders of the two parties: Georg von Schönerer and Dr. Karl Lueger.

From a purely human standpoint they both tower far above the scope and stature of so-called parliamentary figures. Amid the morass of general political corruption their whole life remained pure and unassailable. Nevertheless my personal sympathy lay at first on the side of the Pan-German Schönerer, and turned only little by little toward the Christian Social leader as well.

Compared as to abilities, Schönerer seemed to me even then the better and more profound thinker in questions of principle. He foresaw the inevitable end of the Austrian state more clearly and correctly than anyone else. If, especially in the Reich, people had paid more attention to his warnings against the Habsburg mon-

archy, the calamity of Germany's World War against all Europe
would never have occurred.

But if Schönerer recognized the problems in their innermost
essence, he erred when it came to men.

Here, on the other hand, lay Dr. Lueger's strength.

He had a rare knowledge of men and in particular took good
care not to consider people better than they are. Consequently,
he reckoned more with the real possibilities of life while Schönerer
had but little understanding for them. Theoretically speaking,
all the Pan-German's thoughts were correct, but since he lacked
the force and astuteness to transmit his theoretical knowledge to
the masses — that is, to put it in a form suited to the receptivity
of the broad masses, which is and remains exceedingly limited —
all his knowledge was visionary wisdom, and could never become
practical reality.

And this lack of actual knowledge of men led in the course of
time to an error in estimating the strength of whole movements
as well as age-old institutions.

Finally, Schönerer realized, to be sure, that questions of basic
philosophy were involved, but he did not understand that only
the broad masses of a people are primarily able to uphold such
well-nigh religious convictions.

Unfortunately, he saw only to a limited extent the extraordinary limitation of the will to fight in so-called 'bourgeois'
circles, due, if nothing else, to their economic position which
makes the individual fear to lose too much and thereby holds him
in check.

And yet, on the whole, a philosophy can hope for victory only
if the broad masses adhere to the new doctrine and declare their
readiness to undertake the necessary struggle.

From this deficient understanding of the importance of the
lower strata of the people arose a completely inadequate conception of the social question.

In all this Dr. Lueger was the opposite of Schönerer.

His thorough knowledge of men enabled him to judge the
possible forces correctly, at the same time preserving him from

underestimating existing institutions, and perhaps for this very reason taught him to make use of these institutions as instruments for the achievement of his purposes.

He understood only too well that the political fighting power of the upper bourgeoisie at the present time was but slight and inadequate for achieving the victory of a great movement. He therefore laid the greatest stress in his political activity on winning over the classes whose existence was threatened and therefore tended to spur rather than paralyze the will to fight. Likewise he was inclined to make use of all existing implements of power, to incline mighty existing institutions in his favor, drawing from these old sources of power the greatest possible profit for his own movement.

Thus he adjusted his new party primarily to the middle class menaced with destruction, and thereby assured himself of a following that was difficult to shake, whose spirit of sacrifice was as great as its fighting power. His policy toward the Catholic Church, fashioned with infinite shrewdness, in a short time won over the younger clergy to such an extent that the old Clerical Party was forced either to abandon the field, or, more wisely, to join the new party, in order slowly to recover position after position.

To take this alone as the characteristic essence of the man would be to do him a grave injustice. For in addition to being an astute tactician, he had the qualities of a truly great and brilliant reformer: though here, too, he observed the limits set by a precise knowledge of the existing possibilities as well as his own personal abilities.

It was an infinitely practical goal that this truly significant man had set himself. He wanted to conquer Vienna. Vienna was the heart of the monarchy; from this city the last flush of life flowed out into the sickly, old body of the crumbling empire. The healthier the heart became, the more the rest of the body was bound to revive: an idea, correct in principle, but which could be applied only for a certain limited time.

And herein lay this man's weakness.

What he had done as mayor of Vienna is immortal in the best sense of the word;[1] but he could no longer save the monarchy, it was too late.

His opponent, Schönerer, had seen this more clearly.

All Dr. Lueger's practical efforts were amazingly successful; the hopes he based on them were not realized.

Schönerer's efforts were not successful, but his most terrible fears came true.

Thus neither man realized his ultimate goal. Lueger could no longer save Austria, and Schönerer could no longer save the German people from ruin.

It is infinitely instructive for our present day to study the causes for the failure of both parties. This is particularly useful for my friends, since in many points conditions today are similar to then and errors can thereby be avoided which at that time caused the end of the one movement and the sterility of the other.

To my mind, there were three causes for the collapse of the Pan-German movement in Austria.

In the first place, its unclear conception of the significance of the social problem, especially for a new and essentially revolutionary party.

Since Schönerer and his followers addressed themselves principally to bourgeois circles, the result was bound to be very feeble and tame.

Though some people fail to suspect it, the German bourgeoisie, especially in its upper circles, is pacifistic to the point of positive self-abnegation, where internal affairs of the nation or state are concerned. In good times — that is, in this case, in times of good government — such an attitude makes these classes extremely valuable to the state; but in times of an inferior régime it is positively ruinous. To make possible the waging of any really serious struggle, the Pan-German movement should above all have dedicated itself to winning the masses. That it failed to do so deprived it in advance of the elemental impetus which a wave of its kind simply must have if it is not in a short time to ebb away.

[1] *'unsterblich im besten Sinne des Wortes.'*

Unless this principle is borne in mind and carried out from the very start, the new party loses all possibility of later making up for what has been lost. For, by the admission of numerous moderate bourgeois elements, the basic attitude of the movement will always be governed by them and thus lose any further prospect of winning appreciable forces from the broad masses. As a result, such a movement will not rise above mere grumbling and criticizing. The faith bordering more or less on religion, combined with a similar spirit of sacrifice, will cease to exist; in its place will arise an effort gradually to grind off the edges of struggle by means of 'positive' collaboration; that is, in this case, by acceptance of the existing order, thus ultimately leading to a putrid peace.

And this is what happened to the Pan-German movement because it had not from the outset laid its chief stress on winning supporters from the circles of the great masses. It achieved 'bourgeois respectability and a muffled radicalism.'

From this error arose the second cause of its rapid decline.

At the time of the emergence of the Pan-German movement the situation of the Germans in Austria was already desperate. From year to year the parliament had increasingly become an institution for the slow destruction of the German people. Any attempt at salvation in the eleventh hour could offer even the slightest hope of success only if this institution were eliminated.

Thus the movement was faced with a question of basic importance:

Should its members, to destroy parliament, go into parliament, in order, as people used to say, 'to bore from within,' or should they carry on the struggle from outside by an attack on this institution as such?

They went in and they came out defeated.

To be sure, they couldn't help but go in.

To carry on the struggle against such a power from outside means to arm with unflinching courage and to be prepared for endless sacrifices. You seize the bull by the horns, you suffer many heavy blows, you are sometimes thrown to the earth, sometimes you get up with broken limbs, and only after the hardest contest does victory reward the bold assailant. Only the greatness of the sacrifices will win new fighters for the cause, until at last tenacity is rewarded by success.

But for this the sons of the broad masses are required.

They alone are determined and tough enough to carry through the fight to its bloody end.

And the Pan-German movement did not possess these broad masses; thus no course remained open but to go into parliament.

It would be a mistake to believe that this decision was the result of long soul torments, or even meditations; no, no other idea entered their heads. Participation in this absurdity was only the sediment resulting from general, unclear conceptions regarding the significance and effect of such a participation in an institution which had in principle been recognized as false. In general, the party hoped that this would facilitate the enlightenment of the broad masses, since it would now have an opportunity to speak before the 'forum of the whole nation.' Besides, it seemed plausible that attacking the root of the evil was bound to be more successful than storming it from outside. They thought the security of the individual fighter was increased by the protection of parliamentary immunity, and that this could only enhance the force of the attack.

In reality, it must be said, things turned out very differently.

The forum before which the Pan-German deputies spoke had not become greater but smaller; for each man speaks only to the circle which can hear him, or which obtains an account of his words in the newspapers.

And, not the halls of parliament, but the great public meeting, represents the largest direct forum of listeners.

For, in the latter, there are thousands of people who have come only to hear what the speaker has to say to them, while in the halls of parliament there are only a few hundreds, and most of these are present only to collect their attendance fees, and certainly not to be illuminated by the wisdom of this or that fellow 'representative of the people.'

And above all:

This is always the same public, which will never learn anything new, since, aside from the intelligence, it is lacking in the very rudiments of will.

Never will one of these representatives of the people honor a superior truth of his own accord, and place himself in its service.

No, this is something that not a single one of them will do unless he has reason to hope that by such a shift he may save his mandate for one more session. Only when it is in the air that the party in power will come off badly in a coming election, will these ornaments of virility shift to a party or tendency which they presume will come out better, though you may be confident that this change of position usually occurs amidst a cloudburst of moral justifications. Consequently, when an existing party appears to be falling beneath the disfavor of the people to such an extent that the probability of an annihilating defeat threatens, such a great shift will always begin: then the parliamentary rats leave the party ship.

All this has nothing to do with better knowledge or intentions, but only with that prophetic gift which warns these parliamentary bedbugs at the right moment and causes them to drop, again and again, into another warm party bed.

But to speak to such a 'forum' is really to cast pearls before the well-known domestic beasts. It is truly not worth while. The result can be nothing but zero.

And that is just what it was.

The Pan-German deputies could talk their throats hoarse: the effect was practically nil.

The press either killed them with silence or mutilated their speeches in such a way that any coherence, and often even the sense, was twisted or entirely lost, and public opinion received a very poor picture of the aims of the new movement. What the various gentlemen said was quite unimportant; the important thing was what people read about them. And this was an extract from their speeches, so disjointed that it could — as intended — only seem absurd. The only forum to which they really spoke consisted of five hundred parliamentarians, and that is enough said.

But the worst was the following:

The Pan-German movement could count on success only if it realized from the very first day that what was required was not a new party, but a new philosophy. Only the latter could produce the inward power to fight this gigantic struggle to its end. And for this, only the very best and courageous minds can serve as leaders.

If the struggle for a philosophy is not lead by heroes prepared to make sacrifices, there will, in a short time, cease to be any warriors willing to die. The man who is fighting for his own existence cannot have much left over for the community.

In order to maintain this requirement, every man must know that the new movement can offer the present nothing but honor and fame in posterity.[1] The more easily attainable posts and offices a movement has to hand out, the more inferior stuff [2] it will attract, and in the end these political hangers-on overwhelm a successful party in such number that the honest fighter of former days no longer recognizes the old movement and the new arrivals definitely reject him as an unwelcome intruder. When this happens, the 'mission' of such a movement is done for.

As soon as the Pan-German movement sold its soul to parliament, it attracted 'parliamentarians' instead of leaders and fighters.

Thus it sank to the level of the ordinary political parties of the day and lost the strength to oppose a catastrophic destiny with the defiance of martyrdom. Instead of fighting, it now learned to make speeches and 'negotiate.' And in a short time the new parliamentarian found it a more attractive, because less dangerous, duty to fight for the new philosophy with the 'spiritual' weapons of parliamentary eloquence, than to risk his own life, if

---

[1] '*ausser Ehre und Ruhm der Nachwelt der Gegenwart nichts bieten kann.*' Second edition has improved this jumble to: ' *... dass die neue Bewegung Ehre und Ruhm vor der Nachwelt, in der Gegenwart aber nichts bieten kann.*' (That the new movement can offer honor and fame in the eyes of posterity, but nothing in the present.)

[2] Second edition has: 'the greater the number of inferior characters...'

necessary, by throwing himself into a struggle whose issue was uncertain and which in any case could bring him no profit.

Once they had members in parliament, the supporters outside began to hope and wait for miracles which, of course, did not occur and could not occur. For this reason they soon became impatient, for even what they heard from their own deputies was by no means up to the expectations of the voters. This was perfectly natural, since the hostile press took good care not to give the people any faithful picture of the work of the Pan-German deputies.

The more the new representatives of the people developed a taste for the somewhat gentler variety of 'revolutionary' struggle in parliament and the provincial diets, the less prepared they were to return to the more dangerous work of enlightening the broad masses of the people.

The mass meeting, the only way to exert a truly effective, because personal, influence on large sections of the people and thus possibly to win them, was thrust more and more into the background.

Once the platform of parliament was definitely substituted for the beer table of the meeting hall, and from this forum speeches were poured, not into the people, but on the heads of their so-called 'elect,' the Pan-German movement ceased to be a movement of the people and in a short time dwindled into an academic discussion club to be taken more or less seriously.

Consequently, the bad impression transmitted by the press was in no way corrected by personal agitation at meetings by the individual gentlemen, with the result that finally the word 'Pan-German' began to have a very bad sound in the ears of the broad masses.

For let it be said to all our present-day fops and knights of the pen: the greatest revolutions in this world have never been directed by a goose-quill!

No, to the pen it has always been reserved to provide their theoretical foundations.

But the power which has always started the greatest religious

and political avalanches in history rolling has from time immemorial been the magic power of the spoken word, and that
alone.

Particularly the broad masses of the people can be moved only
by the power of speech. And all great movements are popular
movements, volcanic eruptions of human passions and emotional
sentiments, stirred either by the cruel Goddess of Distress or by
the firebrand of the word hurled among the masses; they are not
the lemonade-like outpourings of literary aesthetes and drawing-
room heroes.

Only a storm of hot passion can turn the destinies of peoples,
and he alone can arouse passion who bears it within himself.

It alone gives its chosen one the words which like hammer
blows can open the gates to the heart of a people.

But the man whom passion fails and whose lips are sealed —
he has not been chosen by Heaven to proclaim its will.

Therefore, let the writer remain by his ink-well, engaging in
'theoretical' activity, if his intelligence and ability are equal to
it; for leadership he is neither born nor chosen.

A movement with great aims must therefore be anxiously on
its guard not to lose contact with the broad masses.

It must examine every question primarily from this standpoint
and make its decisions accordingly.

It must, furthermore, avoid everything which might diminish
or even weaken its ability to move the masses, not for 'demagogic'
reasons, but in the simple knowledge that without the mighty
force of the mass of a people, no great idea, however lofty and
noble it may seem, can be realized.

Hard reality alone must determine the road to the goal; un-
willingness to travel unpleasant roads only too often in this
world means to renounce the goal; which may or may not be
what you want.

As soon as the Pan-German movement by its parliamentary
attitude had shifted the weight of its activity to parliament instead of the people, it lost the future and instead won cheap successes of the moment.

It chose the easier struggle and thereby became unworthy of ultimate victory.

Even in Vienna I pondered this very question with the greatest care, and in the failure to recognize it saw one of the main causes of the collapse of the movement which in those days, in my opinion, was predestined to undertake the leadership of the German element.

The first two mistakes which caused the Pan-German movement to founder were related to each other. Insufficient knowledge of the inner driving forces of great revolutions led to an insufficient estimation of the importance of the broad masses of the people; from this resulted its insufficient interest in the social question, its deficient and inadequate efforts to win the soul of the lower classes of the nation, as well as its over-favorable attitude toward parliament.

If they had recognized the tremendous power which at all times must be attributed to the masses as the repository of revolutionary resistance, they would have worked differently in social and propagandist matters. Then the movement's center of gravity would not have been shifted to parliament, but to the workshop and the street.

Likewise the third error finds its ultimate germ in failure to recognize the value of the masses, which, it is true, need superior minds to set them in motion in a given direction, but which then, like a flywheel, lend the force of the attack momentum and uniform persistence.

The hard struggle which the Pan-Germans fought with the Catholic Church can be accounted for only by their insufficient understanding of the spiritual nature of the people.

The causes for the new party's violent attack on Rome were as follows:

As soon as the House of Habsburg had definitely made up its mind to reshape Austria into a Slavic state, it seized upon every means which seemed in any way suited to this tendency. Even religious institutions were, without the slightest qualms, harnessed to the service of the new 'state idea' by this unscrupulous ruling house.

The use of Czech pastorates and their spiritual shepherds was but one of the many means of attaining this goal, a general Slavization of Austria.

The process took approximately the following form:

Czech pastors were appointed to German communities; slowly but surely they began to set the interests of the Czech people above the interests of the churches, becoming germ-cells of the de-Germanization process.

The German clergy did practically nothing to counter these methods. Not only were they completely useless for carrying on this struggle in a positive German sense; they were even unable to oppose the necessary resistance to the attacks of the adversary. Indirectly, by the misuse of religion on the one hand, and owing to insufficient defense on the other, Germanism was slowly but steadily forced back.

If in small matters the situation was as described, in big things, unfortunately, it was not far different.

Here, too, the anti-German efforts of the Habsburgs did not encounter the resistance they should have, especially on the part of the high clergy, while the defense of German interests sank completely into the background.

The general impression could only be that the Catholic clergy as such was grossly infringing on German rights.

Thus the Church did not seem to feel with the German people, but to side unjustly with the enemy. The root of the whole evil lay, particularly in Schönerer's opinion, in the fact that the directing body of the Catholic Church was not in Germany, and that for this very reason alone it was hostile to the interests of our nationality.

The so-called cultural problems, in this as in virtually every other connection in Austria at that time, were relegated almost entirely to the background. The attitude of the Pan-German movement toward the Catholic Church was determined far less by its position on science, etc., than by its inadequacy in the championing of German rights and, conversely, its continued aid and comfort to Slavic arrogance and greed.

Georg Schönerer was not the man to do things by halves. He took up the struggle toward the Church in the conviction that by it alone he could save the German people. The 'Away-from-Rome'[1] movement seemed the most powerful, though, to be sure, the most difficult, mode of attack, which would inevitably shatter the hostile citadel. If it was successful, the tragic church schism in Germany would be healed, and it was possible that the inner strength of the Empire and the German nation would gain enormously by such a victory.

But neither the premise nor the inference of this struggle was correct.

Without doubt the national force of resistance of the Catholic clergy of German nationality, in all questions connected with Germanism, was less than that of their non-German, particularly Czech, brethren.

Likewise only an ignoramus could fail to see that an offensive in favor of German interests was something that practically never occurred to the German clergyman.

And anyone who was not blind was forced equally to admit that this was due primarily to a circumstance under which all of us Germans have to suffer severely: that is, the objectivity of our attitude toward our nationality as well as everything else.

While the Czech clergyman was subjective in his attitude toward his people and objective only toward the Church, the German pastor was subjectively devoted to the Church and remained objective toward the nation. A phenomenon which, to our misfortune, we can observe equally well in thousands of other cases.

This is by no means a special legacy of Catholicism, but with us it quickly corrodes almost every institution, whether it be governmental or ideal.

Just compare the position which our civil servants, for example, take toward the attempts at a national awakening with the position which in such a case the civil servants of another people would take. Or does anyone believe that an officers' corps anywhere else in the world would subordinate the interests of the

[1] '*Los-von-Rom.*' See note, p. 10.

nation amid mouthings about 'state authority,' in the way that has been taken for granted in our country for the last five years, in fact, has been viewed as especially meritorious? In the Jewish question, for example, do not both denominations today take a standpoint which corresponds neither to the requirements of the nation nor to the real needs of religion? Compare the attitude of a Jewish rabbi in all questions of even the slightest importance for the Jews as a race with the attitude of by far the greatest part of our clergy — of both denominations, if you please!

We always find this phenomenon when it is a question of defending an abstract idea as such.

'State authority,' 'democracy,' 'pacifism,' 'international solidarity,' etc., are all concepts which with us nearly always become so rigid and purely doctrinaire that subsequently all purely national vital necessities are judged exclusively from their standpoint.

This catastrophic way of considering all matters from the angle of a preconceived opinion kills every possibility of thinking oneself subjectively into a matter which is objectively opposed to one's own doctrine, and finally leads to a total reversal of means and ends. People will reject any attempt at a national uprising if it can take place only after the elimination of a bad, ruinous régime, since this would be an offense against 'state authority,' and 'state authority' is not a means to an end, but in the eyes of such a fanatical objectivist rather represents the aim itself, which is sufficient to fill out his whole lamentable life. Thus, for example, they would indignantly oppose any attempt at a dictatorship, even if it was represented by a Frederick the Great and the momentary political comedians of a parliamentary majority were incapable dwarfs or really inferior characters, just because the law of democracy seems holier to such a principle-monger [1] than the welfare of a nation. The one will therefore defend the worst tyranny, a tyranny which is ruining the people, since at

[1] '*Prinzipienbock.*' Second edition has '*Prinzipienblock.*' This rather startling combination would mean a 'bloc' or 'coalition' of men standing on principle.

the moment it embodies 'state authority,' while the other rejects even the most beneficial government as soon as it fails to satisfy his conception of 'democracy.'

In exactly the same way, our German pacifist will accept in silence the bloodiest rape of our nation at the hands of the most vicious military powers if a change in this state of affairs can be achieved only by resistance — that is, force — for this would be contrary to the spirit of his peace society. Let the international German Socialist be plundered in solidarity by the rest of the world, he will accept it with brotherly affection and no thought of retribution or even defense, just because he is — a German.

This may be a sad state of affairs, but to change a thing means to recognize it first.

The same is true of the weak defense of German interests by a part of the clergy.

It is neither malicious ill will in itself, nor is it caused, let us say, by commands from 'above'; no, in such a lack of national determination we see merely the result of an inadequate education in Germanism from childhood up and, on the other hand, an unlimited submission to an idea which has become an idol.

Education in democracy, in socialism of the international variety, in pacifism, etc., is a thing so rigid and exclusive, so purely subjective from these points of view, that the general picture of the remaining world is colored by this dogmatic conception, while the attitude toward Germanism has remained exceedingly objective from early youth. Thus, the pacifist, by giving himself subjectively and entirely to his idea, will, in the presence of any menace to his people, be it ever so grave and unjust, always (in so far as he is a German) seek after the objective right and never from pure instinct of self-preservation join the ranks of his herd and fight with them.

To what extent this is also true of the different religions is shown by the following:

Protestantism as such is a better defender of the interests of Germanism, in so far as this is grounded in its genesis and later tradition; it fails, however, in the moment when this defense of national interests must take place in a province which is either absent from the general line of its ideological world and traditional development, or is for some reason rejected.

Thus, Protestantism will always stand up for the advancement of all Germanism as such, as long as matters of inner purity or national deepening [1] as well as German freedom are involved, since all these things have a firm foundation in its own being; but it combats with the greatest hostility any attempt to rescue the nation from the embrace of its most mortal enemy, since its attitude toward the Jews just happens to be more or less dogmatically established. Yet here we are facing the question without whose solution all other attempts at a German reawakening or resurrection are and remain absolutely senseless and impossible.

In my Vienna period I had leisure and opportunity enough for an unprejudiced examination of this question too, and in my daily contacts was able to establish the correctness of this view a thousand times over.

In this focus of the most varied nationalities, it immediately becomes clearly apparent that the German pacifist is alone in always attempting to view the interests of his own nation objectively, but that the Jew will never regard those of the Jewish people in this way; that only the German Socialist is 'international' in a sense which forbids him to beg justice for his own people except by whimpering and whining in the midst of his international comrades, but never a Czech or a Pole, etc.; in short, I recognized even then that the misfortune lies only partly in these doctrines, and partly in our totally inadequate education in national sentiment and a resultant lack of devotion to our nation.

Thus, the first theoretical foundation for a struggle of the Pan-German movement against Catholicism as such was lacking.

Let the German people be raised from childhood up with that exclusive recognition of the rights of their own nationality, and let not the hearts of children be contaminated with the curse of our 'objectivity,' even in matters regarding the preservation of

[1] '*Nationale Vertiefung.*'

their own ego. Then in a short time it will be seen that (presupposing, of course, a radically national government) in Germany, as in Ireland, Poland, or France, the Catholic will always be a German.

The mightiest proof of this was provided by that epoch which for the last time led our nation into a life-and-death struggle before the judgment seat of history in defense of its own existence.

As long as leadership from above was not lacking, the people fulfilled their duty and obligation overwhelmingly. Whether Protestant pastor or Catholic priest, both together contributed infinitely in maintaining for so long our power to resist, not only at the front but also at home. In these years and particularly at the first flare, there really existed in both camps but a single holy German Reich, for whose existence and future each man turned to his own heaven.

The Pan-German movement in Austria should have asked itself one question: Is the preservation of German-Austrianism possible under a Catholic faith, or is it not? If yes, the political party had no right to concern itself with religious or denominational matters; if not, then what was needed was a religious reformation and never a political party.

Anyone who thinks he can arrive at a religious reformation by the détour of a political organization only shows that he has no glimmer of knowledge of the development of religious ideas or dogmas and their ecclesiastical consequences.

Verily a man cannot serve two masters. And I consider the foundation or destruction of a religion far greater than the foundation or destruction of a state, let alone a party.

And let it not be said that this [1] is only a defense against the attacks from the other side!

It is certain that at all times unscrupulous scoundrels have not shunned to make even religion the instrument of their political bargains (for that is what such rabble almost always and exclusively deal in): but just as certainly it is wrong to make a religious denomination responsible for a number of tramps who

[1] Second edition has 'these attacks' in place of 'this.'

abuse it in exactly the same way as they would probably make anything else serve their low instincts.

Nothing can better suit one of these parliamentarian good-for-nothings and lounge-lizards than when an opportunity is offered to justify his political swindling, even after the fact.

For as soon as religion or even denomination is made responsible for his personal vices and attacked on that ground, this shameless liar sets up a great outcry and calls the whole world to witness that his behavior has been completely justified and that he alone and his eloquence are to be thanked for saving religion of the Church. The public, as stupid as it is forgetful, is, as a rule, prevented by the very outcry from recognizing the real instigator of the struggle or else has forgotten him, and the scoundrel has to all intents and purposes achieved his goal.

The sly fox knows perfectly well that this has nothing to do with religion; and he will silently laugh up his sleeve while his honest but clumsy opponent loses the game and one day, despairing of the loyalty and faith of humanity, withdraws from it all.

And in another sense it would be unjust to make religion as such or even the Church responsible for the failings of individuals. Compare the greatness of the visible organization before our eyes with the average fallibility of man in general, and you will have to admit that in it the relation of good and evil is better than anywhere else. To be sure, even among the priests themselves there are those to whom their holy office is only a means of satisfying their political ambition, yes, who in political struggle forget, in a fashion which is often more than deplorable that they are supposed to be the guardians of a higher truth and not the representatives of lies and slander — but for one such unworthy priest there are a thousand and more honorable ones, shepherds most loyally devoted to their mission, who, in our present false and decadent period, stand out of the general morass like little islands.

No more than I condemn, or would be justified in condemning, the Church as such when a degenerate individual in a cassock obscenely transgresses against morality, do I condemn it when one of the many others besmirches and betrays his nationality at a time when this is a daily occurrence anyway. Particularly today, we must not forget that for one such Ephialtes there are thousands who with bleeding heart feel the misfortune of their people and like the best of our nation long for the hour in which Heaven will smile on us again.

And if anyone replies that here we are not concerned with such everyday problems, but with questions of principle and truth or dogmatic content, we can aptly counter with another question:

If you believe that you have been chosen by Fate to reveal the truth in this matter, do so; but then have the courage to do so, not indirectly through a political party — for this is a swindle; but for today's evil substitute your future good.

But if you lack courage, or if your good is not quite clear even to yourself, then keep your fingers out of the matter; in any case, do not attempt by roundabout sneaking through a political movement to do what you dare not do with an open vizor.

Political parties have nothing to do with religious problems, as long as these are not alien to the nation, undermining the morals and ethics of the race; just as religion cannot be amalgamated with the scheming of political parties.

When Church dignitaries make use of religious institutions or doctrines to injure their nation, we must never follow them on this path and fight with the same methods.

For the political leader the religious doctrines and institutions of his people must always remain inviolable; or else he has no right to be in politics, but should become a reformer, if he has what it takes!

Especially in Germany any other attitude would lead to a catastrophe.

In my study of the Pan-German movement and its struggle against Rome, I then, and even more in the years to come, arrived at the following conviction: This movement's inadequate appreciation of the importance of the social problem cost it the truly militant mass of the people; its entry into parliament took away its mighty impetus and burdened it with all the weaknesses peculiar to this institution; the struggle against the Catholic

Church made it impossible in numerous small and middle circles, and thus robbed it of countless of the best elements that the nation can call its own.

The practical result of the Austrian *Kulturkampf*[1] was next to nil.

To be sure, it succeeded in tearing some hundred thousand members away from the Church, yet without causing it any particular damage. In this case the Church really had no need to shed tears over the lost 'lambs'; for it lost only those who had long ceased to belong to it. The difference between the new reformation and the old one was that in the old days many of the best people in the Church turned away from it through profound religious conviction, while now only those who were lukewarm to begin with departed, and this from 'considerations' of a political nature.

And precisely from the political standpoint the result was just as laughable as it was sad.

Once again a promising political movement for the salvation of the German nation had gone to the dogs because it had not been led with the necessary cold ruthlessness, but had lost itself in fields which could only lead to disintegration.

For one thing is assuredly true:

The Pan-German movement would never have made this mistake but for its insufficient understanding of the psyche of the broad masses. If its leaders had known that to achieve any success one should, on purely psychological grounds, never show the masses two or more opponents, since this leads to a total disintegration of their fighting power, for this reason alone the thrust

---

[1] The struggle carried on by Bismarck against the Catholic Church became known as the *Kulturkampf* from the words of Rudolf Virchow in the Prussian Diet (January 17, 1873): 'The contest has taken on the character of a great cultural struggle.' The struggle was largely released by the proclamation of papal infallibility by the Vatican Council in July, 1870. The Jesuits were expelled from Germany, Church schools were subjected to state control, civil marriage was made obligatory, religious orders were dissolved, etc.

of the Pan-German movement would have been directed at a single adversary. Nothing is more dangerous for a political party than to be led by those jacks-of-all-trades who want everything but can never really achieve anything.

Regardless how much room for criticism there was in any religious denomination a political party must never for a moment lose sight of the fact that in all previous historical experience a purely political party in such situations had never succeeded in producing a religious reformation. And the aim of studying history is not to forget its lessons when occasion arises for its practical application, or to decide that the present situation is different after all, and that therefore its old eternal truths are no longer applicable; no, the purpose of studying history is precisely its lesson for the present. The man who cannot do this must not conceive of himself as a political leader; in reality he is a shallow, though usually very conceited, fool, and no amount of good will can excuse his practical incapacity.

In general the art of all truly great national leaders at all times consists among other things primarily in not dividing the attention of a people, but in concentrating it upon a single foe. The more unified the application of a people's will to fight, the greater will be the magnetic attraction of a movement and the mightier will be the impetus of the thrust. It belongs to the genius of a great leader to make even adversaries far removed from one another seem to belong to a single category, because in weak and uncertain characters the knowledge of having different enemies can only too readily lead to the beginning of doubt in their own right.

Once the wavering mass sees itself in a struggle against too many enemies, objectivity will put in an appearance, throwing open the question whether all others are really wrong and only their own people or their own movement are in the right.

And this brings about the first paralysis of their own power. Hence a multiplicity of different adversaries must always be combined so that in the eyes of the masses of one's own supporters the struggle is directed against only one enemy. This strengthens their faith in their own right and enhances their bitterness against those who attack it.

That the old Pan-German movement failed to understand this deprived it of success.

Its goal had been correct, its will pure, but the road it chose was wrong. It was like a mountain climber who keeps the peak to be climbed in view and who sets out with the greatest determination and energy, but pays no attention to the trail, for his eyes are always on his goal, so that he neither sees nor feels out the character of the ascent and thus comes to grief in the end.

The opposite state of affairs seemed to prevail with its great competitor, the Christian Social Party.

The road it chose was correct and well-chosen, but it lacked clear knowledge of its goal.

In nearly all the matters in which the Pan-German movement was wanting, the attitude of the Christian Social Party was correct and well-planned.

It possessed the necessary understanding for the importance of the masses and from the very first day assured itself of at least a part of them by open emphasis on its social character. By aiming essentially at winning the small and lower middle classes and artisans, it obtained a following as enduring as it was self-sacrificing. It avoided any struggle against a religious institution and thus secured the support of that mighty organization which the Church represents. Consequently, it possessed only a single truly great central opponent. It recognized the value of large-scale propaganda and was a virtuoso in influencing the psychological instincts of the broad masses of its adherents.

If nevertheless it was unable to achieve its goal and dream of saving Austria, this was due to two deficiencies in its method and to its lack of clarity concerning the aim itself.

The anti-Semitism of the new movement was based on religious ideas instead of racial knowledge. The reason for the intrusion of this mistake was the same which brought about the second fallacy.

If the Christian Social Party wanted to save Austria, then in the opinion of its founders it must not operate from the standpoint of the racial principle, for if it did a dissolution of the state would, in a short time, inevitably occur. Particularly the situation in Vienna itself, in the opinion of the party leaders, demanded that all points which would divide their following should be set aside as much as possible, and that all unifying conceptions be emphasized in their stead.

At that time Vienna was so strongly permeated especially with Czech elements that only the greatest tolerance with regard to all racial questions could keep them in a party which was not anti-German to begin with. If Austria were to be saved, this was indispensable. And so they attempted to win over small Czech artisans who were especially numerous in Vienna, by a struggle against liberal Manchesterism, and in the struggle against the Jews on a religious basis they thought they had discovered a slogan transcending all of old Austria's national differences.

It is obvious that combating Jewry on such a basis could provide the Jews with small cause for concern. If the worst came to the worst, a splash of baptismal water could always save the business and the Jew at the same time. With such a superficial motivation, a serious scientific treatment of the whole problem was never achieved, and as a result far too many people, to whom this type of anti-Semitism was bound to be incomprehensible, were repelled. The recruiting power of the idea was limited almost exclusively to intellectually limited circles, unless true knowledge were substituted for purely emotional feeling. The intelligentsia remained aloof as a matter of principle. Thus the whole movement came to look more and more like an attempt at a new conversion of the Jews, or perhaps even an expression of a certain competitive envy. And hence the struggle lost the character of an inner and higher consecration; to many, and not necessarily the worst people, it came to seem immoral and reprehensible. Lacking was the conviction that this was a vital question for all humanity, with the fate of all non-Jewish peoples depending on its solution.

Through this half-heartedness the anti-Semitic line of the Christian Social Party lost its value.

It was a sham anti-Semitism which was almost worse than none at all; for it lulled people into security; they thought they had the foe by the ears, while in reality they themselves were being led by the nose.

In a short time the Jew had become so accustomed to this type of anti-Semitism that he would have missed its disappearance more than its presence inconvenienced him.

If in this the Christian Social Party had to make a heavy sacrifice to the state of nationalities, they had to make an even greater one when it came to championing Germanism as such.

They could not be 'nationalistic' unless they wanted to lose the ground from beneath their feet in Vienna. They hoped that by a pussy-footing evasion of this question they could still save the Habsburg state, and by that very thing they encompassed its ruin. And the movement lost the mighty source of power which alone can fill a political party with inner strength for any length of time.

Through this alone the Christian Social Party became a party like any other.

In those days I followed both movements most attentively. One, by feeling the beat of its innermost heart, the other, carried away by admiration for the unusual man who even then seemed to me a bitter symbol of all Austrian Germanism.

When the mighty funeral procession bore the dead mayor from the City Hall toward the Ring, I was among the many hundred thousands looking on at the tragic spectacle. I was profoundly moved and my feelings told me that the work, even of this man, was bound to be in vain, owing to the fatal destiny which would inevitably lead this state to destruction. If Dr. Karl Lueger had lived in Germany, he would have been ranked among the great minds of our people; that he lived and worked in this impossible state was the misfortune of his work and of himself.

When he died, the little flames in the Balkans were beginning to leap up more greedily from month to month, and it was a gracious fate which spared him from witnessing what he still thought he could prevent.

Out of the failure of the one movement and the miscarriage of the other, I for my part sought to find the causes, and came to the certain conviction that, quite aside from the impossibility of bolstering up the state in old Austria, the errors of the two parties were as follows:

The Pan-German movement was right in its theoretical view about the aim of a German renascence, but unfortunate in its choice of methods. It was nationalistic, but unhappily not socialistic enough to win the masses. But its anti-Semitism was based on a correct understanding of the importance of the racial problem, and not on religious ideas. Its struggle against a definite denomination, however, was actually and tactically false.

The Christian Social movement had an unclear conception of the aim of a German reawakening, but had intelligence and luck in seeking its methods as a party. It understood the importance of the social question, erred in its struggle against the Jews, and had no notion of the power of the national idea.

If, in addition to its enlightened knowledge of the broad masses, the Christian Social Party had had a correct idea of the importance of the racial question, such as the Pan-German movement had achieved; and if, finally, it had itself been nationalistic, or if the Pan-German movement, in addition to its correct knowledge of the aim of the Jewish question, had adopted the practical shrewdness of the Christian Social Party, especially in its attitude toward socialism, there would have resulted a movement which even then in my opinion might have successfully intervened in German destiny.

If this did not come about, it was overwhelmingly due to the nature of the Austrian state.

Since I saw my conviction realized in no other party, I could in the period that followed not make up my mind to enter, let alone fight with, any of the existing organizations. Even then I regarded all political movements as unsuccessful and unable to carry out a national reawakening of the German people on a larger and not purely external scale.

But in this period my inner revulsion toward the Habsburg state steadily grew.

The more particularly I concerned myself with questions of foreign policy, the more my conviction rose and took root that this political formation could result in nothing but the misfortune of Germanism. More and more clearly I saw at last that the fate of the German nation would no longer be decided here, but in the Reich itself. This was true, not only of political questions, but no less for all manifestations of cultural life in general.

Also in the field of cultural or artistic affairs, the Austrian state showed all symptoms of degeneration, or at least of unimportance for the German nation. This was most true in the field of architecture. The new architecture could achieve no special successes in Austria, if for no other reason because since the completion of the Ring its tasks, in Vienna at least, had become insignificant in comparison with the plans arising in Germany.

Thus more and more I began to lead a double life; reason and reality told me to complete a school as bitter as it was beneficial in Austria, but my heart dwelt elsewhere.

An oppressive discontent had seized possession of me, the more I recognized the inner hollowness of this state and the impossibility of saving it, and felt that in all things it could be nothing but the misfortune of the German people.

I was convinced that this state inevitably oppressed and handicapped any really great German as, conversely, it would help every un-German figure.

I was repelled by the conglomeration of races which the capital showed me, repelled by this whole mixture of Czechs, Poles, Hungarians, Ruthenians, Serbs, and Croats, and everywhere, the eternal mushroom of humanity — Jews and more Jews.

To me the giant city seemed the embodiment of racial desecration.

The German of my youth was the dialect of Lower Bavaria; I could neither forget it nor learn the Viennese jargon. The longer I lived in this city, the more my hatred grew for the foreign mixture of peoples which had begun to corrode this old site of German culture.

The idea that this state could be maintained much longer seemed to me positively ridiculous.

Austria was then like an old mosaic; the cement, binding the various little stones together, had grown old and begun to crumble; as long as the work of art is not touched, it can continue to give a show of existence, but as soon as it receives a blow, it breaks into a thousand fragments. The question was only when the blow would come.

Since my heart had never beaten for an Austrian monarchy, but only for a German Reich, the hour of this state's downfall could only seem to me the beginning of the redemption of the German nation.

For all these reasons a longing rose stronger and stronger in me, to go at last whither since my childhood secret desires and secret love had drawn me.

I hoped some day to make a name for myself as an architect and thus, on the large or small scale which Fate would allot me, to dedicate my sincere services to the nation.

But finally I wanted to enjoy the happiness of living and working in the place which some day would inevitably bring about the fulfillment of my most ardent and heartfelt wish: the union of my beloved homeland with the common fatherland, the German Reich.

Even today many would be unable to comprehend the greatness of such a longing, but I address myself to those to whom Fate has either hitherto denied this, or from whom in harsh cruelty it has taken it away; I address myself to all those who, detached from their mother country, have to fight even for the holy treasure of their language, who are persecuted and tortured for their loyalty to the fatherland, and who now, with poignant emotion, long for the hour which will permit them to return to the heart of their faithful mother; I address myself to all these, and I know that they will understand me!

Only he who has felt in his own skin what it means to be a German, deprived of the right to belong to his cherished fatherland, can measure the deep longing which burns at all times in the hearts of children separated from their mother country. It torments those whom it fills and denies them contentment and happiness until the gates of their father's house open, and in the common Reich, common blood gains peace and tranquillity.

Yet Vienna was and remained for me the hardest, though most thorough, school of my life. I had set foot in this town while still half a boy and I left it a man, grown quiet and grave. In it I obtained the foundations for a philosophy in general and a political view in particular which later I only needed to supplement in detail, but which never left me. But not until today have I been able to estimate at their full value those years of study.

That is why I have dealt with this period at some length, because it gave me my first visual instruction in precisely those questions which belonged to the foundations of a party which, arising from smallest beginnings, after scarcely five years is beginning to develop into a great mass movement. I do not know what my attitude toward the Jews, Social Democracy, or rather Marxism as a whole, the social question, etc., would be today if at such an early time the pressure of destiny — and my own study — had not built up a basic stock of personal opinions within me.

For if the misery of the fatherland can stimulate thousands and thousands of men to thought on the inner reasons for this collapse, this can never lead to that thoroughness and deep insight which are disclosed to the man who has himself mastered Fate only after years of struggle.

Munich

IN THE SPRING of 1912 I came at last to Munich.

The city itself was as familiar to me as if I had lived for years within its walls. This is accounted for by my study which at every step had led me to this metropolis of German art. Not only has one not seen Germany if one does not know Munich — no, above all, one does not know German art if one has not seen Munich.

In any case, this period before the War was the happiest and by far the most contented of my life. Even if my earnings were still extremely meager, I did not live to be able to paint, but painted only to be able to secure my livelihood or rather to enable myself to go on studying. I possessed the conviction that I should some day, in spite of all obstacles, achieve the goal I had set myself. And this alone enabled me to bear all other petty cares of daily existence lightly and without anxiety.

In addition to this, there was the heartfelt love which seized me for this city more than for any other place that I knew, almost from the first hour of my sojourn there. A *German* city! What a difference from Vienna! I grew sick to my stomach when I even thought back on this Babylon of races. In addition, the dialect, much closer to me, which particularly in my contacts with Lower Bavarians, reminded me of my former childhood. There were a thousand and more things which were or became inwardly dear and precious to me. But most of all I was attracted by this wonderful marriage of primordial power and fine artistic mood, this single line from the Hofbräuhaus to the Odeon,[1] from the October Festival to the Pinakothek,[2] etc. If today I am more attached to this city than to any other spot of earth in this world, it is partly due to the fact that it is and remains inseparably bound up with the development of my own life; if even then I achieved the happiness of a truly inward contentment, it can be attributed only to the magic which the miraculous residence of the Wittelsbachs exerts on every man who is blessed, not only with a calculating mind but with a feeling soul.

What attracted me most aside from my professional work was, here again, the study of the political events of the day, among them particularly the occurrences in the field of foreign affairs. I came to these latter indirectly through the German alliance policy which from my Austrian days I considered absolutely mistaken. However, the full extent of this self-deception on the part of the Reich had not been clear to me in Vienna. In those days I was inclined to assume — or perhaps I merely talked myself into it as an excuse — that Berlin perhaps knew how weak and unreliable the ally would be in reality, yet, for more or less mysterious reasons, held back this knowledge in order to bolster up an alliance policy which after all Bismarck himself had founded and the sudden cessation of which could not be desirable, if for no other reason lest the lurking foreigner be alarmed in any way, or the shopkeeper at home be worried.

To be sure, my associations, particularly among the people itself, soon made me see to my horror that this belief was false. To my amazement I could not help seeing everywhere that even in otherwise well-informed circles there was not the slightest glimmer of knowledge concerning the nature of the Habsburg monarchy. Particularly the common people were caught in the

---

[1] Odeon. Munich's main concert hall, built by Ludwig I.

[2] Museum of Art in Munich. The Ältere Pinakothek, completed in 1836 by Ludwig I of Bavaria, contains many of the finest works of the old masters. The Neuere Pinakothek, built in 1846–53, was devoted to contemporary art.

mad idea that the ally could be regarded as a serious power which
in the hour of need would surely rise to the situation. Among the
masses the monarchy was still regarded as a 'German' state on
which we could count. They were of the opinion that there, too,
the power could be measured by the millions as in Germany itself,
and completely forgot that, in the first place: Austria had long
ceased to be a German state; and in the second place: the internal
conditions of this Empire were from hour to hour moving closer
to disintegration.

I had come to know this state formation better than the so-
called official 'diplomats,' who blindly, as almost always, rushed
headlong toward catastrophe; for the mood of the people was
always a mere discharge of what was funneled into public opinion
from above. But the people on top made a cult of the 'ally,' as if
it were the Golden Calf. They hoped to replace by cordiality
what was lacking in honesty. And words were always taken for
coin of the realm.

Even in Vienna I had been seized with anger when I reflected
on the disparity appearing from time to time between the speeches
of the official statesmen and the content of the Viennese press.
And yet Vienna, in appearance at least, was still a German city.
How different it was if you left Vienna, or rather German-Austria,
and went to the Slavic provinces of the Empire! You had only to
take up the Prague newspapers to find out what they thought of
the whole exalted hocus-pocus of the Triple Alliance. There
there was nothing but bitter scorn and mockery for this 'master-
piece of statecraft.' In the midst of peace, with both emperors
pressing kisses of friendship on each other's foreheads, the Czechs
made no secret of the fact that this alliance would be done for on
the day when an attempt should be made to translate it from the
moonbeams of the Nibelungen ideal into practical reality.

What excitement seized these same people several years later
when the time finally came for the alliances to show their worth
and Italy leapt out of the triple pact, leaving her two comrades in
the lurch, and in the end even becoming their enemy! That anyone even for a moment should have dared to believe in the possi-

bility of such a miracle — to wit, the miracle that Italy would fight side by side with Austria — could be nothing but incomprehensible to anyone who was not stricken with diplomatic blindness. But in Austria things were not a hair's-breadth different.

In Austria the only exponents of the alliance idea were the Habsburgs and the Germans. The Habsburgs, out of calculation and compulsion; the Germans, from good faith and political — stupidity. From good faith, for they thought that by the Triple Alliance they were performing a great service for the German Reich itself, helping to strengthen and secure it; from political stupidity, because neither did the first-mentioned occur, but on the contrary, they thereby helped to chain the Reich to the corpse of a state which would inevitably drag them both into the abyss, and above all because they themselves, solely by virtue of this alliance, fell more and more a prey to de-Germanization. For by the alliance with the Reich, the Habsburgs thought they could be secure against any interference from this side, which unfortunately was the case, and thus they were able far more easily and safely to carry through their internal policy of slowly eliminating Germanism. Not only that in view of our well-known 'objectivity' they had no need to fear any intervention on the part of the Reich government, but, by pointing to the alliance, they could also silence any embarrassing voice among the Austrian-Germans which might rise in German quarters against Slavization of an excessively disgraceful character.

For what was the German in Austria to do if the Germans of the Reich recognized and expressed confidence in the Habsburg government? Should he offer resistance and be branded by the entire German public as a traitor to his own nationality? When for decades he had been making the most enormous sacrifices precisely for his nationality!

But what value did this alliance have, once Germanism had been exterminated in the Habsburg monarchy? Wasn't the value of the Triple Alliance for Germany positively dependent on the preservation of German predominance in Austria? Or did they really believe that they could live in an alliance with a Slavic-Habsburg Empire?

The attitude of official German diplomacy and of all public opinion toward the internal Austrian problem of nationalities was beyond stupidity, it was positively insane! They banked on an alliance, made the future and security of a people of seventy millions dependent on it — and looked on while the sole basis for this alliance was from year to year, inexorably and by plan, being destroyed in the partner-nation. The day was bound to come when a 'treaty' with Viennese diplomacy would remain, but the aid of an allied empire would be lost.

With Italy this was the case from the very beginning.

If people in Germany had only studied history a little more clearly, and gone into the psychology of nations, they would not have been able to suppose even for an hour that the Quirinal and the Vienna Hofburg would ever stand together in a common fighting front. Sooner would Italy have turned into a volcano than a government have dared to send even a single Italian to the battlefield for the fanatically hated Habsburg state, except as an enemy. More than once in Vienna I saw outbursts of the passionate contempt and bottomless hatred with which the Italian was 'devoted' to the Austrian state. The sins of the House of Habsburg against Italian freedom and independence in the course of the centuries was too great to be forgotten, even if the will to forget them had been present. And it was not present; neither in the people nor in the Italian government. For Italy there were therefore two possibilities for relations with Austria: either alliance or war.

By choosing the first, the Italians were able to prepare, undisturbed, for the second.

Especially since the relation of Austria to Russia had begun to drive closer and closer to a military clash, the German alliance policy was as senseless as it was dangerous.

This was a classic case, bearing witness to the absence of any broad and correct line of thinking.

Why, then, was an alliance concluded? Only in order better to guard the future of the Reich than, reduced to her own resources, she would have been in a position to do. And this future of the Reich was nothing other than the question of preserving the German people's possibility of existence.

Therefore the question could be formulated only as follows:

What form must the life of the German nation assume in the tangible future, and how can this development be provided with the necessary foundations and the required security within the framework of general European relation of forces?

A clear examination of the premises for foreign activity on the part of German statecraft inevitably led to the following conviction:

Germany has an annual increase in population of nearly nine hundred thousand souls. The difficulty of feeding this army of new citizens must grow greater from year to year and ultimately end in catastrophe, unless ways and means are found to forestall the danger of starvation and misery in time.

There were four ways of avoiding so terrible a development for the future:

1. Following the French example, the increase of births could be artificially restricted, thus meeting the problem of overpopulation.

Nature herself in times of great poverty or bad climatic conditions, as well as poor harvest, intervenes to restrict the increase of population of certain countries or races; this, to be sure, by a method as wise as it is ruthless. She diminishes, not the power of procreation as such, but the conservation of the procreated, by exposing them to hard trials and deprivations with the result that all those who are less strong and less healthy are forced back into the womb of the eternal unknown. Those whom she permits to survive the inclemency of existence are a thousandfold tested, hardened, and well adapted to procreate in turn, in order that the process of thoroughgoing selection may begin again from the beginning. By thus brutally proceeding against the individual and immediately calling him back to herself as soon as he shows himself unequal to the storm of life, she keeps the race and species strong, in fact, raises them to the highest accomplishments.

At the same time the diminution of number strengthens the individual and thus in the last analysis fortifies the species.

It is different, however, when man undertakes the limitation of his number. He is not carved of the same wood, he is 'humane.' He knows better than the cruel queen of wisdom. He limits not the conservation of the individual, but procreation itself. This seems to him, who always sees himself and never the race, more human and more justified than the opposite way. Unfortunately, however, the consequences are the reverse:

While Nature, by making procreation free, yet submitting survival to a hard trial, chooses from an excess number of individuals the best as worthy of living, thus preserving them alone and in them conserving their species, man limits procreation, but is hysterically concerned that once a being is born it should be preserved at any price. This correction of the divine will seems to him as wise as it is humane, and he takes delight in having once again gotten the best of Nature and even having proved her inadequacy. The number, to be sure, has really been limited, but at the same time the value of the individual has diminished; this, however, is something the dear little ape of the Almighty does not want to see or hear about.

For as soon as procreation as such is limited and the number of births diminished, the natural struggle for existence which leaves only the strongest and healthiest alive is obviously replaced by the obvious desire to 'save' even the weakest and most sickly at any price, and this plants the seed of a future generation which must inevitably grow more and more deplorable the longer this mockery of Nature and her will continues.

And the end will be that such a people will some day be deprived of its existence on this earth; for man can defy the eternal laws of the will to conservation for a certain time, but sooner or later vengeance comes. A stronger race will drive out the weak, for the vital urge in its ultimate form will, time and again, burst all the absurd fetters of the so-called humanity of individuals, in order to replace it by the humanity of Nature which destroys the weak to give his place to the strong.

Therefore, anyone who wants to secure the existence of the German people by a self-limitation of its reproduction is robbing it of its future.

2. A second way would be one which today we, time and time again, see proposed and recommended: internal colonization. This is a proposal which is well meant by just as many as by most people it is misunderstood, thus doing the greatest conceivable damage that anyone can imagine.[1]

Without doubt the productivity of the soil can be increased up to a certain limit. But only up to a certain limit, and not continuously without end. For a certain time it will be possible to compensate for the increase of the German people without having to think of hunger, by increasing the productivity of our soil. But beside this, we must face the fact that our demands on life ordinarily rise even more rapidly than the number of the population. Man's requirements with regard to food and clothing increase from year to year, and even now, for example, stand in no relation to the requirements of our ancestors, say a hundred years ago. It is, therefore, insane to believe that every rise in production provides the basis for an increase in population: no; this is true only up to a certain degree, since at least a part of the increased production of the soil is spent in satisfying the increased needs of men. But even with the greatest limitation on the one hand and the utmost industry on the other, here again a limit will one day be reached, created by the soil itself. With the utmost toil it will not be possible to obtain any more from it, and then, though postponed for a certain time, catastrophe again manifests itself. First, there will be hunger from time to time, when there is famine, etc. As the population increases, this will happen more and more often, so that finally it will only be absent when rare years of great abundance fill the granaries. But at length the time approaches when even then it will not be possible to satisfy men's needs, and hunger has become the eternal companion of such a people. Then Nature must help again and make a choice among those whom she

[1] '... *um den denkbar grossten Schaden anzurichten, den man sich nur vorzustellen vermag.*'

has chosen for life; but again man helps himself; that is, he turns to artificial restriction of his increase with all the above-indicated dire consequences for race and species.

The objection may still be raised that this future will face the whole of humanity in any case and that consequently the individual nation can naturally not avoid this fate.

At first glance this seems perfectly correct. Yet here the following must be borne in mind:

Assuredly at a certain time the whole of humanity will be compelled, in consequence of the impossibility of making the fertility of the soil keep pace with the continuous increase in population, to halt the increase of the human race and either let Nature again decide or, by self-help if possible, create the necessary balance, though, to be sure, in a more correct way than is done today. But then this will strike all peoples, while today only those races are stricken with such suffering which no longer possess the force and strength to secure for themselves the necessary territories in this world. For as matters stand there are at the present time on this earth immense areas of unusued soil, only waiting for the men to till them. But it is equally true that Nature as such has not reserved this soil for the future possession of any particular nation or race; on the contrary, this soil exists for the people which possesses the force to take it and the industry to cultivate it.

Nature knows no political boundaries. First, she puts living creatures on this globe and watches the free play of forces. She then confers the master's right on her favorite child, the strongest in courage and industry.

When a people limits itself to internal colonization because other races are clinging fast to greater and greater surfaces of this earth, it will be forced to have recourse to self-limitation at a time when the other peoples are still continuing to increase. Some day this situation will arise, and the smaller the living space at the disposal of the people, the sooner it will happen. Since in general, unfortunately, the best nations, or, even more correctly, the only truly cultured races, the standard-bearers of all human progress, all too frequently resolve in their pacifistic blindness to renounce new acquisitions of soil and content themselves with 'internal' colonization, while the inferior races know how to secure immense living areas in this world for themselves — this would lead to the following final result:

The culturally superior, but less ruthless races, would in consequence of their limited soil, have to limit their increase at a time when the culturally inferior but more brutal and more natural [1] peoples, in consequence of their greater living areas, would still be in a position to increase without limit. In other words: some day the world will thus come into possession of the culturally inferior but more active men.

Then, though in a perhaps very distant future, there will be but two possibilities either the world will be governed according to the ideas of our modern democracy, and then the weight of any decision will result in favor of the numerically stronger races, or the world will be dominated in accordance with the laws of the natural order of force, and then it is the peoples of brutal will who will conquer, and consequently once again not the nation of self-restriction.

No one can doubt that this world will some day be exposed to the severest struggles for the existence of mankind. In the end, only the urge for self-preservation can conquer. Beneath it so-called humanity, the expression of a mixture of stupidity, cowardice, and know-it-all conceit, will melt like snow in the March sun. Mankind has grown great in eternal struggle, and only in eternal peace does it perish.

For us Germans the slogan of 'inner colonization' is catastrophic, if for no other reason because it automatically reinforces us in the opinion that we have found a means which, in accordance with the pacifistic tendency, allows us 'to earn' our right to exist by labor in a life of sweet slumbers.[2] Once this doctrine were taken seriously in our country, it would mean the end of every

---

[1] '*brutal-natürlicher.*' Changed in second edition to '*naturhaft-brutaler.*' An English rendition of this subtlety seems impossible.

[2] ' ... *das der pazifistischen Gesinnung entsprechend gestattet, in sanftem Schlummerleben sich das Dasein "erarbeiten" zu können.*'

exertion to preserve for ourselves the place which is our due. Once the average German became convinced that he could secure his life and future in this way, all attempts at an active, and hence alone fertile, defense of German vital necessities would be doomed to failure. In the face of such an attitude on the part of the nation any really beneficial foreign policy could be regarded as buried, and with it the future of the German people as a whole.

Taking these consequences into account, it is no accident that it is always primarily the Jew who tries and succeeds in planting such mortally dangerous modes of thought in our people. He knows his customers too well not to realize that they gratefully let themselves be swindled by any gold-brick salesman who can make them think he has found a way to play a little trick on Nature, to make the hard, inexorable struggle for existence superfluous, and instead, sometimes by work, but sometimes by plain doing nothing, depending on how things 'come out,' to become the lord of the planet.

It cannot be emphasized sharply enough *that any German internal colonization must serve to eliminate social abuses particularly to withdraw the soil from widespread speculation, but can never suffice to secure the future of the nation without the acquisition of new soil.*

If we do not do this, we shall in a short time have arrived, not only at the end of our soil, but also at the end of our strength.

Finally, the following must be stated:

The limitation to a definite small area of soil, inherent in internal colonization, like the same final effect obtained by restriction of procreation, leads to an exceedingly unfavorable politico-military situation in the nation in question.

The size of the area inhabited by a people constitutes in itself an essential factor for determining its outward security. The greater the quantity of space at the disposal of a people, the greater its natural protection; for military decisions against peoples living in a small restricted area have always been obtained more quickly and hence more easily, and in particular more effectively and completely, than can, conversely, be possible against territorially extensive states. In the size of a state's territory there always lies a certain protection against frivolous attacks, since success can be achieved only after hard struggles, and therefore the risk of a rash assault will seem too great unless there are quite exceptional grounds for it. Hence the very size of a state offers in itself a basis for more easily preserving the freedom and independence of a people, while, conversely, the smallness of such a formation is a positive invitation to seizure.

Actually the two first possibilities for creating a balance between the rising population and the stationary amount of soil were rejected in the so-called national circles of the Reich. The reasons for this position were, to be sure, different from those above mentioned: government circles adopted a negative attitude toward the limitation of births out of a certain moral feeling; they indignantly rejected internal colonization because in it they scented an attack against large landholdings and therein the beginning of a wider struggle against private property in general. In view of the form in which particularly the latter panacea was put forward, they may very well have been right in this assumption.

On the whole, the defense against the broad masses was not very skillful and by no means struck at the heart of the problem.

Thus there remained but two ways of securing work and bread for the rising population.

3. Either new soil could be acquired and the superfluous millions sent off each year, thus keeping the nation on a self-sustaining basis; or we could

4. Produce for foreign needs through industry and commerce, and defray the cost of living from the proceeds.

In other words: either a territorial policy, or a colonial and commercial policy.

Both ways were contemplated, examined, recommended, and combated by different political tendencies, and the last was finally taken.

The healthier way of the two would, to be sure, have been the first.

The acquisition of new soil for the settlement of the excess population possesses an infinite number of advantages, particularly if we turn from the present to the future.

For one thing, the possibility of preserving a healthy peasant class as a foundation for a whole nation can never be valued highly enough. Many of our present-day sufferings are only the consequence of the unhealthy relationship between rural and city population. A solid stock of small and middle peasants has at all times been the best defense against social ills such as we possess today. And, moreover, this is the only solution which enables a nation to earn its daily bread within the inner circuit of its economy. Industry and commerce recede from their unhealthy leading position and adjust themselves to the general framework of a national economy of balanced supply and demand. Both thus cease to be the basis of the nation's sustenance and become a mere instrument to that end. Since they now have only a balance [1] between domestic production and demand in all fields, they make the subsistence of the people as a whole more or less independent of foreign countries, and thus help to secure the freedom of the state and the independence of the nation, particularly in difficult periods.

It must be said that such a territorial policy cannot be fulfilled in the Cameroons, but today almost exclusively in Europe. We must, therefore, coolly and objectively adopt the standpoint that it can certainly not be the intention of Heaven to give one people fifty times as much land and soil in this world as another. In this case we must not let political boundaries obscure for us the boundaries of eternal justice. If this earth really has room for all to live in, let us be given the soil we need for our livelihood.

True, they will not willingly do this. But then the law of self-preservation goes into effect; and what is refused to amicable

---

[1] '*Indem sie nur mehr den Ausgleich zwischen eigener Produktion und Bedarf auf allen Gebieten haben...*' Second edition inserts '*zur Aufgabe*' after '*Gebieten.*' The clause now reads: 'Since their sole task now becomes the creation of a balance,' etc. The first version is probably an oversight, perhaps a printer's error.

methods, it is up to the fist to take. If our forefathers had let their decisions depend on the same pacifistic nonsense as our contemporaries, we should possess only a third of our present territory; but in that case there would scarcely be any German people for us to worry about in Europe today. No — it is to our natural determination to fight for our own existence that we owe the two *Ostmarks* [1] of the Reich and hence that inner strength arising from the greatness of our state and national territory which alone has enabled us to exist up to the present.

And for another reason this would have been the correct solution:

Today many European states are like pyramids stood on their heads. Their European area is absurdly small in comparison to their weight of colonies, foreign trade, etc. We may say: summit in Europe, base in the whole world; contrasting with the American Union which possesses its base in its own continent and touches the rest of the earth only with its summit. And from this comes the immense inner strength of this state and the weakness of most European colonial powers.

Nor is England any proof to the contrary, since in consideration of the British Empire we too easily forget the Anglo-Saxon world as such. The position of England, if only because of her linguistic and cultural bond with the American Union, can be compared to no other state in Europe.

For Germany, consequently, the only possibility for carrying out a healthy territorial policy lay in the acquisition of new land in Europe itself. Colonies cannot serve this purpose unless they seem in large part suited for settlement by Europeans. But in the nineteenth century such colonial territories were no longer obtainable by peaceful means. Consequently, such a colonial policy could only have been carried out by means of a hard struggle which, however, would have been carried on to much better purpose, not for territories outside of Europe, but for land on the home continent itself.

[1] The two *Ostmarks* are the Bavarian O., or Austria, and the German O., meaning the territories bordering on Poland. The second came into use among nationalist circles in the late nineteenth century.

Such a decision, it is true, demands undivided devotion. It is not permissible to approach with half measures or even with hesitation a task whose execution seems possible only by the harnessing of the very last possible ounce of energy. This means that the entire political leadership of the Reich should have devoted itself to this exclusive aim; never should any step have been taken, guided by other considerations than the recognition of this task and its requirements. It was indispensable to see clearly that this aim could be achieved only by struggle, and consequently to face the contest of arms with calm and composure.

All alliances, therefore, should have been viewed exclusively from this standpoint and judged according to their possible utilization. If land was desired in Europe, it could be obtained by and large only at the expense of Russia, and this meant that the new Reich must again set itself on the march along the road of the Teutonic Knights of old, to obtain by the German sword sod for the German plow and daily bread for the nation.

For such a policy there was but one ally in Europe: England.

With England alone was it possible, our rear protected, to begin the new Germanic march. Our right to do this would have been no less than the right of our forefathers. None of our pacifists refuses to eat the bread of the East, although the first plowshare in its day bore the name of 'sword'!

Consequently, no sacrifice should have been too great for winning England's willingness. We should have renounced colonies and sea power, and spared English industry our competition.

Only an absolutely clear orientation could lead to such a goal: renunciation of world trade and colonies; renunciation of a German war fleet; concentration of all the state's instruments of power on the land army.

The result, to be sure, would have been a momentary limitation, but a great and mighty future.

There was a time when England would have listened to reason on this point, since she was well aware that Germany as a result of her increased population had to seek some way out and either find it with England in Europe or without England in the world.

And it can primarily be attributed to this realization if at the turn of the century London itself attempted to approach Germany. For the first time a thing became evident which in the last years we have had occasion to observe in a truly terrifying fashion. People were unpleasantly affected by the thought of having to pull England's chestnuts out of the fire; as though there ever could be an alliance on any other basis than a mutual business deal. And with England such a deal could very well have been made. British diplomacy was still clever enough to realize that no service can be expected without a return.

Just suppose that an astute German foreign policy had taken over the rôle of Japan in 1904, and we can scarcely measure the consequences this would have had for Germany.

There would never have been any 'World War.'

The bloodshed in the year 1904 would have saved ten times as much in the years 1914 to 1918.

And what a position Germany would occupy in the world today!

In that light, to be sure, the alliance with Austria was an absurdity.

For this mummy of a state allied itself with Germany, not in order to fight a war to its end, but for the preservation of an eternal peace which could astutely be used for the slow but certain extermination of Germanism in the monarchy.

This alliance was an impossibility for another reason: because we could not expect a state to take the offensive in championing national German interests as long as this state did not possess the power and determination to put an end to the process of de-Germanization on its own immediate borders. If Germany did not possess enough national awareness and ruthless determination to snatch power over the destinies of ten million national comrades from the hands of the impossible Habsburg state, then truly we had no right to expect that she would ever lend her hand to such farseeing and bold plans. The attitude of the old Reich on the Austrian question was the touchstone of its conduct in the struggle for the destiny of the whole nation.

In any case we were not justified in looking on, as year after year Germanism was increasingly repressed, since the value of Austria's fitness for alliance was determined exclusively by the preservation of the German element.

This road, however, was not taken at all.

These people feared nothing so much as struggle, yet they were finally forced into it at the most unfavorable hour.

They wanted to run away from destiny, and it caught up with them. They dreamed of preserving world peace, and landed in the World War.

And this was the most significant reason why this third way of molding the German future was not even considered. They knew that the acquisition of new soil was possible only in the East, they saw the struggle that would be necessary and yet wanted peace at any price; for the watchword of German foreign policy had long ceased to be: preservation of the German nation by all methods; but rather: preservation of world peace by all means. With what success, everyone knows.

I shall return to this point in particular.

Thus there remained the fourth possibility:

Industry and world trade, sea power and colonies.

Such a development, to be sure, was at first easier and also more quickly attainable. The settlement of land is a slow process, often lasting centuries; in fact, its inner strength is to be sought precisely in the fact that it is not a sudden blaze, but a gradual yet solid and continuous growth, contrasting with an industrial development which can be blown up in the course of a few years, but in that case is more like a soapbubble than solid strength. A fleet, to be sure, can be built more quickly than farms can be established in stubborn struggle and settled with peasants, but it is also more rapidly destroyed than the latter.

If, nevertheless, Germany took this road, she should at least have clearly recognized that this development would some day likewise end in struggle. Only children could have thought that they could get their bananas in the 'peaceful contest of nations,' by friendly and moral conduct and constant emphasis on their peaceful intentions, as they so high-soundingly and unctuously babbled; in other words, without ever having to take up arms. No: if we chose this road, England would some day inevitably become our enemy. It was more than senseless — but quite in keeping with our own innocence — to wax indignant over the fact that England should one day take the liberty to oppose our peaceful activity with the brutality of a violent egoist.

It is true that we, I am sorry to say, would never have done such a thing.

If a European territorial policy was only possible against Russia in alliance with England, conversely, a policy of colonies and world trade was conceivable only against England and with Russia. But then we had dauntlessly to draw the consequences — and, above all, abandon Austria in all haste.

Viewed from all angles, this alliance with Austria was real madness by the turn of the century.

But we did not think of concluding an alliance with Russia against England, any more than with England against Russia, for in both cases the end would have been war, and to prevent this we decided in favor of a policy of commerce and industry. In the 'peaceful economic' conquest of the world we possessed a recipe which was expected to break the neck of the former policy of violence once and for all.[1] Occasionally, perhaps, we were not quite sure of ourselves, particularly when from time to time incomprehensible threats came over from England; therefore, we decided to build a fleet, though not to attack and destroy England, but for the 'defense' of our old friend 'world peace' and 'peaceful' conquest of the world. Consequently, it was kept on a somewhat more modest scale in all respects, not only in number but also in the tonnage of the individual ships as well as in armament, so as in the final analysis to let our 'peaceful' intentions shine through after all.

The talk about the 'peaceful economic' conquest of the world was possibly the greatest nonsense which has ever been exalted to be a guiding principle of state policy. What made this non-

---

[1] *'die der bisherigen Gewaltpolitik ein fur allemal das Genick brechen sollte.'*

sense even worse was that its proponents did not hesitate to call upon England as a crown witness for the possibility of such an achievement. The crimes of our academic doctrine and conception of history in this connection can scarcely be made good and are only a striking proof of how many people there are who 'learn' history without understanding or even comprehending it. England, in particular, should have been recognized as the striking refutation of this theory; for no people has ever with greater brutality better prepared its economic conquests with the sword, and later ruthlessly defended them,[1] than the English nation. Is it not positively the distinguishing feature of British statesmanship to draw economic acquisitions from political strength, and at once to recast every gain in economic strength into political power? And what an error to believe that England is personally too much of a *coward* to stake her own blood for her economic policy! The fact that the English people possessed no 'people's army' in no way proved the contrary; for what matters is not the momentary military form of the fighting forces, but rather the will and determination to risk those which do exist. England has always possessed whatever armament she happened to need. She always fought with the weapons which success demanded. She fought with mercenaries as long as mercenaries sufficed; but she reached down into the precious blood of the whole nation when only such a sacrifice could bring victory; but the determination for victory, the tenacity and ruthless pursuit of this struggle, remained unchanged.

In Germany, however, the school, the press, and comic magazines cultivated a conception of the Englishman's character, and almost more so of his empire, which inevitably led to one of the most insidious delusions; for gradually everyone was infected by this nonsense, and the consequence was an underestimation for which we would have to pay most bitterly. This falsification went so deep that people became convinced that in the Englishman they faced a business man as shrewd as personally he was

---

[1] '*hat doch kein Volk mit grösserer Brutalität seine wirtschaftlichen Eroberungen besser vorbereitet und später rücksichtslos verteidigt . . .*'

unbelievably cowardly. The fact that a world empire the size
of the British could not be put together by mere subterfuge and
swindling was unfortunately something that never even occurred
to our exalted professors of academic science. The few who
raised a voice of warning were ignored or killed by silence. I
remember well my comrades' looks of astonishment when we
faced the Tommies in person in Flanders. After the very first
days of battle the conviction dawned on each and every one of
them that these Scotsmen did not exactly jibe with the pictures
they had seen fit to give us in the comic magazines and press
dispatches.

It was then that I began my first reflections about the importance of the form of propaganda.

This falsification, however, did have one good side for those who
spread it: by this example, even though it was incorrect, they
were able to demonstrate the correctness of the economic conquest of the world. If the Englishman had succeeded, we too
were bound to succeed, and our definitely greater honesty, the
absence in us of that specifically English 'perfidy,' was regarded
as a very special plus. For it was hoped that this would enable
us to win the affection, particularly of the smaller nations, and
the confidence of the large ones the more easily.

It did not occur to us that our honesty was a profound horror
to the others, if for no other reason because we ourselves believed all these things seriously while the rest of the world regarded such behavior as the expression of a special slyness and
disingenuousness, until, to their great, infinite amazement, the
revolution gave them a deeper insight into the boundless stupidity of our honest convictions.

However, the absurdity of this 'economic conquest' at once
made the absurdity of the Triple Alliance clear and comprehensible. For with what other state could we ally ourselves? In
alliance with Austria, to be sure, we could not undertake any
military conquest, even in Europe alone. Precisely therein consisted the inner weakness of the alliance from the very first day.
A Bismarck could permit himself this makeshift, but not by a

long shot every bungling successor, least of all at a time when certain essential premises of Bismarck's alliance had long ceased to exist; for Bismarck still believed that in Austria he had to do with a German state. But with the gradual introduction of universal suffrage, this country had sunk to the status of an un-German hodgepodge with a parliamentary government.

Also from the standpoint of racial policy, the alliance with Austria was simply ruinous. It meant tolerating the growth of a new Slavic power on the borders of the Reich, a power which sooner or later would have to take an entirely different attitude toward Germany than, for example, Russia. And from year to year the alliance itself was bound to grow inwardly hollower and weaker in proportion as the sole supporters of this idea in the monarchy lost influence and were shoved out of the most decisive positions.

By the turn of the century the alliance with Austria had entered the very same stage as Austria's pact with Italy.

Here again there were only two possibilities: either we were in a pact with the Habsburg monarchy or we had to lodge protest against the repression of Germanism. But once a power embarks on this kind of undertaking, it usually ends in open struggle.

Even psychologically the value of the Triple Alliance was small, since the stability of an alliance increases in proportion as the individual contracting parties can hope to achieve definite and tangible expansive aims. And, conversely, it will be the weaker the more it limits itself to the preservation of an existing condition. Here, as everywhere else, strength lies not in defense but in attack.

Even then this was recognized in various quarters, unfortunately not by the so-called 'authorities.' Particularly Ludendorff, then a colonel and officer in the great general staff, pointed to these weaknesses in a memorial written in 1912. Of course, none of the 'statesmen' attached any value or significance to the matter; for clear common sense is expected to manifest itself expediently only in common mortals, but may on principle remain absent where 'diplomats' are concerned.

For Germany it was sheer good fortune that in 1914 the war broke out indirectly through Austria, so that the Habsburgs were forced to take part; for if it had happened the other way around Germany would have been alone. Never would the Habsburg state have been able, let alone willing, to take part in a conflict which would have arisen through Germany. What we later so condemned in Italy would then have happened even earlier with Austria: they would have remained 'neutral' in order at least to save the state from a revolution at the very start. Austrian Slavdom would rather have shattered the monarchy even in 1914 than permit aid to Germany.

How great were the dangers and difficulties entailed by the alliance with the Danubian monarchy, only very few realized at that time.

In the first place, Austria possessed too many enemies who were planning to grab what they could from the rotten state to prevent a certain hatred from arising in the course of time against Germany, in whom they saw the cause of preventing the generally hoped and longed-for collapse of the monarchy. They came to the conviction that Vienna could finally be reached only by a detour through Berlin.

In the second place, Germany thus lost her best and most hopeful possibilities of alliance. They were replaced by an ever-mounting tension with Russia and even Italy. For in Rome the general mood was just as pro-German as it was anti-Austrian, slumbering in the heart of the very last Italian and often brightly flaring up.

Now, since we had thrown ourselves into a policy of commerce and industry, there was no longer the slightest ground for war against Russia either. Only the enemies of both nations could still have an active interest in it. And actually these were primarily the Jews and the Marxists, who, with every means, incited and agitated for war between the two states.

Thirdly and lastly, this alliance inevitably involved an infinite peril for Germany, because a great power actually hostile to Bismarck's Reich could at any time easily succeed in mobilizing a whole series of states against Germany, since it was in a position to promise each of them enrichment at the expense of our Austrian ally.

The whole East of Europe could be stirred up against the Danubian monarchy — particularly Russia and Italy. Never would the world coalition which had been forming since the initiating efforts of King Edward have come into existence if Austria as Germany's ally had not represented too tempting a legacy. This alone made it possible to bring states with otherwise so heterogeneous desires and aims into a single offensive front. Each one could hope that in case of a general action against Germany it, too, would achieve enrichment at Austria's expense. The danger was enormously increased by the fact that Turkey seemed to be a silent partner in this unfortunate alliance.

International Jewish world finance needed these lures to enable it to carry out its long-desired plan for destroying the Germany which thus far did not submit to its widespread superstate control of finance and economics. Only in this way could they forge a coalition made strong and courageous by the sheer numbers of the gigantic armies now on the march and prepared to attack the horny [1] Siegfried at last.

The alliance with the Habsburg monarchy, which even in Austria had filled me with dissatisfaction, now became the source of long inner trials which in the time to come reinforced me even more in the opinion I had already conceived.

Even then, among those few people whom I frequented, I made no secret of my conviction that our catastrophic alliance with a state on the brink of ruin would also lead to a fatal collapse of Germany unless we knew enough to release ourselves from it on time. This conviction of mine was firm as a rock, and I did not falter in it for one moment when at last the storm of the World War seemed to have excluded all reasonable thought and a frenzy of enthusiasm had seized even those quarters for

---

[1] In the legend Siegfried kills the dragon by hiding in a pit and stabbing the beast as it passes over him. The dragon's blood pours over Siegfried and makes his skin 'horny,' invulnerable.

which there should have been only the coldest consideration of
reality. And while I myself was at the front, I put forward, whenever these problems were discussed, my opinion that the alliance
had to be broken off, the quicker the better for the German nation,
and that the sacrifice of the Habsburg monarchy would be no
sacrifice at all to make if Germany thereby could achieve a restriction of her adversaries; for it was not for the preservation
of a debauched dynasty that the millions had donned the steel
helmet, but for the salvation of the German nation.

On a few occasions before the War it seemed as though, in
one camp at least, a gentle doubt was arising as to the correctness of the alliance policy that had been chosen. German conservative circles began from time to time to warn against excessive
confidence, but, like everything else that was sensible, this was
thrown to the winds. They were convinced that they were on
the path to a world 'conquest,' whose success would be tremendous and which would entail practically no sacrifices.

There was nothing for those not in authority to do but to
watch in silence why and how the 'authorities' marched straight
to destruction, drawing the dear people behind them like the
Pied Piper of Hamelin.

*   *   *

The deeper cause that made it possible to represent the
absurdity of an 'economic conquest' as a practical political
method, and the preservation of 'world peace' as a political
goal for a whole people, and even to make these things intelligible, lay in the general sickening of our whole political thinking.

With the victorious march of German technology and industry,
the rising successes of German commerce, the realization was
increasingly lost that all this was only possible on the basis of a
strong state. On the contrary, many circles went so far as to put
forward the conviction that the state owed its very existence to
these phenomena, that the state itself primarily represented an

economic institution, that it could be governed according to economic requirements, and that its very existence depended on economics, a state of affairs which was regarded and glorified as by far the healthiest and most natural.

But the state has nothing at all to do with any definite economic conception or development.

It is not a collection of economic contracting parties in a definite delimited living space for the fulfillment of economic tasks, but the organization of a community of physically and psychologically similar living beings for the better facilitation of the maintenance of their species and the achievement of the aim which has been allotted to this species by Providence. This and nothing else is the aim and meaning of a state. Economics is only one of the many instruments required for the achievement of this aim. It is never the cause or the aim of a state unless this state is based on a false, because unnatural, foundation to begin with. Only in this way can it be explained that the state as such does not necessarily presuppose territorial limitation. This will be necessary only among the peoples who want to secure the maintenance of their national comrades by their own resources; in other words, are prepared to fight the struggle for existence by their own labor. Peoples who can sneak their way into the rest of mankind like drones, to make other men work for them under all sorts of pretexts, can form states even without any definitely delimited living space of their own. This applies first and foremost to a people under whose parasitism the whole of honest humanity is suffering, today more than ever: the Jews.

The Jewish state was never spatially limited in itself, but universally unlimited as to space, though restricted in the sense of embracing but one race. Consequently, this people has always formed a state within states. It is one of the most ingenious tricks that was ever devised, to make this state sail under the flag of 'religion,' thus assuring it of the tolerance which the Aryan is always ready to accord a religious creed. For actually the Mosaic religion is nothing other than a doctrine for the preservation of the Jewish race. It therefore embraces almost all sociological, political, and economic fields of knowledge which can have any bearing on this function.

The urge to preserve the species is the first cause for the formation of human communities; thus the state is a national organism and not an economic organization. A difference which is just as large as it is incomprehensible, particularly to our so-called 'statesmen' of today. That is why they think they can build up the state through economics while in reality it results and always will result solely from the action of those qualities which lie in line with the will to preserve the species and race. And these are always heroic virtues and never the egoism of shopkeepers, since the preservation of the existence of a species presupposes a spirit of sacrifice in the individual. The sense of the poet's words, 'If you will not stake your life, you will win no life,' is that the sacrifice of personal existence is necessary to secure the preservation of the species. Thus, the most sensible prerequisite for the formation and preservation of a state is the presence of a certain feeling of cohesion based on similarity of nature and species, and a willingness to stake everything on it with all possible means, something which in peoples with soil of their own will create heroic virtues, but in parasites will create lying hypocrisy and malignant cruelty, or else these qualities must already be present as the necessary and demonstrable basis for their existence as a state so different in form. The formation of a state, originally at least, will occur through the exercise of these qualities, and in the subsequent struggle for self-preservation those nations will be defeated — that is, will fall a prey to subjugation and thus sooner or later die out — which in the mutual struggle possess the smallest share of heroic virtues, or are not equal to the lies and trickery of the hostile parasite. But in this case, too, this must almost always be attributed less to a lack of astuteness than to a lack of determination and courage, which only tries to conceal itself beneath a cloak of humane convictions.

How little the state-forming and state-preserving qualities are connected with economics is most clearly shown by the fact that

the inner strength of a state only in the rarest cases coincides
with so-called economic prosperity, but that the latter, in innumerable cases, seems to indicate the state's approaching
decline. If the formation of human societies were primarily
attributable to economic forces or even impulses, the highest
economic development would have to mean the greatest strength
of the state and not the opposite.

Belief in the state-forming and state-preserving power of economics seems especially incomprehensible when it obtains in a
country which in all things clearly and penetratingly shows the
historic reverse. Prussia, in particular, demonstrates with marvelous sharpness that not material qualities but ideal virtues
alone make possible the formation of a state. Only under their
protection can economic life flourish, until with the collapse of
the pure state-forming faculties the economy collapses too; a
process which we can observe in so terrible and tragic a form
right now. The material interests of man can always thrive
best as long as they remain in the shadow of heroic virtues; but
as soon as they attempt to enter the primary sphere of existence,
they destroy the basis for their own existence.

Always when in Germany there was an upsurge of political
power, the economic conditions began to improve; but always
when economics became the sole content of our people's life,
stifling the ideal virtues, the state collapsed and in a short time
drew economic life along with it.

If, however, we consider the question, what, in reality, are
the state-forming or even state-preserving forces, we can sum
them up under one single head: the ability and will of the individual to sacrifice himself for the totality. That these virtues
have nothing at all to do with economics can be seen from the
simple realization that man never sacrifices himself for the latter,
or, in other words: a man does not die for business, but only for
ideals. Nothing proved the Englishman's superior psychological
knowledge of the popular soul better than the motivation which
he gave to his struggle. While we fought for bread, England
fought for 'freedom'; and not even for her own, no, for that of

the small nations. In our country we laughed at this effrontery, or were enraged at it, and thus only demonstrated how empty-headed and stupid the so-called statesmen of Germany had become even before the War. We no longer had the slightest idea concerning the essence of the force which can lead men to their death of their own free will and decision.

In 1914, as long as the German people thought they were fighting for ideals, they stood firm; but as soon as they were told to fight for their daily bread, they preferred to give up the game.

And our brilliant 'statesmen' were astonished at this change in attitude. It never became clear to them that from the moment when a man begins to fight for an economic interest, he avoids death as much as possible, since death would forever deprive him of his reward for fighting. Anxiety for the rescue of her own child makes a heroine of even the feeblest mother, and only the struggle for the preservation of the species and the hearth, or the state that protects it, has at all times driven men against the spears of their enemies.

The following theorem may be established as an eternally valid truth:

Never yet has a state been founded by peaceful economic means, but always and exclusively by the instincts of preservation of the species regardless whether these are found in the province of heroic virtue or of cunning craftiness; the one results in Aryan states based on work and culture, the other in Jewish colonies of parasites. As soon as economics as such begins to choke out these instincts in a people or in a state, it becomes the seductive cause of subjugation and oppression.

The belief of pre-war days that the world could be peacefully opened up to, let alone conquered for, the German people by a commercial and colonial policy was a classic sign of the loss of real state-forming and state-preserving virtues and of all the insight, will power, and active determination which follow from them; the penalty for this, inevitable as the law of nature, was the World War with its consequences.

For those who do not look more deeply into the matter, this

attitude of the German nation — for it was really as good as general — could only represent an insoluble riddle: for was not Germany above all other countries a marvelous example of an empire which had risen from foundations of pure political power? Prussia, the germ-cell of the Empire, came into being through resplendent heroism and not through financial operations or commercial deals, and the Reich itself in turn was only the glorious reward of aggressive political leadership and the death-defying courage of its soldiers. How could this very German people have succumbed to such a sickening of its political instinct? For here we face, not an isolated phenomenon, but forces of decay which in truly terrifying number soon began to flare up like will-o'-the-wisps, brushing up and down the body politic, or eating like poisonous abscesses into the nation, now here and now there. It seemed as though a continuous stream of poison was being driven into the outermost blood-vessels of this once heroic body by a mysterious power, and was inducing progressively greater paralysis of sound reason and the simple instinct of self-preservation.

As innumerable times I passed in review all these questions, arising through my position on the German alliance policy and the economic policy of the Reich in the years 1912 to 1914 — the only remaining solution to the riddle became to an ever-increasing degree that power which, from an entirely different viewpoint, I had come to know earlier in Vienna: the Marxist doctrine and philosophy, and their organizational results.

For the second time I dug into this doctrine of destruction — this time no longer led by the impressions and effects of my daily associations, but directed by the observation of general processes of political life. I again immersed myself in the theoretical literature of this new world, attempting to achieve clarity concerning its possible effects, and then compared it with the actual phenomena and events it brings about in political, cultural, and economic life.

Now for the first time I turned my attention to the attempts to master this world plague.

ſ studied Bismarck's Socialist legislation [1] in its intention,
struggle, and success.   Gradually I obtained a positively granite
foundation for my own conviction, so that since that time I have
never been forced to undertake a shift in my own inner view on
this question. Likewise the relation of Marxism to the Jews was
submitted to further thorough examination.

Though previously in Vienna, Germany above all had seemed
to me an unshakable colossus, now anxious misgivings sometimes
entered my mind. In silent solitude and in the small circles of
my acquaintance, I was filled with wrath at German foreign
policy and likewise with what seemed to me the incredibly frivolous way in which the most important problem then existing for
Germany, Marxism, was treated. It was really beyond me how
people could rush so blindly into a danger whose effects, pursuant
to the Marxists' own intention, were bound some day to be
monstrous. Even then, among my acquaintance, just as today
on a large scale, I warned against the phrase with which all
wretched cowards comfort themselves: 'Nothing can happen to
us!' This pestilential attitude had once been the downfall of a
gigantic empire. Could anyone believe that Germany alone was
not subject to exactly the same laws as all other human organisms?

In the years 1913 and 1914, I, for the first time in various
circles which today in part faithfully support the National
Socialist movement, expressed the conviction that the question
of the future of the German nation was the question of destroying Marxism.

In the catastrophic German alliance policy I saw only one of
the consequences called forth by the disruptive work of this
doctrine; for the terrible part of it was that this poison almost
invisibly destroyed all the foundations of a healthy conception

[1] Bismarck's Anti-Socialist Law, put through the Reichstag on October
18, 1878, prohibited meetings, collections of funds, and publications of
Social Democrats, Socialists, and Communists; it remained in force until
1890 when the new Emperor, William II, opposed it. Despite the law,
Socialist deputies in the Reichstag retained their parliamentary immunity.

of economy and state, and that often those affected by it did not themselves realize to what an extent their activities and desires emanated from this philosophy which they otherwise sharply rejected.

The internal decline of the German nation had long since begun, yet, as so often in life, people had not achieved clarity concerning the force that was destroying their existence. Sometimes they tinkered around with the disease, but confused the forms of the phenomenon with the virus that had caused it. Since they did not know or want to know the cause, the struggle against Marxism was no better than bungling quackery.

The World War

As a young scamp in my wild years, nothing had so grieved me as having been born at a time which obviously erected its Halls of Fame only to shopkeepers and government officials. The waves of historic events seemed to have grown so smooth that the future really seemed to belong only to the 'peaceful contest of nations'; in other words, a cozy mutual swindling match with the exclusion of violent methods of defense. The various nations began to be more and more like private citizens who cut the ground from under one another's feet, stealing each other's customers and orders, trying in every way to get ahead of one another, and staging this whole act amid a hue and cry as loud as it is harmless. This development seemed not only to endure but was expected in time (as was universally recommended) to remodel the whole world into one big department store in whose vestibules the busts of the shrewdest profiteers and the most lamblike administrative officials would be garnered for all eternity. The English could supply the merchants, the Germans the administrative officials, and the Jews no doubt would have to sacrifice themselves to being the owners, since by their own admission they never make any money, but always 'pay,' and, besides, speak the most languages.

Why couldn't I have been born a hundred years earlier? Say at the time of the Wars of Liberation when a man, even without a 'business,' was really worth something?!

Thus I had often indulged in angry thoughts concerning my
earthly pilgrimage, which, as it seemed to me, had begun too
late, and regarded the period 'of law and order' ahead of me
as a mean and undeserved trick of Fate. Even as a boy I was no
'pacifist,' and all attempts to educate me in this direction came
to nothing.

The Boer War was like summer lightning to me.

Every day I waited impatiently for the newspapers and
devoured dispatches and news reports, happy at the privilege
of witnessing this heroic struggle even at a distance.

The Russo-Japanese War found me considerably more mature,
but also more attentive. More for national reasons I had already taken sides, and in our little discussions at once sided with
the Japanese. In a defeat of the Russians I saw the defeat of
Austrian Slavdom.

Since then many years have passed, and what as a boy had
seemed to me a lingering disease, I now felt to be the quiet before
the storm. As early as my Vienna period, the Balkans were immersed in that livid sultriness which customarily announces the
hurricane, and from time to time a beam of brighter light flared
up, only to vanish again in the spectral darkness. But then came
the Balkan War and with it the first gust of wind swept across
a Europe grown nervous. The time which now followed lay on
the chests of men like a heavy nightmare, sultry as feverish
tropic heat, so that due to constant anxiety the sense of approaching catastrophe turned at last to longing: let Heaven at
last give free rein to the fate which could no longer be thwarted.
And then the first mighty lightning flash struck the earth; the
storm was unleashed and with the thunder of Heaven there
mingled the roar of the World War batteries.

When the news of the murder of Archduke Francis Ferdinand
arrived in Munich (I happened to be sitting at home and heard
of it only vaguely), I was at first seized with worry that the bullets
may have been shot from the pistols of German students, who,
out of indignation at the heir apparent's continuous work of
Slavization, wanted to free the German people from this internal

enemy. What the consequence of this would have been was easy to imagine: a new wave of persecutions which would now have been 'justified' and 'explained' in the eyes of the whole world. But when, soon afterward, I heard the names of the supposed assassins, and moreover read that they had been identified as Serbs, a light shudder began to run through me at this vengeance of inscrutable Destiny.

The greatest friend of the Slavs had fallen beneath the bullets of Slavic fanatics.

Anyone with constant occasion in the last years to observe the relation of Austria to Serbia could not for a moment be in doubt that a stone had been set rolling whose course could no longer be arrested.

Those who today shower the Viennese government with reproaches on the form and content of the ultimatum it issued, do it an injustice. No other power in the world could have acted differently in the same situation and the same position. At her southeastern border Austria possessed an inexorable and mortal enemy who at shorter and shorter intervals kept challenging the monarchy and would never have left off until the moment favorable for the shattering of the Empire had arrived. There was reason to fear that this would occur at the latest with the death of the old Emperor; by then perhaps the old monarchy would no longer be in a position to offer any serious resistance. In the last few years the state had been so bound up with the person of Francis Joseph that the death of this old embodiment of the Empire was felt by the broad masses to be tantamount to the death of the Empire itself. Indeed, it was one of the craftiest artifices, particularly of the Slavic policy, to create the appearance that the Austrian state no longer owed its existence to anything but the miraculous and unique skill of this monarch; this flattery was all the more welcome in the Hofburg, since it corresponded not at all to the real merits of the Emperor. The thorn hidden in these paeans of praise remained undiscovered. The rulers did not see, or perhaps no longer wanted to see, that the more the monarchy depended on the outstanding statecraft, as they put it, of this 'wisest monarch' of all times, the more catastrophic the situation was bound to become if one day Fate were to knock at his door, too, demanding its tribute.

Was old Austria even conceivable without the Emperor?!

Wouldn't the tragedy which had once stricken Maria Theresa have been repeated?

No, it is really doing the Vienna circles an injustice to reproach them with rushing into a war which might otherwise have been avoided. It no longer could be avoided, but at most could have been postponed for one or two years. But this was the curse of German as well as Austrian diplomacy, that it had always striven to postpone the inevitable reckoning, until at length it was forced to strike at the most unfavorable hour. We can be convinced that a further attempt to save peace would have brought war at an even more unfavorable time.

No, those who did not want this war had to have the courage to face the consequences, which could have consisted only in the sacrifice of Austria. Even then the war would have come, but no longer as a struggle of all against ourselves, but in the form of a partition of the Habsburg monarchy. And then they had to make up their minds to join in, or to look on with empty hands and let Fate run its course.

Those very people, however, who today are loudest in cursing the beginning of the war and offer the sagest opinions were those who contributed most fatally to steering us into it.

For decades the Social Democrats had carried on the most scoundrelly war agitation against Russia, and the Center for religious reasons had been most active in making the Austrian state the hinge and pivot of Germany policy. Now we had to suffer the consequences of this lunacy. What came had to come, and could no longer under any circumstances be avoided. The guilt of the German government was that in order to preserve peace it always missed the favorable hours for striking, became entangled in the alliance for the preservation of world peace, and thus finally became the victim of a world coalition which countered the idea of preserving world peace with nothing less than determination for world war.

If the Vienna government had given the ultimatum another, milder form, this would have changed nothing in the situation except at most one thing, that this government would itself have been swept away by the indignation of the people. For in the eyes of the broad masses the tone of the ultimatum was far too gentle and by no means too brutal, let alone too far-reaching. Anyone who today attempts to argue this away is either a forgetful blockhead or a perfectly conscious swindler and liar.

The struggle of the year 1914 was not forced on the masses — no, by the living God — it was desired by the whole people.

People wanted at length to put an end to the general uncertainty. Only thus can it be understood that more than two million German men and boys thronged to the colors for this hardest of all struggles, prepared to defend the flag with the last drop of their blood.

*   *   *

To me those hours seemed like a release from the painful feelings of my youth. Even today I am not ashamed to say that, overpowered by stormy enthusiasm, I fell down on my knees and thanked Heaven from an overflowing heart for granting me the good fortune of being permitted to live at this time.

A fight for freedom had begun, mightier than the earth had ever seen; for once Destiny had begun its course, the conviction dawned on even the broad masses that this time not the fate of Serbia or Austria was involved, but whether the German nation was to be or not to be.

For the last time in many years the people had a prophetic vision of its own future. Thus, right at the beginning of the gigantic struggle the necessary grave undertone entered into the ecstasy of an overflowing enthusiasm; for this knowledge alone made the national uprising more than a mere blaze of straw. The earnestness was only too necessary; for in those days people in general had not the faintest conception of the possible length and duration of the struggle that was now beginning. They dreamed of being home again that winter to continue and renew their peaceful labors.

What a man wants is what he hopes and believes. The overwhelming majority of the nation had long been weary of the eternally uncertain state of affairs; thus it was only too understandable that they no longer believed in a peaceful conclusion of the Austro-Serbian conflict, but hoped for the final settlement.

I, too, was one of these millions.

Hardly had the news of the assassination become known in Munich than at once two thoughts quivered through my brain: first, that at last war would be inevitable; and, furthermore, that now the Habsburg state would be compelled to keep its pact; for what I had always most feared was the possibility that Germany herself would some day, perhaps in consequence of this very alliance, find herself in a conflict not directly caused by Austria, so that the Austrian state for reasons of domestic policy would not muster the force of decision to stand behind her ally. The Slavic majority of the Empire would at once have begun to sabotage any such intention on the part of the state, and would always have preferred to smash the entire state to smithereens than grant its ally the help it demanded. This danger was now eliminated. The old state had to fight whether it wanted to or not.

My own position on the conflict was likewise very simple and clear; for me it was not that Austria was fighting for some Serbian satisfaction, but that Germany was fighting for her existence, the German nation for life or death, freedom and future. The time had come for Bismarck's work to fight; what the fathers had once won in the battles from Weissenburg to Sedan and Paris, young Germany now had to earn once more. If the struggle were carried through to victory, our nation would enter the circle of great nations from the standpoint of external power, and only then could the German Reich maintain itself as a mighty haven of peace without having, for the sake of peace, to cut down on the daily bread of her children.

As a boy and young man I had so often felt the desire to prove at least once by deeds that for me national enthusiasm was no empty whim. It often seemed to me almost a sin to shout hurrah perhaps without having the inner right to do so; for who had the right to use this word without having proved it in the place where all playing is at an end and the inexorable hand of the Goddess of Destiny begins to weigh peoples and men according to the truth and steadfastness of their convictions? Thus my heart, like that of a million others, overflowed with proud joy that at last I would be able to redeem myself from this paralyzing feeling. I had so often sung '*Deutschland über Alles*' and shouted '*Heil*' at the top of my lungs, that it seemed to me almost a be-lated act of grace to be allowed to stand as a witness in the di-vine court of the eternal judge and proclaim the sincerity of this conviction. For from the first hour I was convinced that in case of a war — which seemed to me inevitable — in one way or an-other I would at once leave my books. Likewise I knew that my place would then be where my inner voice directed me.

I had left Austria primarily for political reasons; what was more natural than that, now the struggle had begun, I should really begin to take account of this conviction. I did not want to fight for the Habsburg state, but was ready at any time to die for my people and for the Reich which embodied it.

On the third of August, I submitted a personal petition to His Majesty, King Ludwig III, with a request for permission to enter a Bavarian regiment. The cabinet office certainly had plenty to do in those days; so much the greater was my joy to receive an answer to my request the very next day. With trem-bling hands I opened the document; my request had been ap-proved and I was summoned to report to a Bavarian regiment. My joy and gratitude knew no bounds. A few days later I was wearing the tunic which I was not to doff until nearly six years later.

For me, as for every German, there now began the greatest and most unforgettable time of my earthly existence. Compared to the events of this gigantic struggle, everything past receded to shallow nothingness. Precisely in these days, with the tenth anniversary of the mighty event approaching, I think back with proud sadness on those first weeks of our people's heroic struggle, in which Fate graciously allowed me to take part.

As though it were yesterday, image after image passes before my eyes. I see myself donning the uniform in the circle of my dear comrades, turning out for the first time, drilling, etc., until the day came for us to march off.

A single worry tormented me at that time, me, as so many others: would we not reach the front too late? Time and time again this alone banished all my calm. Thus, in every cause for rejoicing at a new, heroic victory, a slight drop of bitterness was hidden, for every new victory seemed to increase the danger of our coming too late.

At last the day came when we left Munich to begin the fulfillment of our duty. For the first time I saw the Rhine as we rode westward along its quiet waters to defend it, the German stream of streams, from the greed of the old enemy. When through the tender veil of the early morning mist the Niederwald Monument gleamed down upon us in the gentle first rays of the sun, the old *Watch on the Rhine* roared out of the endless transport train into the morning sky, and I felt as though my heart would burst.

And then came a damp, cold night in Flanders, through which we marched in silence, and when the day began to emerge from the mists, suddenly an iron greeting came whizzing at us over our heads, and with a sharp report sent the little pellets flying between our ranks, ripping up the wet ground; but even before the little cloud had passed, from two hundred throats the first hurrah rose to meet the first messenger of death. Then a crackling and a roaring, a singing and a howling began, and with feverish eyes each one of us was drawn forward, faster and faster, until suddenly past turnip fields and hedges the fight began, the fight of man against man. And from the distance the strains of a song reached our ears, coming closer and closer, leaping from company to company, and just as Death plunged a busy hand into our ranks, the song reached us too and we passed it along: '*Deutsch-land, Deutschland über Alles, über Alles in der Welt!*'

Four days later we came back. Even our step had changed. Seventeen-year-old boys now looked like men.

The volunteers of the List Regiment may not have learned to fight properly, but they knew how to die like old soldiers.

This was the beginning.

Thus it went on year after year; but the romance of battle had been replaced by horror. The enthusiasm gradually cooled and the exuberant joy was stifled by mortal fear. The time came when every man had to struggle between the instinct of self-preservation and the admonitions of duty. I, too, was not spared by this struggle. Always when Death was on the hunt, a vague something tried to revolt, strove to represent itself to the weak body as reason, yet it was only cowardice, which in such disguises tried to ensnare the individual. A grave tugging and warning set in, and often it was only the last remnant of conscience which decided the issue. Yet the more this voice admonished one to caution, the louder and more insistent its lures, the sharper resistance grew until at last, after a long inner struggle, consciousness of duty emerged victorious. By the winter of 1915–16, this struggle had for me been decided. At last my will was undisputed master. If in the first days I went over the top with rejoicing and laughter, I was now calm and determined. And this was enduring. Now Fate could bring on the ultimate tests without my nerves shattering or my reason failing.

The young volunteer had become an old soldier.

And this transformation had occurred in the whole army. It had issued old and hard from the eternal battles, and as for those who could not stand up under the storm — well, they were broken.

Now was the time to judge this army. Now, after two or three years, during which it was hurled from one battle into another, forever fighting against superiority in numbers and weapons, suffering hunger and bearing privations, now was the time to test the quality of this unique army.

Thousands of years may pass, but never will it be possible to speak of heroism without mentioning the German army and the World War. Then from the veil of the past the iron front of the gray steel helmet will emerge, unwavering and unflinching, an immortal monument. As long as there are Germans alive, they will remember that these men were sons of their nation.

I was a soldier then, and I didn't want to talk about politics. And really it was not the time for it. Even today I harbor the conviction that the humblest wagon-driver performed more valuable services for the fatherland than the foremost among, let us say, 'parliamentarians.' I had never hated these big-mouths more than now when every red-blooded man with something to say yelled it into the enemy's face or appropriately left his tongue at home and silently did his duty somewhere. Yes, in those days I hated all those politicians. And if it had been up to me, a parliamentary pick-and-shovel battalion would have been formed at once; then they could have chewed the fat to their hearts' content without annoying, let alone harming, honest, decent people.

Thus, at that time I wanted to hear nothing of politics, but I could not help taking a position on certain manifestations which after all did affect the whole nation, and particularly concerned us soldiers.

There were two things which then profoundly angered me and which I regarded as harmful.

After the very first news of victories, a certain section of the press, slowly, and in a way which at first was perhaps unrecognizable to many, began to pour a few drops of wormwood into the general enthusiasm. This was done beneath the mask of a certain benevolence and well-meaning, even of a certain solicitude. They had misgivings about an excess of exuberance in the celebration of the victories. They feared that in this form it was unworthy of so great a nation and hence inappropriate. The bravery and heroic courage of the German soldier were something self-evident, they said, and people should not be carried away too much by thoughtless outbursts of joy, if only for the sake of foreign countries to whom a silent and dignified

form of joy appealed more than unbridled exultation, etc.
Finally, we Germans even now should not forget that the war
was none of our intention and therefore we should not be ashamed
to confess in an open and manly fashion that at any time we
would contribute our part to a reconciliation of mankind. For
that reason it would not be prudent to besmirch the purity of
our army's deeds by too much shouting, since the rest of the
world would have little understanding for such behavior. The
world admired nothing more than the modesty with which a
true hero silently and calmly forgets his deeds, for this was the
gist of the whole argument.

Instead of taking one of these creatures by his long ears, tying
him to a long pole and pulling him up on a long cord, thus making
it impossible for the cheering nation to insult the aesthetic sentiment of this knight of the inkpot, the authorities actually began
to issue remonstrances against 'unseemly' rejoicing over victories.

It didn't occur to them in the least that enthusiasm once
scotched cannot be reawakened at need. It is an intoxication
and must be preserved in this state. And how, without this
power of enthusiasm, should a country withstand a struggle which
in all likelihood would make the most enormous demands on the
spiritual qualities of the nation?

I knew the psyche of the broad masses too well not to be aware
that a high 'aesthetic' tone would not stir up the fire that was
necessary to keep the iron hot. In my eyes it was madness
on the part of the authorities to be doing nothing to intensify
the glowing heat of passion; and when they curtailed what
passion was fortunately present, that was absolutely beyond me.

The second thing that angered me was the attitude which
they thought fit to take toward Marxism. In my eyes, this only
proved that they hadn't so much as the faintest idea concerning
this pestilence. In all seriousness they seemed to believe that,
by the assurance that parties were no longer recognized, they
had brought Marxism to understanding and restraint.

They failed to understand that here no party was involved,
but a doctrine that must lead to the destruction of all humanity,

especially since this cannot be learned in the Jewified universities and, besides, so many, particularly among our higher officials, due to the idiotic conceit that is cultivated in them, don't think it worth the trouble to pick up a book and learn something which was not in their university curriculum. The most gigantic upheaval passes these 'minds' by without leaving the slightest trace, which is why state institutions for the most part lag behind private ones. It is to them, by God, that the popular proverb best applies: 'What the peasant doesn't know, he won't eat.' Here, too, a few exceptions only confirm the rule.

It was an unequaled absurdity to identify the German worker with Marxism in the days of August, 1914. In those hours the German worker had made himself free from the embrace of this venomous plague, for otherwise he would never have been able to enter the struggle. The authorities, however, were stupid enough to believe that Marxism had now become 'national'; a flash of genius which only shows that in these long years none of these official guides of the state had even taken the trouble to study the essence of this doctrine, for if they had, such an absurdity could scarcely have crept in.

Marxism, whose goal is and remains the destruction of all non-Jewish national states, was forced to look on in horror as, in the July days of 1914, the German working class it had ensnared, awakened and from hour to hour began to enter the service of the fatherland with ever-increasing rapidity. In a few days the whole mist and swindle of this infamous betrayal of the people had scattered away, and suddenly the gang of Jewish leaders stood there lonely and forsaken, as though not a trace remained of the nonsense and madness which for sixty years they had been funneling into the masses. It was a bad moment for the betrayers of the German working class, but as soon as the leaders recognized the danger which menaced them, they rapidly pulled the tarn-cap [1] of lies over their ears, and insolently mimicked the national awakening.

[1] Tarn-cap: a cloak conferring invisibility. Occurs frequently in German legends. Siegfried used one in his battle with Brünhilde.

But now the time had come to take steps against the whole treacherous brotherhood of these Jewish poisoners of the people. Now was the time to deal with them summarily without the slightest consideration for any screams and complaints that might arise. In August, 1914, the whole Jewish jabber about international solidarity had vanished at one stroke from the heads of the German working class, and in its stead, only a few weeks later, American shrapnel began to pour down the blessings of brotherhood on the helmets of our march columns. It would have been the duty of a serious government, now that the German worker had found his way back to his nation, to exterminate mercilessly the agitators who were misleading the nation.

If the best men were dying at the front, the least we could do was to wipe out the vermin.

Instead of this, His Majesty the Kaiser himself stretched out his hand to the old criminals, thus sparing the treacherous murderers of the nation and giving them a chance to retrieve themselves.

So now the viper could continue his work, more cautiously than before, but all the more dangerously. While the honest ones were dreaming of peace within their borders,[1] the perjuring criminals were organizing the revolution.

That such terrible half-measures should then be decided upon made me more and more dissatisfied at heart; but at that time I would not have thought it possible that the end of it all would be so frightful.

What, then, should have been done? The leaders of the whole movement should at once have been put behind bars, brought to trial, and thus taken off the nation's neck. All the implements of military power should have been ruthlessly used for the extermination of this pestilence. The parties should have been dissolved, the Reichstag brought to its senses, with bayonets if necessary, but, best of all, dissolved at once. Just as the Re-

[1] *Burgfrieden*, the special legal protection accorded to walled places and cities in the Middle Ages.

public today can dissolve parties, this method should have been used at that time, with more reason. For the life and death of a whole nation was at stake!

One question came to the fore, however: can spiritual ideas be exterminated by the sword? Can 'philosophies' be combated by the use of brute force?

Even at that time I pondered this question more than once:

If we ponder analogous cases, particularly on a religious basis, which can be found in history, the following fundamental principle emerges:

Conceptions and ideas, as well as movements with a definite spiritual foundation, regardless whether the latter is false or true, can, after a certain point in their development, only be broken with technical instruments of power if these physical weapons are at the same time the support of a new kindling thought, idea, or philosophy.

The application of force alone, without the impetus of a basic spiritual idea as a starting point, can never lead to the destruction of an idea and its dissemination, except in the form of a complete extermination of even the very last exponent of the idea and the destruction of the last tradition. This, however, usually means the disappearance of such a state from the sphere of political importance, often for an indefinite time and sometimes forever; for experience shows that such a blood sacrifice strikes the best part of the people, since every persecution which occurs without a spiritual basis seems morally unjustified and whips up precisely the more valuable parts of a people in protest, which results in an adoption of the spiritual content of the unjustly persecuted movement. In many this occurs simply through a feeling of opposition against the attempt to bludgeon down an idea by brute force.

As a result, the number of inward supporters grows in proportion as the persecution increases. Consequently, the complete annihilation of the new doctrine can be carried out only through a process of extermination so great and constantly increasing that in the end all the truly valuable blood is drawn out of the people or state in question. The consequence is that, though a so-called 'inner' purge can now take place, it will only be at the cost of total impotence. Such a method will always prove vain in advance if the doctrine to be combated has overstepped a certain small circle.

Consequently, here, too, as in all growth, the first period of childhood is most readily susceptible to the possibility of extermination, while with the mounting years the power of resistance increases and only with the weakness of approaching old age cedes again to new youth, though in another form and for different reasons.

Indeed, nearly all attempts to exterminate a doctrine and its organizational expression, by force without spiritual foundation, are doomed to failure, and not seldom end with the exact opposite of the desired result for the following reason:

The very first requirement for a mode of struggle with the weapons of naked force is and remains persistence. In other words: only the continuous and steady application of the methods for repressing a doctrine, etc., makes it possible for a plan to succeed. But as soon as force wavers and alternates with forbearance, not only will the doctrine to be repressed recover again and again, but it will also be in a position to draw new benefit from every persecution, since, after such a wave of pressure has ebbed away, indignation over the suffering induced leads new supporters to the old doctrine, while the old ones will cling to it with greater defiance and deeper hatred than before, and even schismatic heretics, once the danger has subsided, will attempt to return to their old viewpoint. Only in the steady and constant application of force lies the very first prerequisite for success. This persistence, however, can always and only arise from a definite spiritual conviction. Any violence which does not spring from a firm, spiritual base, will be wavering and uncertain. It lacks the stability which can only rest in a fanatical outlook. It emanates from the momentary energy and brutal determination of an individual, and is therefore subject to the change of personalities and to their nature and strength.

Added to this there is something else:

Any philosophy, whether of a religious or political nature —
and sometimes the dividing line is hard to determine — fights
less for the negative destruction of the opposing ideology than
for the positive promotion of its own. Hence its struggle is less
defensive than offensive. It therefore has the advantage even in
determining the goal, since this goal represents the victory of its
own idea, while, conversely, it is hard to determine when the
negative aim of the destruction of a hostile doctrine may be
regarded as achieved and assured. For this reason alone, the
philosophy's offensive will be more systematic and also more
powerful than the defensive against a philosophy, since here, too,
as always, the attack and not the defense makes the decision. The fight against a spiritual power with methods of
violence remains defensive, however, until the sword becomes
the support, the herald and disseminator, of a new spiritual doctrine.

Thus, in summing up, we can establish the following:

Any attempt to combat a philosophy with methods of violence
will fail in the end, unless the fight takes the form of attack for
a new spiritual attitude. Only in the struggle between two
philosophies can the weapon of brutal force, persistently and
ruthlessly applied, lead to a decision for the side it supports.

This remained the reason for the failure of the struggle against
Marxism.

This was why Bismarck's Socialist legislation finally failed and
had to fail, in spite of everything. Lacking was the platform of a
new philosophy for whose rise the fight could have been waged.
For only the proverbial wisdom of high government officials will
succeed in believing that drivel about so-called 'state authority'
or 'law and order' could form a suitable basis for the spiritual
impetus of a life-and-death struggle.

Since a real spiritual basis for this struggle was lacking,
Bismarck had to entrust the execution of his Socialist legislation
to the judgment and desires of that institution which itself was a
product of Marxist thinking. By entrusting the fate of his war

on the Marxists to the well-wishing of bourgeois democracy, the Iron Chancellor set the wolf to mind the sheep.

All this was only the necessary consequence of the absence of a basic new anti-Marxist philosophy endowed with a stormy will to conquer.

Hence the sole result of Bismarck's struggle was a grave disillusionment.

Were conditions different during the World War or at its beginning? Unfortunately not.

The more I occupied myself with the idea of a necessary change in the government's attitude toward Social Democracy as the momentary embodiment of Marxism, the more I recognized the lack of a serviceable substitute for this doctrine. What would be given the masses, if, just supposing, Social Democracy had been broken? There was not one movement in existence which could have been expected to succeed in drawing into its sphere of influence the great multitudes of workers grown more or less leaderless. It is senseless and more than stupid to believe that the international fanatic who had left the class party would now at once join a bourgeois party, in other words, a new class organization. For, unpleasant as it may seem to various organizations, it cannot be denied that bourgeois politicians largely take class division quite for granted as long as it does not begin to work out to their political disadvantage.

The denial of this fact only proves the effrontery, and also the stupidity, of the liars.

Altogether, care should be taken not to regard the masses as stupider than they are. In political matters feeling often decides more correctly than reason. The opinion that the stupid international attitude of the masses is sufficient proof of the unsoundness of the masses' sentiments can be thoroughly confuted by the simple reminder that pacifist democracy is no less insane, and that its exponents originate almost exclusively in the bourgeois camp. As long as millions of the bourgeoisie still piously worship their Jewish democratic press every morning, it very ill becomes these gentlemen to make jokes about the stupidity of the 'comrade' who, in the last analysis, only swallows down the same garbage, though in a different form. In both cases the manufacturer is one and the same Jew.

Good care should be taken not to deny things that just happen to be true. The fact that the class question is by no means exclusively a matter of ideal problems, as, particularly before the elections, some people would like to pretend, cannot be denied. The class arrogance of a large part of our people, and to an even greater extent, the underestimation of the manual worker, are phenomena which do not exist only in the imagination of the moonstruck.

Quite aside from this, however, it shows the small capacity for thought of our so-called 'intelligentsia' when, particularly in these circles, it is not understood that a state of affairs which could not prevent the growth of a plague, such as Marxism happens to be, will certainly not be able to recover what has been lost.

The 'bourgeois' parties, as they designate themselves, will never be able to attach the 'proletarian' masses to their camp, for here two worlds oppose each other, in part naturally and in part artificially divided, whose mutual relation [1] can only be struggle. The younger will be victorious — and this is Marxism.

Indeed, a struggle against Social Democracy in the year 1914 was conceivable, but how long this condition would be maintained, in view of the absence of any substitute, remained doubtful.

Here there was a great gap.

I was of this opinion long before the War, and for this reason could not make up my mind to join one of the existing parties. In the course of events of the World War, I was reinforced in this opinion by the obvious impossibility of taking up a ruthless struggle against Social Democracy, owing to this very lack of a movement which would have had to be more than a 'parliamentary' party.

[1] Hitler's word is '*Verhaltungszustand*,' seldom if ever seen before in the language. Literally: 'condition of behavior or attitude.'

With my closer comrades I often expressed myself openly on this point.

And now the first ideas came to me of later engaging in political activity.

Precisely this was what caused me often to assure the small circle of my friends that after the War, I meant to be a speaker in addition to my profession.

I believe that I was very serious about this.

War Propaganda

Ever since I have been scrutinizing political events, I have taken a tremendous interest in propagandist activity. I saw that the Socialist-Marxist organizations mastered and applied this instrument with astounding skill. And I soon realized that the correct use of propaganda is a true art which has remained practically unknown to the bourgeois parties. Only the Christian-Social movement, especially in Lueger's time, achieved a certain virtuosity on this instrument, to which it owed many of its successes.

But it was not until the War that it became evident what immense results could be obtained by a correct application of propaganda. Here again, unfortunately, all our studying had to be done on the enemy side, for the activity on our side was modest, to say the least. The total miscarriage of the German 'enlightenment' service stared every soldier in the face, and this spurred me to take up the question of propaganda even more deeply than before.

There was often more than enough time for thinking, and the enemy offered practical instruction which, to our sorrow, was only too good.

For what we failed to do, the enemy did, with amazing skill and really brilliant calculation. I, myself, learned enormously from this enemy war propaganda. But time passed and left no trace in the minds of all those who should have benefited;

partly because they considered themselves too clever to learn from the enemy, partly owing to lack of good will.

Did we have anything you could call propaganda?

I regret that I must answer in the negative. Everything that actually was done in this field was so inadequate and wrong from the very start that it certainly did no good and sometimes did actual harm.

The form was inadequate, the substance was psychologically wrong: a careful examination of German war propaganda can lead to no other diagnosis.

There seems to have been no clarity on the very first question: Is propaganda a means or an end?

It is a means and must therefore be judged with regard to its end. It must consequently take a form calculated to support the aim which it serves. It is also obvious that its aim can vary in importance from the standpoint of general need, and that the inner value of the propaganda will vary accordingly. The aim for which we were fighting the War was the loftiest, the most overpowering, that man can conceive: it was the freedom and independence of our nation, the security of our future food supply, and — our national honor; a thing which, despite all contrary opinions prevailing today, nevertheless exists, or rather should exist, since peoples without honor have sooner or later lost their freedom and independence, which in turn is only the result of a higher justice, since generations of rabble without honor deserve no freedom. Any man who wants to be a cowardly slave can have no honor, or honor itself would soon fall into general contempt.

The German nation was engaged in a struggle for a human existence, and the purpose of war propaganda should have been to support this struggle; its aim to help bring about victory.

When the nations on this planet fight for existence — when the question of destiny, 'to be or not to be,' cries out for a solution — then all considerations of humanitarianism or aesthetics crumble into nothingness; for all these concepts do not float about in the ether, they arise from man's imagination and are bound up with man. When he departs from this world, these concepts are again dissolved into nothingness, for Nature does not know them. And even among mankind, they belong only to a few nations or rather races, and this in proportion as they emanate from the feeling of the nation or race in question. Humanitarianism and aesthetics would vanish even from a world inhabited by man if this world were to lose the races that have created and upheld these concepts.

But all such concepts become secondary when a nation is fighting for its existence; in fact, they become totally irrelevant to the forms of the struggle as soon as a situation arises where they might paralyze a struggling nation's power of self-preservation. And that has always been their only visible result.

As for humanitarianism, Moltke [1] said years ago that in war it lies in the brevity of the operation, and that means that the most aggressive fighting technique is the most humane.

But when people try to approach these questions with drivel about aesthetics, etc., really only one answer is possible: where the destiny and existence of a people are at stake, all obligation toward beauty ceases. The most unbeautiful thing there can be in human life is and remains the yoke of slavery. Or do these Schwabing [2] decadents view the present lot of the German people as 'aesthetic'? Certainly we don't have to discuss these matters with the Jews, the most modern inventors of this cultural perfume. Their whole existence is an embodied protest against the aesthetics of the Lord's image.

And since these criteria of humanitarianism and beauty must be eliminated from the struggle, they are also inapplicable to propaganda.

Propaganda in the War was a means to an end, and the end was the struggle for the existence of the German people; consequently, propaganda could only be considered in accordance with

---

[1] General Helmuth von Moltke (1800–91) became chief of the Prussian General Staff in 1859. He modernized the Prussian army and was the founder of the German General Staff.

[2] Schwabing: the bohemian quarter of Munich, located near the university.

the principles that were valid for this struggle. In this case the most cruel weapons were humane if they brought about a quicker victory; and only those methods were beautiful which helped the nation to safeguard the dignity of its freedom.

This was the only possible attitude toward war propaganda in a life-and-death struggle like ours.

If the so-called responsible authorities had been clear on this point, they would never have fallen into such uncertainty over the form and application of this weapon: for even propaganda is no more than a weapon, though a frightful one in the hand of an expert.

The second really decisive question was this: To whom should propaganda be addressed? To the scientifically trained intelligentsia or to the less educated masses?

It must be addressed always and exclusively to the masses.

What the intelligentsia — or those who today unfortunately often go by that name — what they need is not propaganda but scientific instruction. The content of propaganda is not science any more than the object represented in a poster is art. The art of the poster lies in the designer's ability to attract the attention of the crowd by form and color. A poster advertising an art exhibit must direct the attention of the public to the art being exhibited; the better it succeeds in this, the greater is the art of the poster itself. The poster should give the masses an idea of the significance of the exhibition, it should not be a substitute for the art on display. Anyone who wants to concern himself with the art itself must do more than study the poster; and it will not be enough for him just to saunter through the exhibition. We may expect him to examine and immerse himself in the individual works, and thus little by little form a fair opinion.

A similar situation prevails with what we today call propaganda.

The function of propaganda does not lie in the scientific training of the individual, but in calling the masses' attention to certain facts, processes, necessities, etc., whose significance is thus for the first time placed within their field of vision.

The whole art consists in doing this so skillfully that everyone will be convinced that the fact is real, the process necessary, the necessity correct, etc. But since propaganda is not and cannot be the necessity in itself, since its function, like the poster, consists in attracting the attention of the crowd, and not in educating those who are already educated or who are striving after education and knowledge, its effect for the most part must be aimed at the emotions and only to a very limited degree at the so-called intellect.

All propaganda must be popular and its intellectual level must be adjusted to the most limited intelligence among those it is addressed to. Consequently, the greater the mass it is intended to reach, the lower its purely intellectual level will have to be. But if, as in propaganda for sticking out a war, the aim is to influence a whole people, we must avoid excessive intellectual demands on our public, and too much caution cannot be exerted in this direction.

The more modest its intellectual ballast, the more exclusively it takes into consideration the emotions of the masses, the more effective it will be. And this is the best proof of the soundness or unsoundness of a propaganda campaign, and not success in pleasing a few scholars or young aesthetes.

The art of propaganda lies in understanding the emotional ideas of the great masses and finding, through a psychologically correct form, the way to the attention and thence to the heart of the broad masses. The fact that our bright boys do not understand this merely shows how mentally lazy and conceited they are.

Once we understand how necessary it is for propaganda to be adjusted to the broad mass, the following rule results:

It is a mistake to make propaganda many-sided, like scientific instruction, for instance.

The receptivity of the great masses is very limited, their intelligence is small, but their power of forgetting is enormous. In consequence of these facts, all effective propaganda must be limited to a very few points and must harp on these in slogans until the last member of the public understands what you want him to understand by your slogan. As soon as you sacrifice this slogan and try to be many-sided, the effect will piddle away, for the crowd can neither digest nor retain the material offered. In this way the result is weakened and in the end entirely cancelled out.

Thus we see that propaganda must follow a simple line and correspondingly the basic tactics must be psychologically sound.

For instance, it was absolutely wrong to make the enemy ridiculous, as the Austrian and German comic papers did. It was absolutely wrong because actual contact with an enemy soldier was bound to arouse an entirely different conviction, and the results were devastating; for now the German soldier, under the direct impression of the enemy's resistance, felt himself swindled by his propaganda service. His desire to fight, or even to stand firm, was not strengthened, but the opposite occurred. His courage flagged.

By contrast, the war propaganda of the English and Americans was psychologically sound. By representing the Germans to their own people as barbarians and Huns, they prepared the individual soldier for the terrors of war, and thus helped to preserve him from disappointments. After this, the most terrible weapon that was used against him seemed only to confirm what his propagandists had told him; it likewise reinforced his faith in the truth of his government's assertions, while on the other hand it increased his rage and hatred against the vile enemy. For the cruel effects of the weapon, whose use by the enemy he now came to know, gradually came to confirm for him the 'Hunnish' brutality of the barbarous enemy, which he had heard all about; and it never dawned on him for a moment that his own weapons possibly, if not probably, might be even more terrible in their effects.

And so the English soldier could never feel that he had been misinformed by his own countrymen, as unhappily was so much the case with the German soldier that in the end he rejected everything coming from this source as 'swindles' and 'bunk.'

All this resulted from the idea that any old simpleton (or even somebody who was intelligent 'in other things') could be assigned to propaganda work, and the failure to realize that the most brilliant psychologists would have been none too good.

And so the German war propaganda offered an unparalleled example of an 'enlightenment' service working in reverse, since any correct psychology was totally lacking.

There was no end to what could be learned from the enemy by a man who kept his eyes open, refused to let his perceptions be ossified, and for four and a half years privately turned the storm-flood of enemy propaganda over in his brain.

What our authorities least of all understood was the very first axiom of all propagandist activity: to wit, the basically subjective and one-sided attitude it must take toward every question it deals with. In this connection, from the very beginning of the War and from top to bottom, such sins were committed that we were entitled to doubt whether so much absurdity could really be attributed to pure stupidity alone.

What, for example, would we say about a poster that was supposed to advertise a new soap and that described other soaps as 'good'?

We would only shake our heads.

Exactly the same applies to political advertising.

The function of propaganda is, for example, not to weigh and ponder the rights of different people, but exclusively to emphasize the one right which it has set out to argue for. Its task is not to make an objective study of the truth, in so far as it favors the enemy, and then set it before the masses with academic fairness; its task is to serve our own right, always and unflinchingly.

It was absolutely wrong to discuss war-guilt from the standpoint that Germany alone could not be held responsible for the outbreak of the catastrophe; it would have been correct to load every bit of the blame on the shoulders of the enemy, even if this had not really corresponded to the true facts, as it actually did.

And what was the consequence of this half-heartedness?

The broad mass of a nation does not consist of diplomats, or even professors of political law, or even individuals capable of forming a rational opinion; it consists of plain mortals, wavering and inclined to doubt and uncertainty. As soon as our own propaganda admits so much as a glimmer of right on the other side, the foundation for doubt in our own right has been laid. The masses are then in no position to distinguish where foreign injustice ends and our own begins. In such a case they become uncertain and suspicious, especially if the enemy refrains from going in for the same nonsense, but unloads every bit of blame on his adversary. Isn't it perfectly understandable that the whole country ends up by lending more credence to enemy propaganda, which is more unified and coherent, than to its own? And particularly a people that suffers from the mania of objectivity as much as the Germans. For, after all this, everyone will take the greatest pains to avoid doing the enemy any injustice, even at the peril of seriously besmirching and even destroying his own people and country.

Of course, this was not the intent of the responsible authorities, but the people never realize that.

The people in their overwhelming majority are so feminine by nature and attitude that sober reasoning determines their thoughts and actions far less than emotion and feeling.

And this sentiment is not complicated, but very simple and all of a piece. It does not have multiple shadings; it has a positive and a negative; love or hate, right or wrong, truth or lie, never half this way and half that way, never partially, or that kind of thing.

English propagandists understood all this most brilliantly — and acted accordingly. They made no half statements that might have given rise to doubts.

Their brilliant knowledge of the primitive sentiments of the broad masses is shown by their atrocity propaganda, which was adapted to this condition. As ruthless as it was brilliant, it created the preconditions for moral steadfastness at the front, even in the face of the greatest actual defeats, and just as strikingly it pilloried the German enemy as the sole guilty party for the outbreak of the War: the rabid, impudent bias and persistence with which this lie was expressed took into account the emotional, always extreme, attitude of the great masses and for this reason was believed.

How effective this type of propaganda was is most strikingly shown by the fact that after four years of war it not only enabled the enemy to stick to its guns, but even began to nibble at our own people.

It need not surprise us that our propaganda did not enjoy this success. In its inner ambiguity alone, it bore the germ of ineffectualness. And finally its content was such that it was very unlikely to make the necessary impression on the masses. Only our feather-brained 'statesmen' could have dared to hope that this insipid pacifistic bilge could fire men's spirits till they were willing to die.

As a result, their miserable stuff [1] was useless, even harmful in fact.

But the most brilliant propagandist technique will yield no success unless one fundamental principle is borne in mind constantly and with unflagging attention. It must confine itself to a few points and repeat them over and over. Here, as so often in this world, persistence is the first and most important requirement for success.

Particularly in the field of propaganda, we must never let ourselves be led by aesthetes or people who have grown blasé: not by the former, because the form and expression of our propaganda would soon, instead of being suitable for the masses, have drawing power only for literary teas; and of the second we must beware, because, lacking in any fresh emotion of their own, they are always on the lookout for new stimulation. These people are quick to weary of everything; they want variety, and they are never able to feel or understand the needs of their fellow men who are not yet so callous. They are always the first to criticize

[1] '*Zeug*.' Second edition has '*Produkt*.'

a propaganda campaign, or rather its content, which seems to them too old-fashioned, too hackneyed, too out-of-date, etc. They are always after novelty, in search of a change, and this makes them mortal enemies of any effective political propaganda. For as soon as the organization and the content of propaganda begin to suit their tastes, it loses all cohesion and evaporates completely.

The purpose of propaganda is not to provide interesting distraction for blasé young gentlemen, but to convince, and what I mean is to convince the masses. But the masses are slow-moving, and they always require a certain time before they are ready even to notice a thing, and only after the simplest ideas are repeated thousands of times will the masses finally remember them.

When there is a change, it must not alter the content of what the propaganda is driving at, but in the end must always say the same thing. For instance, a slogan must be presented from different angles, but the end of all remarks must always and immutably be the slogan itself. Only in this way can the propaganda have a unified and complete effect.

This broadness of outline from which we must never depart, in combination with steady, consistent emphasis, allows our final success to mature. And then, to our amazement, we shall see what tremendous results such perseverance leads to — to results that are almost beyond our understanding.

All advertising, whether in the field of business or politics, achieves success through the continuity and sustained uniformity of its application.

Here, too, the example of enemy war propaganda was typical; limited to a few points, devised exclusively for the masses, carried on with indefatigable persistence. Once the basic ideas and methods of execution were recognized as correct, they were applied throughout the whole War without the slightest change. At first the claims of the propaganda were so impudent that people thought it insane; later, it got on people's nerves; and in the end, it was believed. After four and a half years, a revolution broke out in Germany; and its slogans originated in the enemy's war propaganda.

And in England they understood one more thing: that this spiritual weapon can succeed only if it is applied on a tremendous scale, but that success amply covers all costs.

There, propaganda was regarded as a weapon of the first order, while in our country it was the last resort of unemployed politicians and a comfortable haven for slackers.

And, as was to be expected, its results all in all were zero.

The Revolution

WITH THE YEAR 1915 enemy propaganda began in our country, after 1916 it became more and more intensive, till finally, at the beginning of the year 1918, it swelled to a positive flood. Now the results of this seduction could be seen at every step. The army gradually learned to think as the enemy wanted it to.

And the German counter-action was a complete failure.

In the person of the man whose intellect and will made him its leader, the army had the intention and determination to take up the struggle in this field, too, but it lacked the instrument which would have been necessary. And from the psychological point of view, it was wrong to have this enlightenment work carried on by the troops themselves. If it was to be effective, it had to come from home. Only then was there any assurance of success among the men who, after all, had been performing immortal deeds of heroism and privation for nearly four years for this homeland.

But what came out of the home country?

Was this failure stupidity or crime?

In midsummer of 1918, after the evacuation of the southern bank of the Marne, the German press above all conducted itself with such miserable awkwardness, nay, criminal stupidity, that my wrath mounted by the day, and the question arose within me: Is there really no one who can put an end to this spiritual squandering of the army's heroism?

What happened in France in 1914 when we swept into the country in an unprecedented storm of victory? What did Italy do in the days after her Isonzo front had collapsed? And what again did France do in the spring of 1918 when the attack of the German divisions seemed to lift her positions off their hinges and the far-reaching arm of the heavy long-range batteries began to knock at the doors of Paris?

How they whipped the fever heat of national passion into the faces of the hastily retreating regiments in those countries! What propaganda and ingenious demagogy were used to hammer the faith in final victory back into the hearts of the broken fronts!

Meanwhile, what happened in our country?

Nothing, or worse than nothing.

Rage and indignation often rose up in me when I looked at the latest newspapers, and came face to face with the psychological mass murder that was being committed.

More than once I was tormented by the thought that if Providence had put me in the place of the incapable or criminal incompetents or scoundrels in our propaganda service, our battle with Destiny would have taken a different turn.

In these months I felt for the first time the whole malice of Destiny which kept me at the front in a position where every nigger might accidentally shoot me to bits, while elsewhere I would have been able to perform quite different services for the fatherland!

For even then I was rash enough to believe that I would have succeeded in this.

But I was a nameless soldier, one among eight million!

And so it was better to hold my tongue and do my duty in the trenches as best I could.

*   *   *

In the summer of 1915, the first enemy leaflets fell into our hands.

Aside from a few changes in the form of presentation, their content was almost always the same, to wit: that the suffering was growing greater and greater in Germany; that the War was going to last forever while the hope of winning it was gradually vanishing; that the people at home were, therefore, longing for peace, but that 'militarism' and the 'Kaiser' did not allow it; that the whole world — to whom this was very well known — was, therefore, not waging a war on the German people, but exclusively against the sole guilty party, the Kaiser; that, therefore, the War would not be over before this enemy of peaceful humanity should be eliminated; that when the War was ended, the libertarian and democratic nations would take the German people into the league of eternal world peace, which would be assured from the hour when 'Prussian militarism' was destroyed.

The better to illustrate these claims, 'letters from home' were often reprinted whose contents seemed to confirm these assertions.

On the whole, we only laughed in those days at all these efforts. The leaflets were read, then sent back to the higher staffs, and for the most part forgotten until the wind again sent a load of them sailing down into the trenches; for, as a rule, the leaflets were brought over by airplanes.

In this type of propaganda there was one point which soon inevitably attracted attention: in every sector of the front where Bavarians were stationed, Prussia was attacked with extraordinary consistency, with the assurance that not only was Prussia on the one hand the really guilty and responsible party for the whole war, but that on the other hand there was not the slightest hostility against Bavaria in particular; however, there was no helping Bavaria as long as she served Prussian militarism and helped to pull its chestnuts out of the fire.

Actually this kind of propaganda began to achieve certain effects in 1915. The feeling against Prussia grew quite visibly among the troops — yet not a single step was taken against it from above. This was more than a mere sin of omission, and sooner or later we were bound to suffer most catastrophically for it; and not just the 'Prussians,' but the whole German people, to which Bavaria herself is not the last to belong.

In this direction enemy propaganda began to achieve unquestionable successes from 1916 on.

Likewise the complaining letters direct from home had long been having their effect. It was no longer necessary for the enemy to transmit them to the frontline soldiers by means of leaflets, etc. And against this, aside from a few psychologically idiotic 'admonitions' on the part of the 'government,' nothing was done. Just as before, the front was flooded with this poison dished up by thoughtless women at home, who, of course, did not suspect that this was the way to raise the enemy's confidence in victory to the highest pitch, thus consequently to prolong and sharpen the sufferings of their men at the fighting front. In the time that followed, the senseless letters of German women cost hundreds of thousands of men their lives.

Thus, as early as 1916, there appeared various phenomena that would better have been absent.[1] The men at the front complained and 'beefed'; they began to be dissatisfied in many ways and sometimes were even righteously indignant. While they starved and suffered, while their people at home lived in misery, there was abundance and high-living in other circles. Yes, even at the fighting front all was not in order in this respect.

Even then a slight crisis was emerging — but these were still 'internal' affairs. The same man, who at first had cursed and grumbled, silently did his duty a few minutes later as though this was a matter of course. The same company, which at first was discontented, clung to the piece of trench it had to defend as though Germany's fate depended on these few hundred yards of mudholes. It was still the front of the old, glorious army of heroes!

I was to learn the difference between it and the homeland in a glaring contrast.

At the end of September, 1916, my division moved into the

---

[1] '*Erscheinungen die besser nicht vorhanden gewesen waren.*' Second edition has: '*bedenkliche Erscheinungen*': disquieting phenomena.

Battle of the Somme. For us it was the first of the tremendous
battles of matériel which now followed, and the impression was
hard to describe — it was more like hell than war.

Under a whirlwind of drumfire that lasted for weeks, the
German front held fast, sometimes forced back a little, then again
pushing forward, but never wavering.

On October 7, 1916, I was wounded.

I was brought safely to the rear, and from there was to return
to Germany with a transport.

Two years had now passed since I had seen the homeland,
under such conditions an almost endless time. I could scarcely
imagine how Germans looked who were not in uniform. As I
lay in the field hospital at Hermies, I almost collapsed for fright
when suddenly the voice of a German woman serving as a nurse
addressed a man lying beside me.

For the first time in two years to hear such a sound!

The closer our train which was to bring us home approached
the border, the more inwardly restless each of us became. All
the towns passed by, through which we had ridden two years
previous as young soldiers: Brussels, Louvain, Liège, and at last
we thought we recognized the first German house by its high
gable and beautiful shutters.

The fatherland!

In October, 1914, we had burned with stormy enthusiasm as
we crossed the border; now silence and emotion reigned. Each
of us was happy that Fate again permitted him to see what he
had had to defend so hard with his life, and each man was well-
nigh ashamed to let another look him in the eye.

It was almost on the anniversary of the day when I left for
the front that I reached the hospital at Beelitz near Berlin.

What a change! From the mud of the Battle of the Somme
into the white beds of this miraculous building! In the beginning
we hardly dared to lie in them properly. Only gradually could
we reaccustom ourselves to this new world.

Unfortunately, this world was new in another respect as well.
The spirit of the army at the front seemed no longer to be a

guest here.[1] Here for the first time I heard a thing that was still unknown at the front; men bragging about their own cowardice! For the cursing and 'beefing' you could hear at the front were never an incitement to shirk duty or a glorification of the coward. No! The coward still passed as a coward and as nothing else; and the contempt which struck him was still general, just like the admiration that was given to the real hero. But here in the hospital it was partly almost the opposite: the most unscrupulous agitators did the talking and attempted with all the means of their contemptible eloquence to make the conceptions of the decent soldiers ridiculous and hold up the spineless coward as an example. A few wretched scoundrels in particular set the tone. One boasted that he himself had pulled his hand through a barbed-wire entanglement in order to be sent to the hospital; in spite of this absurd wound he seemed to have been here for an endless time, and for that matter he had only gotten into the transport to Germany by a swindle. This poisonous fellow went so far in his insolent effrontery as to represent his own cowardice as an emanation [2] of higher bravery than the hero's death of an honest soldier. Many listened in silence, others went away, but a few assented.

Disgust mounted to my throat, but the agitator was calmly tolerated in the institution. What could be done? The management couldn't help knowing, and actually did know, exactly who and what he was. But nothing was done.

When I could again walk properly, I obtained permission to go to Berlin.

Clearly there was dire misery everywhere. The big city was suffering from hunger. Discontent was great. In various soldiers' homes the tone was like that in the hospital. It gave you the impression that these scoundrels were intentionally frequenting such places in order to spread their views.

But much, much worse were conditions in Munich itself!

When I was discharged from the hospital as cured and transferred to the replacement battalion, I thought I could no longer

---

[1] *'schien hier kein Gast mehr zu sein.'*   [2] *'Ausfluss.'*

recognize the city. Anger, discontent, cursing, wherever you went! In the replacement battalion itself the mood was beneath all criticism. Here a contributing factor was the immeasurably clumsy way in which the field soldiers were treated by old training officers who hadn't spent a single hour in the field and for this reason alone were only partially able to create a decent relationship with the old soldiers. For it had to be admitted that the latter possessed certain qualities which could be explained by their service at the front, but which remained totally incomprehensible to the leaders of these replacement detachments, while the officer who had come from the front was at least able to explain them. The latter, of course, was respected by the men quite differently than the rear commander. But aside from this, the general mood was miserable: to be a slacker passed almost as a sign of higher wisdom, while loyal steadfastness was considered a symptom of inner weakness and narrow-mindedness. The offices were filled with Jews. Nearly every clerk was a Jew and nearly every Jew was a clerk. I was amazed at this plethora of warriors of the chosen people and could not help but compare them with their rare representatives at the front.

As regards economic life, things were even worse Here the Jewish people had become really 'indispensable.' The spider was slowly beginning to suck the blood out of the people's pores. Through the war corporations, they had found an instrument with which, little by little, to finish off the national free economy.

The necessity of an unlimited centralization was emphasized.

Thus, in the year 1916–17 nearly the whole of production was under the control of Jewish finance.

But against whom was the hatred of the people directed?

At this time I saw with horror a catastrophe approaching which, unless averted in time, would inevitably lead to collapse.

While the Jew robbed the whole nation and pressed it beneath his domination, an agitation was carried on against the 'Prussians.' At home, as at the front, nothing was done against this poisonous propaganda. No one seemed to suspect that the collapse of Prussia would not by a long shot bring with it a resurgence of Bavaria; no, that on the contrary any fall of the one would inevitably carry the other along with it into the abyss.

I felt very badly about this behavior. In it I could only see the craftiest trick of the Jew, calculated to distract the general attention from himself and to others. While the Bavarian and the Prussian fought, he stole the existence of both of them from under their nose; while the Bavarians were cursing the Prussians, the Jew organized the revolution and smashed Prussia and Bavaria at once.

I could not bear this accursed quarrel among German peoples, and was glad to return to the front, for which I reported at once after my arrival in Munich.

At the beginning of March, 1917, I was back with my regiment.

*   *   *

Toward the end of 1917, the low point of the army's dejection seemed to have passed. The whole army took fresh hope and fresh courage after the Russian collapse. The conviction that the War would end with the victory of Germany, after all, began to seize the troops more and more. Again singing could be heard and the Calamity Janes became rarer. Again people believed in the future of the fatherland.

Especially the Italian collapse of autumn, 1917, had had the most wonderful effect; in this victory we saw a proof of the possibility of breaking through the front, even aside from the Russian theater of war. A glorious faith flowed again into the hearts of the millions, enabling them to await spring, 1918, with relief and confidence. The foe was visibly depressed. In this winter he remained quieter than usual. This was the lull before the storm.

But, while those at the front were undertaking the last preparations for the final conclusion of the eternal struggle, while endless transports of men and matériel were rolling toward the West Front, and the troops were being trained for the great attack — the biggest piece of chicanery in the whole war broke out in Germany.

Germany must not be victorious; in the last hour, with victory already threatening to be with the German banners, a means was chosen which seemed suited to stifle the German spring attack in the germ with one blow, to make victory impossible:

The munitions strike was organized.

If it succeeded, the German front was bound to collapse, and the *Vorwärts'* [1] desire that this time victory should not be with the German banners would inevitably be fulfilled. Owing to the lack of munitions, the front would inevitably be pierced in a few weeks; thus the offensive was thwarted, the Entente saved, international capital was made master of Germany, and the inner aim of the Marxist swindle of nations achieved.

To smash the national economy and establish the rule of international capital — a goal which actually was achieved, thanks to the stupidity and credulity of the one side and the bottomless cowardice of the other.

To be sure, the munitions strike did not have all the hoped-for success with regard to starving the front of arms; it collapsed too soon for the lack of munitions as such — as the plan had been — to doom the army to destruction.

But how much more terrible was the moral damage that had been done!

In the first place: What was the army fighting for if the homeland itself no longer wanted victory? For whom the immense sacrifices and privations? The soldier is expected to fight for victory and the homeland goes on strike against it!

And in the second place: What was the effect on the enemy?

In the winter of 1917 to 1918, dark clouds appeared for the first time in the firmament of the Allied world. For nearly four years they had been assailing the German warrior and had been unable to encompass his downfall; and all this while the German had only his shield arm free for defense, while his sword was

---

[1] *Vorwärts.* Official organ of the Social Democratic Party of Germany, founded in 1884 as the *Berliner Volksblatt.* It was renamed *Vorwärts* in 1890. Wilhelm Liebknecht directed it from then until his death in 1900. It continued to appear until 1933.

obliged to strike, now in the East, now in the South. But now at last the giant's back was free. Streams of blood had flown before he administered final defeat to one of his foes. Now in the West his shield was going to be joined by his sword; up till then the enemy had been unable to break his defense, and now he himself was facing attack.

The enemy feared him and trembled for their victory.

In London and Paris one deliberation followed another, but at the front sleepy silence prevailed. Suddenly their high mightinesses lost their effrontery. Even enemy propaganda was having a hard time of it; it was no longer so easy to prove the hopelessness of German victory.

But this also applied to the Allied troops at the fronts. A ghastly light began to dawn slowly even on them. Their inner attitude toward the German soldier had changed. Until then he may have seemed to them a fool destined to defeat; but now it was the destroyer of the Russian ally that stood before them. The limitation of the German offensives to the East, though born of necessity, now seemed to them brilliant tactics. For three years these Germans had stormed the Russian front, at first it seemed without the slightest success. The Allies almost laughed over this aimless undertaking; for in the end the Russian giant with his overwhelming number of men was sure to remain the victor while Germany would inevitably collapse from loss of blood. Reality seemed to confirm this hope.

Since the September days of 1914, when for the first time the endless hordes of Russian prisoners from the Battle of Tannenberg began moving into Germany over the roads and railways, this stream was almost without end — but for every defeated and destroyed army a new one arose. Inexhaustibly the gigantic Empire gave the Tsar more and more new soldiers and the War its new victims. How long could Germany keep up this race? Would not the day inevitably come when the Germans would win their last victory and still the Russian armies would not be marching to their last battle? And then what? In all human probability the victory of Russia could be postponed, but it was bound to come.

Now all these hopes were at an end: the ally who had laid the greatest blood sacrifices on the altar of common interests was at the end of his strength, and lay prone at the feet of the inexorable assailant. Fear and horror crept into the hearts of the soldiers who had hitherto believed so blindly. They feared the coming spring. For if up until then they had not succeeded in defeating the German when he was able to place only part of his forces on the Western Front, how could they count on victory now that the entire power of this incredible heroic state seemed to be concentrating for an attack on the West?

The shadows of the South Tyrolean Mountains lay oppressive on the fantasy; as far as the mists of Flanders, the defeated armies of Cadorna conjured up gloomy faces, and faith in victory ceded to fear of coming defeat.

Then — when out of the cool nights the Allied soldiers already seemed to hear the dull rumble of the advancing storm units of the German army, and with eyes fixed in fear and trepidation awaited the approaching judgment, suddenly a flaming red light arose in Germany, casting its glow into the last shell-hole of the enemy front: at the very moment when the German divisions were receiving their last instructions for the great attack, the general strike broke out in Germany.

At first the world was speechless. But then enemy propaganda hurled itself with a sigh of relief on this help that came in the eleventh hour. At one stroke the means was found to restore the sinking confidence of the Allied soldiers, once again to represent the probability of victory as certain,[1] and transform dread anxiety in the face of coming events into determined confidence. Now the regiments awaiting the German attack could be sent into the greatest battle of all time with the conviction that, not the boldness of the German assault would decide the end of this war, but the perseverance of the defense. Let the Germans achieve as many victories as they pleased; at home the revolution was before the door, and not the victorious army.

English, French, and American newspapers began to implant

[1] '*die Wahrscheinlichkeit als sicher hinzustellen . . .*'

this faith in the hearts of their readers while an infinitely shrewd propaganda raised the spirits of the troops at the front.

'Germany facing revolution! Victory of the Allies inevitable!' This was the best medicine to help the wavering poilu and Tommy back on their feet. Now rifles and machine guns could again be made to fire, and a headlong flight in panic fear was replaced by hopeful resistance.

This was the result of the munitions strike. It strengthened the enemy peoples' belief in victory and relieved the paralyzing despair of the Allied front — in the time that followed, thousands of German soldiers had to pay for this with their blood. The instigators of this vilest of all scoundrelly tricks were the aspirants to the highest state positions of revolutionary Germany.

On the German side, it is true, the visible reaction to this crime could at first apparently be handled; on the enemy side, however, the consequences did not fail to appear. The resistance had lost the aimlessness of an army giving up all as lost, and took on the bitterness of a struggle for victory.

For now, in all human probability, victory was inevitable if the Western Front could stand up under a German attack for only a few months. The parliaments of the Entente, however, recognized the possibilities for the future and approved unprecedented expenditures for continuing the propaganda to disrupt Germany.

*   *   *

I had the good fortune to fight in the first two offensives and in the last.

These became the most tremendous impressions of my life; tremendous because now for the last time, as in 1914, the fight lost the character of defense and assumed that of attack. A sigh of relief passed through the trenches and the dugouts of the German army when at length, after more than three years' endurance in the enemy hell, the day of retribution came. Once

again the victorious battalions cheered and hung the last wreaths
of immortal laurel on their banners rent by the storm of victory.
Once again the songs of the fatherland roared to the heavens
along the endless marching columns, and for the last time the
Lord's grace smiled on His ungrateful children.

*   *   *

In midsummer of 1918, oppressive sultriness lay over the front.
At home there was fighting. For what? In the different detachments of the field army all sorts of things were being said: that
the war was now hopeless and only fools could believe in victory.
That not the people but only capital and the monarchy had an
interest in holding out any longer — all this came from the
homeland and was discussed even at the front.

At first the front reacted very little. What did we care about
universal suffrage? Had we fought four years for that? It was
vile banditry to steal the war aim of the dead heroes from their
very graves. The young regiments had not gone to their death
in Flanders crying: 'Long live universal suffrage and the secret
ballot,' but crying: '*Deutschland über Alles in der Welt.*' A small,
yet not entirely insignificant, difference. But most of those who
cried out for suffrage hadn't ever been in the place where they
now wanted to fight for it. The front was unknown to the whole
political party rabble. Only a small fraction of the Parliamentarian gentlemen could be seen where all decent Germans with
sound limbs left were sojourning at that time.

And so the old personnel at the front was not very receptive to
this new war aim of Messrs. Ebert, Scheidemann,[1] Barth, Lieb-

---

[1] Friedrich Ebert and Philip Scheidemann were leaders of the majority
Socialists who took over the German government on the abdication of
William II on November 9, 1918. On November 25, representatives of the
new provincial governments met in Berlin and decided on the election of a
National Assembly. Elections took place on January 19. The Assembly,
which met in Weimar on February 6, was controlled by a coalition of the

nitz, etc. They couldn't for the life of them see why suddenly the slackers should have the right to arrogate to themselves control of the state over the heads of the army.

My personal attitude was established from the very start. I hated the whole gang of miserable party scoundrels and betrayers of the people in the extreme. It had long been clear to me that this whole gang was not really concerned with the welfare of the nation, but with filling empty pockets. For this they were ready to sacrifice the whole nation, and if necessary to let Germany be destroyed; and in my eyes this made them ripe for hanging. To take consideration of their wishes was to sacrifice the interests of the working people for the benefit of a few pickpockets; these wishes could only be fulfilled by giving up Germany.

And the great majority of the embattled army still thought the same. Only the reinforcements coming from home rapidly grew worse and worse, so that their arrival meant, not a reinforcement but a weakening of our fighting strength. Especially the young reinforcements were mostly worthless. It was often hard to believe that these were sons of the same nation which had once sent its youth out to the battle for Ypres.

In August and September, the symptoms of disorganization increased more and more rapidly, although the effect of the enemy attack was not to be compared with the terror of our former defensive battles. The past Battle of Flanders and the Battle of the Somme had been awesome by comparison.

At the end of September, my division arrived for the third time at the positions which as young volunteer regiments we had once stormed.

What a memory!

Socialists, the Center, and the Democrats, led by Scheidemann. On February 11, it chose Friedrich Ebert President of Germany. The Scheidemann Cabinet resigned on June 20 because it was unwilling to sign the peace treaty. The treaty was signed by the succeeding Cabinet of Gustav Bauer after the Assembly had voted acceptance. The Socialist and Democrat majority were attacked by both Right and Left for accepting this 'national disgrace.'

In October and November of 1914, we had there received our baptism of fire. Fatherland love in our heart and songs on our lips, our young regiments had gone into the battle as to a dance. The most precious blood there sacrificed itself joyfully, in the faith that it was preserving the independence and freedom of the fatherland.

In July, 1917, we set foot for the second time on the ground that was sacred to all of us. For in it the best comrades slumbered, still almost children, who had run to their death with gleaming eyes for the one true fatherland.

We old soldiers, who had then marched out with the regiment, stood in respectful emotion at this shrine of 'loyalty and obedience to the death.'

Now in a hard defensive battle the regiment was to defend this soil which it had stormed three years earlier.

With three weeks of drumfire the Englishman prepared the great Flanders offensive. The spirits of the dead seemed to quicken; the regiment clawed its way into the filthy mud, bit into the various holes and craters, and neither gave ground nor wavered. As once before in this place, it grew steadily smaller and thinner, until the British attack finally broke loose on July 13, 1917.

In the first days of August we were relieved.

The regiment had turned into a few companies: crusted with mud they tottered back, more like ghosts than men. But aside from a few hundred meters of shell holes, the Englishman had found nothing but death.

Now, in the fall of 1918, we stood for the third time on the storm site of 1914. The little city of Comines where we then rested had now become our battlefield. Yet, though the battlefield was the same, the men had changed: for now 'political discussions' went on even among the troops. As everywhere, the poison of the hinterland began, here too, to be effective. And the younger recruit fell down completely — for he came from home.

In the night of October 13, the English gas attack on the

southern front before Ypres burst loose; they used yellow-cross gas, whose effects were still unknown to us as far as personal experience was concerned. In this same night I myself was to become acquainted with it. On a hill south of Wervick, we came on the evening of October 13 into several hours of drumfire with gas shells which continued all night more or less violently. As early as midnight, a number of us passed out, a few of our comrades forever. Toward morning I, too, was seized with pain which grew worse with every quarter hour, and at seven in the morning I stumbled and tottered back with burning eyes; taking with me my last report of the War.

A few hours later, my eyes had turned into glowing coals; it had grown dark around me.

Thus I came to the hospital at Pasewalk in Pomerania, and there I was fated to experience — the greatest villainy of the century.[1]

*   *   *

For a long time there had been something indefinite but repulsive in the air. People were telling each other that in the next few weeks it would 'start in' — but I was unable to imagine what was meant by this. First I thought of a strike like that of the spring. Unfavorable rumors were constantly coming from the navy, which was said to be in a state of ferment. But this, too, seemed to me more the product of the imagination of individual scoundrels than an affair involving real masses. Even in the hospital, people were discussing the end of the War which they hoped would come soon, but no one counted on anything immediate. I was unable to read the papers.

In November the general tension increased.

And then one day, suddenly and unexpectedly, the calamity descended. Sailors arrived in trucks and proclaimed the revolu-

---

[1] 'greatest villainy of the century' changed to 'revolution' in second edition.

tion; a few Jewish youths were the 'leaders' in this struggle for the 'freedom, beauty, and dignity' of our national existence. None of them had been at the front. By way of a so-called 'gonorrhoea hospital,' the three Orientals had been sent back home from their second-line base. Now they raised the red rag in the homeland.

In the last few days I had been getting along better. The piercing pain in my eye sockets was diminishing; slowly I succeeded in distinguishing the broad outlines of the things about me. I was given grounds for hoping that I should recover my eyesight at least well enough to be able to pursue some profession later. To be sure, I could no longer hope that I would ever be able to draw again. In any case, I was on the road to improvement when the monstrous thing happened.

My first hope was still that this high treason might still be a more or less local affair. I also tried to bolster up a few comrades in this view. Particularly my Bavarian friends in the hospital were more than accessible to this. The mood there was anything but 'revolutionary.' I could not imagine that the madness would break out in Munich, too. Loyalty to the venerable House of Wittelsbach [1] seemed to me stronger, after all, than the will of a few Jews. Thus I could not help but believe that this was merely a *Putsch* on the part of the navy and would be crushed in the next few days.

The next few days came and with them the most terrible certainty of my life. The rumors became more and more oppressive. What I had taken for a local affair was now said to be a general revolution. To this was added the disgraceful news from the front. They wanted to capitulate. Was such a thing really possible?

On November 10, the pastor came to the hospital for a short address: now we learned everything.

In extreme agitation, I, too, was present at the short speech. The dignified old gentleman seemed all a-tremble as he informed

[1] Wittelsbach. Family of the Kings of Bavaria, dating back to the tenth century.

us that the House of Hollenzollern should no longer bear the German imperial crown; that the fatherland had become a 'republic'; that we must pray to the Almighty not to refuse His blessing to this change and not to abandon our people in the times to come. He could not help himself, he had to speak a few words in memory of the royal house. He began to praise its services in Pomerania, in Prussia, nay, to the German fatherland, and — here he began to sob gently to himself — in the little hall the deepest dejection settled on all hearts, and I believe that not an eye was able to restrain its tears. But when the old gentleman tried to go on, and began to tell us that we must now end the long War, yes, that now that it was lost and we were throwing ourselves upon the mercy of the victors, our fatherland would for the future be exposed to dire oppression, that the armistice should be accepted with confidence in the magnanimity of our previous enemies — I could stand it no longer. It became impossible for me to sit still one minute more. Again everything went black before my eyes; I tottered and groped my way back to the dormitory, threw myself on my bunk, and dug my burning head into my blanket and pillow.

Since the day when I had stood at my mother's grave, I had not wept. When in my youth Fate seized me with merciless hardness, my defiance mounted. When in the long war years Death snatched so many a dear comrade and friend from our ranks, it would have seemed to me almost a sin to complain — after all, were they not dying for Germany? And when at length the creeping gas — in the last days of the dreadful struggle — attacked me, too, and began to gnaw at my eyes, and beneath the fear of going blind forever, I nearly lost heart for a moment, the voice of my conscience thundered at me: Miserable wretch, are you going to cry when thousands are a hundred times worse off than you! And so I bore my lot in dull silence. But now I could not help it. Only now did I see how all personal suffering vanishes in comparison with the misfortune of the fatherland.

And so it had all been in vain. In vain all the sacrifices and privations; in vain the hunger and thirst of months which were often endless; in vain the hours in which, with mortal fear clutching at our hearts, we nevertheless did our duty; and in vain the death of two millions who died. Would not the graves of all the hundreds of thousands open, the graves of those who with faith in the fatherland had marched forth never to return? Would they not open and send the silent mud- and blood-covered heroes back as spirits of vengeance to the homeland which had cheated them with such mockery of the highest sacrifice which a man can make to his people in this world? Had they died for this, the soldiers of August and September, 1914? Was it for this that in the autumn of the same year the volunteer regiments marched after their old comrades? Was it for this that these boys of seventeen sank into the earth of Flanders? Was this the meaning of the sacrifice which the German mother made to the fatherland when with sore heart she let her best-loved boys march off, never to see them again? Did all this happen only so that a gang of wretched criminals could lay hands on the fatherland?

Was it for this that the German soldier had stood fast in the sun's heat and in snowstorms, hungry, thirsty, and freezing, weary from sleepless nights and endless marches? Was it for this that he had lain in the hell of the drumfire and in the fever of gas attacks without wavering, always thoughtful of his one duty to preserve the fatherland from the enemy peril?

Verily these heroes deserved a headstone: 'Thou Wanderer who comest to Germany, tell those at home that we lie here, true to the fatherland and obedient to duty.'

And what about those at home — ?

And yet, was it only our own sacrifice that we had to weigh in the balance? Was the Germany of the past less precious? Was there no obligation toward our own history? Were we still worthy to relate the glory of the past to ourselves? And how could this deed be justified to future generations?

Miserable and degenerate criminals!

The more I tried to achieve clarity on the monstrous event in this hour, the more the shame of indignation and disgrace burned my brow. What was all the pain in my eyes compared to this misery?

There followed terrible days and even worse nights — I knew that all was lost. Only fools, liars, and criminals could hope in the mercy of the enemy. In these nights hatred grew in me, hatred for those responsible for this deed.

In the days that followed, my own fate became known to me.

I could not help but laugh at the thought of my own future which only a short time before had given me such bitter concern. Was it not ridiculous to expect to build houses on such ground? At last it became clear to me that what had happened was what I had so often feared but had never been able to believe with my emotions.

Kaiser William II was the first German Emperor to hold out a conciliatory hand to the leaders of Marxism, without suspecting that scoundrels have no honor. While they still held the imperial hand in theirs, their other hand was reaching for the dagger.

There is no making pacts with Jews; there can only be the hard: either — or.

I, for my part, decided to go into politics.

The Beginning of My Political Activity

AT THE END of November, 1918, I returned to Munich. Again I went to the replacement battalion of my regiment, which was in the hands of 'soldiers' councils.' Their whole activity was so repellent to me that I decided at once to leave again as soon as possible. With Schmiedt Ernst, a faithful war comrade, I went to Traunstein and remained there till the camp was broken up.

In March, 1919, we went back to Munich.

The situation was untenable and moved inevitably toward a further continuation of the revolution. Eisner's death only hastened the development and finally led to a dictatorship of the Councils,[1] or, better expressed, to a passing rule of the Jews, as had been the original aim of the instigators of the whole revolution.

At this time endless plans chased one another through my head. For days I wondered what could be done, but the end of every meditation was the sober realization that I, nameless as I was, did not possess the least basis for any useful action. I shall come back to speak of the reasons why then, as before, I could not decide to join any of the existing parties.

In the course of the new revolution of the Councils I for the

[1] Kurt Eisner (1867–1919). Edited the *Vorwärts* from 1899 to 1905. In 1917 he went over from the Majority to the Independent Socialists. On November 7, 1918, he led the revolution in Munich and headed a government of Majority and Independent Socialists. He was assassinated on February 21. 1919

first time acted in such a way as to arouse the disapproval of the
Central Council. Early in the morning of April 27, 1919, I was
to be arrested, but, faced with my leveled carbine, the three
scoundrels lacked the necessary courage and marched off as they
had come.

A few days after the liberation of Munich, I was ordered to report to the examining commission concerned with revolutionary
occurrences in the Second Infantry Regiment.

This was my first more or less purely political activity.

Only a few weeks afterward I received orders to attend a
'course' that was held for members of the armed forces. In it the
soldier was supposed to learn certain fundamentals of civic thinking. For me the value of the whole affair was that I now obtained
an opportunity of meeting a few like-minded comrades with
whom I could thoroughly discuss the situation of the moment.
All of us were more or less firmly convinced that Germany could
no longer be saved from the impending collapse by the parties of
the November crime, the Center and the Social Democracy, and
that the so-called 'bourgeois-national' formations, even with the
best of intentions, could never repair what had happened. A whole
series of preconditions were lacking, without which such a task
simply could not succeed. The following period confirmed the
opinion we then held. Thus, in our own circle we discussed the
foundation of a new party. The basic ideas which we had in mind
were the same as those later realized in the 'German Workers'
Party.' The name of the movement to be founded would from
the very beginning have to offer the possibility of approaching
the broad masses; for without this quality the whole task seemed
aimless and superfluous. Thus we arrived at the name of 'Social
Revolutionary Party'; this because the social views of the new
organization did indeed mean a revolution.

But the deeper ground for this lay in the following: however
much I had concerned myself with economic questions at an
earlier day, my efforts had remained more or less within the
limits resulting from the contemplation of social questions as
such. Only later did this framework broaden through examina-

tion of the German alliance policy. This in very great part was the outcome of a false estimation of economics as well as unclarity concerning the possible basis for sustaining the German people in the future. But all these ideas were based on the opinion that capital in any case was solely the result of labor and, therefore, like itself was subject to the correction of all those factors which can either advance or thwart human activity; and the national importance of capital was that it depended so completely on the greatness, freedom, and power of the state, hence of the nation, that this bond in itself would inevitably cause capital to further the state and the nation owing to its simple instinct of self-preservation or of reproduction. This dependence of capital on the independent free state would, therefore, force capital in turn to champion this freedom, power, strength, etc., of the nation.

Thus, the task of the state toward capital was comparatively simple and clear: it only had to make certain that capital remain the handmaiden of the state and not fancy itself the mistress of the nation. This point of view could then be defined between two restrictive limits: preservation of a solvent, national, and independent economy on the one hand, assurance of the social rights of the workers on the other.

Previously I had been unable to recognize with the desired clarity the difference between this pure capital as the end result of productive labor and a capital whose existence and essence rests exclusively on speculation. For this I lacked the initial inspiration, which had simply not come my way.

But now this was provided most amply by one of the various gentlemen lecturing in the above-mentioned course: Gottfried Feder.[1]

---

[1] Gottfried Feder: born in Würzburg in 1883. An engineer by profession. In 1917, he founded the *Deutscher Kampfbund zur Brechung der Zinsknechtschaft* (German Fighting League for the breaking of interest slavery). Became a National Socialist member of the Reichstag in 1924. Later, he became head of *Hauptabteilung 1 (Wirtschaftsabteilung) der Reichsleitung der NSDAP*, an economic body with elaborate plans for socialization. This was

For the first time in my life I heard a principled discussion of
international stock exchange and loan capital.

Right after listening to Feder's first lecture, the thought ran
through my head that I had now found the way to one of the
most essential premises for the foundation of a new party.

*   *   *

In my eyes Feder's merit consisted in having established with
ruthless brutality the speculative and economic character of
stock exchange and loan capital, and in having exposed its eternal
and age-old presupposition which is interest. His arguments were
so sound in all fundamental questions that their critics from the
start questioned the theoretical correctness of the idea less than
they doubted the practical possibility of its execution. But what
in the eyes of others was a weakness of Feder's arguments, in my
eyes constituted their strength.

*   *   *

It is not the task of a theoretician to determine the varying
degrees in which a cause can be realized, but to establish the
cause as such: that is to say: he must concern himself less with the
road than with the goal. In this, however, the basic correctness
of an idea is decisive and not the difficulty of its execution. As
soon as the theoretician attempts to take account of so-called
'utility' and 'reality' instead of the absolute truth, his work will
cease to be a polar star of seeking humanity and instead will become a prescription for everyday life. The theoretician of a movement must lay down its goal, the politician strive for its fulfill-

dissolved by Hitler in 1932 to please the industrialists. From 1933 to 1934
Feder was Under-Secretary of State for Labor. Then he became professor
of economics at the Technische Hochschule in Charlottenburg and virtually
disappeared from public life.

ment. The thinking of the one, therefore, will be determined by eternal truth, the actions of the other more by the practical reality of the moment. The greatness of the one lies in the absolute abstract soundness of his idea, that of the other in his correct attitude toward the given facts and their advantageous application; and in this the theoretician's aim must serve as his guiding star. While the touchstone for the stature of a politician may be regarded as the success of his plans and acts — in other words, the degree to which they become reality — the realization of the theoretician's ultimate purpose can never be realized, since, though human thought can apprehend truths and set up crystal-clear aims, complete fulfillment will fail due to the general imperfection and inadequacy of man. The more abstractly correct and hence powerful the idea will be, the more impossible remains its complete fulfillment as long as it continues to depend on human beings. Therefore, the stature of the theoretician must not be measured by the fulfillment of his aims, but by their soundness and the influence they have had on the development of humanity. If this were not so, the founders of religion could not be counted among the greatest men of this earth, since the fulfillment of their ethical purposes will never be even approximately complete. In its workings, even the religion of love is only the weak reflection of the will of its exalted founder; its significance, however, lies in the direction which it attempted to give to a universal human development of culture, ethics, and morality.

The enormous difference between the tasks of the theoretician and the politician is also the reason why a union of both in one person is almost never found. This is especially true of the so-called 'successful' politician of small format, whose activity for the most part is only an 'art of the possible,' as Bismarck rather modestly characterized politics in general. The freer such a 'politician' keeps himself from great ideas, the easier and often the more visible, but always the more rapid, his successes will be. To be sure, they are dedicated to earthly transitoriness and sometimes do not survive the death of their fathers. The work of such politicians, by and large, is unimportant for posterity, since their successes in the present are based solely on keeping at a distance all really great and profound problems and ideas, which as such would only have been of value for later generations.

The execution of such aims, which have value and significance for the most distant times, usually brings little reward to the man who champions them and rarely finds understanding among the great masses, who for the moment have more understanding for beer and milk regulations than for farsighted plans for the future, whose realization can only occur far hence, and whose benefits will be reaped only by posterity.

Thus, from a certain vanity, which is always a cousin of stupidity, the great mass of politicians will keep far removed from all really weighty plans for the future, in order not to lose the momentary sympathy of the great mob. The success and significance of such a politician lie then exclusively in the present, and do not exist for posterity. But small minds are little troubled by this; they are content.

With the theoretician conditions are different. His importance lies almost always solely in the future, for not seldom he is what is described by the world as 'unworldly.' For if the art of the politician is really the art of the possible, the theoretician is one of those of whom it can be said that they are pleasing to the gods only if they demand and want the impossible. He will almost always have to renounce the recognition of the present, but in return, provided his ideas are immortal, will harvest the fame of posterity.

In long periods of humanity, it may happen once that the politician is wedded to the theoretician. The more profound this fusion, however, the greater are the obstacles opposing the work of the politician. He no longer works for necessities which will be understood by the first best shopkeeper, but for aims which only the fewest comprehend. Therefore, his life is torn by love and hate. The protest of the present which does not understand the man, struggles with the recognition of posterity — for which he works.

For the greater a man's works for the future, the less the present can comprehend them; the harder his fight, and the rarer success. If, however, once in centuries success does come to a man, perhaps in his latter days a faint beam of his coming glory may shine upon him. To be sure, these great men are only the Marathon runners of history; the laurel wreath of the present touches only the brow of the dying hero.

Among them must be counted the great warriors in this world who, though not understood by the present, are nevertheless prepared to carry the fight for their ideas and ideals to their end. They are the men who some day will be closest to the heart of the people; it almost seems as though every individual feels the duty of compensating in the past for the sins which the present once committed against the great.[1] Their life and work are followed with admiring gratitude and emotion, and especially in days of gloom they have the power to raise up broken hearts and despairing souls.

To them belong, not only the truly great statesmen, but all other great reformers as well. Beside Frederick the Great stands Martin Luther as well as Richard Wagner.

As I listened to Gottfried Feder's first lecture about the 'breaking of interest slavery,' I knew at once that this was a theoretical truth which would inevitably be of immense importance for the future of the German people. The sharp separation of stock exchange capital from the national economy offered the possibility of opposing the internationalization of the German economy without at the same time menacing the foundations of an independent national self-maintenance by a struggle against all capital. The development of Germany was much too clear in my eyes for me not to know that the hardest battle would have to be fought, not against hostile nations, but against international capital. In Feder's lecture I sensed a powerful slogan for this coming struggle.

And here again later developments proved how correct our sentiment of those days was. Today the know-it-alls among our

[1] *'Nun in der Vergangenheit gut zu machen was die Gegenwart einst an den Grossen gesündigt hatte.'*

bourgeois politicians no longer laugh at us: today even they, in so far as they are not conscious liars, see that international stock exchange capital was not only the greatest agitator for the War, but that especially, now that the fight is over, it spares no effort to turn the peace into a hell.

The fight against international finance and loan capital became the most important point in the program of the German nation's struggle for its economic independence and freedom.

As regards the objections of so-called practical men, they can be answered as follows: All fears regarding the terrible economic consequences of the 'breaking of interest slavery' are superfluous; for, in the first place, the previous economic prescriptions have turned out very badly for the German people, and your positions on the problems of national self-maintenance remind us strongly of the reports of similar experts in former times, for example, those of the Bavarian medical board on the question of introducing the railroad. It is well known that none of the fears of this exalted corporation were later realized: the travelers in the trains of the new 'steam horse' did not get dizzy, the onlookers did not get sick, and the board fences to hide the new invention from sight were given up — only the board fences around the brains of all so-called 'experts' were preserved for posterity.

In the second place, the following should be noted: every idea, even the best, becomes a danger if it parades as a purpose in itself, being in reality only a means to one. For me and all true National Socialists there is but one doctrine: people and fatherland.

What we must fight for is to safeguard the existence and reproduction of our race and our people, the sustenance of our children and the purity of our blood, the freedom and independence of the fatherland, so that our people may mature for the fulfillment of the mission allotted it by the creator of the universe.

Every thought and every idea, every doctrine and all knowledge, must serve this purpose. And everything must be examined from this point of view and used or rejected according to its utility. Then no theory will stiffen into a dead doctrine, since it is life alone that all things must serve.

Thus, it was the conclusions of Gottfried Feder that caused me to delve into the fundamentals of this field with which I had previously not been very familiar.

I began to study again, and now for the first time really achieved an understanding of the content of the Jew Karl Marx's life effort. Only now did his *Kapital* become really intelligible to me, and also the struggle of the Social Democracy against the national economy, which aims only to prepare the ground for the domination of truly international finance and stock exchange capital.

*   *   *

But also in another respect these courses were of the greatest consequence to me.

One day I asked for the floor. One of the participants felt obliged to break a lance for the Jews and began to defend them in lengthy arguments. This aroused me to an answer. The overwhelming majority of the students present took my standpoint. The result was that a few days later I was sent into a Munich regiment as a so-called 'educational officer.'

Discipline among the men was still comparatively weak at that time. It suffered from the after-effects of the period of soldiers' councils. Only very slowly and cautiously was it possible to replace voluntary obedience — the pretty name that was given to the pig-sty under Kurt Eisner — by the old military discipline and subordination. Accordingly, the men were now expected to learn to feel and think in a national and patriotic way. In these two directions lay the field of my new activity.

I started out with the greatest enthusiasm and love. For all at once I was offered an opportunity of speaking before a larger audience; and the thing that I had always presumed from pure feeling without knowing it was now corroborated: I could 'speak.' My voice, too, had grown so much better that I could be sufficiently understood at least in every corner of the small squad rooms.

No task could make me happier than this, for now before being discharged I was able to perform useful services to the institution which had been so close to my heart: the army.

And I could boast of some success: in the course of my lectures I led many hundreds, indeed thousands, of comrades back to their people and fatherland. I 'nationalized' the troops and was thus also able to help strengthen the general discipline.

Here again I became acquainted with a number of like-minded comrades, who later began to form the nucleus of the new movement.

The 'German Workers' Party'

ONE DAY I received orders from my head-
quarters to find out what was behind an apparently political
organization which was planning to hold a meeting within the
next few days under the name of 'German Workers' Party' —
with Gottfried Feder as one of the speakers. I was told to go and
take a look at the organization and then make a report.

The curiosity of the army toward political parties in those days
was more than understandable. The revolution had given the
soldiers the right of political activity, and it was just the most
inexperienced among them who made the most ample use of it.
Not until the moment when the Center and the Social Democracy
were forced to recognize, to their own grief, that the sympathies
of the soldiers were beginning to turn away from the revolutionary parties toward the national movement and reawakening, did
they see fit to deprive the troops of suffrage again and prohibit
their political activity.

It was illuminating that the Center and the Marxists should
have taken this measure, for if they had not undertaken this curtailment of 'civil rights' — as the political equality of the soldiers
after the revolution was called — within a few years there would
have been no revolution, and hence no more national dishonor
and disgrace. The troops were then well on their way toward
ridding the nation of its leeches and the stooges of the Entente
within our walls. The fact that the so-called 'national' parties

voted enthusiastically for the correction of the previous views of the November criminals, and thus helped to blunt the instrument of a national rising, again showed what the eternally doctrinaire ideas of these innocents among innocents can lead to. This bourgeoisie was really suffering from mental senility; in all seriousness they harbored the opinion that the army would again become what it had been, to wit, a stronghold of German military power; while the Center and Marxism planned only to tear out its dangerous national poison fang, without which, however, an army remains forever a police force, but is not a troop capable of fighting an enemy — as has been amply proved in the time that followed.

Or did our 'national politicians' believe that the development of the army could have been other than national? That would have been confoundedly like the gentlemen and is what comes of not being a soldier in war but a big-mouth; in other words, a parliamentarian with no notion of what goes on in the hearts of men who are reminded by the most colossal past that they were once the best soldiers in the world.

And so I decided to attend the above-mentioned meeting of this party which up till then had been entirely unknown to me too.

In the evening when I entered the 'Leiber Room' of the former Sterneckerbräu in Munich, I found some twenty to twenty-five people present, chiefly from the lower classes of the population.

Feder's lecture was known to me from the courses, so I was able to devote myself to an inspection of the organization itself.

My impression was neither good nor bad; a new organization like so many others. This was a time in which anyone who was not satisfied with developments and no longer had any confidence in the existing parties felt called upon to found a new party. Everywhere these organizations sprang out of the ground, only to vanish silently after a time. The founders for the most part had no idea what it means to make a party — let alone a movement — out of a club. And so these organizations nearly always stifle automatically in their absurd philistinism.

I judged the 'German Workers' Party' no differently. When Feder finally stopped talking, I was happy. I had seen enough and wanted to leave when the free discussion period, which was now announced, moved me to remain, after all. But here, too, everything seemed to run along insignificantly until suddenly a 'professor' took the floor; he first questioned the soundness of Feder's arguments and then — after Feder replied very well — suddenly appealed to 'the facts,' but not without recommending most urgently that the young party take up the 'separation' of Bavaria from 'Prussia' as a particularly important program-matic point. With bold effrontery the man maintained that in this case German-Austria would at once join Bavaria, that the peace would then become much better, and more similar non-sense. At this point I could not help demanding the floor and giving the learned gentleman my opinion on this point — with the result that the previous speaker, even before I was finished, left the hall like a wet poodle. As I spoke, the audience had lis-tened with astonished faces, and only as I was beginning to say good night to the assemblage and go away did a man come leaping after me, introduce himself (I had not quite understood his name), and press a little booklet into my hand, apparently a political pamphlet, with the urgent request that I read it.

This was very agreeable to me, for now I had reason to hope that I might become acquainted with this dull organization in a simpler way, without having to attend any more such interesting meetings. Incidentally this apparent worker had made a good impression on me. And with this I left the hall.

At that time I was still living in the barracks of the Second Infantry Regiment in a little room that still very distinctly bore the traces of the revolution. During the day I was out, mostly with the Forty-First Rifle Regiment, or at meetings, or lectures in some other army unit, etc. Only at night did I sleep in my quarters. Since I regularly woke up before five o'clock in the morning, I had gotten in the habit of putting a few left-overs or crusts of bread on the floor for the mice which amused themselves in my little room, and watching the droll little beasts chasing around after these choice morsels. I had known so much poverty in my life that I was well able to imagine the hunger, and hence also the pleasure, of the little creatures.

At about five o'clock in the morning after this meeting, I thus lay awake in my cot, watching the chase and bustle. Since I could no longer fall asleep, I suddenly remembered the past evening and my mind fell on the booklet which the worker had given me. I began to read. It was a little pamphlet in which the author, this same worker, described how he had returned to national thinking out of the Babel of Marxist and trade-unionist phrases; hence also the title: *My Political Awakening*.[1] Once I had begun, I read the little book through with interest; for it reflected a process similar to the one which I myself had gone through twelve years before. Involuntarily I saw my own development come to life before my eyes. In the course of the day I reflected a few times on the matter and was finally about to put it aside when, less than a week later, much to my surprise, I received a postcard saying that I had been accepted in the German Workers' Party; I was requested to express myself on the subject and for this purpose to attend a committee meeting of this party on the following Wednesday.

I must admit that I was astonished at this way of 'winning' members and I didn't know whether to be angry or to laugh. I had no intention of joining a ready-made party, but wanted to found one of my own. What they asked of me was presumptuous and out of the question.

I was about to send the gentlemen my answer in writing when curiosity won out and I decided to appear on the appointed day to explain my reasons by word of mouth.

[1] Anton Drexler, *Mein Politisches Erwachen* (München, E. Boepple, 1920). Drexler was the founder and leading spirit of the *Deutsche Arbeiterpartei*, which Hitler transformed into the National Socialist Party. A mechanic in the Munich railroad workshop, he was sickly, uneducated, a poor speaker. His claim to leadership was his idea of winning the German worker for the nationalist idea. Soon after Hitler assumed leadership of the party, Drexler was expelled. From 1924 to 1928, he was Vice-President of the Bavarian Diet. In 1930, he became reconciled with Hitler, but never returned to active politics.

Wednesday came. The tavern in which the said meeting was to take place was the 'Altes Rosenbad' in the Herrenstrasse, a very run-down place that no one seemed to stray into more than once in a blue moon. No wonder, in the year 1919 when the menu of even the larger restaurants could offer only the scantiest and most modest allurements. Up to this time this tavern had been totally unknown to me.

I went through the ill-lit dining room in which not a soul was sitting, opened the door to the back room, and the 'session' was before me. In the dim light of a broken-down gas lamp four young people sat at a table, among them the author of the little pamphlet, who at once greeted me most joyfully and bade me welcome as a new member of the German Workers' Party.

Really, I was somewhat taken aback. As I was now informed that the actual 'national chairman' had not yet arrived, I decided to wait with my declaration. This gentleman finally appeared. It was the same who had presided at the meeting in the Sterneckerbräu on the occasion of Feder's lecture.

Meanwhile, I had again become very curious, and waited expectantly for what was to come. Now at least I came to know the names of the individual gentlemen. The chairman of the 'national organization' was a Herr Harrer,[1] that of the Munich District, Anton Drexler.

The minutes of the last meeting were read and the secretary was given a vote of confidence. Next came the treasury report — all in all the association possessed seven marks and fifty pfennigs — for which the treasurer received a vote of general confidence. This, too, was entered in the minutes. Then the first chairman read the answers to a letter from Kiel, one from Düsseldorf, and one from Berlin, and everyone expressed approval. Next a report was given on the incoming mail: a letter from Berlin, one from Düsseldorf and one from Kiel, whose arrival seemed to be received

---

[1] Karl Harrer, reporter on the formerly liberal, at that time German Nationalist *Münchener-Augsburger Abendzeitung*. He bore the title of Reichs Chairman of the Party. He is described as a club-footed, ungainly figure, proletarian and shabbily dressed.

with great satisfaction. This growing correspondence was interpreted as the best and most visible sign of the spreading importance of the German Workers' Party, and then — then there was a long deliberation with regard to the answers to be made.

Terrible, terrible! This was club life of the worst manner and sort. Was I to join this organization?

Next, new memberships were discussed; in other words, my capture was taken up.

I now began to ask questions — but, aside from a few directives, there was nothing, no program, no leaflet, no printed matter at all, no membership cards, not even a miserable rubber stamp, only obvious good faith and good intentions.

I had stopped smiling, for what was this if not a typical sign of the complete helplessness and total despair of all existing parties, their programs, their purposes, and their activity? The thing that drove these few young people to activity that was outwardly so absurd was only the emanation of their inner voice, which more instinctively than consciously showed them that all parties up till then were suited neither for raising up the German nation nor for curing its inner wounds. I quickly read the typed 'directives' and in them I saw more seeking than knowledge. Much was vague or unclear, much was missing, but nothing was present which could not have passed as a sign of a struggling realization.

I knew what these men felt: it was the longing for a new movement which should be more than a party in the previous sense of the woid.

That evening when I returned to the barracks I had formed my judgment of this association.

I was facing the hardest question of my life: should I join or should I decline?

Reason could advise me only to decline, but my feeling left me no rest, and as often as I tried to remember the absurdity of this whole club, my feeling argued for it.

I was restless in the days that followed.

I began to ponder back and forth. I had long been resolved to engage in political activity; that this could be done only in a new movement was likewise clear to me; only the impetus to act had hitherto been lacking. I am not one of those people who begin something today and lay it down tomorrow, if possible taking up something else again. This very conviction among others was the main reason why it was so hard for me to make up my mind to join such a new organization. I knew that for me a decision would be for good, with no turning back. For me it was no passing game, but grim earnest. Even then I had an instinctive revulsion toward men who start everything and never carry anything out. These jacks-of-all-trades were loathsome to me. I regarded the activity of such people as worse than doing nothing.

And this way of thinking constituted one of the main reasons why I could not make up my mind as easily as some others do to found a cause which either had to become everything or else would do better not to exist at all.

Fate itself now seemed to give me a hint. I should never have gone into one of the existing large parties, and later on I shall go into the reasons for this more closely. This absurd little organization with its few members seemed to me to possess the one advantage that it had not frozen into an 'organization,' but left the individual an opportunity for real personal activity. Here it was still possible to work, and the smaller the movement, the more readily it could be put into the proper form. Here the content, the goal, and the road could still be determined, which in the existing great parties was impossible from the outset.

The longer I tried to think it over, the more the conviction grew in me that through just such a little movement the rise of the nation could some day be organized, but never through the political parliamentary parties which clung far too greatly to the old conceptions or even shared in the profits of the new régime. For it was a new philosophy and not a new election slogan that had to be proclaimed.

Truly a very grave decision — to begin transforming this intention into reality!

What prerequisites did I myself bring to this task?

That I was poor and without means seemed to me the most bearable part of it, but it was harder that I was numbered among the nameless, that I was one of the millions whom chance permits to live or summons out of existence without even their closest neighbors condescending to take any notice of it. In addition, there was the difficulty which inevitably arose from my lack of schooling.

The so-called 'intelligentsia' always looks down with a really limitless condescension on anyone who has not been dragged through the obligatory schools and had the necessary knowledge pumped into him. The question has never been: What are the man's abilities? but: What has he learned? To these 'educated' people the biggest empty-head, if he is wrapped in enough diplomas, is worth more than the brightest boy who happens to lack these costly envelopes. And so it was easy for me to imagine how this 'educated' world would confront me, and in this I erred only in so far as even then I still regarded people as better than in cold reality they for the most part unfortunately are. As they are, to be sure, the exceptions, as everywhere else, shine all the more brightly. Thereby, however, I learned always to distinguish between the eternal students and the men of real ability.

After two days of agonized pondering and reflection, I finally came to the conviction that I had to take this step.

It was the most decisive resolve of my life. From here there was and could be no turning back.

And so I registered as a member of the German Workers' Party and received a provisional membership card with the number 7.

CHAPTER

X

Causes of the Collapse

T HE EXTENT of the fall of a body is always measured by the distance between its momentary position and the one it originally occupied. The same is true of nations and states. A decisive significance must be ascribed to their previous position or rather elevation. Only what is accustomed to rise above the common limit can fall and crash to a manifest low. This is what makes the collapse of the Reich so hard and terrible for every thinking and feeling man, since it brought a crash from heights which today, in view of the depths of our present degradation, are scarcely conceivable.

The very founding of the Reich seemed gilded by the magic of an event which uplifted the entire nation. After a series of incomparable victories, a Reich was born for the sons and grandsons — a reward for immortal heroism. Whether consciously or unconsciously, it matters not, the Germans all had the feeling that this Reich, which did not owe its existence to the trickery of parliamentary fractions, towered above the measure of other states by the very exalted manner of its founding; for not in the cackling of a parliamentary battle of words, but in the thunder and rumbling of the front surrounding Paris was the solemn act performed: a proclamation of our will, declaring that the Germans, princes and people, were resolved in the future to constitute a Reich and once again to raise the imperial crown to symbolic heights. And this was not done by cowardly murder; no deserters and slackers were the founders of the Bismarckian state, but the regiments at the front.

This unique birth and baptism of fire in themselves surrounded the Reich with a halo of historic glory such as only the oldest states — and they but seldom — could boast.

And what an ascent now began!

Freedom on the outside provided daily bread within. The nation became rich in numbers and earthly goods. The honor of the state, and with it that of the whole people, was protected and shielded by an army which could point most visibly to the difference from the former German Union.

So deep is the downfall of the Reich and the German people that everyone, as though seized by dizziness, seems to have lost feeling and consciousness; people can scarcely remember the former height, so dreamlike and unreal do the old greatness and glory seem compared to our present-day misery. Thus it is understandable that people are so blinded by the sublime that they forget to look for the omens of the gigantic collapse which must after all have been somehow present.

Of course, this applies only to those for whom Germany was more than a mere stop-over for making and spending money, since they alone can feel the present condition as a collapse, while to the others it is the long-desired fulfillment of their hitherto unsatisfied desires.

The omens were then present and visible, though but very few attempted to draw a certain lesson from them.

Yet today this is more necessary than ever.

The cure of a sickness can only be achieved if its cause is known, and the same is true of curing political evils. To be sure, the outward form of a sickness, its symptom which strikes the eye, is easier to see and discover than the inner cause. And this is the reason why so many people never go beyond the recognition of external effects and even confuse them with the cause, attempting, indeed, to deny the existence of the latter. Thus most of us primarily see the German collapse only in the general economic misery and the consequences arising therefrom. Nearly every one of us must personally suffer these — a cogent ground for every individual to understand the catastrophe. Much less does the great mass see the collapse in its political, cultural, ethical, and moral aspect. In this the feeling and understanding of many fail completely.

That this should be so among the broad masses may still pass, but for even the circles of the intelligentsia to regard the German collapse as primarily an 'economic catastrophe,' which can therefore be cured by economic means, is one of the reasons why a recovery has hitherto been impossible. Only when it is understood that here, too, economics is only of second or third-rate importance, and the primary rôle falls to factors of politics, ethics, morality, and blood, will we arrive at an understanding of the present calamity, and thus also be able to find the ways and means for a cure.

The question of the causes of the German collapse is, therefore, of decisive importance, particularly for a political movement whose very goal is supposed to be to quell the defeat.

But, in such research into the past, we must be very careful not to confuse the more conspicuous effects with the less visible causes.

The easiest and hence most widespread explanation of the present misfortune is that it was brought about by the consequences of the lost War and that therefore the War is the cause of the present evil.

There may be many who will seriously believe this nonsense, but there are still more from whose mouth such an explanation can only be a lie and conscious falsehood. This last applies to all those who today feed at the government's cribs. For didn't the prophets of the revolution again and again point out most urgently to the people that it was a matter of complete indifference to the broad masses how this War turned out? Did they not, on the contrary, gravely assure us that at most the 'big capitalist' could have an interest in a victorious end of the gigantic struggle of nations, but never the German people as such, let alone the German worker? Indeed, didn't these apostles of world conciliation maintain the exact opposite: didn't they say that by a German defeat 'militarism' would be destroyed, but that the German nation would celebrate its most glorious resurrection? Didn't these circles glorify the benevolence of the Entente, and didn't they shove the blame for the whole bloody struggle on Germany? And could they have done this without declaring that even military defeat would be without special consequences for the nation? Wasn't the whole revolution embroidered with the phrase that it would prevent the victory of the German flag, but that through it the German people would at last begin advancing toward freedom at home and abroad?

Will you claim that this was not so, you wretched, lying scoundrels?

It takes a truly Jewish effrontery to attribute the blame for the collapse solely to the military defeat when the central organ of all traitors to the nation, the Berlin *Vorwärts*, wrote that this time the German people must not bring its banner home victorious!

And now this is supposed to be the cause of our collapse?

Of course, it would be perfectly futile to fight with such forgetful liars. I wouldn't waste my words on them if unfortunately this nonsense were not parroted by so many thoughtless people, who do not seem inspired by malice or conscious insincerity. Furthermore, these discussions are intended to give our propaganda fighters an instrument which is very much needed at a time when the spoken word is often twisted in our mouths.

Thus we have the following to say to the assertion that the lost War is responsible for the German collapse:

Certainly the loss of the War was of terrible importance for the future of our fatherland; however, its loss is not a cause, but itself only a consequence of causes. It was perfectly clear to everyone with insight and without malice that an unfortunate end of this struggle for life and death would inevitably lead to extremely devastating consequences. But unfortunately there were also people who seemed to lack this insight at the right time or who, contrary to their better knowledge, contested and denied this truth. Such for the most part were those who, after the fulfill-

ment of their secret wish, suddenly and belatedly became aware
of the catastrophe which had been brought about by themselves
among others. They are guilty of the collapse — not the lost
War as it suddenly pleases them to say and believe. For its loss
was, after all, only the consequence of their activity and not, as
they now try to say, the result of 'bad' leadership. The foe did
not consist of cowards either; he, too, knew how to die. His number from the first day was greater than that of the German army,
for he could draw on the technical armament and the arsenals of
the whole world; hence the German victories, won for four years
against a whole world, must regardless of all heroic courage and
'organization,' be attributed solely to superior leadership, and
this is a fact which cannot be denied out of existence. The organization and leadership of the German army were the mightiest
that the earth had ever seen. Their deficiencies lay in the limits of
all human adequacy in general.

The collapse of this army was not the cause of our present-day
misfortune, but only the consequence of other crimes, a consequence which itself again, it must be admitted, ushered in the
beginning of a further and this time visible collapse.

The truth of this can be seen from the following:

Must a military defeat lead to so complete a collapse of a nation and a state? Since when is this the result of an unfortunate
war? Do peoples perish in consequence of lost wars as such?

The answer to this can be very brief: always, when military
defeat is the payment meted out to peoples for their inner rottenness, cowardice, lack of character, in short, unworthiness. If this
is not the case, the military defeat will rather be the inspiration
of a great future resurrection than the tombstone of a national
existence.

History offers innumerable examples for the truth of this
assertion.

Unfortunately, the military defeat of the German people is
not an undeserved catastrophe, but the deserved chastisement
of eternal retribution. We more than deserved this defeat. It is
only the greatest outward symptom of decay amid a whole series

of inner symptoms, which perhaps had remained hidden and invisible to the eyes of most people, or which like ostriches people did not want to see.

Just consider the attendant circumstances amid which the German people accepted this defeat. Didn't many circles express the most shameless joy at the misfortune of the fatherland? And who would do such a thing if he does not really deserve such a punishment? Why, didn't they go even further and brag of having finally caused the front to waver? And it was not the enemy that did this — no, no, it was Germans who poured such disgrace upon their heads! Can it be said that misfortune struck them unjustly? Since when do people step forward and take the guilt for a war on themselves? And against better knowledge and better judgment!

No, and again no. In the way in which the German people received its defeat, we can recognize most clearly that the true cause of our collapse must be sought in an entirely different place from the purely military loss of a few positions or in the failure of an offensive; for if the front as such had really flagged and if its downfall had really encompassed the doom of the fatherland, the German people would have received the defeat quite differently. Then they would have borne the ensuing misfortune with gritted teeth or would have mourned it, overpowered by grief; then all hearts would have been filled with rage and anger toward the enemy who had become victorious through a trick of chance or the will of fate; then, like the Roman Senate, the nation would have received the defeated divisions with the thanks of the fatherland for the sacrifices they had made and besought them not to despair of the Reich. The capitulation would have been signed only with the reason, while the heart even then would have beaten for the resurrection to come.

This is how a defeat for which only fate was responsible would have been received. Then people would not have laughed and danced, they would not have boasted of cowardice and glorified the defeat, they would not have scoffed at the embattled troops and dragged their banner and cockade in the mud. But above all:

then we should never have had the terrible state of affairs which prompted a British officer, Colonel Repington, to make the contemptuous statement: 'Of the Germans, every third man is a traitor.' No, this plague would never have been able to rise into the stifling flood which for five years now has been drowning the very last remnant of respect for us on the part of the rest of the world.

This most of all shows the assertion that the lost War was the cause of the German collapse to be a lie. No, this military collapse was itself only the consequence of a large number of symptoms of disease and their causes, which even in peacetime were with the German nation. This was the first consequence, catastrophic and visible to all, of an ethical and moral poisoning, of a diminution in the instinct of self-preservation and its preconditions, which for many years had begun to undermine the foundations of the people and the Reich.

It required the whole bottomless falsehood of the Jews and their Marxist fighting organization to lay the blame for the collapse on that very man who alone, with superhuman energy and will power, tried to prevent the catastrophe he foresaw and save the nation from its time of deepest humiliation and disgrace. By branding Ludendorff as guilty for the loss of the World War, they took the weapon of moral right from the one dangerous accuser who could have risen against the traitors to the fatherland. In this they proceeded on the sound principle that the magnitude of a lie always contains a certain factor of credibility, since the great masses of the people in the very bottom of their hearts tend to be corrupted rather than consciously and purposely evil, and that, therefore, in view of the primitive simplicity of their minds, they more easily fall a victim to a big lie than to a little one, since they themselves lie in little things, but would be ashamed of lies that were too big. Such a falsehood will never enter their heads, and they will not be able to believe in the possibility of such monstrous effrontery and infamous misrepresentation in others; yes, even when enlightened on the subject, they will long doubt and waver, and continue to accept at least one of these causes as true. Therefore, something of even the most insolent lie will always remain and stick — a fact which all the great lie-virtuosi and lying-clubs in this world know only too well and also make the most treacherous use of.

The foremost connoisseurs of this truth regarding the possibilities in the use of falsehood and slander have always been the Jews; for after all, their whole existence is based on one single great lie, to wit, that they are a religious community while actually they are a race — and what a race! One of the greatest minds of humanity has nailed them forever as such in an eternally correct phrase of fundamental truth: he called them 'the great masters of the lie.' And anyone who does not recognize this or does not want to believe it will never in this world be able to help the truth to victory.

For the German people it must almost be considered a great good fortune that its period of creeping sickness was suddenly cut short by so terrible a catastrophe, for otherwise the nation would have gone to the dogs more slowly perhaps, but all the more certainly. The disease would have become chronic, while in the acute form of the collapse it at least became clearly and distinctly recognizable to a considerable number of people. It was no accident that man mastered the plague more easily than tuberculosis. The one comes in terrible waves of death that shake humanity to the foundations, the other slowly and stealthily; the one leads to terrible fear, the other to gradual indifference. The consequence is that man opposed the one with all the ruthlessness of his energy, while he tries to control consumption with feeble means. Thus he mastered the plague, while tuberculosis masters him.

Exactly the same is true of diseases of national bodies. If they do not take the form of catastrophe, man slowly begins to get accustomed to them and at length, though it may take some time, perishes all the more certainly of them. And so it is a good fortune — though a bitter one, to be sure — when Fate resolves to take a hand in this slow process of putrefaction and with a sudden blow makes the victim visualize the end of his disease.

For more than once, that is what such a catastrophe amounts to. Then it can easily become the cause of a recovery beginning with the utmost determination.

But even in such a case, the prerequisite is again the recognition of the inner grounds which cause the disease in question.

Here, too, the most important thing remains the distinction between the causes and the conditions they call forth. This will be all the more difficult, the longer the toxins remain in the national body and the more they become an ingredient of it which is taken for granted. For it is easily possible that after a certain time unquestionably harmful poisons will be regarded as an ingredient of one's own nation or at best will be tolerated as a necessary evil, so that a search for the alien virus is no longer regarded as necessary.

Thus, in the long peace of the pre-War years, certain harmful features had appeared and been recognized as such, though next to nothing was done against their virus, aside from a few exceptions. And here again these exceptions were primarily manifestations of economic life, which struck the consciousness of the individual more strongly than the harmful features in a number of other fields.

There were many symptoms of decay which should have aroused serious reflection.

*   *   *

With respect to economics, the following should be said:

Through the amazing increase in the German population before the War, the question of providing the necessary daily bread stepped more and more sharply into the foreground of all political and economic thought and action. Unfortunately, those in power could not make up their minds to choose the only correct solution, but thought they could reach their goal in an easier way. When they renounced the acquisition of new soil and replaced it by the lunacy of world economic conquest, the result was bound to be an industrialization as boundless as it was harmful.

The first consequence of gravest importance was the weakening of the peasant class. Proportionately as the peasant class diminished, the mass of the big city proletariat increased more and more, until finally the balance was completely upset.

Now the abrupt alternation [1] between rich and poor became really apparent. Abundance and poverty lived so close together that the saddest consequences could and inevitably did arise. Poverty and frequent unemployment began to play havoc with people, leaving behind them a memory of discontent and embitterment. The consequence of this seemed to be political class division. Despite all the economic prosperity, dissatisfaction became greater and deeper; in fact, things came to such a pass that the conviction that 'it can't go on like this much longer' became general, yet without people having or being able to have any definite idea of what ought to have been done.

These were the typical symptoms of deep discontent which sought to express themselves in this way.

But worse than this were other consequences induced by the economization of the nation.

In proportion as economic life grew to be the dominant mistress of the state, money became the god whom all had to serve and to whom each man had to bow down. More and more, the gods of heaven were put into the corner as obsolete and outmoded, and in their stead incense was burned to the idol Mammon. A truly malignant degeneration set in; what made it most malignant was that it began at a time when the nation, in a presumably menacing and critical hour, needed the highest heroic attitude. Germany had to accustom herself to the idea that some day her attempt to secure her daily bread by means of 'peaceful economic labor' would have to be defended by the sword.

Unfortunately, the domination of money was sanctioned even by that authority which should have most opposed it: His Majesty the Kaiser acted most unfortunately by drawing the aristocracy into the orbit of the new finance capital. It must be said to his credit, however, that unfortunately even Bismarck himself

[1] *'der schroffe Wechsel.'*

did not recognize the menacing danger in this respect. Thereby the ideal virtues for all practical purposes had taken a position second to the value of money, for it was clear that once a beginning had been made in this direction, the aristocracy of the sword would in a short time inevitably be overshadowed by the financial aristocracy. Financial operations succeed more easily than battles. It was no longer inviting for the real hero or statesman to be brought into relations with some old bank Jew: the man of true merit could no longer have an interest in the bestowal of cheap decorations; he declined them with thanks. But regarded purely from the standpoint of blood, such a development was profoundly unfortunate: more and more, the nobility lost the racial basis for its existence, and in large measure the designation of 'ignobility' would have been more suitable for it.

A grave economic symptom of decay was the slow disappearance of the right of private property, and the gradual transference of the entire economy to the ownership of stock companies.

Now for the first time labor had sunk to the level of an object of speculation for unscrupulous Jewish business men; the alienation of property from the wage-worker was increased *ad infinitum.* The stock exchange began to triumph and prepared slowly but surely to take the life of the nation into its guardianship and control.

The internationalization of the German economic life had been begun even before the War through the medium of stock issues. To be sure, a part of German industry still attempted with resolution to ward off this fate. At length, however, it, too, fell a victim to the united attack of greedy finance capital which carried on this fight, with the special help of its most faithful comrade, the Marxist movement.

The lasting war against German 'heavy industry' was the visible beginning of the internationalization of German economy toward which Marxism was striving, though this could not be carried to its ultimate end until the victory of Marxism and the revolution. While I am writing these words, the general attack against the German state railways has finally succeeded, and they are now being handed over to international finance capital.[1] 'International' Social Democracy has thus realized one of its highest goals.

How far this 'economization' of the German people had succeeded is most visible in the fact that after the War one of the leading heads of German industry, and above all of commerce, was finally able to express the opinion that economic effort as such was alone in a position to re-establish Germany. This nonsense was poured forth at a moment when France was primarily bringing back the curriculum of her schools to humanistic foundations in order to combat the error that the nation and the state owed their survival to economics and not to eternal ideal values. These words pronounced by a Stinnes created the most incredible confusion; they were picked up at once, and with amazing rapidity became the *leitmotiv* of all the quacks and big-mouths that since the revolution Fate has let loose on Germany in the capacity of 'statesmen.'

*   *   *

One of the worst symptoms of decay in Germany of the pre-War era was the steadily increasing habit of doing things by halves. This is always a consequence of uncertainty on some matter and of the cowardice resulting from this and other grounds. This disease was further promoted by education.

[1] On April 16, 1924, the German government accepted the Dawes Plan providing for reparations payments beginning at 1,000,000,000 marks annually and increasing to 2,500,000,000 marks at the end of five years. It provided for a reorganization of the Reichsbank under Allied supervision.

By the terms of the Plan, the German railway system had to pay 11,000,-000,000 marks of the reparations debt, or, at 6 per cent, 660,000,000 marks annually. The railways, which had been nationalized by the Republic, were turned over to a private corporation with a capital of 26,000,000,000 marks. Eleven billions of this capital was mortgaged to the Dawes Plan trustees. Moreover, Allied representatives were placed on the administrative board of the railways.

German education before the War was afflicted with an extraordinary number of weaknesses. It was extremely one-sided and adapted to breeding pure 'knowledge,' with less attention to 'ability.' Even less emphasis was laid on the development of the character of the individual — in so far as this is possible; exceedingly little on the sense of joy in responsibility, and none at all on the training of will and force of decision. Its results, you may be sure, were not strong men, but compliant 'walking encyclopedias,' as we Germans were generally looked upon and accordingly estimated before the War. People liked the German because he was easy to make use of, but respected him little, precisely because of his weakness of will. It was not for nothing that more than almost any other people he was prone to lose his nationality and fatherland. The lovely proverb, 'with hat in hand, he travels all about the land,' [1] tells the whole story.

This compliance became really disastrous, however, when it determined the sole form in which the monarch could be ap- proached; that is, never to contradict him, but agree to anything and everything that His Majesty condescends to do. Precisely in this place was free, manly dignity most necessary; otherwise the monarchic institution was one day bound to perish from all this crawling; for crawling it was and nothing else! And only miserable crawlers and sneaks — in short, all the decadents who have always felt more at ease around the highest thrones than sincere, decent, honorable souls — can regard this as the sole proper form of intercourse with the bearers of the crown! These 'most humble' creatures, to be sure, despite all their humility before their master and source of livelihood, have always demonstrated the greatest arrogance toward the rest of humanity, and worst of all when they pass themselves off with shameful effrontery on their sinful fellow men as the only 'monarchists'; this is real gall such as only these ennobled or even unennobled tapeworms are capable of! For in reality these people remained the gravediggers of the monarchy and particularly the monarchistic idea. Nothing else is conceivable: a man who is prepared to stand up for a cause will

[1] '*Gefügigkeit.*' In second edition, '*Gesellschaft*': society.

never and can never be a sneak and a spineless lickspittle. Anyone who is really serious about the preservation and furtherance
of an institution will cling to it with the last fiber of his heart and
will not be able to abandon it if evils of some sort appear in this
institution. To be sure, he will not cry this out to the whole
public as the democratic 'friends' of the monarchy did in the
exact same lying way; he will most earnestly warn and attempt
to influence His Majesty, the bearer of the crown himself. He
will not and must not adopt the attitude that His Majesty remains free to act according to his own will anyway, even if this
obviously must and will lead to a catastrophe, but in such a case
he will have to protect the monarchy against the monarch, and
this despite all perils. If the value of this institution lay in the
momentary person of the monarch, it would be the worst institution that can be imagined; for monarchs only in the rarest cases
are the cream of wisdom and reason or even of character, as some
people like to claim. This is believed only by professional lickspittles and sneaks, but all straightforward men — and these remain the most valuable men in the state despite everything —
will only feel repelled by the idea of arguing such nonsense. For
them history remains history and the truth the truth even where
monarchs are concerned. No, the good fortune to possess a great
monarch who is also a great man falls to peoples so seldom that
they must be content if the malice of Fate abstains at least from
the worst possible mistakes.

Consequently, the value and importance of the monarchic
idea cannot reside in the person of the monarch himself except if
Heaven decides to lay the crown on the brow of a heroic genius
like Frederick the Great or a wise character like William I. This
happens once in centuries and hardly more often. Otherwise the
idea takes precedence over the person and the meaning of this
institution must lie exclusively in the institution itself. With this
the monarch himself falls into the sphere of service. Then he, too,
becomes a mere cog in this work, to which he is obligated as such.
Then he, too, must comply with a higher purpose, and the 'monarchist' is then no longer the man who in silence lets the bearer

of the crown profane it, but the man who prevents this. Otherwise, it would not be permissible to depose an obviously insane prince, if the sense of the institution lay not in the idea, but in the 'sanctified' person at any price.

Today it is really necessary to put this down, for in recent times more and more of these creatures, to whose wretched attitude the collapse of the monarchy must not least of all be attributed are rising out of obscurity. With a certain naïve gall, these people have started in again to speak of nothing but 'their King' — whom only a few years ago they left in the lurch in the critical hour and in the most despicable fashion — and are beginning to represent every person who is not willing to agree to their lying tirades as a bad German. And in reality they are the very same poltroons who in 1919 scattered and ran from every red armband, abandoned their King, in a twinkling exchanged the halberd for the walking stick, put on noncommittal neckties, and vanished without trace as peaceful 'citizens.' At one stroke they were gone, these royal champions, and only after the revolutionary storm, thanks to the activity of others, had subsided enough so that a man could again roar his 'Hail, hail to the King' into the breezes, these 'servants and counselors' of the crown began again cautiously to emerge. And now they are all here again, looking back longingly to the fleshpots of Egypt; they can hardly restrain themselves in their loyalty to the King and their urge to do great things, until the day when again the first red arm-band will appear, and the whole gang of ghosts profiting from the old monarchy will again vanish like mice at the sight of a cat!

If the monarchs were not themselves to blame for these things, they could be most heartily pitied because of their present defenders. In any case, they might as well know that with such knights a crown can be lost, but no crowns gained.

This servility, however, was a flaw in our whole education, for which we suffered most terribly in this connection. For, as its consequence, these wretched creatures were able to maintain themselves at all the courts and gradually undermine the foundations of the monarchy. And when the structure finally began to totter, they evaporated. Naturally: cringers and lickspittles do not let themselves be knocked dead for their master. That monarchs never know this and fail to learn it almost on principle has from time immemorial been their undoing.

*   *   *

One of the worst symptoms of decay was the increasing cowardice in the face of responsibility, as well as the resultant half-heartedness in all things.

To be sure, the starting point of this plague in our country lies in large part in the parliamentary institution in which irresponsibility of the purest breed is cultivated. Unfortunately, this plague slowly spread to all other domains of life, most strongly to state life. Everywhere responsibility was evaded and inadequate half-measures were preferred as a result; for in the use of such measures personal responsibility seems reduced to the smallest dimensions.

Just examine the attitude of the various governments toward a number of truly injurious manifestations of our public life, and you will easily recognize the terrible significance of this general half-heartedness and cowardice in the face of responsibility.

I shall take only a few cases from the mass of existing examples:

Journalistic circles in particular like to describe the press as a 'great power' in the state. As a matter of fact, its importance really is immense. It cannot be overestimated, for the press really continues education in adulthood.

Its readers, by and large, can be divided into three groups:

First, into those who believe everything they read;

second, into those who have ceased to believe anything;

third, into the minds which critically examine what they read, and judge accordingly.

Numerically, the first group is by far the largest. It consists of the great mass of the people and consequently represents the simplest-minded part of the nation. It cannot be listed in terms of professions, but at most in general degrees of intelligence. To it belong all those who have neither been born nor trained to think independently, and who partly from incapacity and partly from incompetence believe everything that is set before them in black and white. To them also belongs the type of lazybones who could perfectly well think, but from sheer mental laziness seizes gratefully on everything that someone else has thought, with the modest assumption that the someone else has exerted himself considerably. Now, with all these types, who constitute the great masses, the influence of the press will be enormous. They are not able or willing themselves to examine what is set before them, and as a result their whole attitude toward all the problems of the day can be reduced almost exclusively to the outside influence of others. This can be advantageous when their enlightenment is provided by a serious and truth-loving party, but it is catastrophic when scoundrels and liars provide it.

The second group is much smaller in number. It is partly composed of elements which previously belonged to the first group, but after long and bitter disappointments shifted to the opposite and no longer believe anything that comes before their eyes in print. They hate every newspaper; either they don't read it at all, or without exception fly into a rage over the contents, since in their opinion they consist only of lies and falsehoods. These people are very hard to handle, since they are suspicious even in the face of the truth. Consequently, they are lost for all positive, political work.

The third group, finally, is by far the smallest; it consists of the minds with real mental subtlety, whom natural gifts and education have taught to think independently, who try to form their own judgment on all things, and who subject everything they read to a thorough examination and further development of their own. They will not look at a newspaper without always collaborating in their minds, and the writer has no easy time of it. Journalists love such readers with the greatest reserve.

For the members of this third group, it must be admitted, the

nonsense that newspaper scribblers can put down is not very
dangerous or even very important. Most of them in the course
of their lives have learned to regard every journalist as a rascal
on principle, who tells the truth only once in a blue moon. Unfortunately, however, the importance of these splendid people
lies only in their intelligence and not in their number — a misfortune at a time when wisdom is nothing and the majority is
everything! Today, when the ballot of the masses decides, the
chief weight lies with the most numerous group, and this is the
first: the mob of the simple or credulous.

It is of paramount interest to the state and the nation to prevent these people from falling into the hands of bad, ignorant, or
even vicious educators. The state, therefore, has the duty of
watching over their education and preventing any mischief. It
must particularly exercise strict control over the press; for its
influence on these people is by far the strongest and most penetrating, since it is applied, not once in a while, but over and over
again. In the uniformity and constant repetition of this instruction lies its tremendous power. If anywhere, therefore, it is here
that the state must not forget that all means must serve an end;
it must not let itself be confused by the drivel about so-called
'freedom of the press' and let itself be talked into neglecting its
duty and denying the nation the food which it needs and which is
good for it; with ruthless determination it must make sure of this
instrument of popular education, and place it in the service of
the state and the nation.

But what food did the German press of the pre-War period dish
out to the people? Was it not the worst poison that can even be
imagined? Wasn't the worst kind of pacifism injected into the
heart of our people at a time when the rest of the world was preparing to throttle Germany, slowly but surely? Even in peacetime didn't the press inspire the minds of the people with doubt
in the right of their own state, thus from the outset limiting them
in the choice of means for its defense? Was it not the German
press which knew how to make the absurdity of 'Western democracy' palatable to our people until finally, ensnared by all

the enthusiastic tirades, they thought they could entrust their future to a League of Nations? Did it not help to teach our people a miserable immorality? Did it not ridicule morality and ethics as backward and petty-bourgeois, until our people finally became 'modern'? Did it not with its constant attacks undermine the foundations of the state's authority until a single thrust sufficed to make the edifice collapse? Did it not fight with all possible means against every effort to give unto the state that which is the state's? Did it not belittle the army with constant criticism, sabotage universal conscription, demand the refusal of military credits, etc., until the result became inevitable?

The so-called liberal press was actively engaged in digging the grave of the German people and the German Reich. We can pass by the lying Marxist sheets in silence; to them lying is just as vitally necessary as catching mice for a cat; their function is only to break the people's national and patriotic backbone and make them ripe for the slave's yoke of international capital and its masters, the Jews.

And what did the state do against this mass poisoning of the nation? Nothing, absolutely nothing. A few ridiculous decrees, a few fines for villainy that went too far, and that was the end of it. Instead, they hoped to curry favor with this plague by flattery, by recognition of the 'value' of the press, its 'importance,' its 'educational mission,' and more such nonsense — as for the Jews, they took all this with a crafty smile and acknowledged it with sly thanks.

The reason, however, for this disgraceful failure on the part of the state was not that it did not recognize the danger, but rather in a cowardice crying to high Heaven and the resultant half-heartedness of all decisions and measures. No one had the courage to use thoroughgoing radical methods, but in this as in everything else they tinkered about with a lot of halfway pre-scriptions, and instead of carrying the thrust to the heart, they at most irritated the viper — with the result that not only did every-thing remain as before, but on the contrary the power of the in-stitutions which should have been combated increased from year to year.

The defensive struggle of the German government at that time against the press — mainly that of Jewish origin — which was slowly ruining the nation was without any straight line, irresolute and above all without any visible goal. The intelligence of the privy councilors failed completely when it came to estimating the importance of this struggle, to choosing means or drawing up a clear plan. Planlessly they fiddled about; sometimes, after being bitten too badly, they locked up one of the journalistic vipers for a few weeks or months, but they left the snakes' nest as such perfectly unmolested.

True — this resulted partly from the infinitely wily tactics of the Jews, on the one hand, and from a stupidity and innocence such as only privy councilors are capable of, on the other. The Jew was much too clever to allow his entire press to be attacked uniformly. No, one part of it existed in order to cover the other. While the Marxist papers assailed in the most dastardly way everything that can be holy to man; while they infamously attacked the state and the government and stirred up large sections of the people against one another, the bourgeois-democratic papers knew how to give an appearance of their famous objectivity, painstakingly avoided all strong words, well knowing that empty heads can judge only by externals and never have the faculty of penetrating the inner core, so that for them the value of a thing is measured by this exterior instead of by the content; a human weakness to which they owe what esteem they themselves enjoy.

For these people the *Frankfurter Zeitung* was the embodiment of respectability. For it never uses coarse expressions, it rejects all physical brutality and keeps appealing for struggle with 'intellectual' weapons, a conception, strange to say, to which especially the least intelligent people are most attached. This is a result of our half-education which removes people from the instinct of Nature and pumps a certain amount of knowledge into them, but cannot create full understanding, since for this industry and good will alone are no use; the necessary intelligence must be present, and what is more, it must be inborn. The ultimate wisdom is always the understanding of the instinct [1] — that is: a man must never fall into the lunacy of believing that he has really risen to be lord and master of Nature — which is so easily induced by the conceit of half-education; he must understand the fundamental necessity of Nature's rule, and realize how much his existence is subjected to these laws of eternal fight and upward struggle. Then he will feel that in a universe where planets revolve around suns, and moons turn about planets, where force alone forever masters weakness, compelling it to be an obedient slave or else crushing it, there can be no special laws for man. For him, too, the eternal principles of this ultimate wisdom hold sway. He can try to comprehend them; but escape them, never.

And it is precisely for our intellectual *demi-monde* that the Jew writes his so-called intellectual press. For them the *Frankfurter Zeitung* and the *Berliner Tageblatt* are made; for them their tone is chosen, and on them they exercise their influence. Seemingly they all most sedulously avoid any outwardly crude forms, and meanwhile from other vessels they nevertheless pour their poison into the hearts of their readers. Amid a *Gezeires* [2] of fine sounds and phrases they lull their readers into believing that pure science or even morality is really the motive of their acts, while in reality it is nothing but a wily, ingenious trick for stealing the enemy's weapon against the press from under his nose. The one variety oozes respectability, so all soft-heads are ready to believe them when they say that the faults of others are only trivial abuses, which should never lead to an infringement of the 'freedom of the press' — their term for poisoning and lying to the people. And so the authorities shy away from taking measures against these bandits, for they fear that, if they did, they would at once have the 'respectable' press against them, a fear which is only too justified. For as soon as they attempt to proceed against one of these shameful rags, all the others will at once take its part, but

[1] '*das Verstehen des Instinktes.*' Second edition has: '. . . *der Instinktursachen*': of the instinctive causes.

[2] *Gezeires:* Yiddish, meaning arbitrary decrees. What Hitler thinks it means is not clear.

by no means to sanction its mode of struggle, God forbid! — but only to defend the principle of freedom of the press and freedom of public opinion; these alone must be defended. But in the face of all this shouting, the strongest men grow weak, for does it not issue from the mouths of 'respectable' papers?

This poison was able to penetrate the bloodstream of our people unhindered and do its work, and the state did not possess the power to master the disease. In the laughable half-measures which it used against the poison, the menacing decay of the Reich was manifest. *For an institution which is no longer resolved to defend itself with all weapons has for practical purposes abdicated.* Every half-measure is a visible sign of inner decay which must and will be followed sooner or later by outward collapse.

I believe that the present generation, properly led, will more easily master this danger. It has experienced various things which had the power somewhat to strengthen the nerves of those who did not lose them entirely. In future days the Jew will certainly continue to raise a mighty uproar in his newspapers if a hand is ever laid on his favorite nest, if an end is put to the mischief of the press and this instrument of education is put into the service of the state and no longer left in the hands of aliens and enemies of the people. But I believe that this will bother us younger men less than our fathers. A thirty-centimeter shell has always hissed more loudly than a thousand Jewish newspaper vipers — so let them hiss!

*   *   *

A further example of the half-heartedness and weakness of the leaders of pre-War Germany in meeting the most important vital questions of the nation is the following: running parallel to the political, ethical, and moral contamination of the people, there had been for many years a no less terrible poisoning of the health of the national body. Especially in the big cities, syphilis was beginning to spread more and more, while tuberculosis steadily reaped its harvest of death throughout nearly the whole country.

Though in both cases the consequences were terrible for the nation, the authorities could not summon up the energy to take decisive measures.

Particularly with regard to syphilis, the attitude of the leadership of the nation and the state can only be designated as total capitulation. To fight it seriously, they would have had to take somewhat broader measures than was actually the case. The invention of a remedy of questionable character and its commercial exploitation can no longer help much against this plague. Here again it was only the fight against causes that mattered and not the elimination of the symptoms. The cause lies, primarily, in our prostitution of love. Even if its result were not this frightful plague, it would nevertheless be profoundly injurious to man, since the moral devastations which accompany this degeneracy suffice to destroy a people slowly but surely. This Jewification of our spiritual life and mammonization of our mating instinct will sooner or later destroy our entire offspring, for the powerful children of a natural emotion will be replaced by the miserable creatures of financial expediency which is becoming more and more the basis and sole prerequisite of our marriages. Love finds its outlet elsewhere.

Here, too, of course, Nature can be scorned for a certain time, but her vengeance will not fail to appear, only it takes a time to manifest itself, or rather: it is often recognized too late by man.

But the devastating consequences of a lasting disregard of the natural requirements for marriage can be seen in our nobility. Here we have before us the results of procreation based partly on purely social compulsion and partly on financial grounds. The one leads to a general weakening, the other to a poisoning of the blood, since every department store Jewess is considered fit to augment the offspring of His Highness — and, indeed, the offspring look it. In both cases complete degeneration is the consequence.

Today our bourgeoisie strive to go the same road, and they will end up at the same goal.

Hastily and indifferently, people tried to pass by the unpleasant truths, as though by such an attitude events could be undone. No, the fact that our big city population is growing more and more prostituted in its love life cannot just be denied out of existence; it simply is so. The most visible results of this mass contamination can, on the one hand, be found in the insane asylums, and on the other, unfortunately, in our — children. They in particular are the sad product of the irresistibly spreading contamination of our sexual life; the vices of the parents are revealed in the sicknesses of the children.

There are different ways of reconciling oneself to this unpleasant, yes, terrible fact: the ones see nothing at all or rather want to see nothing; this, of course, is by far the simplest and easiest 'position.' The others wrap themselves in a saint's cloak of prudishness as absurd as it is hypocritical; they speak of this whole field as if it were a great sin, and above all express their profound indignation against every sinner caught in the act, then close their eyes in pious horror to this godless plague and pray God to let sulphur and brimstone — preferably after their own death — rain down on this whole Sodom and Gomorrah, thus once again making an instructive example of this shameless humanity. The third, finally, are perfectly well aware of the terrible consequences which this plague must and will some day induce, but only shrug their shoulders, convinced that nothing can be done against the menace, so the only thing to do is to let things slide.

All this, to be sure, is comfortable and simple, but it must not be forgotten that a nation will fall victim to such comfortableness. The excuse that other peoples are no better off, it goes without saying, can scarcely affect the fact of our own ruin, except that the feeling of seeing others stricken by the same calamity might for many bring a mitigation of their own pains. But then more than ever the question becomes: Which people will be the first and only one to master this plague by its own strength, and which nations will perish from it? And this is the crux of the whole matter. Here again we have a touchstone of a race's value — the race which cannot stand the test will simply die out, making place for healthier or tougher and more resisting races. For since this question primarily regards the offspring, it is one of those concerning which it is said with such terrible justice that the sins of the fathers are avenged down to the tenth generation. But this applies only to profanation of the blood and the race.

Blood sin and desecration of the race are the original sin in this world and the end of a humanity which surrenders to it.

How truly wretched was the attitude of pre-War Germany on this one very question! What was done to check the contamination of our youth in the big cities? What was done to attack the infection and mammonization of our love life? What was done to combat the resulting syphilization of our people?

This can be answered most easily by stating what should have been done.

First of all, it was not permissible to take this question frivolously; it had to be understood that the fortune or misfortune of generations would depend on its solution; yes, that it could, if not had to be, decisive for the entire future of our people. Such a realization, however, obligated us to ruthless measures and surgical operations. What we needed most was the conviction that first of all the whole attention of the nation had to be concentrated upon this terrible danger, so that every single individual could become inwardly conscious of the importance of this struggle. Truly incisive and sometimes almost unbearable obligations and burdens can only be made generally effective if, in addition to compulsion, the realization of necessity is transmitted to the individual. But this requires a tremendous enlightenment excluding all other problems of the day which might have a distracting effect.

In all cases where the fulfillment of apparently impossible demands or tasks is involved, the whole attention of a people must be focused and concentrated on this one question, as though life and death actually depended on its solution. Only in this way will a people be made willing and able to perform great tasks and exertions.

This principle applies also to the individual man in so far as he wants to achieve great goals. He, too, will be able to do this only in steplike sections, and he, too, will always have to unite his entire energies on the achievement of a definitely delimited task, until this task seems fulfilled and a new section can be marked out. Anyone who does not so divide the road to be conquered into separate stages and does not try to conquer these one by one, systematically with the sharpest concentration of all his forces, will never be able to reach the ultimate goal, but will be left lying somewhere along the road, or perhaps even off it. This gradual working up to a goal is an art, and to conquer the road step by step in this way you must throw in your last ounce of energy.

The very first prerequisite needed for attacking such a difficult stretch of the human road is for the leadership to succeed in representing to the masses of the people the partial goal which now has to be achieved, or rather conquered, as the one which is solely and alone worthy of attention, on whose conquest everything depends. The great mass of the people cannot see the whole road ahead of them without growing weary and despairing of the task. A certain number of them will keep the goal in mind, but will only be able to see the road in small, partial stretches, like the wanderer, who likewise knows and recognizes the end of his journey, but is better able to conquer the endless highway if he divides it into sections and boldly attacks each one as though it represented the desired goal itself. Only in this way does he advance without losing heart.

Thus, by the use of all propagandist means, the question of combating syphilis should have been made to appear as *the* task of the nation. Not just *one more* task. To this end, its injurious effects should have been thoroughly hammered into people as the most terrible misfortune, and this by the use of all available means, until the entire nation arrived at the conviction that everything — future or ruin — depended upon the solution of this question.

Only after such a preparation, if necessary over a period of years, will the attention, and consequently the determination, of the entire nation be aroused to such an extent that we can take exceedingly hard measures exacting the greatest sacrifices without running the risk of not being understood or of suddenly being left in the lurch by the will of the masses.

For, seriously to attack this plague, tremendous sacrifices and equally great labors are necessary.

The fight against syphilis demands a fight against prostitution, against prejudices, old habits, against previous conceptions, general views among them not least the false prudery of certain circles.

The first prerequisite for even the moral right to combat these things is the facilitation of earlier marriage for the coming generation. In late marriage alone lies the compulsion to retain an institution which, twist and turn as you like, is and remains a disgrace to humanity, an institution which is damned ill-suited to a being who with his usual modesty likes to regard himself as the 'image' of God.

Prostitution is a disgrace to humanity, but it cannot be eliminated by moral lectures, pious intentions, etc.; its limitation and final abolition presuppose the elimination of innumerable preconditions. The first is and remains the creation of an opportunity for early marriage as compatible with human nature — particularly for the man, as the woman in any case is only the passive part.

How lost, how incomprehensible a part of humanity has become today can be seen from the fact that mothers in so-called 'good' society can not seldom be heard to say that they are glad to have found their child a husband who has sown his wild oats, etc. Since there is hardly any lack of these, but rather the contrary, the poor girl will be happy to find one of these worn-out Siegfrieds,[1] and the children will be the visible result of this

[1] The German here has an untranslatable and rather elaborate pun. To sow wild oats is '*sich die Hörner abstossen*,' to butt off one's horns. The word I have rendered as 'worn-out' is *enthörnt*, literally de-horned. Siegfried did not have horns, the reference is to the horny skin which made him invulnerable.

'sensible' marriage. If we bear in mind that, aside from this, propagation as such is limited as much as possible, so that Nature is prevented from making any choice, since naturally every creature, regardless how miserable, must be preserved, the only question that remains is why such an institution exists at all any more and what purpose it is supposed to serve? Isn't it exactly the same as prostitution itself? Hasn't duty toward posterity passed completely out of the picture? Or do people fail to realize what a curse on the part of their children and children's children they are heaping on themselves by such criminal frivolity in observing the ultimate natural law as well as our ultimate natural obligation?[1]

Thus, the civilized peoples degenerate and gradually perish.

And marriage cannot be an end in itself, but must serve the one higher goal, the increase and preservation of the species and the race. This alone is its meaning and its task.

Under these conditions its soundness can only be judged by the way in which it fulfills this task. For this reason alone early marriage is sound, for it gives the young marriage that strength from which alone a healthy and resistant offspring can arise. To be sure, it can be made possible only by quite a number of social conditions without which early marriage is not even thinkable. Therefore, a solution of this question, small as it is, cannot occur without incisive measures of a social sort. The importance of these should be most understandable at a time when the 'social' republic, if only by its incompetence in the solution of the housing question, simply prevents numerous marriages and thus encourages prostitution.

Our absurd way of regulating salaries, which concerns itself much too little with the question of the family and its sustenance, is one more reason that makes many an early marriage impossible.

Thus, a real fight against prostitution can only be undertaken if a basic change in social conditions makes possible an earlier

[1] '... in der Wahrung des letzten Naturrechtes, aber auch der letzten Natur-verpflichtung?'

marriage than at present can generally take place. This is the very first premise for a solution of this question.

In the second place, education and training must eradicate a number of evils about which today no one bothers at all. Above all, in our present education a balance must be created between mental instruction and physical training. The institution that is called a *Gymnasium* today is a mockery of the Greek model. In our educational system it has been utterly forgotten that in the long run a healthy mind can dwell only in a healthy body. Especially if we bear in mind the mass of the people, aside from a few exceptions, this statement becomes absolutely valid.

In pre-War Germany there was a period in which no one concerned himself in the least about this truth. They simply went on sinning against the body and thought that in the one-sided training of the 'mind,' they possessed a sure guaranty for the greatness of the nation. A mistake whose consequences began to be felt sooner than was expected. It is no accident that the Bolshevistic wave never found better soil than in places inhabited by a population degenerated by hunger and constant undernourishment: in Central Germany, Saxony, and the Ruhr. But in all these districts the so-called intelligentsia no longer offers any serious resistance to this Jewish disease, for the simple reason that this intelligentsia is itself completely degenerate physically, though less for reasons of poverty than for reasons of education. In times when not the mind but the fist decides, the purely intellectual emphasis of our education in the upper classes makes them incapable of defending themselves, let alone enforcing their will. Not infrequently the first reason for personal cowardice lies in physical weaknesses.

The excessive emphasis on purely intellectual instruction and the neglect of physical training also encourage the emergence of sexual ideas at a much too early age. The youth who achieves the hardness of iron by sports and gymnastics succumbs to the need of sexual satisfaction less than the stay-at-home fed exclusively on intellectual fare. And a sensible system of education must bear this in mind. It must, moreover, not fail to consider that the healthy young man will expect different things from the woman than a prematurely corrupted weakling.

Thus, the whole system of education must be so organized as to use the boy's free time for the useful training of his body. He has no right to hang about in idleness during these years, to make the streets and movie-houses unsafe; after his day's work he should steel and harden his young body, so that later life will not find him too soft. To begin this and also carry it out, to direct and guide it, is the task of education, and not just the pumping of so-called wisdom. We must also do away with the conception that the treatment of the body is the affair of every individual. There is no freedom to sin at the cost of posterity and hence of the race.

Parallel to the training of the body, a struggle against the poisoning of the soul must begin. Our whole public life today is like a hothouse for sexual ideas and stimulations. Just look at the bill of fare served up in our movies, vaudeville and theaters, and you will hardly be able to deny that this is not the right kind of food, particularly for the youth. In shop windows and billboards the vilest means are used to attract the attention of the crowd. Anyone who has not lost the ability to think himself into their soul must realize that this must cause great damage in the youth. This sensual, sultry atmosphere leads to ideas and stimulations at a time when the boy should have no understanding of such things. The result of this kind of education can be studied in present-day youth, and it is not exactly gratifying. They mature too early and consequently grow old before their time. Sometimes the public learns of court proceedings which permit shattering insights into the emotional life of our fourteen- and fifteen-year-olds. Who will be surprised that even in these age-groups syphilis begins to seek its victims? And is it not deplorable to see a good number of these physically weak, spiritually corrupted young men obtaining their introduction to marriage through big-city whores?

No, anyone who wants to attack prostitution must first of all help to eliminate its spiritual basis. He must clear away the filth of the moral plague of big-city 'civilization' and he must do this ruthlessly and without wavering in the face of all the shouting and screaming that will naturally be let loose. If we do not lift the youth out of the morass of their present-day environment, they will drown in it. Anyone who refuses to see these things supports them, and thereby makes himself an accomplice in the slow prostitution of our future which, whether we like it or not, lies in the coming generation. This cleansing of our culture must be extended to nearly all fields. Theater, art, literature, cinema, press, posters, and window displays must be cleansed of all manifestations of our rotting world and placed in the service of a moral, political, and cultural idea. Public life must be freed from the stifling perfume of our modern eroticism, just as it must be freed from all unmanly, prudish hypocrisy. In all these things the goal and the road must be determined by concern for the preservation of the health of our people in body and soul. The right of personal freedom recedes before the duty to preserve the race.

Only after these measures are carried out can the medical struggle against the plague itself be carried through with any prospect of success. But here, too, there must be no half-measures; the gravest and most ruthless decisions will have to be made. It is a half-measure to let incurably sick people steadily contaminate the remaining healthy ones. This is in keeping with the humanitarianism which, to avoid hurting one individual, lets a hundred others perish. The demand that defective people be prevented from propagating equally defective offspring is a demand of the clearest reason and if systematically executed represents the most humane act of mankind. It will spare millions of unfortunates undeserved sufferings, and consequently will lead to a rising improvement of health as a whole. The determination to proceed in this direction will oppose a dam to the further spread of venereal diseases. For, if necessary, the incurably sick will be pitilessly segregated — a barbaric measure for the unfortunate who is struck by it, but a blessing for his fellow men and posterity. The passing pain of a century can and will redeem millenniums from sufferings.

The struggle against syphilis and the prostitution which prepares the way for it is one of the most gigantic tasks of humanity, gigantic because we are facing, not the solution of a single question, but the elimination of a large number of evils which bring about this plague as a resultant manifestation. For in this case the sickening of the body is only the consequence of a sickening of the moral, social, and racial instincts.

But if out of smugness, or even cowardice, this battle is not fought to its end, then take a look at the peoples five hundred years from now. I think you will find but few images of God, unless you want to profane the Almighty.

But how did they try to deal with this plague in old Germany? Viewed calmly, the answer is really dismal. Assuredly, government circles well recognized the terrible evils, though perhaps they were not quite able to ponder the consequences; but in the struggle against it they failed totally, and instead of thoroughgoing reforms preferred to take pitiful measures. They tinkered with the disease and left the causes untouched. They submitted the individual prostitute to a medical examination, supervised her as best they could, and, in case they established disease, put her in some hospital from which after a superficial cure they again let her loose on the rest of humanity.

To be sure, they had introduced a 'protective paragraph' according to which anyone who was not entirely healthy or cured must avoid sexual intercourse under penalty of the law. Surely this measure is sound in itself, but in its practical application it was almost a total failure. In the first place, the woman, in case she is smitten by misfortune — if only due to our, or rather her, education — will in most cases refuse to be dragged into court as a witness against the wretched thief of her health — often under the most embarrassing attendant circumstances. She, in particular, has little to gain from it; in most cases she will be the one to suffer most — for she will be struck much harder by the contempt of her loveless fellow creatures than would be the case with a man. Finally, imagine the situation if the conveyor of the disease is her own husband. Should she accuse him? Or what else should she do?

In the case of the man, there is the additional fact that unfortunately he often runs across the path of this plague after ample consumption of alcohol, since in this condition he is least able to judge the qualities of his 'fair one,' a fact which is only too well known to the diseased prostitute, and always causes her to angle after men in this ideal condition. And the upshot of it all is that the man who gets an unpleasant surprise later can, even by thoroughly racking his brains, not remember his kind benefactress, which should not be surprising in a city like Berlin or even Munich. In addition, it must be considered that often we have to deal with visitors from the provinces who are completely befuddled by all the magic of the big city.

Finally, however: who can know whether he is sick or healthy? Are there not numerous cases in which a patient apparently cured relapses and causes frightful mischief without himself suspecting it at first?

Thus, the practical effect of this protection by legal punishment of a guilty infection is in reality practically nil. Exactly the same is true of the supervision of prostitutes; and finally, the cure itself, even today, is dubious. Only one thing is certain: despite all measures the plague spread more and more, giving striking confirmation of their ineffectualness.

The fight against the prostitution of the people's soul was a failure all along the line, or rather, that is, nothing at all was done.

Let anyone who is inclined to take this lightly just study the basic statistical facts on the dissemination of this plague, compare its growth in the last hundred years, and then imagine its further development — and he would really need the simplicity of an ass to keep an unpleasant shudder from running down his back.

The weakness and half-heartedness of the position taken in old Germany toward so terrible a phenomenon may be evaluated as a visible sign of a people's decay. *If the power to fight for one's own health is no longer present, the right to live in this world of struggle ends.* This world belongs only to the forceful 'whole' man and not to the weak 'half' man.

One of the most obvious manifestations of decay in the old Reich was the slow decline of the cultural level, and by culture I do not mean what today is designated by the word 'civilization.' The latter, on the contrary, rather seems hostile to a truly high standard of thinking and living.

Even before the turn of the century an element began to intrude into our art which up to that time could be regarded as entirely foreign and unknown. To be sure, even in earlier times there were occasional aberrations of taste, but such cases were rather artistic derailments, to which posterity could attribute at least a certain historical value, than products no longer of an artistic degeneration, but of a spiritual degeneration that had reached the point of destroying the spirit. In them the political collapse, which later became more visible, was culturally indicated.

Art Bolshevism is the only possible cultural form and spiritual expression of Bolshevism as a whole.

Anyone to whom this seems strange need only subject the art of the happily Bolshevized states to an examination, and, to his horror, he will be confronted by the morbid excrescences of insane and degenerate men, with which, since the turn of the century, we have become familiar under the collective concepts of cubism and dadaism, as the official and recognized art of those states. Even in the short period of the Bavarian Republic of Councils, this phenomenon appeared. Even here it could be seen that all the official posters, propagandist drawings in the newspapers, etc., bore the imprint, not only of political but of cultural decay.

No more than a political collapse of the present magnitude would have been conceivable sixty years ago was a cultural collapse such as began to manifest itself in futurist and cubist works since 1900 thinkable. Sixty years ago an exhibition of so-called dadaistic 'experiences' would have seemed simply impossible and its organizers would have ended up in the madhouse, while today they even preside over art associations. This plague could not appear at that time, because neither would public opinion have tolerated it nor the state calmly looked on. For it is the business of the state, in other words, of its leaders, to prevent a people from being driven into the arms of spiritual madness. And this is where such a development would some day inevitably end. For on the day when this type of art really corresponded to the general view of things, one of the gravest transformations of humanity would have occurred: the regressive development of the human mind would have begun and the end would be scarcely conceivable.

Once we pass the development of our cultural life in the last twenty-five years in review from this standpoint, we shall be horrified to see how far we are already engaged in this regression. Everywhere we encounter seeds which represent the beginnings of parasitic growths which must sooner or later be the ruin of our culture. In them, too, we can recognize the symptoms of decay of a slowly rotting world. Woe to the peoples who can no longer master this disease!

Such diseases could be seen in Germany in nearly every field of art and culture. Everything seemed to have passed the high point and to be hastening toward the abyss. The theater was sinking manifestly lower and even then would have disappeared completely as a cultural factor if the Court Theaters at least had not turned against the prostitution of art. If we disregard them and a few other praiseworthy examples, the offerings of the stage were of such a nature that it would have been more profitable for the nation to keep away from them entirely. It was a sad sign of inner decay that the youth could no longer be sent into most of these so-called 'abodes of art' — a fact which was admitted with shameless frankness by a general display of the penny-arcade warning: 'Young people are not admitted!'

Bear in mind that such precautionary measures had to be taken in the places which should have existed primarily for the education of the youth and not for the delectation of old and jaded sections of the population. What would the great dramatists of all times have said to such a regulation, and what, above all, to the circumstances which caused it? How Schiller would have flared up, how Goethe would have turned away in indignation!

But after all, what are Schiller, Goethe, or Shakespeare com-

pared to the heroes of the newer German poetic art? Old, out-
worn, outmoded, nay, obsolete. For that was the characteristic
thing about that period: not that the period itself produced
nothing but filth, but that in the bargain it befouled everything
that was really great in the past. This, to be sure, is a phenomenon that can always be observed at such times. The baser and
more contemptible the products of the time and its people, the
more it hates the witnesses to the greater nobility and dignity of
a former day. In such times the people would best like to efface
the memory of mankind's past completely, so that by excluding
every possibility of comparison they could pass off their own
trash as 'art.' Hence every new institution, the more wretched
and miserable it is, will try all the harder to extinguish the last
traces of the past time, whereas every true renascence of humanity can start with an easy mind from the good achievements of
past generations; in fact, can often make them truly appreciated
for the first time. It does not have to fear that it will pale before
the past; no, of itself it contributes so valuable an addition to the
general store of human culture that often, in order to make this
culture fully appreciated, it strives to keep alive the memory of
former achievements, thus making sure that the present will fully
understand the new gift. Only those who can give nothing valuable to the world, but try to act as if they were going to give it
God knows what, will hate everything that was previously given
and would best like to negate or even destroy it.

The truth of this is by no means limited to the field of general
culture, but applies to politics as well. Revolutionary new movements will hate the old forms in proportion to their own inferiority. Here, too, we can see how eagerness to make their own
trash appear to be something noteworthy leads to blind hatred
against the superior good of the past. As long, for example, as
the historical memory of Frederick the Great is not dead, Friedrich Ebert can arouse nothing but limited amazement. The
hero of Sans-Souci is to the former Bremen saloon keeper [1] ap-

---

[1] Before going into politics, Ebert was a saddle-maker. The source of
the legend about his being a tavern-keeper is that, as secretary of his trade
union in Bremen, he helped to administer the union's restaurant and bar.

proximately as the sun to the moon; only when the rays of the sun die can the moon shine. Consequently, the hatred of all new moons of humanity for the fixed stars is only too comprehensible. In political life, such nonentities, if Fate temporarily casts power in their lap, not only besmirch and befoul the past with untiring zeal, but also remove themselves from general criticism by the most extreme methods. The new German Reich's legislation for the defense of the Republic may pass as an example of this.

Therefore, if any new idea, a doctrine, a new philosophy, or even a political or economic movement tries to deny the entire past, tries to make it bad or worthless, for this reason alone we must be extremely cautious and suspicious. As a rule the reason for such hatred is either its own inferiority or even an evil intention as such. A really beneficial renascence of humanity will always have to continue building where the last good foundation stops. It will not have to be ashamed of using already existing truths. For the whole of human culture, as well as man himself, is only the result of a single long development in which every generation contributed and fitted in its stone. Thus the meaning and purpose of revolutions is not to tear down the whole building, but to remove what is bad or unsuitable and to continue building on the sound spot that has been laid bare.

Thus alone can we and may we speak of the progress of humanity. Otherwise the world would never be redeemed from chaos, since every generation would be entitled to reject the past and hence destroy the works of the past as the presupposition for its own work.

Thus, the saddest thing about the state of our whole culture of the pre-War period was not only the total impotence of artistic and cultural creative power in general, but the hatred with which the memory of the greater past was besmirched and effaced. In nearly all fields of art, especially in the theater and literature, we began around the turn of the century to produce less that was new and significant, but to disparage the best of the old work and represent it as inferior and surpassed; as though this epoch of the most humiliating inferiority could surpass anything at all. And from this effort to remove the past from the eyes of the present, the evil intent of the apostles of the future could clearly and distinctly be seen. By this it should have been recognized that these were no new, even if false, cultural conceptions, but a process of destroying all culture, paving the way for a stultification of healthy artistic feeling: the spiritual preparation of political Bolshevism. For if the age of Pericles seems embodied in the Parthenon, the Bolshevistic present is embodied in a cubist monstrosity.

In this connection we must also point to the cowardice which here again was manifest in the section of our people which on the basis of its education and position should have been obligated to resist this cultural disgrace. But from pure fear of the clamor raised by the apostles of Bolshevistic art, who furiously attacked anyone who didn't want to recognize the crown of creation in them and pilloried him as a backward philistine, they renounced all serious resistance and reconciled themselves to what seemed after all inevitable. They were positively scared stiff that these half-wits or scoundrels would accuse them of lack of understanding; as though it were a disgrace not to understand the products of spiritual degenerates or slimy swindlers. These cultural disciples, it is true, possessed a very simple means of passing off their nonsense as something God knows how important: they passed off all sorts of incomprehensible and obviously crazy stuff on their amazed fellow men as a so-called inner experience, a cheap way of taking any word of opposition out of the mouths of most people in advance. For beyond a doubt this could be an inner experience; the doubtful part was whether it is permissible to dish up the hallucinations of lunatics or criminals to the healthy world. The works of a Moritz von Schwind, or of a Böcklin, were also an inner experience, but of artists graced by God and not of clowns.

Here was a good occasion to study the pitiful cowardice of our so-called intelligentsia, which dodged any serious resistance to this poisoning of the healthy instinct of our people and left it to the people themselves to deal with this insolent nonsense. In order not to be considered lacking in artistic understanding, people stood for every mockery of art and ended up by becoming really uncertain in the judgment of good and bad.

All in all, these were tokens of times that were getting very bad.

* * *

As another disquieting attribute, the following must yet be stated:

In the nineteenth century our cities began more and more to lose the character of cultural sites and to descend to the level of mere human settlements. The small attachment of our present big-city proletariat for the town they live in is the consequence of the fact that it is only the individual's accidental local stopping place, and nothing more. This is partly connected with the frequent change of residence caused by social conditions, which do not give a man time to form a closer bond with the city, and another cause is to be found in the general cultural insignificance and poverty of our present-day cities *per se*.

At the time of the wars of liberation, the German cities were not only small in number, but also modest as to size. The few really big cities were mostly princely residences, and as such nearly always possessed a certain cultural value and for the most part also a certain artistic picture. The few places with more than fifty thousand inhabitants were, compared to present-day cities with the same population, rich in scientific and artistic treasures. When Munich numbered sixty thousand souls, it was already on its way to becoming one of the first German art centers; today nearly every factory town has reached this number, if not many times surpassed it, yet some cannot lay claim to the slightest real values. Masses of apartments and tenements, and nothing more. How, in view of such emptiness, any special bond could be expected to arise with such a town must remain a mystery. No one will be particularly attached to a city which has nothing more to offer than every other, which lacks every individual note and in which everything has been carefully avoided which might even look like art or anything of the sort.

But, as if this were not enough, even the really big cities grow relatively poorer in real art treasures with the mounting increase in the population. They seem more and more standardized and give entirely the same picture as the poor little factory towns, though in larger dimensions. What recent times have added to the cultural content of our big cities is totally inadequate. All our cities are living on the fame and treasures of the past. For instance, take from present-day Munich everything that was created under Ludwig I,[1] and you will note with horror how poor the addition of significant artistic creations has been since that time. The same is true of Berlin and most other big cities.

The essential point, however, is the following: our big cities of today possess no monuments dominating the city picture, which might somehow be regarded as the symbols of the whole epoch. This was true in the cities of antiquity, since nearly every one possessed a special monument in which it took pride. The characteristic aspect of the ancient city did not lie in private buildings, but in the community monuments which seemed made, not for the moment, but for eternity, because they were intended to reflect, not the wealth of an individual owner, but the greatness and wealth of the community. Thus arose monuments which were very well suited to unite the individual inhabitant with his city in a way which today sometimes seems almost incomprehensible to us. For what the ancient had before his eyes was less the humble houses of private owners than the magnificent edifices of the whole community. Compared to them the dwelling house really sank to the level of an insignificant object of secondary importance.

---

[1] Ludwig I, King of Bavaria from 1825 to 1848. Hitler apparently overlooks the fact that from 1806 to 1809 he fought on the side of Napoleon against Austria and Prussia. Ludwig was a man of scholarly and artistic leanings; he made Munich a home of the arts. He was an enthusiast for Greek independence, and his son Otto became the first King of independent Greece.

Only if we compare the dimensions of the ancient state structures with contemporary dwelling houses can we understand the
overpowering sweep and force of this emphasis on the principle of giving first place to public works. The few still towering
colossuses which we admire in the ruins and wreckage of the
ancient world are not former business palaces, but temples and
state structures; in other words, works whose owner was the
community. Even in the splendor of late Rome the first place
was not taken by the villas and palaces of individual citizens, but
by the temples and baths, the stadiums, circuses, aqueducts,
basilicas, etc., of the state, hence of the whole people.

Even the Germanic Middle Ages upheld the same guiding
principle, though amid totally different conceptions of art. What
in antiquity found its expression in the Acropolis or the Pantheon
now cloaked itself in the forms of the Gothic Cathedral. Like
giants these monumental structures towered over the swarming
frame, wooden, and brick buildings of the medieval city, and thus
became symbols which even today, with the tenements climbing
higher and higher beside them, determine the character and picture of these towns. Cathedrals, town halls, grain markets, and
battlements are the visible signs of a conception which in the last
analysis was the same as that of antiquity.

Yet how truly deplorable the relation between state buildings
and private buildings has become today! If the fate of Rome
should strike Berlin, future generations would some day admire
the department stores of a few Jews as the mightiest works of our
era and the hotels of a few corporations as the characteristic expression of the culture of our times. Just compare the miserable
discrepancy prevailing in a city like even Berlin between the
structures of the Reich and those of finance and commerce.

Even the sum of money spent on state buildings is usually
laughable and inadequate. Works are not built for eternity, but
at most for the need of the moment. And in them there is no
dominant higher idea. At the time of its construction, the Berlin
Schloss was a work of different stature than the new library, for
instance, in the setting of the present time. While a single battle-

ship represented a value of approximately sixty millions, hardly half of this sum was approved for the first magnificent building of the Reich, intended to stand for eternity, the Reichstag Building. Indeed, when the question of interior furnishings came up for decision, the exalted house voted against the use of stone and ordered the walls trimmed with plaster; this time, I must admit, the parliamentarians did right for a change: stone walls are no place for plaster heads.

Thus, our cities of the present lack the outstanding symbol of national community which, we must therefore not be surprised to find, sees no symbol of itself in the cities. The inevitable result is a desolation whose practical effect is the total indifference of the big-city dweller to the destiny of his city.

This, too, is a sign of our declining culture and our general collapse. The epoch is stifling in the pettiest utilitarianism or better expressed in the service of money. And we have no call for surprise if under such a deity little sense of heroism remains. The present time is only harvesting what the immediate past has sown.

*   *   *

All these symptoms of decay are in the last analysis only the consequences of the absence of a definite, uniformly acknowledged philosophy and the resultant general uncertainty in the judgment and attitude toward the various great problems of the time. That is why, beginning in education, everyone is half-hearted and vacillating, shunning responsibility and thus ending in cowardly tolerance of even recognized abuses. Humanitarian bilge becomes stylish and, by weakly yielding to cankers and sparing individuals, the future of millions is sacrificed.

How widespread the general disunity was growing is shown by an examination of religious conditions before the War. Here, too, a unified and effective philosophical conviction had long since been lost in large sections of the nation. In this the members officially breaking away from the churches play a less important rôle than those who are completely indifferent. While both denominations maintain missions in Asia and Africa in order to win new followers for their doctrine — an activity which can boast but very modest success compared to the advance of the Mohammedan faith in particular — right here in Europe they lose millions and millions of inward adherents who either are alien to all religious life or simply go their own ways. The consequences, particularly from the moral point of view, are not favorable.

Also noteworthy is the increasingly violent struggle against the dogmatic foundations of the various churches without which in this human world the practical existence of a religious faith is not conceivable. The great masses of people do not consist of philosophers; precisely for the masses, faith is often the sole foundation of a moral attitude. The various substitutes have not proved so successful from the standpoint of results that they could be regarded as a useful replacement for previous religious creeds. But if religious doctrine and faith are really to embrace the broad masses, the unconditional authority of the content of this faith is the foundation of all efficacy. What the current *mores*, without which assuredly hundreds of thousands of well-bred people would live sensibly and reasonably but millions of others would not, are for general living, state principles are for the state, and dogmas for the current religion. Only through them is the wavering and infinitely interpretable, purely intellectual idea delimited and brought into a form without which it could never become faith. Otherwise the idea would never pass beyond a metaphysical conception; in short, a philosophical opinion. The attack against dogmas as such, therefore, strongly resembles the struggle against the general legal foundations of a state, and, as the latter would end in a total anarchy of the state, the former would end in a worthless religious nihilism.

For the political man, the value of a religion must be estimated less by its deficiencies than by the virtue of a visibly better substitute. As long as this appears to be lacking, what is present can be demolished only by fools or criminals.

Not the smallest blame for the none too delectable religious conditions must be borne by those who encumber the religious idea with too many things of a purely earthly nature and thus often bring it into a totally unnecessary conflict with so-called exact science. In this victory will almost always fall to the latter, though perhaps after a hard struggle, and religion will suffer serious damage in the eyes of all those who are unable to raise themselves above a purely superficial knowledge.

Worst of all, however, is the devastation wrought by the misuse of religious conviction for political ends. In truth, we cannot sharply enough attack those wretched crooks who would like to make religion an implement to perform political or rather business services for them. These insolent liars, it is true, proclaim their creed in a stentorian voice to the whole world for other sinners to hear; but their intention is not, if necessary, to die for it, but to live better. For a single political swindle, provided it brings in enough, they are willing to sell the heart of a whole religion; for ten parliamentary mandates they would ally themselves with the Marxistic mortal enemies of all religions — and for a minister's chair they would even enter into marriage with the devil, unless the devil were deterred by a remnant of decency.

If in Germany before the War religious life for many had an unpleasant aftertaste, this could be attributed to the abuse of Christianity on the part of a so-called 'Christian' party and the shameless way in which they attempted to identify the Catholic faith with a political party.

This false association was a calamity which may have brought parliamentary mandates to a number of good-for-nothings but injury to the Church.

The consequence, however, had to be borne by the whole nation, since the outcome of the resultant slackening of religious life occurred at a time when everyone was beginning to waver and vacillate anyway, and the traditional foundations of ethics and morality were threatening to collapse.

This, too, created cracks and rifts in our nation which might present no danger as long as no special strain arose, but which inevitably became catastrophic when by the force of great events the question of the inner solidity of the nation achieved decisive importance.

*   *   *

Likewise in the field of politics the observant eye could discern evils which, if not remedied or altered within a reasonable time, could be and had to be regarded as signs of the Reich's coming decay. The aimlessness of German domestic and foreign policy was apparent to everyone who was not purposely blind. The régime of compromise seemed to be most in keeping with Bismarck's conception that 'politics is an art of the possible.' But between Bismarck and the later German chancellors there was a slight difference which made it permissible for the former to let fall such an utterance on the nature of politics while the same view from the mouths of his successors could not but take on an entirely different meaning. For Bismarck with this phrase only wanted to say that for the achievement of a definite political goal all possibilities should be utilized, or, in other words, that all possibilities should be taken into account; in the view of his successors, however, this utterance solemnly released them from the necessity of having any political ideas or goals whatever. And the leadership of the Reich at this time really had no more political goals; for the necessary foundation of a definite philosophy was lacking, as well as the necessary clarity on the inner laws governing the development of all political life.

There were not a few who saw things blackly in this respect and flayed the planlessness and heedlessness of the Reich's policies, and well recognized their inner weakness and hollowness, but these were only outsiders in political life; the official government authorities passed by the observations of a Houston Stewart Chamberlain [1] with the same indifference as still occurs today.

[1] Houston Stewart Chamberlain (1855–1927). Son of an English general, studied in Geneva. Went to Germany in 1885 and remained there until his death. He became a staunch German patriot and pan-German, a close friend of William II. In 1908, he married Eva Wagner, Richard's daughter, and went to live in Bayreuth. His chief work is *Die Grundlagen des neun-*

These people are too stupid to think anything for themselves
and too conceited to learn what is necessary from others — an
age-old truth which caused Oxenstierna to cry out: 'The world
is governed by a mere fraction of wisdom';[1] and indeed nearly
every ministerial secretary embodies only an atom of this fraction. Only since Germany has become a republic, this no longer
applies. That is why it has been forbidden by the Law for the
Defense of the Republic[2] to believe, let alone discuss, any such
thought. Oxenstierna was lucky to live when he did, and not
in this wise republic of ours.

Even in the pre-War period that institution which was supposed to embody the strength of the Reich was recognized by
many as its greatest weakness: the parliament or Reichstag.
Cowardice and irresponsibility were here completely wedded.

One of the foolish remarks which today we not infrequently
hear is that parliamentarism in Germany has 'gone wrong since
the revolution.' This too easily gives the impression that it was

zehnten Jahrhunderts (Foundations of the Nineteenth Century). Shortly before
his death he met Hitler, by whom he was greatly impressed. Many of Hitler's
ideas seem to originate in Chamberlain. Alfred Rosenberg is said to have
prepared excerpts from Chamberlain's ponderous work for *Der Führer* to
read.

[1] The usual version of this saying is, *'Weisst du nicht mein Sohn mit wie
wenig Verstand die Welt regiert wird?'* (Do you not know, my son, with how
little understanding the world is governed?) Hitler's cryptic formulation
seems to be his own. The saying has been attributed to Pope Julius III
(1550–1555) and to Axel Oxenstierna (1583–1654), prime minister to
Gustavus Adolphus of Sweden. It does not occur in any of Oxenstierna's
works, but was attributed to him by Johann Arkenholtz, an eighteenth-
century writer, in his *Schwedische Merkwürdigkeiten (Swedish Curiosities).*
It has also been attributed to various other people. The phrase apparently
had considerable currency at the beginning of the century, for the *Frankfurter Zeitung* for October 26, 1910, carries an article on its origin.

[2] *'Gesetz zum Schutz der Republik,'* passed as a result of the murder of
Walter Rathenau on July 21, 1922. Contains penal provisions against
membership in any organization plotting to assassinate members of the
government or to overthrow the government by force; it also punishes insults
to deceased members of the government, to the republican state form, or the
flag. It includes provisions governing meetings and the press. Passage of
the law was opposed by the extreme Right and Left.

different before the revolution. In reality the effect of this institution can be nothing else than devastating — and this was true even in those days when most people wore blinders and saw nothing and wanted to see nothing. For if Germany was crushed, it was owing not least to this institution; no thanks are owing to the Reichstag that the catastrophe did not occur earlier; this must be attributed to the resistance to the activity of this gravedigger of the German nation and the German Reich, which persisted in the years of peace.

Out of the vast number of devastating evils for which this institution was directly or indirectly responsible, I shall pick only a single one which is most in keeping with the inner essence of this most irresponsible institution of all times: the terrible half-heartedness and weakness of the political leaders of the Reich both at home and abroad, which, primarily attributable to the activities of the Reichstag, developed into one of the chief reasons for the political collapse.

Half-hearted was everything that was subject in any way to the influence of this parliament, regardless which way you look.

Half-hearted and weak was the alliance policy of the Reich in its foreign relations. By trying to preserve peace it steered inevitably toward war.

Half-hearted was the Polish policy. It consisted in irritating without ever seriously going through with anything. The result was neither a victory for the Germans nor conciliation of the Poles, but hostility with Russia instead.

Half-hearted was the solution of the Alsace-Lorraine question. Instead of crushing the head of the French hydra once and for all with a brutal fist, and then granting the Alsatian equal rights, neither of the two was done. Nor could it be, for in the ranks of the biggest parties sat the biggest traitors — in the Center, for example, Herr Wetterlé.[1]

---

[1] Emile Wetterlé, Alsatian Catholic politician. From 1898 to 1914 he belonged to the Reichstag, and in 1910 he founded the Alsatian Nationalist Party. Bitterly hostile to Germany, he fled to France in 1914. In 1919 he became a French deputy. His books include *Les Coulisses du Reichstag* (1918) and *L'Alsace et La Guerre* (1919).

All this, however, would have been bearable if the general half-heartedness had not taken possession of that power on whose existence the survival of the Reich ultimately depended: the army.

The sins of the so-called 'German Reichstag' would alone suffice to cover it for all times with the curse of the German nation. For the most miserable reasons, these parliamentary rabble stole and struck from the hand of the nation its weapon of self-preservation, the only defense of our people's freedom and independence. If today the graves of Flanders field were to open, from them would arise the bloody accusers, hundreds of thousands of the best young Germans who, due to the unscrupulousness of these parliamentarian criminals, were driven, poorly trained and half-trained, into the arms of death; the fatherland lost them and millions of crippled and dead, solely and alone so that a few hundred misleaders of the people could perpetrate their political swindles and blackmail, or merely rattle off their doctrinaire theories.

While the Jews in their Marxist and democratic press proclaimed to the whole world the lie about 'German militarism' and sought to incriminate Germany by all means, the Marxist and democratic parties were obstructing any comprehensive training of the German national man-power. The enormous crime that was thus committed could not help but be clear to everyone who just considered that, in case of a coming war, the entire nation would have to take up arms, and that, therefore, through the rascality of these savory representatives of their own so-called 'popular representation,' millions of Germans were driven to face the enemy half-trained and badly trained. But even if the consequences resulting from the brutal and savage unscrupulousness of these parliamentary pimps were left entirely out of consideration: this lack of trained soldiers at the beginning of the War could easily lead to its loss, and this was most terribly confirmed in the great World War.

The loss of the fight for the freedom and independence of the German nation is the result of the half-heartedness and weakness

manifested even in peacetime as regards drafting the entire
national man-power for the defense of the fatherland.

*   *   *

If too few recruits were trained on the land, the same half-
heartedness was at work on the sea, making the weapon of
national self-preservation more or less worthless. Unfortunately,
the navy leadership was itself infected with the spirit of half-
heartedness. The tendency to build all ships a little smaller than
the English ships which were being launched at the same time
was hardly farsighted, much less brilliant. Especially a fleet
which from the beginning can in point of pure numbers not be
brought to the same level as its presumable adversary must seek
to compensate for the lack of numbers by the superior fighting
power of its individual ships. It is the superior fighting power
which matters and not any legendary superiority in 'quality.'
Actually modern technology is so far advanced and has achieved
so much uniformity in the various civilized countries that it
must be held impossible to give the ships of one power an appreci-
ably larger combat value than the ships of like tonnage of another state. And it is even less conceivable to achieve a superiority with smaller deplacement as compared to larger.

In actual fact, the smaller tonnage of the German ships was
possible only at the cost of speed and armament. The phrase
with which people attempted to justify this fact showed a very
serious lack of logic in the department responsible for this in
peacetime. They declared, for instance, that the material of the
German guns was so obviously superior to the British that the
German 28-centimeter gun was not behind the British 30.5-
centimeter gun in performance!!

But for this very reason it would have been our duty to change
over to the 30.5-centimeter gun, for the goal should have been
the achievement, not of equal but of superior fighting power.
Otherwise it would have been superfluous for the army to order

the 42-centimeter mortar, since the German 21-centimeter mortar
was in itself superior to any then existing high trajectory French
cannon, and the fortresses would have likewise fallen to the
30.5-centimeter mortar. The leadership of the land army, however, thought soundly, while that of the navy unfortunately
did not.

The neglect of superior artillery power and superior speed lay
entirely in the absolutely erroneous so-called 'idea of risk.' The
navy leadership by the very form in which it expanded the fleet
renounced attack and thus from the outset inevitably assumed
the defensive. But in this they also renounced the ultimate success which is and can only be forever in attack.

A ship of smaller speed and weaker armament will as a rule be
sent to the bottom by a speedier and more heavily armed enemy
at the firing distance favorable for the latter. A number of our
cruisers were to find this out to their bitter grief. The utter
mistakenness of the peacetime opinion of the navy staff was
shown by the War, which forced the introduction, whenever
possible, of modified armament in old ships and better armament
in newer ones. If in the battle of Skagerrak the German ships
had had the tonnage, the armament, the same speed as the
English ships, the British navy would have found a watery grave
beneath the hurricane of the more accurate and more effective
German 38-centimeter shells.

Japan carried on a different naval policy in those days. There,
on principle, the entire emphasis was laid on giving every single
new ship superior fighting power over the presumable adversary.
The result was a greater possibility of offensive utilization of the
navy.

While the staff of the land army still kept free of such basically
false trains of thought, the navy, which unfortunately had better
'parliamentary' representation, succumbed to the spirit of parliament. It was organized on the basis of half-baked ideas and was
later used in a similar way. What immortal fame the navy nevertheless achieved could only be set to the account of the skill of
the German armaments worker and the ability and incomparable

heroism of the individual officers and crews. If the previous naval high command had shown corresponding intelligence, these sacrifices would not have been in vain.

Thus perhaps it was precisely the superior parliamentary dexterity of the navy's peacetime head that resulted in its misfortune, since, even in its building, parliamentary instead of purely military criteria unfortunately began to play the decisive rôle. The half-heartedness and weakness as well as the meager logic in thinking,[1] characteristic of the parliamentary institution, began to color the leadership of the navy.

The land army, as already emphasized, still refrained from such basically false trains of thought. Particularly the colonel in the great General Staff of that time, Ludendorff, carried on a desperate struggle against the criminal half-heartedness and weakness with which the Reichstag approached the vital problems of the nation, and for the most part negated them. If the struggle which this officer then carried on was nevertheless in vain, the blame was borne half by parliament and half by the attitude and weakness — even more miserable, if possible — of Reich Chancellor Bethmann Hollweg. Yet today this does not in the least prevent those who were responsible for the German collapse from putting the blame precisely on him who alone combated this neglect of national interests — one swindle more or less is nothing to these born crooks.

Anyone who contemplates all the sacrifices which were heaped on the nation by the criminal frivolity of these most irresponsible among irresponsibles, who passes in review all the uselessly sacrificed dead and maimed, as well as the boundless shame and disgrace, the immeasurable misery which has now struck us, and knows that all this happened only to clear the path to ministers' chairs for a gang of unscrupulous climbers and job-hunters — anyone who contemplates all this will understand that these creatures can, believe me, be described only by words such as 'scoundrel,' 'villain,' 'scum,' and 'criminal,' otherwise the meaning and purpose of having these expressions in our lin-

---

[1] *'Geringe Logik im Denken.'*

guistic usage would be incomprehensible. For compared to these traitors to the nation, every pimp is a man of honor.

*   *   *

Strangely enough, all the really seamy sides of old Germany attracted attention only when the inner solidarity of the nation would inevitably suffer thereby. Yes, indeed, in such cases the unpleasant truths were positively bellowed to the broad masses, while otherwise the same people preferred modestly to conceal many things and in part simply to deny them. This was the case when the open discussion of a question might have led to an improvement. At the same time, the government offices in charge knew next to nothing of the value and nature of propaganda. The fact that by clever and persevering use of propaganda even heaven can be represented as hell to the people, and conversely the most wretched life as paradise, was known only to the Jew, who acted accordingly; the German, or rather his government, hadn't the faintest idea of this.

During the War we were to suffer most gravely for all this.

*   *   *

Along with all the evils of German life before the War here indicated, and many more, there were also many advantages. In a fair examination, we must even recognize that most of our weaknesses were largely shared by other countries and peoples, and in some, indeed, we were put completely in the shade, while they did not possess many of our own actual advantages.

At the head of these advantages we can, among other things, set the fact that, of nearly all European peoples, the German people still made the greatest attempt to preserve the national character of its economy and despite certain evil omens was least subject to international financial control. A dangerous advantage, to be sure, which later became the greatest instigator of the World War. But aside from this and many other things, we must, from the vast number of healthy sources of national strength, pick three institutions which in their kind were exemplary and in part unequaled.

First, the state form as such and the special stamp which it had received in modern Germany.

Here we may really disregard the individual monarchs who as men are subject to all the weaknesses which are customarily visited upon this earth and its children; if we were not lenient in this, we would have to despair of the present altogether, for are not the representatives of the present régime, considered as personalities, intellectually and morally of the most modest proportions that we can conceive of even racking our brains for a long time? Anyone who measures the 'value' of the German revolution by the value and stature of the personalities which it has given the German people since November, 1919, will have to hide his head for shame before the judgment of future generations, whose tongue it will no longer be possible to stop by protective laws, etc., and which therefore will say what today all of us know to be true, to wit, that brains and virtue in our modern German leaders are inversely proportionate to their vices and the size of their mouths.

To be sure, the monarchy had grown alien to many, to the broad masses above all. This was the consequence of the fact that the monarchs were not always surrounded by the brightest — to put it mildly — and above all not by the sincerest minds. Unfortunately, a number of them liked flatterers better than straightforward natures, and consequently it was the flatterers who 'instructed' them. A very grave evil at a time when many of the world's old opinions had undergone a great change, spreading naturally to the estimation in which many old-established traditions of the courts were held.

Thus, at the turn of the century the common man in the street could no longer find any special admiration for the princess who rode along the front in uniform. Apparently those in authority were incapable of correctly judging the effect of such a parade in the eyes of the people, for if they had, such unfortunate performances would doubtless not have occurred. Moreover, the humanitarian bilge — not always entirely sincere — that these circles went in for repelled more than it attracted. If, for example, Princess X condescended to taste a sample of food in a people's kitchen, in former days it might have looked well, but now the result was the opposite. We may be justified in assuming that Her Highness really had no idea that the food on the day she sampled it was a little different from what it usually was; but it was quite enough that the people knew it.

Thus, what may possibly have been the best intention became ridiculous, if not actually irritating.

Stories about the monarch's proverbial frugality, his much too early rising and his slaving away until late into the night, amid the permanent peril of threatening undernourishment, aroused very dubious comments. People did not ask to know what food and how much of it the monarch deigned to consume; they did not begrudge him a 'square' meal; nor were they out to deprive him of the sleep he needed; they were satisfied if in other things, as a man and character, he was an honor to the name of his house and to the nation, and if he fulfilled his duties as a ruler. Telling fairy tales helped little, but did all the more harm.

This and many similar things were mere trifles, however. What had a worse effect on sections of the nation, that were unfortunately very large, was the mounting conviction that people were ruled from the top no matter what happened, and that, therefore, the individual had no need to bother about anything. As long as the government was really good, or at least had the best intentions, this was bearable. But woe betide if the old government whose intentions were after all good were replaced by a new one which was not so decent; then spineless compliance and childlike faith were the gravest calamity that could be conceived of.

But along with these and many other weaknesses, there were unquestionable assets.

For one thing, the stability of the entire state leadership, brought about by the monarchic form of state and the removal of the highest state posts from the welter of speculation by ambitious politicians. Furthermore, the dignity of the institution as such and the authority which this alone created: likewise the raising of the civil service and particularly the army above the level of party obligations. One more advantage was the personal embodiment of the state's summit in the monarch as a person, and the example of responsibility which is bound to be stronger in a monarch than in the accidental rabble of a parliamentary majority — the proverbial incorruptibility of the German administration could primarily be attributed to this. Finally, the cultural value of the monarchy for the German people was high and could very well compensate for other drawbacks. The German court cities were still the refuge of an artistic state of mind, which is increasingly threatening to die out in our materialistic times. What the German princes did for art and science, particularly in the nineteenth century, was exemplary. The present period in any case cannot be compared with it.

*   *   *

As the greatest credit factor, however, in this period of incipient and slowly spreading decomposition of our nation, we must note the army. It was the mightiest school of the German nation, and not for nothing was the hatred of all our enemies directed against this buttress of national freedom and independence. No more glorious monument can be dedicated to this unique institution than a statement of the truth that it was slandered, hated, combated, and also feared by all inferior peoples. The fact that the rage of the international exploiters of our people in Versailles was directed primarily against the old German army permits us to recognize it as the bastion of our national freedom against the power of the stock exchange. Without this warning power, the intentions of Versailles would long since have been carried out

against our people. What the German people owes to the army
can be briefly summed up in a single word, to wit: everything.

The army trained men for unconditional responsibility at a
time when this quality had grown rare and evasion of it was becoming more and more the order of the day, starting with the
model prototype of all irresponsibility, the parliament; it trained
men in personal courage in an age when cowardice threatened
to become a raging disease and the spirit of sacrifice, the willingness to give oneself for the general welfare, was looked on almost
as stupidity, and the only man regarded as intelligent was the
one who best knew how to indulge and advance his own ego; it
was the school that still taught the individual German not to
seek the salvation of the nation in lying phrases about an international brotherhood between Negroes, Germans, Chinese,
French, etc., but in the force and solidarity of our own nation.

The army trained men in resolution while elsewhere in life
indecision and doubt were beginning to determine the actions
of men. In an age when everywhere the know-it-alls were setting
the tone, it meant something to uphold the principle that some
command is always better than none. In this sole principle there
was still an unspoiled robust health which would long since have
disappeared from the rest of our life if the army and its training
had not provided a continuous renewal of this primal force.
We need only see the terrible indecision of the Reich's present
leaders, who can summon up the energy for no action unless it
is the forced signing of a new decree for plundering the people;
in this case, to be sure, they reject all responsibility and with the
agility of a court stenographer sign everything that anyone may
see fit to put before them. In this case the decision is easy to
take; for it is dictated.

The army trained men in idealism and devotion to the fatherland and its greatness while everywhere else greed and material-
ism had spread abroad. It educated a single people in contrast
to the division into classes and in this perhaps its sole mistake
was the institution of voluntary one-year enlistment. A mistake, because through it the principle of unconditional equality

was broken, and the man with higher education was removed from the setting of his general environment, while precisely the exact opposite would have been advantageous. In view of the great unworldliness of our upper classes and their constantly mounting estrangement from their own people, the army could have exerted a particularly beneficial effect if in its own ranks, at least, it had avoided any segregation of the so-called intelligentsia. That this was not done was a mistake; but what institution in this world makes no mistakes? In this one, at any rate, the good was so predominant that the few weaknesses lay far beneath the average degree of human imperfection.[1]

It must be attributed to the army of the old Reich as its highest merit that at a time when heads were generally counted by majorities, it placed heads above the majority. Confronted with the Jewish-democratic idea of a blind worship of numbers, the army sustained belief in personality. And thus it trained what the new epoch most urgently needed: men. In the morass of a universally spreading softening and effeminization, each year three hundred and fifty thousand vigorous young men sprang from the ranks of the army, men who in their two years' training had lost the softness of youth and achieved bodies hard as steel. The young man who practiced obedience during this time could then learn to command. By his very step you could recognize the soldier who had done his service.

This was the highest school of the German nation, and it was not for nothing that the bitterest hatred of those who from envy and greed needed and desired the impotence of the Reich and the defenselessness of its citizens was concentrated on it. What many Germans in their blindness or ill will did not want to see was recognized by the foreign world: the German army was the mightiest weapon serving the freedom of the German nation and the sustenance of its children.

*   *   *

[1] '... weit unter dem Durchschnittsgrade der menschlichen Unzulänglichkeit ...'

The third in the league, along with the state form and the army, was the incomparable civil service of the old Reich.

Germany was the best organized and best administered country in the world. The German government official might well be accused of bureaucratic red tape, but in the other countries things were no better in this respect; they were worse. But what the other countries did not possess was the wonderful solidity of this apparatus and the incorruptible honesty of its members. It was better to be a little old-fashioned, but honest and loyal, than enlightened and modern, but of inferior character and, as is often seen today, ignorant and incompetent. For if today people like to pretend that the German administration of the pre-War period, though bureaucratically sound, was bad from a business point of view, only the following answer can be given: what country in the world had an institution better directed and better organized in a business sense than Germany's state railways? It was reserved to the revolution to go on wrecking this exemplary apparatus until at last it seemed ripe for being taken out of the hands of the nation and socialized according to the lights of this Republic's founders; in other words, made to serve international stock exchange capital, the power behind the German revolution.

What especially distinguished the German civil service and administrative apparatus was their independence from the individual governments whose passing political views could have no effect on the job of German civil servant. Since the revolution, it must be admitted, this has completely changed. Ability and competence were replaced by party ties and a self-reliant, independent character became more of a hindrance than a help.

The state form, the army. and the civil service formed the basis for the old Reich's wonderful power and strength. These first and foremost were the reasons for a quality which is totally lacking in the present-day state: state's authority! For this is not based on bull-sessions in parliaments or provincial diets, or on laws for its protection, or court sentences to frighten those who insolently deny it, etc., but on the general confidence which may and can be placed in the leadership and administration of a commonwealth. This confidence, in turn, results only from an unshakable inner faith in the selflessness and honesty of the government and administration of a country and from an agreement between the spirit of the laws and the general ethical view. For in the long run government systems are not maintained by the pressure of violence, but by faith in their soundness and in the truthfulness with which they represent and advance the interests of a people.

*   *   *

Gravely as certain evils of the pre-War period corroded and threatened to undermine the inner strength of the nation, it must not be forgotten that other states suffered even more than Germany from most of these ailments and yet in the critical hour of danger did not flag and perish. But if we consider that the German weaknesses before the War were balanced by equally great strengths, the ultimate cause of the collapse can and must lie in a different field; and this is actually the case.

The deepest and ultimate reason for the decline of the old Reich lay in its failure to recognize the racial problem and its importance for the historical development of peoples. For events in the lives of peoples are not expressions of chance, but processes related to the self-preservation and propagation of the species and the race and subject to the laws of Nature, even if people are not conscious of the inner reason for their actions.

Nation and Race

T HERE are some truths which are so obvious that for this very reason they are not seen or at least not recognized by ordinary people. They sometimes pass by such truisms as though blind and are most astonished when someone suddenly discovers what everyone really ought to know. Columbus's eggs lie around by the hundreds of thousands, but Columbuses are met with less frequently.

Thus men without exception wander about in the garden of Nature; they imagine that they know practically everything and yet with few exceptions pass blindly by one of the most patent principles of Nature's rule: the inner segregation of the species of all living beings on this earth.

Even the most superficial observation shows that Nature's restricted form of propagation and increase is an almost rigid basic law of all the innumerable forms of expression of her vital urge. Every animal mates only with a member of the same species. The titmouse seeks the titmouse, the finch the finch, the stork the stork, the field mouse the field mouse, the dormouse the dormouse, the wolf the she-wolf, etc.

Only unusual circumstances can change this, primarily the compulsion of captivity or any other cause that makes it impossible to mate within the same species. But then Nature begins to resist this with all possible means, and her most visible protest consists either in refusing further capacity for propagation to bastards or in limiting the fertility of later offspring; in most cases, however, she takes away the power of resistance to disease or hostile attacks.

This is only too natural.

Any crossing of two beings not at exactly the same level produces a medium between the level of the two parents. This means: the offspring will probably stand higher than the racially lower parent, but not as high as the higher one. Consequently, it will later succumb in the struggle against the higher level. Such mating is contrary to the will of Nature for a higher breeding of all life. The precondition for this does not lie in associating superior and inferior, but in the total victory of the former. The stronger must dominate and not blend with the weaker, thus sacrificing his own greatness. Only the born weakling can view this as cruel, but he after all is only a weak and limited man; for if this law did not prevail, any conceivable higher development of organic living beings would be unthinkable.

The consequence of this racial purity,[1] universally valid in Nature, is not only the sharp outward delimitation of the various races, but their uniform character in themselves. The fox is always a fox, the goose a goose, the tiger a tiger, etc., and the difference can lie at most in the varying measure of force, strength, intelligence, dexterity, endurance, etc., of the individual specimens. But you will never find a fox who in his inner attitude might, for example, show humanitarian tendencies toward geese, as similarly there is no cat with a friendly inclination toward mice.

Therefore, here, too, the struggle among themselves arises less from inner aversion than from hunger and love. In both cases, Nature looks on calmly, with satisfaction, in fact. In the struggle for daily bread all those who are weak and sickly or less determined succumb, while the struggle of the males for the female grants the right or opportunity to propagate only to the healthiest. And struggle is always a means for improving a species' health and power of resistance and, therefore, a cause of its higher development.

[1] Second edition inserts 'urge toward' before 'racial purity.'

If the process were different, all further and higher development would cease and the opposite would occur. For, since the
inferior always predominates numerically over the best, if both
had the same possibility of preserving life and propagating, the
inferior would multiply so much more rapidly that in the end
the best would inevitably be driven into the background, unless
a correction of this state of affairs were undertaken. Nature does
just this by subjecting the weaker part to such severe living
conditions that by them alone the number is limited, and by not
permitting the remainder to increase promiscuously, but making
a new and ruthless choice according to strength and health.

No more than Nature desires the mating of weaker with
stronger individuals, even less does she desire the blending of a
higher with a lower race, since, if she did, her whole work of
higher breeding, over perhaps hundreds of thousands of years,
night be ruined with one blow.

Historical experience offers countless proofs of this. It shows
with terrifying clarity that in every mingling of Aryan blood
with that of lower peoples the result was the end of the cultured
people. North America, whose population consists in by far the
largest part of Germanic elements who mixed but little with the
lower colored peoples, shows a different humanity and culture
from Central and South America, where the predominantly Latin
immigrants often mixed with the aborigines on a large scale.
By this one example, we can clearly and distinctly recognize the
effect of racial mixture. The Germanic inhabitant of the American continent, who has remained racially pure and unmixed,
rose to be master of the continent; he will remain the master as
long as he does not fall a victim to defilement of the blood.

The result of all racial crossing is therefore in brief always
the following:

(a) Lowering of the level of the higher race;

(b) Physical and intellectual regression and hence the beginning of a slowly but surely progressing sickness.

To bring about such a development is, then, nothing else but
to sin against the will of the eternal creator.

And as a sin this act is rewarded.

When man attempts to rebel against the iron logic of Nature, he comes into struggle with the principles to which he himself owes his existence as a man. And this attack [1] must lead to his own doom.

Here, of course, we encounter the objection of the modern pacifist, as truly Jewish in its effrontery as it is stupid! 'Man's rôle is to overcome Nature!'

Millions thoughtlessly parrot this Jewish nonsense and end up by really imagining that they themselves represent a kind of conqueror of Nature; though in this they dispose of no other weapon than an idea, and at that such a miserable one, that if it were true no world at all would be conceivable.

But quite aside from the fact that man has never yet conquered Nature in anything, but at most has caught hold of and tried to lift one or another corner of her immense gigantic veil of eternal riddles and secrets, that in reality he invents nothing but only discovers everything, that he does not dominate Nature, but has only risen on the basis of his knowledge of various laws and secrets of Nature to be lord over those other living creatures who lack this knowledge — quite aside from all this, an idea cannot overcome the preconditions for the development and being of humanity, since the idea itself depends only on man. Without human beings there is no human idea in this world, therefore, the idea as such is always conditioned by the presence of human beings and hence of all the laws which created the precondition for their existence.

And not only that! Certain ideas are even tied up with certain men. This applies most of all to those ideas whose content originates, not in an exact scientific truth, but in the world of emotion, or, as it is so beautifully and clearly expressed today, reflects an 'inner experience.' All these ideas, which have nothing to do with cold logic as such, but represent only pure expressions of feeling, ethical conceptions, etc., are chained to the existence of men, to whose intellectual imagination and creative power

[1] Second edition: 'so his action against Nature' instead of 'this attack.'

they owe their existence. Precisely in this case the preservation of these definite races and men is the precondition for the existence of these ideas. Anyone, for example, who really desired the victory of the pacifistic idea in this world with all his heart would have to fight with all the means at his disposal for the conquest of the world by the Germans; for, if the opposite should occur, the last pacifist would die out with the last German, since the rest of the world has never fallen so deeply as our own people, unfortunately, has for this nonsense so contrary to Nature and reason. Then, if we were serious, whether we liked it or not, we would have to wage wars in order to arrive at pacifism. This and nothing else was what Wilson, the American world savior, intended, or so at least our German visionaries believed — and thereby his purpose was fulfilled.

In actual fact the pacifistic-humane idea is perfectly all right perhaps when the highest type of man has previously conquered and subjected the world to an extent that makes him the sole ruler of this earth. Then this idea lacks the power of producing evil effects in exact proportion as its practical application becomes rare and finally impossible. Therefore, first struggle and then we shall see what can be done.[1] Otherwise mankind has passed the high point of its development and the end is not the domination of any ethical idea but barbarism and consequently chaos. At this point someone or other may laugh, but this planet once moved through the ether for millions of years without human beings and it can do so again some day if men forget that they owe their higher existence, not to the ideas of a few crazy ideologists, but to the knowledge and ruthless application of Nature's stern and rigid laws.

Everything we admire on this earth today — science and art, technology and inventions — is only the creative product of a few peoples and originally perhaps of *one* race. On them depends the existence of this whole culture. If they perish, the beauty of this earth will sink into the grave with them.

However much the soil, for example, can influence men, the

[1] Second edition: 'struggle and then perhaps pacifism.'

result of the influence will always be different depending on the races in question. The low fertility of a living space may spur the one race to the highest achievements; in others it will only be the cause of bitterest poverty and final undernourishment with all its consequences. The inner nature of peoples is always determining for the manner in which outward influences will be effective. What leads the one to starvation trains the other to hard work.

All great cultures of the past perished only because the originally creative race died out from blood poisoning.

The ultimate cause of such a decline was their forgetting that all culture depends on men and not conversely; hence that to preserve a certain culture the man who creates it must be preserved. This preservation is bound up with the rigid law of necessity and the right to victory of the best and stronger in this world.[1]

Those who want to live, let them fight, and those who do not want to fight in this world of eternal struggle do not deserve to live.

Even if this were hard — that is how it is! Assuredly, however, by far the harder fate is that which strikes the man who thinks he can overcome Nature, but in the last analysis only mocks her. Distress, misfortune, and diseases are her answer.

The man who misjudges and disregards the racial laws actually forfeits the happiness that seems destined to be his. He thwarts the triumphal march of the best race and hence also the precondition for all human progress, and remains, in consequence, burdened with all the sensibility of man, in the animal realm of helpless misery.[2]

*   *   *

[1] Second edition omits: 'in this world.'

[2] '*und verbleibt in der Folge dann, belastet mit der Empfindlichkeit des Menschen, im Bereich des hilflosen Jammers der Tiere.*' Second edition has: '*Er begibt sich in der Folge, belastet mit der Empfindlichkeit des Menschen, ins Bereich des hilflosen Tieres.*' This would read: 'In consequence, burdened with all the sensibility of man, he moves into the realm of the helpless beast.'

It is idle to argue which race or races were the original representative of human culture and hence the real founders of all that we sum up under the word 'humanity.' It is simpler to raise this question with regard to the present, and here an easy, clear answer results. All the human culture, all the results of art, science, and technology that we see before us today, are almost exclusively the creative product of the Aryan. This very fact admits of the not unfounded inference that he alone was the founder of all higher humanity, therefore representing the prototype of all that we understand by the word 'man.' He is the Prometheus of mankind from whose bright forehead the divine spark of genius has sprung at all times, forever kindling anew that fire of knowledge which illumined the night of silent mysteries and thus caused man to climb the path to mastery over the other beings of this earth. Exclude him — and perhaps after a few thousand years darkness will again descend on the earth, human culture will pass, and the world turn to a desert.

If we were to divide mankind into three groups, the founders of culture, the bearers of culture, the destroyers of culture, only the Aryan could be considered as the representative of the first group. From him originate the foundations and walls of all human creation, and only the outward form and color are determined by the changing traits of character of the various peoples. He provides the mightiest building stones and plans for all human progress and only the execution corresponds to the nature of the varying men and races. In a few decades, for example, the entire east of Asia will possess a culture whose ultimate foundation will be Hellenic spirit and Germanic technology, just as much as in Europe. Only the *outward* form — in part at least — will bear the features of Asiatic character. It is not true, as some people think, that Japan adds European technology to its culture; no, European science and technology are trimmed with Japanese characteristics. The foundation of actual life is no longer the special Japanese culture, although it determines the color of life — because outwardly, in consequence of its inner difference, it is more conspicuous to the European — but the gigantic scientific-technical achievements of Europe and America; that is, of Aryan peoples. Only on the basis of these achievements can the Orient follow general human progress. They furnish the basis of the struggle for daily bread, create weapons and implements for it, and only the outward form is gradually adapted to Japanese character.

If beginning today all further Aryan influence on Japan should stop, assuming that Europe and America should perish, Japan's present rise in science and technology might continue for a short time; but even in a few years the well would dry up, the Japanese special character would gain, but the present culture would freeze and sink back into the slumber from which it was awakened seven decades ago by the wave of Aryan culture. Therefore, just as the present Japanese development owes its life to Aryan origin, long ago in the gray past foreign influence and foreign spirit awakened the Japanese culture of that time. The best proof of this is furnished by the fact of its subsequent sclerosis and total petrifaction. This can occur in a people only when the original creative racial nucleus has been lost, or if the external influence which furnished the impetus and the material for the first development in the cultural field was later lacking. But if it is established that a people receives the most essential basic materials of its culture from foreign races, that it assimilates and adapts them, and that then, if further external influence is lacking, it rigidifies again and again, such a race may be designated as '*culture-bearing*,' but never as '*culture-creating*.' An examination of the various peoples from this standpoint points to the fact that practically none of them were originally *culture-founding*, but almost always *culture-bearing*.

Approximately the following picture of their development always results:

Aryan races — often absurdly small numerically — subject foreign peoples, and then, stimulated by the special living conditions of the new territory (fertility, climatic conditions, etc.) and assisted by the multitude of lower-type beings standing at their disposal as helpers, develop the intellectual and organizational capacities dormant within them. Often in a few millenniums or even centuries they create cultures which originally bear all the inner characteristics of their nature, adapted to the above-indicated special qualities of the soil and subjected beings. In the end, however, the conquerors transgress against the principle of blood purity, to which they had first adhered; they begin to mix with the subjugated inhabitants and thus end their own existence; for the fall of man in paradise has always been followed by his expulsion.

After a thousand years and more, the last visible trace of the former master people is often seen in the lighter skin color which its blood left behind in the subjugated race, and in a petrified culture which it had originally created. For, once the actual and spiritual conqueror lost himself in the blood of the subjected people, the fuel for the torch of human progress was lost! Just as, through the blood of the former masters, the color preserved a feeble gleam in their memory, likewise the night of cultural life is gently illumined by the remaining creations of the former light-bringers. They shine through all the returned barbarism and too often inspire the thoughtless observer of the moment with the opinion that he beholds the picture of the present people before him, whereas he is only gazing into the mirror of the past.

It is then possible that such a people will a second time, or even more often in the course of its history, come into contact with the race of those who once brought it culture, and the memory of former encounters will not necessarily be present. Unconsciously the remnant of the former master blood will turn toward the new arrival, and what was first possible only by compulsion can now succeed through the people's own will. A new cultural wave makes its entrance and continues until those who have brought it are again submerged in the blood of foreign peoples.

It will be the task of a future cultural and world history to carry on researches in this light and not to stifle in the rendition of external facts, as is so often, unfortunately, the case with our present historical science.

This mere sketch of the development of 'culture-bearing' nations gives a picture of the growth, of the activity, and — the decline — of the true culture-founders of this earth, the Aryans themselves.

As in daily life the so-called genius requires a special cause, indeed, often a positive impetus, to make him shine, likewise the genius-race in the life of peoples. In the monotony of everyday life even significant men often seem insignificant, hardly rising above the average of their environment; as soon, however, as they are approached by a situation in which others lose hope or go astray, the genius rises manifestly from the inconspicuous average child, not seldom to the amazement of all those who had hitherto seen him in the pettiness of bourgeois life — and that is why the prophet seldom has any honor in his own country. Nowhere have we better occasion to observe this than in war. From apparently harmless children, in difficult hours when others lose hope, suddenly heroes shoot up with death-defying determination and an icy cool presence of mind.[1] If this hour of trial had not come, hardly anyone would ever have guessed that a young hero was hidden in this beardless boy. It nearly always takes some stimulus to bring the genius on the scene. The hammer-stroke of Fate which throws one man to the ground suddenly strikes steel in another, and when the shell of everyday life is broken, the previously hidden kernel lies open before the eyes of the astonished world. The world then resists and does not want to believe that the type which is apparently identical with it is suddenly a very different being; a process which is repeated with every eminent son of man.

Though an inventor, for example, establishes his fame only on the day of his invention, it is a mistake to think that genius as such entered into the man only at this hour — the spark of genius exists in the brain of the truly creative man from the hour of his birth. True genius is always inborn and never cultivated, let alone learned.

As already emphasized, this applies not only to the individual man but also to the race. Creatively active peoples always have

[1] '*eisige Kühle der Uberlegung.*'

a fundamental creative gift, even if it should not be recognizable to the eyes of superficial observers. Here, too, outward recognition is possible only in consequence of accomplished deeds, since the rest of the world is not capable of recognizing genius in itself, but sees only its visible manifestations in the form of inventions, discoveries, buildings, pictures, etc.; here again it often takes a long time before the world can fight its way through to this knowledge. Just as in the life of the outstanding individual, genius or extraordinary ability strives for practical realization only when spurred on by special occasions, likewise in the life of nations the creative forces and capacities which are present can often be exploited only when definite preconditions invite.

We see this most distinctly in connection with the race which has been and is the bearer of human cultural development — the Aryans. As soon as Fate leads them toward special conditions, their latent abilities begin to develop in a more and more rapid sequence and to mold themselves into tangible forms. The cultures which they found in such cases are nearly always decisively determined by the existing soil, the given climate, and — the subjected people. This last item, to be sure, is almost the most decisive. The more primitive the technical foundations for a cultural activity, the more necessary is the presence of human helpers who, organizationally assembled and employed, must replace the force of the machine. Without this possibility of using lower human beings, the Aryan would never have been able to take his first steps toward his future culture; just as without the help of various suitable beasts which he knew how to tame, he would not have arrived at a technology which is now gradually permitting him to do without these beasts. The saying, 'The Moor has worked off his debt, the Moor can go,'[1] unfortunately has only too deep a meaning. For thousands of years the horse had to serve man and help him lay the founda-

[1] Schiller. *Die Verschwörung des Fiesko*, Act III, Scene 4, spoken by the Moor. It properly reads: '*Der Moor hat seine Arbeit getan*,' etc. (The Moor has done his work), but is often quoted in the altered version with '*Schuldigkeit*' in place of '*Arbeit*.'

tions of a development which now, in consequence of the motor car, is making the horse superfluous. In a few years his activity will have ceased, but without his previous collaboration man might have had a hard time getting where he is today.

Thus, for the formation of higher cultures the existence of lower human types was one of the most essential preconditions, since they alone were able to compensate for the lack of technical aids without which a higher development is not conceivable. It is certain that the first culture of humanity was based less on the tamed animal than on the use of lower human beings.

Only after the enslavement of subjected races did the same fate strike beasts, and not the other way around, as some people would like to think. For first the conquered warrior drew the plow — and only after him the horse. Only pacifistic fools can regard this as a sign of human depravity, failing to realize that this development had to take place in order to reach the point where today these sky-pilots could force their drivel on the world.

The progress of humanity is like climbing an endless ladder; it is impossible to climb higher without first taking the lower steps. Thus, the Aryan had to take the road to which reality directed him and not the one that would appeal to the imagination of a modern pacifist. The road of reality is hard and difficult, but in the end it leads where our friend would like to bring humanity by dreaming, but unfortunately removes more than bringing it closer.

Hence it is no accident that the first cultures arose in places where the Aryan, in his encounters with lower peoples, subjugated them and bent them to his will. They then became the first technical instrument in the service of a developing culture.

Thus, the road which the Aryan had to take was clearly marked out. As a conqueror he subjected the lower beings and regulated their practical activity under his command, according to his will and for his aims. But in directing them to a useful, though arduous activity, he not only spared the life of those he subjected; perhaps he gave them a fate that was better than their previous so-called 'freedom.' As long as he ruthlessly upheld the master attitude, not only did he really remain master, but also the preserver and increaser of culture. For culture was based exclusively on his abilities and hence on his actual survival. As soon as the subjected people began to raise themselves up and probably approached the conqueror in language, the sharp dividing wall between master and servant fell. The Aryan gave up the purity of his blood and, therefore, lost his sojourn in the paradise which he had made for himself. He became submerged in the racial mixture, and gradually, more and more, lost his cultural capacity, until at last, not only mentally but also physically, he began to resemble the subjected aborigines more than his own ancestors. For a time he could live on the existing cultural benefits, but then petrifaction set in and he fell a prey to oblivion.

Thus cultures and empires collapsed to make place for new formations.

Blood mixture and the resultant drop in the racial level is the sole cause of the dying out of old cultures; for men do not perish as a result of lost wars, but by the loss of that force of resistance which is contained only in pure blood.

All who are not of good race in this world are chaff.

And all occurrences in world history are only the expression of the races' instinct of self-preservation, in the good or bad sense.

*   *   *

The question of the inner causes of the Aryan's importance can be answered to the effect that they are to be sought less in a natural instinct of self-preservation than in the special type of its expression. The will to live, subjectively viewed, is everywhere equal and different only in the form of its actual expression. In the most primitive living creatures the instinct of self-preservation does not go beyond concern for their own ego. Egoism, as we designate this urge, goes so far that it even embraces time; the moment itself claims everything, granting nothing to the coming hours. In this condition the animal lives only for himself, seeks food only for his present hunger, and fights only for his own life. As long as the instinct of self-preservation expresses itself in this way, every basis is lacking for the formation of a group, even the most primitive form of family. Even a community between male and female beyond pure mating, demands an extension of the instinct of self-preservation, since concern and struggle for the ego are now directed toward the second party; the male sometimes seeks food for the female, too, but for the most part both seek nourishment for the young. Nearly always one comes to the defense of the other, and thus the first, though infinitely simple, forms of a sense of sacrifice result. As soon as this sense extends beyond the narrow limits of the family, the basis for the formation of larger organisms and finally formal states is created.

In the lowest peoples of the earth this quality is present only to a very slight extent, so that often they do not go beyond the formation of the family. The greater the readiness to subordinate purely personal interests, the higher rises the ability to establish comprehensive communities.

This self-sacrificing will to give one's personal labor and if necessary one's own life for others is most strongly developed in the Aryan. The Aryan is not greatest in his mental qualities as such, but in the extent of his willingness to put all his abilities in the service of the community. In him the instinct of self-preservation has reached the noblest form, since he willingly subordinates his own ego to the life of the community and, if the hour demands, even sacrifices it.

Not in his intellectual gifts lies the source of the Aryan's capacity for creating and building culture. If he had just this alone, he could only act destructively, in no case could he organize; for the innermost essence of all organization requires that the individual renounce putting forward his personal opinion and interests and sacrifice both in favor of a larger group. Only by way of this general community does he again recover his share. Now, for example, he no longer works directly for himself, but with his activity articulates himself with the community, not only for his own advantage, but for the advantage of all. The most wonderful elucidation of this attitude is provided by his word 'work,' by which he does not mean an activity for maintaining life in itself, but exclusively a creative effort that does not conflict with the interests of the community. Otherwise he designates human activity, in so far as it serves the instinct of self-preservation without consideration for his fellow men, as theft, usury, robbery, burglary, etc.

This state of mind, which subordinates the interests of the ego to the conservation of the community, is really the first premise for every truly human culture. From it alone can arise all the great works of mankind, which bring the founder little reward, but the richest blessings to posterity. Yes, from it alone can we understand how so many are able to bear up faithfully under a scanty life which imposes on them nothing but poverty and frugality, but gives the community the foundations of its existence. Every worker, every peasant, every inventor, official, etc., who works without ever being able to achieve any happiness or prosperity for himself, is a representative of this lofty idea, even if the deeper meaning of his activity remains hidden in him.

What applies to work as the foundation of human sustenance and all human progress is true to an even greater degree for the defense of man and his culture. In giving one's own life for the existence of the community lies the crown of all sense of sacrifice. It is this alone that prevents what human hands have built from being overthrown by human hands or destroyed by Nature.

Our own German language possesses a word which magnificently designates this kind of activity: *Pflichterfüllung* (fulfillment of duty); it means not to be self-sufficient but to serve the community.

The basic attitude from which such activity arises, we call — to distinguish it from egoism and selfishness — idealism. By this we understand only the individual's capacity to make sacrifices for the community, for his fellow men.

How necessary it is to keep realizing that idealism does not represent a superfluous expression of emotion, but that in truth it has been, is, and will be, the premise for what we designate as human culture, yes, that it alone created the concept of 'man'! It is to this inner attitude that the Aryan owes his position in this world, and to it the world owes man; for it alone formed from pure spirit the creative force which, by a unique pairing of the brutal fist and the intellectual genius, created the monuments of human culture.

Without his idealistic attitude all, even the most dazzling faculties of the intellect, would remain mere intellect as such — outward appearance without inner value, and never creative force.

But, since true idealism is nothing but the subordination of the interests and life of the individual to the community, and this in turn is the precondition for the creation of organizational forms of all kinds, it corresponds in its innermost depths to the ultimate will of Nature. It alone leads men to voluntary recognition of the privilege of force and strength, and thus makes them into a dust particle of that order which shapes and forms the whole universe.

The purest idealism is unconsciously equivalent to the deepest knowledge.

How correct this is, and how little true idealism has to do with playful flights of the imagination, can be seen at once if we let the unspoiled child, a healthy boy, for example, judge. The same boy who feels like throwing up [1] when he hears the tirades of a pacifist 'idealist' is ready to give his young life for the ideal of his nationality.

Here the instinct of knowledge unconsciously obeys the deeper necessity of the preservation of the species, if necessary at the cost of the individual, and protests against the visions of the pacifist windbag who in reality is nothing but a cowardly, though camouflaged, egoist, transgressing the laws of development; for development requires willingness on the part of the individual to sacrifice himself for the community, and not the sickly imaginings of cowardly know-it-alls and critics of Nature.

[1] Second edition has: 'who is hostile and without understanding.'

Especially, therefore, at times when the ideal attitude threatens to disappear, we can at once recognize a diminution of that force which forms the community and thus creates the premises of culture. As soon as egoism becomes the ruler of a people, the bands of order are loosened and in the chase after their own happiness men fall from heaven into a real hell.

Yes, even posterity forgets the men who have only served their own advantage and praises the heroes who have renounced their own happiness.

*   *   *

The mightiest counterpart to the Aryan is represented by the Jew. In hardly any people in the world is the instinct of self-preservation developed more strongly than in the so-called 'chosen.' Of this, the mere fact of the survival of this race may be considered the best proof. Where is the people which in the last two thousand years has been exposed to so slight changes of inner disposition, character, etc., as the Jewish people? What people, finally, has gone through greater upheavals than this one — and nevertheless issued from the mightiest catastrophes of mankind unchanged? What an infinitely tough will to live and preserve the species speaks from these facts!

The mental qualities of the Jew have been schooled in the course of many centuries. Today he passes as 'smart,' and this in a certain sense he has been at all times. But his intelligence is not the result of his own development, but of visual instruction through foreigners. For the human mind cannot climb to the top without steps; for every step upward he needs the foundation of the past, and this in the comprehensive sense in which it can be revealed only in general culture. All thinking is based only in small part on man's own knowledge, and mostly on the experience of the time that has preceded. The general cultural level provides the individual man, without his noticing it as a rule, with such a profusion of preliminary knowledge that, thus armed, he can more easily take further steps of his own. The boy of today, for example, grows up among a truly vast number of technical acquisitions of the last centuries, so that he takes for granted and no longer pays attention to much that a hundred years ago was a riddle to even the greatest minds, although for following and understanding our progress in the field in question it is of decisive importance to him. If a very genius from the twenties of the past century should suddenly leave his grave today, it would be harder for him even intellectually to find his way in the present era than for an average boy of fifteen today. For he would lack all the infinite preliminary education which our present contemporary unconsciously, so to speak, assimilates while growing up amidst the manifestations of our present general civilization.

Since the Jew — for reasons which will at once become apparent — was never in possession of a culture of his own, the foundations of his intellectual work were always provided by others. His intellect at all times developed through the cultural world surrounding him.

The reverse process never took place.

For if the Jewish people's instinct of self-preservation is not smaller but larger than that of other peoples, if his intellectual faculties can easily arouse the impression that they are equal to the intellectual gifts of other races, he lacks completely the most essential requirement for a cultured people, the idealistic attitude.

In the Jewish people the will to self-sacrifice does not go beyond the individual's naked instinct of self-preservation. Their apparently great sense of solidarity is based on the very primitive herd instinct that is seen in many other living creatures in this world. It is a noteworthy fact that the herd instinct leads to mutual support only as long as a common danger makes this seem useful or inevitable. The same pack of wolves which has just fallen on its prey together disintegrates when hunger abates into its individual beasts. The same is true of horses which try to defend themselves against an assailant in a body, but scatter again as soon as the danger is past.

It is similar with the Jew. His sense of sacrifice is only apparent. It exists only as long as the existence of the individual makes it absolutely necessary. However, as soon as the common enemy is conquered, the danger threatening all averted and the booty hidden, the apparent harmony of the Jews among themselves ceases, again making way for their old causal [1] tendencies. The Jew is only united when a common danger forces him to be or a common booty entices him; if these two grounds are lacking, the qualities of the crassest egoism come into their own, and in the twinkling of an eye the united people turns into a horde of rats, fighting bloodily among themselves.

If the Jews were alone in this world, they would stifle in filth and offal; they would try to get ahead of one another in hate-filled struggle and exterminate one another, in so far as the absolute absence of all sense of self-sacrifice, expressing itself in their cowardice, did not turn battle into comedy here too.

So it is absolutely wrong to infer any ideal sense of sacrifice in the Jews from the fact that they stand together in struggle, or, better expressed, in the plundering of their fellow men.

Here again the Jew is led by nothing but the naked egoism of the individual.

That is why the Jewish state — which should be the living organism for preserving and increasing a race — is completely unlimited as to territory. For a state formation to have a definite spatial setting always presupposes an idealistic attitude on the part of the state-race, and especially a correct interpretation of the concept of work. In the exact measure in which this attitude is lacking, any attempt at forming, even of preserving, a spatially delimited state fails. And thus the basis on which alone culture can arise is lacking.

Hence the Jewish people, despite all apparent intellectual qualities, is without any true culture, and especially without any culture of its own. For what sham culture the Jew today pos-

---

[1] *'ursächlich vorhandene Anlagen.'* *'Ursächlich'* is no doubt intended as a refinement of *'ursprünglich'* (originally). The phrase would then read: 'their originally existing tendencies.'

sesses is the property of other peoples, and for the most part it is ruined in his hands.

In judging the Jewish people's attitude on the question of human culture, the most essential characteristic we must always bear in mind is that there has never been a Jewish art and accordingly there is none today either; that above all the two queens of all the arts, architecture and music, owe nothing original to the Jews. What they do accomplish in the field of art is either patchwork or intellectual theft. Thus, the Jew lacks those qualities which distinguish the races that are creative and hence culturally blessed.

To what an extent the Jew takes over foreign culture, imitating or rather ruining it, can be seen from the fact that he is mostly found in the art which seems to require least original invention, the art of acting. But even here, in reality, he is only a 'juggler,' or rather an ape; for even here he lacks the last touch that is required for real greatness; even here he is not the creative genius, but a superficial imitator, and all the twists and tricks that he uses are powerless to conceal the inner lifelessness of his creative gift. Here the Jewish press most lovingly helps him along by raising such a roar of hosannahs about even the most mediocre bungler, just so long as he is a Jew, that the rest of the world actually ends up by thinking that they have an artist before them, while in truth it is only a pitiful comedian.

No, the Jew possesses no culture-creating force of any sort, since the idealism, without which there is no true higher development of man, is not present in him and never was present. Hence his intellect will never have a constructive effect, but will be destructive, and in very rare cases perhaps will at most be stimulating, but then as the prototype of the 'force which always wants evil and nevertheless creates good.' [1] Not through him does any progress of mankind occur, but in spite of him.

Since the Jew never possessed a state with definite territorial limits and therefore never called a culture his own, the conception arose that this was a people which should be reckoned among

[1] Goethe's *Faust*, lines 1336–1337: Mephistopheles to Faust.

the ranks of the *nomads*. This is a fallacy as great as it is dangerous. The nomad does possess a definitely limited living space,
only he does not cultivate it like a sedentary peasant, but lives
from the yield of his herds with which he wanders about in his
territory. The outward reason for this is to be found in the small
fertility of a soil which simply does not permit of settlement.
The deeper cause, however, lies in the disparity between the
technical culture of an age or people and the natural poverty of
a living space. There are territories in which even the Aryan is
enabled only by his technology, developed in the course of more
than a thousand years, to live in regular settlements, to master
broad stretches of soil and obtain from it the requirements of
life. If he did not possess this technology, either he would have
to avoid these territories or likewise have to struggle along as
a nomad in perpetual wandering, provided that his thousand-
year-old education and habit of settled residence did not make
this seem simply unbearable to him. We must bear in mind
that in the time when the American continent was being opened
up, numerous Aryans fought for their livelihood as trappers,
hunters, etc., and often in larger troops with wife and children,
always on the move, so that their existence was completely like
that of the nomads. But as soon as their increasing number
and better implements permitted them to clear the wild soil and
make a stand against the natives, more and more settlements
sprang up in the land.

Probably the Aryan was also first a nomad, settling in the
course of time, but for that very reason he was never a Jew!
No, the Jew is no nomad; for the nomad had also a definite attitude toward the concept of work which could serve as a basis
for his later development in so far as the necessary intellectual
premises were present. In him the basic idealistic view is present,
even if in infinite dilution, hence in his whole being he may seem
strange to the Aryan peoples, but not unattractive. In the Jew,
however, this attitude is not at all present; for that reason he
was never a nomad, but only and always a *parasite* in the body of
other peoples. That he sometimes left his previous living space

has nothing to do with his own purpose, but results from the fact that from time to time he was thrown out by the host nations he had misused. His spreading is a typical phenomenon for all parasites; he always seeks a new feeding ground for his race.

This, however, has nothing to do with nomadism, for the reason that a Jew never thinks of leaving a territory that he has occupied, but remains where he is, and he sits so fast that even by force it is very hard to drive him out. His extension to ever-new countries occurs only in the moment in which certain conditions for his existence are there present, without which — unlike the nomad — he would not change his residence. He is and remains the typical parasite, a sponger who like a noxious bacillus keeps spreading as soon as a favorable medium invites him. And the effect of his existence is also like that of spongers: wherever he appears, the host people dies out after a shorter or longer period.

Thus, the Jew of all times has lived in the states of other peoples, and there formed his own state, which, to be sure, habitually sailed under the disguise of 'religious community' as long as outward circumstances made a complete revelation of his nature seem inadvisable. But as soon as he felt strong enough to do without the protective cloak, he always dropped the veil and suddenly became what so many of the others previously did not want to believe and see: the Jew.

The Jew's life as a parasite in the body of other nations and states explains a characteristic which once caused Schopenhauer, as has already been mentioned, to call him the 'great master in lying.' Existence impels the Jew to lie, and to lie perpetually, just as it compels the inhabitants of the northern countries to wear warm clothing.

His life within other peoples can only endure for any length of time if he succeeds in arousing the opinion that he is not a people but a 'religious community,' though of a special sort.

And this is the first great lie.

In order to carry on his existence as a parasite on other peoples, he is forced to deny his inner nature. The more intelligent the

individual Jew is, the more he will succeed in this deception.
Indeed, things can go so far that large parts of the host people
will end by seriously believing that the Jew is really a Frenchman or an Englishman, a German or an Italian, though of a
special religious faith. Especially state authorities, which always
seem animated by the historical fraction of wisdom, most easily
fall a victim to this infinite deception. Independent thinking
sometimes seems to these circles a true sin against holy advancement, so that we may not be surprised if even today a Bavarian
state ministry, for example, still has not the faintest idea that
the Jews are members of a *people* and not of a *'religion'* though a
glance at the Jew's own newspapers should indicate this even to
the most modest mind. The *Jewish Echo* is not yet an official
organ, of course, and consequently is unauthoritative as far as
the intelligence of one of these government potentates is concerned.

The Jew has always been a people with definite racial characteristics and never a religion; only in order to get ahead he
early sought for a means which could distract unpleasant attention from his person. And what would have been more expedient
and at the same time more innocent than the 'embezzled' concept of a religious community? For here, too, everything is borrowed or rather stolen. Due to his own original special nature,
the Jew cannot possess a religious institution, if for no other
reason because he lacks idealism in any form, and hence belief
in a hereafter is absolutely foreign to him. And a religion in the
Aryan sense cannot be imagined which lacks the conviction of
survival after death in some form. Indeed, the Talmud is not a
book to prepare a man for the hereafter, but only for a practical
and profitable life in this world.

The Jewish religious doctrine consists primarily in prescriptions
for keeping the blood of Jewry pure and for regulating the relation
of Jews among themselves, but even more with the rest of the
world; in other words, with non-Jews. But even here it is by
no means ethical problems that are involved, but extremely
modest economic ones. Concerning the moral value of Jewish

religious instruction, there are today and have been at all times rather exhaustive studies (not by Jews; the drivel of the Jews themselves on the subject is, of course, adapted to its purpose) which make this kind of religion seem positively monstrous according to Aryan conceptions. The best characterization is provided by the product of this religious education, the Jew himself. His life is only of this world, and his spirit is inwardly as alien to true Christianity as his nature two thousand years previous was to the great founder of the new doctrine. Of course, the latter made no secret of his attitude toward the Jewish people, and when necessary he even took to the whip to drive from the temple of the Lord this adversary of all humanity, who then as always saw in religion nothing but an instrument for his business existence. In return, Christ was nailed to the cross, while our present-day party Christians debase themselves to begging for Jewish votes at elections and later try to arrange political swindles with atheistic Jewish parties — and this against their own nation.

On this first and greatest lie, that the Jews are not a race but a religion, more and more lies are based in necessary consequence. Among them is the lie with regard to the language of the Jew. For him it is not a means for expressing his thoughts, but a means for concealing them. When he speaks French, he thinks Jewish, and while he turns out German verses, in his life he only expresses the nature of his nationality. As long as the Jew has not become the master of the other peoples, he must speak their languages whether he likes it or not, but as soon as they became his slaves, they would all have to learn a universal language (Esperanto, for instance!), so that by this additional means the Jews could more easily dominate them!

To what an extent the whole existence of this people is based on a continuous lie is shown incomparably by the *Protocols of the Wise Men of Zion*, so infinitely hated by the Jews. They are based on a forgery, the *Frankfurter Zeitung* moans and screams once every week: the best proof that they are authentic. What many Jews may do unconsciously is here consciously exposed.

And that is what matters. It is completely indifferent from what Jewish brain these disclosures originate; the important thing is that with positively terrifying certainty they reveal the nature and activity of the Jewish people and expose their inner contexts as well as their ultimate final aims. The best criticism applied to them, however, is reality. Anyone who examines the historical development of the last hundred years from the standpoint of this book will at once understand the screaming of the Jewish press. For once this book has become the common property of a people, the Jewish menace may be considered as broken.

*   *   *

The best way to know the Jew is to study the road which he has taken within the body of other peoples in the course of the centuries. It suffices to follow this up in only one example, to arrive at the necessary realizations. As his development has always and at all times been the same, just as that of the peoples corroded by him has also been the same, it is advisable in such an examination to divide his development into definite sections which in this case for the sake of simplicity I designate alphabetically. The first Jews came to ancient Germany in the course of the advance of the Romans, and as always they came as merchants. In the storms of the migrations, however, they seem to have disappeared again, and thus the time of the first Germanic state formation may be viewed as the beginning of a new and this time lasting Jewification of Central and Northern Europe. A development set in which has always been the same or similar wherever the Jews encountered Aryan peoples.

*   *   *

(a) With the appearance of the first fixed settlement, the Jew is suddenly 'at hand.' He comes as a merchant and at first attaches little importance to the concealment of his nationality. He is still a Jew, partly perhaps among other reasons because the outward racial difference between himself and the host people is too great, his linguistic knowledge still too small, and the cohesion of the host people too sharp for him to dare to try to appear as anything else than a foreign merchant. With his dexterity and the inexperience of his host people, the retention of his character as a Jew represents no disadvantage for him, but rather an advantage; the stranger is given a friendly reception.

(b) Gradually he begins slowly [1] to become active in economic life, not as a producer, but exclusively as a middleman. With his thousand-year-old mercantile dexterity he is far superior to the still helpless, and above all boundlessly honest, Aryans, so that in a short time commerce threatens to become his monopoly. He begins to lend money and as always at usurious interest. As a matter of fact, he thereby introduces interest. The danger of this new institution is not recognized at first, but because of its momentary advantages is even welcomed.

(c) The Jew has now become a steady resident; that is, he settles special sections of the cities and villages and more and more constitutes a state within a state. He regards commerce as well as all financial transactions as his own special privilege which he ruthlessly exploits.

(d) Finance and commerce have become his complete monopoly. His usurious rates of interest finally arouse resistance, the rest of his increasing effrontery indignation, his wealth envy. The cup is full to overflowing when he draws the soil into the sphere of his commercial objects and degrades it to the level of a commodity to be sold or rather traded. Since he himself never cultivates the soil, but regards it only as a property to be exploited on which the peasant can well remain, though amid the most miserable extortions on the part of his new master, the aversion against him gradually increases to open hatred. His blood-sucking tyranny becomes so great that excesses against him occur. People begin to look at the foreigner more and more

---

[1] *'Allmählich beginnt er sich langsam...'*

closely and discover more and more repulsive traits and characteristics in him until the cleft becomes unbridgeable.

At times of the bitterest distress, fury against him finally breaks out, and the plundered and ruined masses begin to defend themselves against the scourge of God. In the course of a few centuries they have come to know him, and now they feel that the mere fact of his existence is as bad as the plague.

(e) Now the Jew begins to reveal his true qualities. With repulsive flattery he approaches the governments, puts his money to work, and in this way always manages to secure new license to plunder his victims. Even though the rage of the people sometimes flares high against the eternal blood-sucker, it does not in the least prevent him from reappearing in a few years in the place he had hardly left and beginning the old life all over again. No persecution can deter him from his type of human exploitation, none can drive him away; after every persecution he is back again in a short time, and just the same as before.

To prevent the very worst, at least, the people begin to withdraw the soil from his usurious hands by making it legally impossible for him to acquire soil.

(f) Proportionately as the power of the princes begins to mount, he pushes closer and closer to them. He begs for 'patents' and 'privileges,' which the lords, always in financial straits, are glad to give him for suitable payment. However much this may cost him, he recovers the money he has spent in a few years through interest and compound interest. A true blood-sucker that attaches himself to the body of the unhappy people and cannot be picked off until the princes themselves again need money and with their own exalted hand tap off the blood he has sucked from them.

This game is repeated again and again, and in it the rôle of the so-called 'German princes' is just as miserable as that of the Jews themselves. These lords were really God's punishment for their beloved peoples and find their parallels only in the various ministers of the present time.

It is thanks to the German princes that the German nation

was unable to redeem itself for good from the Jewish menace. In this, too, unfortunately, nothing changed as time went on; all they obtained from the Jew was the thousandfold reward for the sins they had once committed against their peoples. They made a pact with the devil and landed in hell.

(g) And so, his ensnarement of the princes leads to their ruin. Slowly but surely their relation to the peoples loosens in the measure in which they cease to serve the people's interests and instead become mere exploiters of their subjects. The Jew well knows what their end will be and tries to hasten it as much as possible. He himself adds to their financial straits by alienating them more and more from their true tasks, by crawling around them with the vilest flattery, by encouraging them in vices, and thus making himself more and more indispensable to them. With his deftness, or rather unscrupulousness, in all money matters he is able to squeeze, yes, to grind, more and more money out of the plundered subjects, who in shorter and shorter intervals go the way of all flesh. Thus every court has its 'court Jew' — as the monsters are called who torment the 'beloved people' to despair and prepare eternal pleasures for the princes. Who then can be surprised that these ornaments of the human race ended up by being ornamented, or rather decorated, in the literal sense, and rose to the hereditary nobility, helping not only to make this institution ridiculous, but even to poison it?

Now, it goes without saying, he can really make use of his position for his own advancement.

Finally he needs only to have himself baptized to possess himself of all the possibilities and rights of the natives of the country. Not seldom he concludes this deal to the joy of the churches over the son they have won and of Israel over the successful swindle.

(h) Within Jewry a change now begins to take place. Up till now they have been Jews; that is, they attach no importance to appearing to be something else, which they were unable to do, anyway, because of the very distinct racial characteristics on both sides. At the time of Frederick the Great it still entered no one's head to regard the Jew as anything else but a 'foreign' people, and Goethe was still horrified at the thought that in future marriage between Christians and Jews would no longer be forbidden by law. And Goethe, by God, was no reactionary, let alone a helot;[1] what spoke out of him was only the voice of the blood and of reason. Thus — despite all the shameful actions of the courts — the people instinctively saw in the Jew a foreign element and took a corresponding attitude toward him.

But now all this was to change. In the course of more than a thousand years he has learned the language of the host people to such an extent that he now thinks he can venture in future to emphasize his Judaism less and place his 'Germanism' more in the foreground; for ridiculous, nay, insane, as it may seem at first, he nevertheless has the effrontery to turn 'Germanic,' in this case a 'German.' With this begins one of the most infamous deceptions that anyone could conceive of. Since of Germanism he possesses really nothing but the art of stammering its language — and in the most frightful way — but apart from this has never mixed with the Germans, his whole Germanism rests on the language alone. Race, however, does not lie in the language, but exclusively in the blood, which no one knows better than the Jew, who attaches very little importance to the preservation of his language, but all importance to keeping his blood pure. A man can change his language without any trouble — that is, he can use another language; but in his new language he will express the old ideas; his inner nature is not changed. This is best shown by the Jew who can speak a thousand languages and nevertheless remains a Jew. His traits of character have remained the same, whether two thousand years ago as a grain dealer in Ostia, speaking Roman, or whether as a flour profiteer of today, jabbering German with a Jewish accent. It is always the same Jew. That this obvious fact is not understood by a ministerial secretary or higher police official is also self-evident, for there is scarcely any creature with less instinct and intelligence running around in the world today than these servants of our present model state authority.

[1] Corrected in second edition to 'zealot.'

The reason why the Jew decides suddenly to become a 'German' is obvious. He feels that the power of the princes is slowly tottering and therefore tries at an early time to get a platform beneath his feet. Furthermore, his financial domination of the whole economy has advanced so far that without possession of all 'civil' rights he can no longer support the gigantic edifice, or at any rate, no further increase of his influence is possible. And he desires both of these; for the higher he climbs, the more alluring his old goal that was once promised him rises from the veil of the past, and with feverish avidity his keenest minds see the dream of world domination tangibly approaching. And so his sole effort is directed toward obtaining full possession of 'civil' rights.

This is the reason for his emancipation from the ghetto.

(i) So from the court Jew there gradually develops the people's Jew, which means, of course: the Jew remains as before in the entourage of the high lords; in fact, he tries to push his way even more into their circle; but at the same time another part of his race makes friends with the 'beloved people.' If we consider how greatly he has sinned against the masses in the course of the centuries, how he has squeezed and sucked their blood again and again; if furthermore, we consider how the people gradually learned to hate him for this, and ended up by regarding his existence as nothing but a punishment of Heaven for the other peoples, we can understand how hard this shift must be for the Jew. Yes, it is an arduous task suddenly to present himself to his flayed victims as a 'friend of mankind.'

First, therefore, he goes about making up to the people for his previous sins against them. He begins his career as the 'benefactor' of mankind. Since his new benevolence has a practical foundation, he cannot very well adhere to the old Biblical recommendation, that the left hand should not know what the right hand giveth; no, whether he likes it or not, he must reconcile himself to letting as many people as possible know how deeply he feels the sufferings of the masses and all the sacrifices that he himself is making to combat them. With this 'modesty' which is inborn in him, he blares out his merits to the rest of the world

until people really begin to believe in them. Anyone who does
not believe in them is doing him a bitter injustice. In a short time
he begins to twist things around to make it look as if all the injustice in the world had always been done to him and not the
other way around. The very stupid believe this and then they
just can't help but pity the poor 'unfortunate.'

In addition, it should be remarked here that the Jew, despite
all his love of sacrifice, naturally never becomes personally im-
poverished. He knows how to manage; sometimes, indeed, his
charity is really comparable to fertilizer, which is not strewn on
the field for love of the field, but with a view to the farmer's own
future benefit. In any case, everyone knows in a comparatively
short time that the Jew has become a 'benefactor and friend
of mankind.' What a strange transformation!

But what is more or less taken for granted in others arouses
the greatest astonishment and in many distinct admiration for
this very reason. So it happens that he gets much more credit
for every such action than the rest of mankind, in whom it is
taken for granted.

But even more: all at once the Jew also becomes liberal and
begins to rave about the necessary progress of mankind.

Slowly he makes himself the spokesman of a new era.

Also, of course, he destroys more and more thoroughly the
foundations of any economy that will really benefit the people.
By way of stock shares he pushes his way into the circuit of national production which he turns into a purchasable or rather
tradable object, thus robbing the enterprises of the foundations
of a personal ownership. Between employer and employee there
arises that inner estrangement which later leads to political class
division.

Finally, the Jewish influence on economic affairs grows with
terrifying speed through the stock exchange. He becomes
the owner, or at least the controller, of the national labor
force.

To strengthen his political position he tries to tear down the
racial and civil barriers which for a time continue to restrain him

at every step. To this end he fights with all the tenacity innate
in him for religious tolerance — and in Freemasonry, which has
succumbed to him completely, he has an excellent instrument
with which to fight for his aims and put them across. The governing circles and the higher strata of the political and economic
bourgeoisie are brought into his nets by the strings of Freemasonry, and never need to suspect what is happening.

Only the deeper and broader strata of the people as such, or
rather that class which is beginning to wake up and fight for its
rights and freedom, cannot yet be sufficiently taken in by these
methods. But this is more necessary than anything else; for the
Jew feels that the possibility of his rising to a dominant rôle
exists only if there is someone ahead of him to clear the way; and
this someone he thinks he can recognize in the bourgeoisie, in
their broadest strata in fact. The glovemakers and linen weavers,
however, cannot be caught in the fine net of Freemasonry; no, for
them coarser but no less drastic means must be employed. Thus,
Freemasonry is joined by a second weapon in the service of the
Jews: the *press*. With all his perseverance and dexterity he seizes
possession of it. With it he slowly begins to grip and ensnare, to
guide and to push all public life, since he is in a position to create
and direct that power which, under the name of 'public opinion,'
is better known today than a few decades ago.

In this he always represents himself personally as having an
infinite thirst for knowledge, praises all progress, mostly, to be
sure, the progress that leads to the ruin of others; for he judges
all knowledge and all development only according to its possibilities for advancing his nation, and where this is lacking, he is
the inexorable mortal enemy of all light, a hater of all true culture.
He uses all the knowledge he acquires in the schools of other peoples, exclusively for the benefit of his race.

And this nationality he guards as never before. While he seems
to overflow with 'enlightenment,' 'progress,' 'freedom,' 'humanity,' etc., he himself practices the severest segregation of his race.
To be sure, he sometimes palms off his women on influential
Christians, but as a matter of principle he always keeps his male

line pure. He poisons the blood of others, but preserves his own. The Jew almost never marries a Christian woman; it is the Christian who marries a Jewess. The bastards, however, take after the Jewish side. Especially a part of the high nobility degenerates completely. The Jew is perfectly aware of this, and therefore systematically carries on this mode of 'disarming' the intellectual leader class of his racial adversaries. In order to mask his activity and lull his victims, however, he talks more and more of the equality of all men without regard to race and color. The fools begin to believe him.

Since, however, his whole being still has too strong a smell of the foreign for the broad masses of the people in particular to fall readily into his nets, he has his press give a picture of him which is as little in keeping with reality as conversely it serves his desired purpose. His comic papers especially strive to represent the Jews as a harmless little people, with their own peculiarities, of course — like other peoples as well — but even in their gestures, which seem a little strange, perhaps, giving signs of a possibly ludicrous, but always thoroughly honest and benevolent, soul. And the constant effort is to make him seem almost more 'insignificant' than *dangerous*.

His ultimate goal in this stage is the victory of 'democracy,' or, as he understands it: the rule of parliamentarianism. It is most compatible with his requirements; for it excludes the personality — and puts in its place the majority characterized by stupidity, incompetence, and last but not least, cowardice.

The final result will be the overthrow of the monarchy, which is now sooner or later bound to occur.

(j) The tremendous economic development leads to a change in the social stratification of the people. The small craftsman slowly dies out, and as a result the worker's possibility of achieving an independent existence becomes rarer and rarer; in consequence the worker becomes visibly proletarianized. There arises the industrial 'factory worker' whose most essential characteristic is to be sought in the fact that he hardly ever is in a position to found an existence of his own in later life. He is propertyless in the truest sense of the word. His old age is a torment and can scarcely be designated as living.

Once before, a similar situation was created, which pressed urgently for a solution and also found one. The peasants and artisans had slowly been joined by the officials and salaried workers — particularly of the state — as a new class. They, too, were propertyless in the truest sense of the word. The state finally found a way out of this unhealthy condition by assuming the care of the state employee who could not himself provide for his old age; it introduced the pension. Slowly, more and more enterprises followed this example, so that nearly every regularly employed brain-worker draws a pension in later life, provided the concern he works in has achieved or surpassed a certain size. Only by safeguarding the state official in his old age could he be taught the selfless devotion to duty which in the pre-War period was the most eminent quality of German officialdom.

In this way a whole class that had remained propertyless was wisely snatched away from social misery and articulated with the body of the people.

Now this question again, and this time on a much larger scale, faced the state and the nation. More and more masses of people, numbering millions, moved from peasant villages to the larger cities to earn their bread as factory workers in the newly established industries. The working and living conditions of the new class were more than dismal. If nothing else, the more or less mechanical transference of the old artisan's or even peasant's working methods to the new form was by no means suitable. The work done by these men could not be compared with the exertions which the industrial factory worker has to perform. In the old handicraft, this may not have been very important, but in the new working methods it was all the more so. The formal transference of the old working hours to the industrial large-scale enterprise was positively catastrophic, for the actual work done before was but little in view of the absence of our present intensive working methods. Thus, though previously the fourteen- or even fifteen-hour working day had been bearable, it certainly ceased to be bearable at a time when every minute was exploited to the fullest. The result of this senseless transference of the old working hours to the new industrial activity was really unfortunate in two respects: the worker's health was undermined and his faith in a higher justice destroyed. To this finally was added the miserable wages on the one hand and the employer's correspondingly and obviously so vastly superior position on the other.

In the country there could be no social question, since master and hired hand did the same work and above all ate out of the same bowls. But this, too, changed.

The separation of worker and employer now seems complete in all fields of life. How far the inner Judaization of our people has progressed can be seen from the small respect, if not contempt, that is accorded to manual labor. This is not German. It took the foreignization of our life, which was in truth a Jewification, to transform the old respect for manual work into a certain contempt for all physical labor.

Thus, there actually comes into being a new class enjoying very little respect, and one day the question must arise whether the nation would possess the strength to articulate the new class into general society, or whether the social difference would broaden into a classlike cleavage.

But one thing is certain: the new class did not count the worst elements in its ranks, but on the contrary definitely the most energetic elements. The overrefinements of so-called culture had not yet exerted their disintegrating and destructive effects. The broad mass of the new class was not yet infected with the poison of pacifist weakness; it was robust and if necessary even brutal.

While the bourgeoisie is not at all concerned about this all-important question, but indifferently lets things slide, the Jew seizes the unlimited opportunity it offers for the future; while on the one hand he organizes capitalistic methods of human exploitation to their ultimate consequence, he approaches the very victims of his spirit and his activity and in a short time becomes the leader of their struggle against himself. 'Against himself' is only figuratively speaking; for the great master of lies understands as always how to make himself appear to be the pure one and to load the blame on others. Since he has the gall to lead the masses, it never even enters their heads that this might be the most infamous betrayal of all times.

And yet it was.

Scarcely has the new class grown out of the general economic shift than the Jew, clearly and distinctly, realizes that it can open the way for his own further advancement. First, he used the bourgeoisie as a battering-ram against the feudal world, then the worker against the bourgeois world. If formerly he knew how to swindle his way to civil rights in the shadow of the bourgeoisie, now he hopes to find the road to his own domination in the worker's struggle for existence.

From now on the worker has no other task but to fight for the future of the Jewish people. Unconsciously he is harnessed to the service of the power which he thinks he is combating. He is seemingly allowed to attack capital, and this is the easiest way of making him fight for it. In this the Jew keeps up an outcry against international capital and in truth he means the national economy which must be demolished in order that the international stock exchange can triumph over its dead body.

Here the Jew's procedure is as follows:

He approaches the worker, simulates pity with his fate, or even indignation at his lot of misery and poverty, thus gaining his confidence. He takes pains to study all the various real or imaginary hardships of his life — and to arouse his longing for a change in such an existence. With infinite shrewdness he fans the need for social justice, somehow slumbering in every Aryan man, into hatred against those who have been better favored by fortune, and thus gives the struggle for the elimination of social evils a very definite philosophical stamp. He establishes the Marxist doctrine.

By presenting it as inseparably bound up with a number of socially just demands, he promotes its spread and conversely the aversion of decent people to fulfill demands which, advanced in such form and company, seem from the outset unjust and impossible to fulfill. For under this cloak of purely social ideas truly diabolic purposes are hidden, yes, they are publicly proclaimed with the most insolent frankness. This theory represents an inseparable mixture of reason and human madness, but always in such a way that only the lunacy can become reality and never the reason. By the categorical rejection of the personality and hence of the nation and its racial content, it destroys the elementary foundations of all human culture which is dependent on just these factors. This is the true inner kernel of the Marxist philosophy in so far as this figment of a criminal brain can be designated as a 'philosophy.' With the shattering of the personality and the race, the essential obstacle is removed to the domination of the inferior being — and this is the Jew.

Precisely in political and economic madness lies the sense of this doctrine. For this prevents all truly intelligent people from entering its service, while those who are intellectually less active and poorly educated in economics hasten to it with flying colors. The intellectuals for this movement — for even this movement needs intellectuals for its existence — are 'sacrificed' by the Jew from his own ranks.

Thus there arises a pure movement entirely of manual workers under Jewish leadership, apparently aiming to improve the situation of the worker, but in truth planning the enslavement and with it the destruction of all non-Jewish peoples.

The general pacifistic paralysis of the national instinct of self-preservation begun by Freemasonry in the circles of the so-called intelligentsia is transmitted to the broad masses and above all to the bourgeoisie by the activity of the big papers which today are always Jewish. Added to these two weapons of disintegration comes a third and by far the most terrible, the organization of brute force. As a shock and storm troop, Marxism is intended to finish off what the preparatory softening up with the first two weapons has made ripe for collapse.

Here we have teamwork that is positively brilliant — and we need really not be surprised if in confronting it those very institutions which always like to represent themselves as the pillars

of a more or less legendary state authority hold up least. It is in
our high and highest state officialdom that the Jew has at all times
(aside from a few exceptions) found the most compliant abettor of
his work of disintegration. Cringing submissiveness to superiors
and high-handed arrogance to inferiors distinguish this class to
the same degree as a narrow-mindedness that often cries to high
Heaven and is only exceeded by a self-conceit that is sometimes
positively amazing.

And these are qualities that the Jew needs in our authorities
and loves accordingly.

The practical struggle which now begins, sketched in broad
outlines, takes the following course:

In keeping with the ultimate aims of the Jewish struggle,
which are not exhausted in the mere economic conquest of the
world, but also demand its political subjugation, the Jew divides
the organization of his Marxist world doctrine into two halves
which, apparently separate from one another, in truth form an
inseparable whole: the political and the trade-union movement.

The trade-union movement does the recruiting. In the hard
struggle for existence which the worker must carry on, thanks
to the greed and shortsightedness of many employers, it offers
him aid and protection, and thus the possibility of winning better
living conditions. If, at a time when the organized national community, the state, concerns itself with him little or not at all, the
worker does not want to hand over the defense of his vital human
rights to the blind caprice of people who in part have little sense
of responsibility and are often heartless to boot, he must take
their defense into his own hands. In exact proportion as the so-
called national bourgeoisie, blinded by financial interests, sets
the heaviest obstacles in the path of this struggle for existence
and not only resists all attempts at shortening the inhumanly
long working day, abolishing child labor, safeguarding and protecting the woman, improving sanitary conditions in the work-
shops and homes, but often actually sabotages them, the shrewder
Jew takes the oppressed people under his wing. Gradually he be-

comes the leader of the trade-union movement, all the more easily as he is not interested in really eliminating social evils in an honest sense, but only in training an economic storm troop, blindly devoted to him, with which to destroy the national economic independence. For while the conduct of a healthy social policy will consistently move between the aims of preserving the national health on the one hand and safeguarding an independent national economy on the other, for the Jew in his struggle these two criteria not only cease to exist, but their elimination, among other things, is his life goal. He desires, not the preservation of an independent national economy, but its destruction. Consequently, no pangs of conscience can prevent him as a leader of the trade-union movement from raising demands which not only overshoot the goal, but whose fulfillment is either impossible for practical purposes or means the ruin of the national economy. Moreover, he does not want to have a healthy, sturdy race before him, but a rickety herd capable of being subjugated. This desire again permits him to raise demands of the most senseless kind whose practical fulfillment he himself knows to be impossible and which, therefore, could not lead to any change in things, but at most to a wild incitement of the masses. And that is what he is interested in and not a true and honest improvement of social conditions.

Hence the Jewish leadership in trade-union affairs remains uncontested until an enormous work of enlightenment influences the broad masses and sets them right about their never-ending misery, or else the state disposes of the Jew and his work. For as long as the insight of the masses remains as slight as now and the state as indifferent as today, these masses will always be first to follow the man who in economic matters offers the most shameless promises. And in this the Jew is a master. For in his entire activity he is restrained by no moral scruples!

And so he inevitably drives every competitor in this sphere from the field in a short time. In keeping with all his inner rapacious brutality, he at once teaches the trade-union movement the most brutal use of violence. If anyone by his intelligence resists the Jewish lures, his defiance and understanding are broken by terror. The success of such an activity is enormous.

Actually the Jew by means of the trade union, which could be a blessing for the nation, shatters the foundations of the national economy.

Parallel with this, the political organization advances.

It plays hand in glove with the trade-union movement, for the latter prepares the masses for political organization, in fact, lashes them into it with violence and coercion. Furthermore, it is the permanent financial source from which the political organization feeds its enormous apparatus. It is the organ controlling the political activity of the individual and does the pandering in all big demonstrations of a political nature. In the end it no longer comes out for political interests at all, but places its chief instrument of struggle, the cessation of work in the form of a mass and general strike, in the service of the political idea.

By the creation of a press whose content is adapted to the intellectual horizon of the least educated people, the political and trade-union organization finally obtains the agitational institution by which the lowest strata of the nation are made ripe for the most reckless acts. Its function is not to lead people out of the swamp of a base mentality to a higher stage, but to cater to their lowest instincts. Since the masses are as mentally lazy as they are sometimes presumptuous, this is a business as speculative as it is profitable.

It is this press, above all, which wages a positively fanatical and slanderous struggle, tearing down everything which can be regarded as a support of national independence, cultural elevation, and the economic independence of the nation.

Above all, it hammers away at the characters of all those who will not bow down to the Jewish presumption to dominate, or whose ability and genius in themselves seem a danger to the Jew. For to be hated by the Jew it is not necessary to combat him; no, it suffices if he suspects that someone might even conceive the idea of combating him some time or that on the strength of his superior genius he is an augmenter of the power and greatness of a nationality hostile to the Jew.

His unfailing instinct in such things scents the original soul [1] in everyone, and his hostility is assured to anyone who is not spirit of his spirit. Since the Jew is not the attacked but the attacker, not only anyone who attacks passes as his enemy, but also anyone who resists him. But the means with which he seeks to break such reckless but upright souls is not honest warfare, but lies and slander.

Here he stops at nothing, and in his vileness he becomes so gigantic that no one need be surprised if among our people the personification of the devil as the symbol of all evil assumes the living shape of the Jew.

The ignorance of the broad masses about the inner nature of the Jew, the lack of instinct and narrow-mindedness of our upper classes, make the people an easy victim for this Jewish campaign of lies.

While from innate cowardice the upper classes turn away from a man whom the Jew attacks with lies and slander, the broad masses from stupidity or simplicity believe everything. The state authorities either cloak themselves in silence or, what usually happens, in order to put an end to the Jewish press campaign, they persecute the unjustly attacked, which, in the eyes of such an official ass, passes as the preservation of state authority and the safeguarding of law and order.

Slowly fear of the Marxist weapon of Jewry descends like a nightmare on the mind and soul of decent people.

They begin to tremble before the terrible enemy and thus have become his final victim.

The Jew's domination in the state seems so assured that now not only can he call himself a Jew again, but he ruthlessly admits his ultimate national and political designs. A section of his race openly owns itself to be a foreign people, yet even here they lie. For while the Zionists try to make the rest of the world believe that the national consciousness of the Jew finds its satisfaction

[1] *'die ursprüngliche Seele.'*

in the creation of a Palestinian state, the Jews again slyly dupe the dumb *Goyim*.[1] It doesn't even enter their heads to build up a Jewish state in Palestine for the purpose of living there; all they want is a central organization for their international world swindle, endowed with its own sovereign rights and removed from the intervention of other states: a haven for convicted scoundrels and a university for budding crooks.

It is a sign of their rising confidence and sense of security that at a time when one section is still playing the German, Frenchman, or Englishman, the other with open effrontery comes out as the Jewish race.

How close they see approaching victory can be seen by the hideous aspect which their relations with the members of other peoples takes on.

With satanic joy in his face, the black-haired Jewish youth lurks in wait for the unsuspecting girl whom he defiles with his blood, thus stealing her from her people. With every means he tries to destroy the racial foundations of the people he has set out to subjugate. Just as he himself systematically ruins women and girls, he does not shrink back from pulling down the blood barriers for others, even on a large scale. It was and it is Jews who bring the Negroes into the Rhineland, always with the same secret thought and clear aim of ruining the hated white race by the necessarily resulting bastardization, throwing it down from its cultural and political height, and himself rising to be its master.

For a racially pure people which is conscious of its blood can never be enslaved by the Jew. In this world he will forever be master over bastards and bastards alone.

And so he tries systematically to lower the racial level by a continuous poisoning of individuals.

And in politics he begins to replace the idea of democracy by the dictatorship of the proletariat.

In the organized mass of Marxism he has found the weapon which lets him dispense with democracy and in its stead allows

[1] Yiddish for Gentiles.

him to subjugate and govern the peoples with a dictatorial and brutal fist.

He works systematically for revolutionization in a twofold sense: economic and political.

Around peoples who offer too violent a resistance to attack from within he weaves a net of enemies, thanks to his international influence, incites them to war, and finally, if necessary, plants the flag of revolution on the very battlefields.

In economics he undermines the states until the social enterprises which have become unprofitable are taken from the state and subjected to his financial control.

In the political field he refuses the state the means for its self-preservation, destroys the foundations of all national self-maintenance and defense, destroys faith in the leadership, scoffs at its history and past, and drags everything that is truly great into the gutter.

Culturally he contaminates art, literature, the theater, makes a mockery of natural feeling, overthrows all concepts of beauty and sublimity, of the noble and the good, and instead drags men down into the sphere of his own base nature.

Religion is ridiculed, ethics and morality represented as outmoded, until the last props of a nation in its struggle for existence in this world have fallen.

(e) [1] Now begins the great last revolution. In gaining political power the Jew casts off the few cloaks that he still wears. The democratic people's Jew becomes the blood-Jew and tyrant over peoples. In a few years he tries to exterminate the national intelligentsia and by robbing the peoples of their natural intellectual leadership makes them ripe for the slave's lot of permanent subjugation.

The most frightful example of this kind is offered by Russia, where he killed or starved about thirty million people with positively fanatical savagery, in part amid inhuman tortures, in order to give a gang of Jewish journalists and stock exchange bandits domination over a great people.

[1] The writer has lost track of his letters.   Second edition has (i).

The end is not only the end of the freedom of the peoples oppressed by the Jew, but also the end of this parasite upon the nations. After the death of his victim, the vampire sooner or later dies too.

*   *   *

If we pass all the causes of the German collapse in review, the ultimate and most decisive remains the failure to recognize the racial problem and especially the Jewish menace.

The defeats on the battlefield in August, 1918, would have been child's play to bear. They stood in no proportion to the victories of our people. It was not they that caused our downfall; no, it was brought about by that power which prepared these defeats by systematically over many decades robbing our people of the political and moral instincts and forces which alone make nations capable and hence worthy of existence.

In heedlessly ignoring the question of the preservation of the racial foundations of our nation, the old Reich disregarded the sole right which gives life in this world. Peoples which bastardize themselves, or let themselves be bastardized, sin against the will of eternal Providence, and when their ruin is encompassed by a stronger enemy it is not an injustice done to them, but only the restoration of justice. If a people no longer wants to respect the Nature-given qualities of its being which root in its blood, it has no further right to complain over the loss of its earthly existence.

Everything on this earth is capable of improvement. Every defeat can become the father of a subsequent victory, every lost war the cause of a later resurgence, every hardship the fertilization of human energy, and from every oppression the forces for a new spiritual rebirth can come — as long as the blood is preserved pure.

The lost purity of the blood alone destroys inner happiness forever, plunges man into the abyss for all time, and the consequences can never more be eliminated from body and spirit.

Only by examining and comparing all other problems of life in the light of this one question shall we see how absurdly petty they are by this standard. They are all limited in time — but the question of preserving or not preserving the purity of the blood will endure as long as there are men.

All really significant symptoms of decay of the pre-War period can in the last analysis be reduced to racial causes.

Whether we consider questions of general justice or cankers of economic life, symptoms of cultural decline or processes of political degeneration, questions of faulty schooling or the bad influence exerted on grown-ups by the press, etc., everywhere and always it is fundamentally the disregard of the racial needs of our own people or failure to see a foreign racial menace.

And that is why all attempts at reform, all works for social relief and political exertions, all economic expansion and every apparent increase of intellectual knowledge were futile as far as their results were concerned. The nation, and the organism which enables [1] and preserves its life on this earth, the state, did not grow inwardly healthier, but obviously languished more and more. All the illusory prosperity of the old Reich could not hide its inner weakness, and every attempt really to strengthen the Reich failed again and again, due to disregarding the most important question.

It would be a mistake to believe that the adherents of the various political tendencies which were tinkering around on the German national body — yes, even a certain section of the leaders — were bad or malevolent men in themselves. Their activity was condemned to sterility only because the best of them saw at most the forms of our general disease and tried to combat them, but blindly ignored the virus. Anyone who systematically follows the old Reich's line of political development is bound to arrive, upon calm examination, at the realization that even at the time of the unification, hence the rise of the German nation, the inner decay was already in full swing, and that despite all ap-

[1] *'Die Nation und ihr das Leben auf dieser Erde befähigender und erhaltender Organismus.'*

parent political successes and despite increasing economic wealth, the general situation was deteriorating from year to year. If nothing else, the elections for the Reichstag announced, with their outward swelling of the Marxist vote, the steadily approaching inward and hence also outward collapse. All the successes of the so-called bourgeois parties were worthless, not only because even with so-called bourgeois electoral victories they were unable to halt the numerical growth of the Marxist flood, but because they themselves above all now bore the ferments of decay in their own bodies. Without suspecting it, the bourgeois world itself was inwardly infected with the deadly poison of Marxist ideas and its resistance often sprang more from the competitor's envy of ambitious leaders than from a fundamental rejection of adversaries determined to fight to the utmost. In these long years there was only one who kept up an imperturbable, unflagging fight, and this was the *Jew*. His Star of David [1] rose higher and higher in proportion as our people's will for self-preservation vanished.

Therefore, in August, 1914, it was not a people resolved to attack which rushed to the battlefield; no, it was only the last flicker of the national instinct of self-preservation in face of the progressing pacifist-Marxist paralysis of our national body. Since even in these days of destiny, our people did not recognize the inner enemy, all outward resistance was in vain and Providence did not bestow her reward on the victorious sword, but followed the law of eternal retribution.

On the basis of this inner realization, there took form in our new movement the leading principles as well as the tendency, which in our conviction were alone capable, not only of halting the decline of the German people, but of creating the granite foundation upon which some day a state will rest which represents, not an alien mechanism of economic concerns and interests, but a national organism:

A Germanic State of the
German Nation

---

[1] Typical Hitlerian metaphor. The Star of David, it will be remembered, is not a star, but a shield.

The First Period of Development of the National Socialist German Workers' Party

I<small>F AT THE END</small> of this volume I describe the first period in the development of our movement and briefly discuss a number of questions it raises, my aim is not to give a dissertation on the spiritual aims of the movement. The aims and tasks of the new movement are so gigantic that they can only be treated in a special volume. In a second volume, therefore, I shall discuss the programmatic foundations of the movement in detail and attempt to draw a picture of what we conceive of under the word 'state.' By 'us' I mean all the hundreds of thousands who fundamentally long for the same thing without as individuals finding the words to describe outwardly [1] what they inwardly visualize; for the noteworthy fact about all reforms is that at first they possess but a single champion yet many million supporters. Their aim has often been for centuries the inner longing of hundreds of thousands, until one man stands up to proclaim such a general will, and as a standard-bearer guides the old longing to victory in the form of the new idea.

The fact that millions bear in their hearts the desire for a basic change in the conditions obtaining today proves the deep discontent under which they suffer. It expresses itself in thousandfold manifestations, with one in despair and hopelessness, with another in ill will, anger, and indignation; with this man in indifference, and with that man in furious excesses. As witnesses

[1] 'Outwardly' omitted in second edition.

to this inner dissatisfaction we may consider those who are weary of elections as well as the many who tend to the most fanatical extreme of the Left.

The young movement was intended primarily to appeal to these last. It is not meant to constitute an organization of the contented and satisfied, but to embrace those tormented by suffering, those without peace, the unhappy and the discontented, and above all it must not swim on the surface of a national body, but strike roots deep within it.

*   *   *

In purely political terms, the following picture presented itself in 1918: a people torn into two parts. The one, by far the smaller, includes the strata of the national intelligentsia, excluding all the physically active. It is outwardly national, yet under this word can conceive of nothing but a very insipid and weak-kneed defense of so-called state interests, which in turn seem identical with dynastic interests. They attempt to fight for their ideas and aims with spiritual weapons which are as fragmentary as they are superficial, and which fail completely in the face of the enemy's brutality. With a single frightful blow this class, which only a short time before was still governing, is stretched on the ground and with trembling cowardice suffers every humiliation at the hands of the ruthless victor.

Confronting it is a second class, the broad mass of the laboring population. It is organized in more or less radical Marxist movements, determined to break all spiritual resistance by the power of violence. It does not want to be national, but consciously rejects any promotion of national interests, just as, conversely, it aids and abets all foreign oppression. It is numerically the stronger and above all comprises all those elements of the nation without which a national resurrection is unthinkable and impossible.

For in 1918 this much was clear: no resurrection of the German people can occur except through the recovery of outward power. But the prerequisites for this are not arms, as our bourgeois 'statesmen' keep prattling, but the forces of the will. The German people had more than enough arms before. They were not able to secure freedom because the energies of the national instinct of self-preservation, the will for self-preservation, were lacking. The best weapon is dead, worthless material as long as the spirit is lacking which is ready, willing, and determined to use it. Germany became defenseless, not because arms were lacking, but because the will was lacking to guard the weapon for national survival.

If today more than ever our Left politicians are at pains to point out the lack of arms as the necessary cause of their spineless, compliant, actually treasonous policy, we must answer only one thing: no, the reverse is true. Through your anti-national, criminal policy of abandoning national interests, you surrendered our arms. Now you attempt to represent the lack of arms as the underlying cause of your miserable villainy. This, like everything you do, is *lies* and falsification.

But this reproach applies just as much to the politicians on the Right. For, thanks to their miserable cowardice, the Jewish rabble that had come to power was able in 1918 to steal the nation's arms. They, too, have consequently no ground and no right to palm off our present lack of arms as the compelling ground for their wily caution (read 'cowardice'); on the contrary, our defenselessness is the consequence of their cowardice.

Consequently the question of regaining German power is not: How shall we manufacture arms? but: How shall we manufacture the spirit which enables a people to bear arms? If this spirit dominates a people, the will finds a thousand ways, every one of which ends in a weapon! But give a coward ten pistols and if attacked he will not be able to fire a single shot. And so for him they are more worthless than a knotted stick for a courageous man.

The question of regaining our people's political power is primarily a question of recovering our national instinct of self-preservation, if for no other reason because experience shows that any preparatory foreign policy, as well as any evaluation of a state as such, takes its cue less from the existing weapons than from a nation's recognized or presumed moral capacity for resistance. A nation's ability to form alliances is determined much less by dead stores of existing arms than by the visible presence of an ardent national will for self-preservation and heroic death-defying courage. For an alliance is not concluded with arms but with men. Thus, the English nation will have to be considered the most valuable ally in the world as long as its leadership and the spirit of its broad masses justify us in expecting that brutality and perseverance which is determined to fight a battle once begun to a victorious end, with every means and without consideration of time and sacrifices; and what is more, the military armament existing at any given moment does not need to stand in any proportion to that of other states.

If we understand that the resurrection of the German nation represents a question of regaining our political will for self-preservation, it is also clear that this cannot be done by winning elements which in point of will at least are already national, but only by the nationalization of the consciously anti-national masses.

A young movement which, therefore, sets itself the goal of resurrecting a German state with its own sovereignty will have to direct its fight entirely to winning the broad masses. Wretched as our so-called 'national bourgeoisie' is on the whole, inadequate as its national attitude seems, certainly from this side no serious resistance is to be expected against a powerful domestic and foreign policy in the future. Even if the German bourgeoisie, for their well-known narrow-minded and short-sighted reasons, should, as they once did toward Bismarck, maintain an obstinate attitude of passive resistance in the hour of coming liberation — an active resistance, in view of their recognized and proverbial cowardice, is never to be feared.

It is different with the masses of our internationally minded comrades. In their natural primitiveness, they are more inclined to the idea of violence, and, moreover, their Jewish leadership is more brutal and ruthless. They will crush any German resurrection just as they once broke the backbone of the German army. But above all: in this state with its parliamentary government they will, thanks to their majority in numbers, not only obstruct any national foreign policy, but also make impossible any higher estimation of the German strength, thus making us seem undesirable as an ally. For not only are we ourselves aware of the element of weakness lying in our fifteen million Marxists, democrats, pacifists, and Centrists; it is recognized even more by foreign countries, which measure the value of a possible alliance with us according to the weight of this burden. No one allies himself with a state in which the attitude of the active part of the population toward any determined foreign policy is passive, to say the least.

To this we must add the fact that the leaderships of these parties of national treason must and will be hostile to any resurrection, out of mere instinct of self-preservation. Historically it is just not conceivable that the German people could recover its former position without settling accounts with those who were the cause and occasion of the unprecedented collapse which struck our state. For before the judgment seat of posterity November, 1918, will be evaluated, not as high treason, but as treason against the fatherland.[1]

Thus, any possibility of regaining outward German independence is bound up first and foremost with the recovery of the inner unity of our people's will.

But regarded even from the purely technical point of view, the idea of an outward German liberation seems senseless as long as the broad masses are not also prepared to enter the service of this liberating idea. From the purely military angle, every officer above all will realize after a moment's thought that a

---

[1] *'Hochverrat'* (high treason) is an offense against the government or constitution; *'Landesverrat'* (treason against the fatherland) consists of military betrayal (mutiny, desertion, betrayal of secrets to the enemy, etc.), and is a far graver offense.

foreign struggle cannot be carried on with student battalions, that in addition to the brains of a people, the fists are also needed. In addition, we must bear in mind that a national defense, which is based only on the circles of the so-called intelligentsia, would squander irreplaceable treasures. The absence of the young German intelligentsia which found its death on the fields of Flanders in the fall of 1914 was sorely felt later on. It was the highest treasure that the German nation possessed and during the War its loss could no longer be made good. Not only is it impossible to carry on the struggle itself if the storming battal- :ons do not find the masses of the workers in their ranks; the technical preparations are also impracticable without the inner unity of our national will. Especially our people, doomed to languish along unarmed beneath the thousand eyes of the Versailles peace treaty, can only make technical preparations for the achievement of freedom and human independence if the army of domestic stoolpigeons is decimated down to those whose inborn lack of character permits them to betray anything and everything for the well-known thirty pieces of silver.[1] For with these we can deal. Unconquerable by comparison seem the millions who oppose the national resurrection out of political conviction — unconquerable as long as the inner cause of their opposition, the international Marxist philosophy of life, is not combated and torn out of their hearts and brains.

Regardless, therefore, from what standpoint we examine the possibility of regaining our state and national independence, whether from the standpoint of preparations in the sphere of foreign policy, from that of technical armament or that of battle itself, in every case the presupposition for everything remains the previous winning of the broad masses of our people for the idea of our national independence.

Without the recovery of our external freedom, however, any

[1] As the thought seems too complicated to be true, I cite the German:
‘ ... wenn das Heer innerer Spitzel auf diejenigen dezimiert wird, denen ange-borene Charakterlosignkeit gestattet, fur die bekannten drei Silberlinge alles und jedes zu verraten.’

internal reform, even in the most favorable case, means only the increase of our productivity as a colony. The surplus of all so-called economic improvements falls to the benefit of our international control commissions, and every social improvement at best raises the productivity of our work for them. No cultural advances will fall to the share of the German nation; they are too contingent on the political independence and dignity of our nation.

Thus, if a favorable solution of the German future requires a national attitude on the part of the broad masses of our people, this must be the highest, mightiest task of a movement whose activity is not intended to exhaust itself in the satisfaction of the moment, but which must examine all its commissions and omissions solely with a view to their presumed consequences in the future.

Thus, by 1919 we clearly realized that, as its highest aim, the new movement must first accomplish the nationalization of the masses.

From a tactical standpoint a number of demands resulted from this.

(1) To win the masses for a national resurrection, no social sacrifice is too great.

Whatever economic concessions are made to our working class today, they stand in no proportion to the gain for the entire nation if they help to give the broad masses back to their nation. Only pigheaded short-sightedness, such as is often unfortunately found in our employer circles, can fail to recognize that in the long run there can be no economic upswing for them and hence no economic profit, unless the inner national solidarity of our people is restored.

If during the War the German unions had ruthlessly guarded the interests of the working class, if even during the War they had struck a thousand times over and forced approval of the demands of the workers they represented on the dividend-hungry employers of those days; but if in matters of national defense they had avowed their Germanism with the same fanaticism;

and if with equal ruthlessness they had given to the fatherland that which is the fatherland's, the War would not have been lost. And how trifling all economic concessions, even the greatest, would have been, compared to the immense importance of winning the War!

Thus a movement which plans to give the German worker back to the German people must clearly realize that in this question economic sacrifices are of no importance whatever as long as the preservation and independence of the national economy are not threatened by them.

(2) The national education of the broad masses can only take place indirectly through a social uplift, since thus exclusively can those general economic premises be created which permit the individual to partake of the cultural goods of the nation.

(3) The nationalization of the broad masses can never be achieved by half-measures, by weakly emphasizing a so-called objective standpoint, but only by a ruthless and fanatically one-sided orientation toward the goal to be achieved. That is to say, a people cannot be made 'national' in the sense understood by our present-day bourgeoisie, meaning with so and so many limitations, but only nationalistic with the entire vehemence that is inherent in the extreme. Poison is countered only by an antidote, and only the shallowness of a bourgeois mind can regard the middle course as the road to heaven.

The broad masses of a people consist neither of professors nor of diplomats. The scantiness of the abstract knowledge they possess directs their sentiments more to the world of feeling. That is where their positive or negative attitude lies. It is receptive only to an expression of force in one of these two directions and never to a half-measure hovering between the two. Their emotional attitude at the same time conditions their extraordinary stability. Faith is harder to shake than knowledge, love succumbs less to change than respect, hate is more enduring than aversion, and the impetus to the mightiest upheavals on this earth has at all times consisted less in a scientific knowledge dominating the masses than in a fanaticism which inspired them and sometimes in a hysteria which drove them forward.

Anyone who wants to win the broad masses must know the key that opens the door to their heart. Its name is not objectivity (read weakness), but will and power.

(4) The soul of the people can only be won if along with carrying on a positive struggle for our own aims, we destroy the opponent of these aims.

The people at all times see the proof of their own right in ruthless attack on a foe, and to them renouncing the destruction of the adversary seems like uncertainty with regard to their own right if not a sign of their own unright.

The broad masses are only a piece of Nature and their sentiment does not understand the mutual handshake of people who claim that they want the opposite things. What they desire is the victory of the stronger and the destruction of the weak or his unconditional subjection.

The nationalization of our masses will succeed only when, aside from all the positive struggle for the soul of our people, their international poisoners are exterminated.

(5) All great questions of the day are questions of the moment and represent only consequences of definite causes. Only one among all of them, however, possesses causal importance,[1] and that is the question of the racial preservation of the nation. In the blood alone resides the strength as well as the weakness of man. As long as peoples do not recognize and give heed to the importance of their racial foundation, they are like men who would like to teach poodles the qualities of greyhounds, failing to realize that the speed of the greyhound like the docility of the poodle are not learned, but are qualities inherent in the race. Peoples which renounce the preservation of their racial purity renounce with it the unity of their soul in all its expressions. The divided state of their nature is the natural consequence of the divided state of their blood, and the change in their intellectual and creative force is only the effect of the change in their racial foundations.

Anyone who wants to free the German blood from the mani-

[1] '*ursähliche Bedeutung.*'

festations and vices of today, which were originally alien to its nature, will first have to redeem it from the foreign virus of these manifestations.

Without the clearest knowledge of the racial problem and hence of the Jewish problem there will never be a resurrection of the German nation.

The racial question gives the key not only to world history, but to all human culture.

(6) Organizing the broad masses of our people which are today in the international camp into a national people's community does not mean renouncing the defense of justified class interests. Divergent class and professional interests are not synonymous with class cleavage, but are natural consequences of our economic life. Professional grouping is in no way opposed to a true national community, for the latter consists in the unity of a nation in all those questions which affect this nation as such.

The integration of an occupational group which has become a class with the national community, or merely with the state, is not accomplished by the lowering of higher classes but by uplifting the lower classes. This process in turn can never be upheld by the higher class, but only by the lower class fighting for its equal rights. The present-day bourgeoisie was not organized into the state by measures of the nobility, but by its own energy under its own leadership.

The German worker will not be raised to the framework of the German national community via feeble scenes of fraternization, but by a conscious raising of his social and cultural situation until the most serious differences may be viewed as bridged. A movement which sets this development as its goal will have to take its supporters primarily from this camp.[1] It may fall back on the intelligentsia only in so far as the latter has completely understood the goal to be achieved. This process of transformation and equalization will not be completed in ten or twenty years; experience shows that it comprises many generations.

The severest obstacle to the present-day worker's approach to

[1] Changed in second edition to 'the workers' camp.'

the national community lies not in the defense of his class interests, but in his international leadership and attitude which are
hostile to the people and the fatherland. The same unions with a
fanatical national leadership in political and national matters
would make millions of workers into the most valuable members
of their nation regardless of the various struggles that took place
over purely economic matters.

A movement which wants honestly to give the German worker
back to his people and tear him away from the international delusion must sharply attack a conception dominant above all in
employer circles, which under national community understands
the unresisting economic surrender of the employee to the employer and which chooses to regard any attempt at safeguarding
even justified interests regarding the employee's economic existence as an attack on the national community. Such an assertion
is not only untrue, but a conscious lie, because the national community imposes its obligations not only on one side but also on
the other.

Just as surely as a worker sins against the spirit of a real
national community when, without regard for the common welfare and the survival of a national economy, he uses his power to
raise extortionate demands, an employer breaks this community
to the same extent when he conducts his business in an inhuman,
exploiting way, misuses the national labor force and makes millions out of its sweat. He then has no right to designate himself
as national, no right to speak of a national community; no, he is
a selfish scoundrel who induces social unrest and provokes future
conflicts which whatever happens must end in harming the
nation.

Thus, the reservoir from which the young movement must
gather its supporters will primarily be the masses of our workers.
Its work will be to tear these away from the international delusion, to free them from their social distress, to raise them out of
their cultural misery and lead them to the national community as
a valuable, united factor, national in feeling and desire.

If, in the circles of the national intelligentsia, there are found men with the warmest hearts for their people and its future, imbued with the deepest knowledge of the importance of this struggle for the soul of these masses, they will be highly welcome in the ranks of this movement, as a valuable spiritual backbone. But winning over the bourgeois voting cattle can never be the aim of this movement. If it were, it would burden itself with a dead weight which by its whole nature would paralyze our power to recruit from the broad masses. For regardless of the theoretical beauty of the idea of leading together the broadest masses from below and from above within the framework of the movement, there is the opposing fact that by psychological propagandizing of bourgeois masses in general meetings, it may be possible to create moods and even to spread insight, but not to do away with qualities of character or, better expressed, vices, whose development and origin embrace centuries. The difference with regard to the cultural level on both sides and the attitude on both sides toward questions raised by economic interests is at present still so great that, as soon as the intoxication of the meetings has passed, it would at once manifest itself as an obstacle.

Finally, the goal is not to undertake a restratification in the camp that is national to begin with, but to win over the antinational camp.

And this point of view, finally, is determining for the tactical attitude of the whole movement.

(7) This one-sided but thereby clear position must express itself in the propaganda of the movement and on the other hand in turn is required on propagandist grounds.

If propaganda is to be effective for the movement, it must be addressed to only one quarter, since otherwise, in view of the difference in the intellectual training of the two camps in question, either it will not be understood by the one group, or by the other it would be rejected as obvious and therefore uninteresting.

Even the style and the tone of its individual products cannot be equally effective for two such extreme groups. If propaganda renounces primitiveness of expression, it does not find its way to

the feeling of the broad masses. If, however, in word and ges-
ture, it uses the masses' harshness of sentiment and expression,
it will be rejected by the so-called intelligentsia as coarse and
vulgar. Among a hundred so-called speakers there are hardly
ten capable of speaking with equal effect today before a public
consisting of street-sweepers, locksmiths, sewer-cleaners, etc.,
and tomorrow holding a lecture with necessarily the same thought
content in an auditorium full of university professors and students. But among a thousand speakers there is perhaps only a
single one who can manage to speak to locksmiths and university
professors at the same time, in a form which not only is suitable
to the receptivity of both parties, but also influences both parties
with equal effect or actually lashes them into a wild storm of
applause. We must always bear in mind that even the most
beautiful idea of a sublime theory in most cases can be disseminated only through the small and smallest minds. The important
thing is not what the genius who has created an idea has in
mind, but what, in what form, and with what success the prophets of this idea transmit it to the broad masses.

The strong attractive power of the Social Democracy, yes, of
the whole Marxist movement, rested in large part on the homogeneity and hence one-sidedness of the public it addressed. The
more seemingly limited, indeed, the narrower its ideas were, the
more easily they were taken up and assimilated by a mass whose
intellectual level corresponded to the material offered.

Likewise for the new movement a simple and clear line thus
resulted:

Propaganda must be adjusted to the broad masses in content
and in form, and its soundness is to be measured exclusively by
its effective result.

In a mass meeting of all classes it is not that speaker who is
mentally closest to the intellectuals present who speaks best, but
the one who conquers the heart of the masses.

A member of the intelligentsia present at such a meeting, who
carps at the intellectual level of the speech despite the speaker's
obvious effect on the lower strata he has set out to conquer,

proves the complete incapacity of his thinking and the worthlessness of his person for the young movement. It can use only that intellectual who comprehends the task and goal of the movement to such an extent that he has learned to judge the activity of propaganda according to its success and not according to the impressions which it leaves behind in himself. For propaganda is not intended to provide entertainment for people who are national-minded to begin with, but to win the enemies of our nationality, in so far as they are of our blood.

In general those trends of thought which I have briefly summed up under the heading of war propaganda should be determining and decisive for our movement in the manner and execution of its own enlightenment work.

That it was right was demonstrated by its success.

(8) The goal of a political reform movement will never be reached by enlightenment work or by influencing ruling circles, but only by the achievement of political power. Every world-moving idea has not only the right, but also the duty, of securing, those means which make possible the execution of its ideas. Success is the one earthly judge concerning the right or wrong of such an effort, and under success we must not understand, as in the year 1918, the achievement of power in itself, but an exercise of that power that will benefit the nation. Thus, a coup d'état must not be regarded as successful if, as senseless state's attorneys in Germany think today, the revolutionaries have succeeded in possessing themselves of the state power, but only if, by the realization of the purposes and aims underlying such a revolutionary action, more benefit accrues to the nation than under the past régime. Something which cannot very well be claimed for the German revolution, as the gangster job of autumn, 1918, calls itself.

If the achievement of political power constitutes the precondition for the practical execution of reform purposes, the movement with reform purposes must from the first day of its existence feel itself a movement of the masses and not a literary tea-club or a shopkeepers' bowling society.

(9) The young movement is in its nature and inner organization anti-parliamentarian; that is, it rejects, in general and in its own inner structure, a principle of majority rule in which the leader is degraded to the level of a mere executant of other people's will and opinion. In little as well as big things, the movement advocates the principle of a Germanic democracy: the leader is elected, but then enjoys unconditional authority.

The practical consequences of this principle in the movement are the following:

The first chairman of a local group is elected, but then he is the responsible leader of the local group. All committees are subordinate to him and not, conversely, he to a committee. There are no electoral committees, but only committees for work. The responsible leader, the first chairman, organizes the work. The first principle applies to the next higher organization, the precinct, the district or county. The leader is always elected, but thereby he is vested with unlimited powers and authority. And, finally, the same applies to the leadership of the whole party. The chairman is elected, but he is the exclusive leader of the movement.[1] All committees are subordinate to him and not

---

[1] This is one of the few passages the sense of which has been radically changed in the second edition. By the time of the appearance of the second edition, Hitler had emerged victorious from the factional conflicts within the party. His authority was now uncontested. In the second edition the passage reads:

'The young movement is in its nature and inner organization anti-parliamentarian; that is, it rejects in general and in its own inner structure, a principle of majority rule in which the leader is degraded to the level of a mere executant of other people's will and opinion. In little as well as big things, the movement advocates the principle of unconditional authority of the leader, coupled with the highest responsibility.

'The practical consequences of this principle in the movement are the following:

'The first chairman of a local group is appointed by the next highest leader; he is the responsible leader of the local group. All committees are subordinate to him and not, conversely, he to a committee. There are no electoral committees, but only committees for work. The responsible leader, the first chairman, organizes the work. The first principle applies to

he to the committees. He makes the decisions and hence bears the responsibility on his shoulders. Members of the movement are free to call him to account before the forum of a new election, to divest him of his office in so far as he has infringed on the principles of the movement or served its interests badly. His place is then taken by an abler, new man, enjoying, however, the same authority and the same responsibility.

It is one of the highest tasks of the movement to make this principle determining, not only within its own ranks, but for the entire state.

Any man who wants to be leader bears, along with the highest unlimited authority, also the ultimate and heaviest responsibility.

Anyone who is not equal to this or is too cowardly to bear the consequences of his acts is not fit to be leader; only the hero is cut out for this.

The progress and culture of humanity are not a product of the majority, but rest exclusively on the genius and energy of the personality.

To cultivate the personality and establish it in its rights is one of the prerequisites for recovering the greatness and power of our nationality.

Hence the movement is anti-parliamentarian, and even its participation in a parliamentary institution can only imply activity for its destruction, for eliminating an institution in which we must see one of the gravest symptoms of mankind's decay.

(10) The movement decisively rejects any position on questions which either lie outside the frame of its political work or, being not of basic importance, are irrelevant for it. Its task is not a religious reformation, but a political reorganization of our people. In both religious denominations it sees equally valuable pillars

the next higher organization, the precinct, the district or county. The leader is always appointed from above and at the same time vested with unlimited powers and authority. Only the leader of the whole party is elected, in a general membership meeting compatible with the laws governing associations. But he is the exclusive leader of the movement.'

for the existence of our people and therefore combats those parties which want to degrade this foundation of an ethical, moral, and religious consolidation of our national body to the level of an instrument of their party interests.

The movement finally sees its task, not in the restoration of a definite state form and in the struggle against another, but in the creation of those basic foundations without which neither republic nor monarchy can endure for any length of time. Its mission lies not in the foundation of a monarchy or in the reinforcement of a republic, but in the creation of a Germanic state.

The question of the outward shaping of this state, its crowning, so to speak, is not of basic importance, but is determined only by questions of practical expediency.

For a people that has once understood the great problems and tasks of its existence, the questions of outward formalities will no longer lead to inner struggle.

(11) The question of the movement's inner organization is one of expediency and not of principle.

The best organization is not that which inserts the greatest, but that which inserts the smallest, intermediary apparatus between the leadership of a movement and its individual adherents. For the function of organization is the transmission of a definite idea — which always first arises from the brain of an individual — to a larger body of men and the supervision of its realization.

Hence organization is in all things only a necessary evil. In the best case it is a means to an end, in the worst case an end in itself.

Since the world produces more mechanical than ideal natures, the forms of organization are usually created more easily than ideas as such.

The practical development of every idea striving for realization in this world, particularly of one possessing a reform character, is in its broad outlines as follows:

Some idea of genius arises in the brain of a man who feels called upon to transmit his knowledge to the rest of humanity. He preaches his view and gradually wins a certain circle of adherents. This process of the direct and personal transmittance of a man's ideas to the rest of his fellow men [1] is the most ideal and natural. With the rising increase in the adherents of the new doctrine, it gradually becomes impossible for the exponent of the idea to go on exerting a personal, direct influence on the innumerable supporters, to lead and direct them. Proportionately as, in consequence of the growth of the community, the direct and shortest communication is excluded, the necessity of a connecting organization arises: thus, the ideal condition is ended and is replaced by the necessary evil of organization. Little sub-groups are formed which in the political movement, for example, call themselves local groups and constitute the germ-cells of the future organization.

If the unity of the doctrine is not to be lost, however, this sub-division must not take place until the authority of the spiritual founder and of the school trained by him can be regarded as unconditional. The geo-political significance of a focal center in a movement cannot be overemphasized. Only the presence of such a place, exerting the magic spell of a Mecca or a Rome, can in the long run give the movement a force which is based on inner unity and the recognition of a summit representing this unity.

Thus, in forming the first organizational germ-cells we must never lose sight of the necessity, not only of preserving the importance of the original local source of the idea, but of making it paramount. This intensification of the ideal, moral, and factual immensity of the movement's point of origin and direction must take place in exact proportion as the movement's germ-cells, which have now become innumerable, demand new links in the shape of organizational forms.

For, as the increasing number of individual adherents makes it impossible to continue direct communication with them for the formation of the lowest bodies, the ultimate innumerable increase [2] of these lowest organizational forms compels in turn

[1] *'die andere Mitwelt.'*

[2] *'die zahllose Vermehrung.'*

creation of higher associations which politically can be designated roughly as county or district groups.

Easy as it still may be to maintain the authority of the original center toward the lowest local groups, it will be equally difficult to maintain this position toward the higher organizational forms which now arise. But this is the precondition for the unified existence of the movement and hence for carrying out an idea.

If, finally, these larger intermediary divisions are also combined into new organizational forms, the difficulty is further increased of safeguarding, even toward them, the unconditional leading character of the original founding site, its school, etc.

Therefore, the mechanical forms of an organization may only be developed to the degree in which the spiritual ideal authority of a center seems unconditionally secured. In political formations this guaranty can often seem provided only by practical power.

From this the following directives for the inner structure of the movement resulted:

(a) Concentration for the time being of all activity in a single place: Munich. Training of a community of unconditionally reliable supporters and development of a school for the subsequent dissemination of the idea. Acquisition of the necessary authority for the future by the greatest possible visible successes in this one place.

To make the movement and its leaders known, it was necessary, not only to shake the belief in the invincibility of the Marxist doctrine in one place for all to see, but to demonstrate the possibility of an opposing movement.

(b) Formation of local groups only when the authority of the central leadership in Munich may be regarded as unquestionably recognized.

(c) Likewise the formation of district, county, or provincial groups depends, not only on the need for them, but also on certainty that an unconditional recognition of the center has been achieved.

Furthermore, the creation of organizational forms is dependent on the men who are available and can be considered as leaders.

This may occur in two ways:

(a) The movement disposes of the necessary financial means for the training and schooling of minds capable of future leadership. It then distributes the material thus acquired systematically according to criteria of tactical and other expediency.

This way is the easier and quicker; however, it demands great financial means, since this leader material is only able to work for the movement when paid.

(b) The movement, owing to the lack of financial means, is not in a position to appoint official leaders, but for the present must depend on honorary officers.

This way is the slower and more difficult.

Under certain circumstances the leadership of a movement must let large territories lie fallow, unless there emerges from the adherents a man able and willing to put himself at the disposal of the leadership, and organize and lead the movement in the district in question.

It may happen that in large territories there will be no one, in other places, however, two or even three almost equally capable. The difficulty that lies in such a development is great and can only be overcome in the course of years.

The prerequisite for the creation of an organizational form is and remains the man necessary for its leadership.

As worthless as an army in all its organizational forms is without officers, equally worthless is a political organization without the suitable leader.

Not founding a local group is more useful to the movement when a suitable leader personality is lacking than to have its organization miscarry due to the absence of a leader to direct and drive it forward.

Leadership itself requires not only will but also ability, and a greater importance must be attached to will and energy than to intelligence as such, and most valuable of all is a combination of ability, determination, and perseverance.

(12) The future of a movement is conditioned by the fanati-

cism, yes, the intolerance, with which its adherents uphold it as the sole correct movement, and push it past other formations of a similar sort.

It is the greatest error to believe that the strength of a movement increases through a union with another of similar character. It is true that every enlargement of this kind at first means an increase in outward dimensions, which to the eyes of superficial observers means power; in truth, however, it only takes over the germs of an inner weakening that will later become effective.

For whatever can be said about the like character of two movements, in reality it is never present. For otherwise there would actually be not two movements but one. And regardless wherein the differences lie — even if they consisted only in the varying abilities of the leadership — they exist. But the natural law of all development demands, not the coupling of two formations which are simply not alike, but the victory of the stronger and the cultivation of the victor's force and strength made possible alone by the resultant struggle.

Through the union of two more or less equal political party formations momentary advantages may arise, but in the long run any success won in this way is the cause of inner weaknesses which appear later.

The greatness of a movement is exclusively guaranteed by the unrestricted development of its inner strength and its steady growth up to the final victory over all competitors.

Yes, we can say that its strength and hence the justification of its existence increases only so long as it recognizes the principle of struggle as the premise of its development, and that it has passed the high point of its strength in the moment when complete victory inclines to its side.

Therefore, it is only profitable for a movement to strive for this victory in a form which does not lead to an early momentary success, but which in a long struggle occasioned by absolute intolerance also provides long growth.

Movements which increase only by the so-called fusion of similar formations, thus owing their strength to compromises, are like hothouse plants. They shoot up, but they lack the strength to defy the centuries and withstand heavy storms.

The greatness of every mighty organization embodying an idea in this world lies in the religious fanaticism and intolerance with which, fanatically convinced of its own right, it intolerantly imposes its will against all others. If an idea in itself is sound and, thus armed, takes up a struggle on this earth, it is unconquerable and every persecution will only add to its inner strength.

The greatness of Christianity did not lie in attempted negotiations for compromise with any similar philosophical opinions in the ancient world, but in its inexorable fanaticism in preaching and fighting for its own doctrine.

The apparent head start which movements achieve by fusions is amply caught up with by the steady increase in the strength of a doctrine and organization that remain independent and fight their own fight.

(13) On principle the movement must so educate its members that they do not view the struggle as something idly cooked up, but as the thing that they themselves are striving for.[1] Therefore, they must not fear the hostility of their enemies, but must feel that it is the presupposition for their own right to exist. They must not shun the hatred of the enemies of our nationality and our philosophy and its manifestations; they must long for them. And among the manifestations of this hate are lies and slander.

Any man who is not attacked in the Jewish newspapers, not slandered and vilified, is no decent German and no true National Socialist. The best yardstick for the value of his attitude, for the sincerity of his conviction, and the force of his will is the hostility he receives from the mortal enemy of our people.

It must, over and over again, be pointed out to the adherents of the movement and in a broader sense to the whole people that the Jew and his newspapers always lie and that even an occasional truth is only intended to cover a bigger falsification and is therefore itself in turn a deliberate untruth. The Jew is the great

---

[1] '*das selbst Erstrebte.*' Second edition has: '*das selbst Erlebte*': 'something they themselves experience.'

master in lying, and lies and deception are his weapons in struggle.

Every Jewish slander and every Jewish lie is a scar of honor on the body of our warriors.

The man they have most reviled stands closest to us and the man they hate worst is our best friend.

Anyone who picks up a Jewish newspaper in the morning and does not see himself slandered in it has not made profitable use of the previous day; for if he had, he would be persecuted, reviled, slandered, abused, befouled. And only the man who combats this mortal enemy of our nation and of all Aryan humanity and culture most effectively may expect to see the slanders of this race and the struggle of this people directed against him.

When these principles enter the flesh and blood of our supporters, the movement will become unshakable and invincible.

(14) The movement must promote respect for personality by all means; it must never forget that in personal worth lies the worth of everything human; that every idea and every achievement is the result of one man's creative force and that the admiration of greatness constitutes, not only a tribute of thanks to the latter, but casts a unifying bond around the grateful.

Personality cannot be replaced; especially when it embodies not the mechanical but the cultural and creative element. No more than a famous master can be replaced and another take over the completion of the half-finished painting he has left behind can the great poet and thinker, the great statesman and the great soldier, be replaced. For their activity lies always in the province of art. It is not mechanically trained, but inborn by God's grace.

The greatest revolutionary changes and achievements of this earth, its greatest cultural accomplishments, the immortal deeds in the field of statesmanship, etc., are forever inseparably bound up with a name and are represented by it. To renounce doing homage to a great spirit means the loss of an immense strength which emanates from the names of all great men and women.

The Jew knows this best of all. He, whose great men are only great in the destruction of humanity and its culture, makes sure that they are idolatrously admired. He attempts only to represent the admiration of the nations for their own spirits as unworthy and brands it as a 'personality cult.'

As soon as a people becomes so cowardly that it succumbs to this Jewish arrogance and effrontery, it renounces the mightiest power that it possesses; for this is based, not on respect for the masses, but on the veneration of genius and on uplift and enlightenment by his example.

When human hearts break and human souls despair, then from the twilight of the past the great conquerors of distress and care, of disgrace and misery, of spiritual slavery and physical compulsion, look down on them and hold out their eternal hands to the despairing mortals!

Woe to the people that is ashamed to take them!

* * *

In the first period of our movement's development we suffered from nothing so much as from the insignificance, the unknownness of our names, which in themselves made our success questionable. The hardest thing in this first period, when often only six, seven, or eight heads met together, to use the words of an opponent, was to arouse and preserve in this tiny circle faith in the mighty future of the movement.

Consider that six or seven men, all nameless poor devils, had joined together with the intention of forming a movement, hoping to succeed — where the powerful great mass parties had hitherto failed — in restoring a German Reich of greater power and glory. If people had attacked us in those days, yes, even if they had laughed at us, in both cases we should have been happy. For the oppressive thing was neither the one nor the other; it was the complete lack of attention we found in those days.[1]

When I entered the circle of these few men, there could be no question of a party or a movement. I have already described my

[1] Second edition adds: 'from which I suffered most.'

impressions regarding my first meeting with this little formation. In the weeks that followed, I had time and occasion to study this so-called 'party' which at first looked so impossible. And, by God, the picture was depressing and discouraging. There was nothing here, really positively nothing. The name of a party whose committee constituted practically the whole membership, which, whether we liked it or not, was exactly what it was trying to combat, a parliament on a small scale. Here, too, the vote ruled; if big parliaments yelled their throats hoarse for months at a time, it was about important problems at least, but in this little circle the answer to a safely arrived letter let loose an interminable argument!

The public, of course, knew nothing at all about this. Not a soul in Munich knew the party even by name, except for its few supporters and their few friends.

Every Wednesday a so-called committee meeting took place in a Munich café, and once a week an evening lecture. Since the whole membership of the 'movement' was at first represented in the committee, the faces of course were always the same. Now the task was at last to burst the bonds of the small circle, to win new supporters, but above all to make the name of the movement known at any price.

In this we used the following technique:

Every month, and later every two weeks, we tried to hold a 'meeting.' The invitations to it were written on the typewriter or sometimes by hand on slips of paper and the first few times were distributed, or handed out,[1] by us personally. Each one of us turned to the circle of his friends, and tried to induce someone or other to attend one of these affairs.

The result was miserable.

I still remember how I myself in this first period once distributed about eighty of these slips of paper, and how in the evening we sat waiting for the masses who were expected to appear.

An hour late, the 'chairman' finally had to open the 'meeting.' We were again seven men, the old seven.

[1] '*verteilt bzw. ausgetragen.*'

We changed over to having the invitation slips written on a machine and mimeographed in a Munich stationery store. The result at the next meeting was a few more listeners. Thus the number rose slowly from eleven to thirteen, finally to seventeen, to twenty-three, to thirty-four listeners.

By little collections among us poor devils the funds were raised with which at last to advertise the meeting by notices in the then independent *Münchener Beobachter* in Munich. And this time the success was positively amazing. We had organized the meeting in the Munich Hofbräuhauskeller (not to be confused with the Munich Hofbräuhaus-Festsaal), a little room with a capacity of barely one hundred and thirty people. To me personally the room seemed like a big hall and each of us was worried whether we would succeed in filling this 'mighty' edifice with people.

At seven o'clock one hundred and eleven people were present and the meeting was opened.

A Munich professor made the main speech, and I, for the first time, in public, was to speak second.

In the eyes of Herr Harrer, then first chairman of the party, the affair seemed a great adventure. This gentleman, who was certainly otherwise honest, just happened to be convinced that I might be capable of doing certain things, but not of speaking. And even in the time that followed he could not be dissuaded from this opinion.

Things turned out differently. In this first meeting that could be called public I had been granted twenty minutes' speaking time.

I spoke for thirty minutes, and what before I had simply felt within me, without in any way knowing it, was now proved by reality: I could speak! After thirty minutes the people in the small room were electrified and the enthusiasm was first expressed by the fact that my appeal to the self-sacrifice of those present led to the donation of three hundred marks. This relieved us of a great worry. For at this time the financial stringency was so great that we were not even in a position to have slogans printed for the movement, or even distribute leaflets. Now the foundation was laid for a little fund from which at least our barest needs and most urgent necessities could be defrayed. But in another respect as well, the success of this first larger meeting was considerable.

At that time I had begun to bring a number of fresh young forces into the committee. During my many years in the army I had come to know a great number of faithful comrades who now slowly, on the basis of my persuasion, began to enter the movement. They were all energetic young people, accustomed to discipline, and from their period of service raised in the principle: nothing at all is impossible, everything can be done if you only want it.

How necessary such a transfusion of new blood was, I myself could recognize after only a few weeks of collaboration.

Herr Harrer, then first chairman of the party, was really a journalist and as such he was certainly widely educated. But for a party leader he had one exceedingly serious drawback: he was no speaker for the masses. As scrupulously conscientious and precise as his work in itself was, it nevertheless lacked — perhaps because of this very lack of a great oratorical gift — the great sweep. Herr Drexler, then chairman of the Munich local group, was a simple worker, likewise not very significant as a speaker, and moreover he was no soldier. He had not served in the army, even during the War he had not been a soldier, so that feeble and uncertain as he was in his whole nature, he lacked the only schooling which was capable of turning uncertain and soft natures into men. Thus both men were not made of stuff which would have enabled them not only to bear in their hearts fanatical faith in the victory of a movement, but also with indomitable energy and will, and if necessary with brutal ruthlessness, to sweep aside any obstacles which might stand in the path of the rising new idea. For this only beings were fitted in whom spirit and body had acquired those military virtues which can perhaps best be described as follows: swift as greyhounds, tough as leather, and hard as Krupp steel.

At that time I myself was still a soldier. My exterior and interior had been whetted and hardened for well-nigh six years, so that at first I must have seemed strange in this circle. I, too, had forgotten how to say: 'that's impossible,' or 'it won't work'; 'we can't risk that,' 'that is too dangerous,' etc.

For of course the business was dangerous. Little attention as the Reds paid to one of your bourgeois gossip clubs whose inner innocence and hence harmlessness for themselves they knew better than its own members, they were determined to use every means to get rid of a movement which did seem dangerous to them. Their most effective method in such cases has at all times been terror or violence.

In the year 1920, in many regions of Germany, a national meeting that dared to address its appeal to the broad masses and publicly invite attendance was simply impossible. The participants in such a meeting were dispersed and driven away with bleeding heads. Such an accomplishment, to be sure, did not require much skill: for after all the biggest so-called bourgeois mass meeting would scatter at the sight of a dozen Communists like hares running from a hound.

Most loathsome to the Marxist deceivers of the people was inevitably a movement whose explicit aim was the winning of those masses which had hitherto stood exclusively in the service of the international Marxist Jewish stock exchange parties. The very name of 'German Workers' Party' had the effect of goading them. Thus one could easily imagine that on the first suitable occasion the conflict would begin with the Marxist inciters who were then still drunk with victory.

In the small circle that the movement then was a certain fear of such a fight prevailed. The members wanted to appear in public as little as possible, for fear of being beaten up. In their mind's eye they already saw the first great meeting smashed and the movement finished for good. I had a hard time putting forward my opinion that we must not dodge this struggle, but prepare for it, and for this reason acquire the armament which alone offers protection against violence. Terror is not broken by the mind, but by terror. The success of the first meeting strengthened my position in this respect. We gained courage for a second meeting on a somewhat larger scale.

About October, 1919, the second, larger meeting took place in the Eberlbräukeller. Topic: Brest-Litovsk and Versailles. Four gentlemen appeared as speakers. I myself spoke for almost an hour and the success was greater than at the first rally. The audience had risen to more than one hundred and thirty. An attempted disturbance was at once nipped in the bud by my comrades. The disturbers flew down the stairs with gashed heads.

Two weeks later another meeting took place in the same hall. The attendance had risen to over one hundred and seventy and the room was well filled. I had spoken again, and again the success was greater than at the previous meeting.

I pressed for a larger hall. At length we found one at the other end of town in the 'Deutsches Reich' on Dachauer Strasse. The first meeting in the new hall was not so well attended as the previous one: barely one hundred and forty persons. In the committee, hopes began to sink and the eternal doubters felt that the excessive repetition of our 'demonstrations' had to be considered the cause of the bad attendance. There were violent arguments in which I upheld the view that a city of seven hundred thousand inhabitants could stand not one meeting every two weeks, but ten every week, that we must not let ourselves be misled by failures, that the road we had taken was the right one, and that sooner or later, with steady perseverance, success was bound to come. All in all, this whole period of winter 1919–20 was a single struggle to strengthen confidence in the victorious might of the young movement and raise it to that fanaticism of faith which can move mountains.

The next meeting in the same hall showed me to be right. The attendance had risen to over two hundred; the public as well as financial success was brilliant.

I urged immediate preparations for another meeting. It took place barely two weeks later and the audience rose to over two hundred and seventy heads.

Two weeks later, for the seventh time, we called together the supporters and friends of the new movement and the same hall could barely hold the people who had grown to over four hundred.

It was at this time that the young movement received its inner form. In the small circle there were sometimes more or less violent disputes. Various quarters — then as today — carped at designating the young movement as a party. In such a conception I have always seen proof of the critics' practical incompetence and intellectual smallness. They were and always are the men who cannot distinguish externals from essentials, and who try to estimate the value of a movement according to the most bombastic-sounding titles, most of which, sad to say, the vocabulary of our forefathers must provide.

It was hard, at that time, to make it clear to people that every movement, as long as it has not achieved the victory of its ideas, hence its goal, is a party even if it assumes a thousand different names.

If any man wants to put into practical effect a bold idea whose realization seems useful in the interests of his fellow men, he will first of all have to seek supporters who are ready to fight for his intentions. And if this intention consists only in destroying the existing parties, of ending the fragmentation, the exponents of this view and propagators of this determination are themselves a party, as long as this goal has not been achieved. It is hair-splitting and shadow-boxing when some antiquated folkish theoretician, whose practical successes stand in inverse proportion to his wisdom, imagines that he can change the party character which every young movement possesses by changing this term.

On the contrary.

If anything is unfolkish, it is this tossing around of old Germanic expressions which neither fit into the present period nor represent anything definite, but can easily lead to seeing the significance of a movement in its outward vocabulary. This is a real menace which today can be observed on countless occasions.

Altogether then, and also in the period that followed, I had to warn again and again against those *deutschvölkisch*[1] wandering scholars whose positive accomplishment is always practically nil, but whose conceit can scarcely be excelled. The young movement had and still has to guard itself against an influx of people whose sole recommendation for the most part lies in their declaration that they have fought for thirty and even forty years for the same idea. Anyone who fights for forty years for a so-called idea without being able to bring about even the slightest success, in fact, without having prevented the victory of the opposite, has, with forty years of activity, provided proof of his own incapacity. The danger above all lies in the fact that such natures do not want to fit into the movement as links, but keep shooting off their mouths about leading circles in which alone, on the strength of their age-old activity, they can see a suitable place for further activity. But woe betide if a young movement is surrended to the mercies of such people. No more than a business man who in forty years of activity has steadily run a big business into the ground is fitted to be the founder of a new one, is a folkish Methuselah, who in exactly the same time has gummed up and petrified a great idea, fit for the leadership of a new, young movement!

Besides, only a fragment of all these people come into the new movement to serve it, but in most cases, under its protection or

---

[1] The *Deutschvölkische Partei* was founded in 1914 by a union of various anti-Semitic groups. After the revolution of November, 1918, it dissolved into the *Deutschnationale Volkspartei* (German National People's Party) which was the strongest party of the extreme Right under the Weimar Republic, until overshadowed by the National Socialists. At the same time a *Deutschvölkischer Bund* was founded, which in 1922 assumed the name of *Deutschvölkischer Schutz- und Trutzbund* and was suppressed after the murder of Rathenau. A *Deutschvölkische Freiheitspartei* was also founded in 1922, but soon merged with the Nazis. Finally, there was a *Deutschvölkische Freiheitsbewegung* under the leadership of Ludendorff, which broke away from the Nazis in 1925.

Hitler's attacks here are directed against all his Rightist competitors, those belonging to rival movements and those struggling for leadership within his own movement.

through the possibilities it offers, to warm over their old cabbage.

They do not want to benefit the idea of the new doctrine, they only expect it to give them a chance to make humanity miserable with their own ideas. For what kind of ideas they often are, it is hard to tell.

The characteristic thing about these people is that they rave about old Germanic heroism, about dim prehistory, stone axes, spear and shield, but in reality are the greatest cowards that can be imagined. For the same people who brandish scholarly imitations of old German tin swords, and wear a dressed bearskin with bull's horns over their bearded heads, preach for the present nothing but struggle with spiritual weapons, and run away as fast as they can from every Communist blackjack. Posterity will have little occasion to glorify their own heroic existence in a new epic.

I came to know these people too well not to feel the profoundest disgust at their miserable play-acting. But they make a ridiculous impression on the broad masses, and the Jew has every reason to spare these folkish comedians, even to prefer them to the true fighters for a coming German state. With all this, these people are boundlessly conceited; despite all the proofs of their complete incompetence, they claim to know everything better and become a real plague for all straightforward and honest fighters to whom heroism seems worth honoring, not only in the past, but who also endeavor to give posterity a similar picture by their own actions.

And often it can be distinguished only with difficulty which of these people act out of inner stupidity or incompetence and which only pretend to for certain reasons. Especially with the so-called religious reformers on an old Germanic basis, I always have the feeling that they were sent by those powers which do not want the resurrection of our people. For their whole activity leads the people away from the common struggle against the common enemy, the Jew, and instead lets them waste their strength on inner religious squabbles as senseless as they are disastrous. For these very reasons the establishment of a strong central power implying the unconditional authority of a leader-ship is necessary in the movement. By it alone can such ruinous elements be squelched. And for this reason the greatest enemies of a uniform, strictly led and conducted movement are to be found in the circles of these folkish wandering Jews. In the move-ment they hate the power that checks their mischief.

Not for nothing did the young movement establish a definite program in which it did not use the word 'folkish.' The concept folkish, in view of its conceptual boundlessness, is no possible basis for a movement and offers no standard for membership in one. The more indefinable this concept is in practice, the more and broader interpretations it permits, the greater becomes the possibility of invoking its authority. The insertion of such an indefinable and variously interpretable concept into the politi-cal struggle leads to the destruction of any strict fighting soli-darity, since the latter does not permit leaving to the individual the definition of his faith and will.

And it is disgraceful to see all the people who run around today with the word 'folkish' on their caps and how many have their own interpretation of this concept. A Bavarian professor by the name of Bayer,[1] a famous fighter with spiritual weapons, rich in equally spiritual marches on Berlin, thinks that the concept folkish consists only in a monarchistic attitude. This learned mind, however, has thus far forgotten to give a closer explana-tion of the identity of our German monarchs of the past with the folkish opinion of today. And I fear that in this the gentleman would not easily succeed. For anything less folkish than most of the Germanic monarchic state formations can hardly be imag-ined. If this were not so, they would never have disappeared, or their disappearance would offer proof of the unsoundness of the folkish outlook.

And so everyone shoots off his mouth about this concept as he happens to understand it. As a basis for a movement of political

---

[1] Second edition has: 'a well-known professor in Bavaria.' The reason for the change seems to have been nothing more than the professor's unimpor-tance.

struggle, such a multiplicity of opinions is out of the question.

I shall not even speak of the unworldliness of these folkish Saint Johns of the twentieth century or their ignorance of the popular soul. It is sufficiently illustrated by the ridicule with which they are treated by the Left, which lets them talk and laughs at them.

Anyone in this world who does not succeed in being hated by his adversaries does not seem to me to be worth much as a friend. And thus the friendship of these people for our young movement was not only worthless, but solely and always harmful, and it was also the main reason why, first of all, we chose the name of 'party' — we had grounds for hoping that by this alone a whole swarm of these folkish sleepwalkers would be frightened away from us — and why in the second place we termed ourselves *National Socialist German Workers' Party*.

The first expression kept away the antiquity enthusiasts, the big-mouths and superficial proverb-makers of the so-called 'folkish idea,' and the second freed us from the entire host of knights of the 'spiritual sword,' all the poor wretches who wield the 'spiritual weapon' as a protecting shield to hide their actual cowardice.

It goes without saying that in the following period we were attacked hardest especially by these last, not actively, of course, but only with the pen, just as you would expect from such folkish goose-quills. For them our principle, 'Against those who attack us with force we will defend ourselves with force,' had something terrifying about it. They persistently reproached us, not only with brutal worship of the blackjack, but with lack of spirit as such. The fact that in a public meeting a Demosthenes can be brought to silence if only fifty idiots, supported by their voices and their fists, refuse to let him speak, makes no impression whatever on such a quack. His inborn cowardice never lets him get into such danger. For he does not work 'noisily' and 'obtrusively,' but in 'silence.'

Even today I cannot warn our young movement enough against falling into the net of these so-called 'silent workers.' They are not only cowards, but they are also always incompetents and do-nothings. A man who knows a thing, who is aware of a given danger, and sees the possibility of a remedy with his own eyes, has the duty and obligation, by God, not to work 'silently,' but to stand up before the whole public against the evil and for its cure. If he does not do so, he is a disloyal, miserable weakling who fails either from cowardice or from laziness and inability. To be sure, this does not apply at all to most of these people, for they know absolutely nothing, but behave as though they knew God knows what; they can do nothing but try to swindle the whole world with their tricks; they are lazy, but with the 'silent' work they claim to do, they arouse the impression of an enormous and conscientious activity; in short, they are swindlers, political crooks who hate the honest work of others. As soon as one of these folkish moths praises the darkness [1] of silence, we can bet a thousand to one that by it he produces nothing, but steals, steals from the fruits of other people's work.

To top all this, there is the arrogance and conceited effrontery with which this lazy, light-shunning rabble fall upon the work of others, trying to criticize it from above, thus in reality aiding the mortal enemies of our nationality.

Every last agitator who possesses the courage to stand on a tavern table among his adversaries, to defend his opinions with manly forthrightness, does more than a thousand of these lying, treacherous sneaks. He will surely be able to convert one man or another and win him for the movement. It will be possible to examine his achievement and establish the effect of his activity by its results. Only the cowardly swindlers who praise their 'silent' work and thus wrap themselves in the protective cloak of a despicable anonymity, are good for nothing and may in the truest sense of the word be considered drones in the resurrection of our people.

*　　　*　　　*

[1] '*sich immer auf das Dunkel der Stille beruft.*'　Second edition has: '*den Wert*' for '*das Dunkel.*' The meaning would then be: 'cites the value of silence.'

At the beginning of 1920, I urged the holding of the first great mass meeting. Differences of opinion arose. A few leading party members regarded the affair as premature and hence disastrous in effect. The Red press had begun to concern itself with us and we were fortunate enough gradually to achieve its hatred. We had begun to speak in the discussions at other meetings. Of course, each of us was at once shouted down. There was, however, some success. People got to know us and proportionately as their knowledge of us deepened, the aversion and rage against us grew. And thus we were entitled to hope that in our first great mass meeting we would be visited by a good many of our friends from the Red camp.

I, too, realized that there was great probability of the meeting being broken up. But the struggle had to be carried through, if not now, a few months later. It was entirely in our power to make the movement eternal on the very first day by blindly and ruthlessly fighting for it. I knew above all the mentality of the adherents of the Red side far too well, not to know that resistance to the utmost not only makes the biggest impression, but also wins supporters. And so we just had to be resolved to put up this resistance.

Herr Harrer,[1] then first chairman of the party, felt he could not support my views with regard to the time chosen and consequently, being an honest, upright man, he withdrew from the leadership of the party. His place was taken by Herr Anton Drexler. I had reserved for myself the organization of propaganda and began ruthlessly to carry it out.

And so, the date of February 4, 1920, was set for the holding of this first great mass meeting of the still unknown movement.

I personally conducted the preparations. They were very brief.

---

[1] Harrer was for the masses whom he claimed Hitler was alienating. He believed in telling the truth and opposed Hitler's unbridled propaganda. He also opposed anti-Semitism and the wealthy backers whom Hitler was trying to bring into the movement. After the acceptance of the 25 Points which he considered demagogic and which included the anti-Semitic Point 4, he resigned.

Altogether the whole apparatus was adjusted to make lightning decisions. Its aim was to enable us to take a position on current questions in the form of mass meetings within twenty-four hours. They were to be announced by posters and leaflets whose content was determined according to those guiding principles which in rough outlines I have set down in my treatise on propaganda. Effect on the broad masses, concentration on a few points, constant repetition of the same, self-assured and self-reliant framing of the text in the forms of an apodictic statement, greatest perseverance in distribution and patience in awaiting the effect.

On principle, the color red was chosen; it is the most exciting; we knew it would infuriate and provoke our adversaries the most and thus bring us to their attention and memory whether they liked it or not.

In the following period the inner fraternization in Bavaria between the Marxists and the Center as a political party was most clearly shown in the concern with which the ruling Bavarian People's Party tried to weaken the effect of our posters on the Red working masses and later to prohibit them. If the police found no other way to proceed against them, 'considerations of traffic' had to do the trick, till finally, to please the inner, silent Red ally, these posters, which had given back hundreds of thousands of workers, incited and seduced by internationalism, to their German nationality, were forbidden entirely with the helping hand of a so-called German National People's Party.[1] As an appendix and example to our young movement, I am adding a number of these proclamations. They come from a period embracing nearly three years; they can best illustrate the mighty struggle which the young movement fought at this time. They will also bear witness to posterity of the will and honesty of our convictions and the despotism of the so-called national authorities in prohibiting, just because they personally found it uncomfortable, a nationalization which would have won back broad masses of our nationality.

They will also help to destroy the opinion that there had been

[1] See page 360.

a national government as such in Bavaria and also document for posterity the fact that the national Bavaria of 1919, 1920, 1921, 1922, 1923 was not forsooth the result of a national government, but that the government was merely forced to take consideration of a people that was gradually feeling national.[1]

The governments themselves did everything to eliminate this process of recovery and to make it impossible.

Here only two men must be excluded:

Ernst Pöhner, the police president at that time, and Chief Deputy *Frick*,[2] his faithful advisor, were the only higher state officials who even then had the courage to be first Germans and then officials. Ernst Pöhner was the only man in a responsible post who did not curry favor with the masses, but felt responsible to his nationality and was ready to risk and sacrifice everything, even if necessary his personal existence, for the resurrection of the German people whom he loved above all things. And for this reason he was always a troublesome thorn in the eyes of those

---

[1] '*auf ein allmählich national fühlendes Volk.*'

[2] Ernst Pöhner, former Bavarian police president, became a justice of the Bavarian Supreme Court in 1921. It was he who induced the Munich police not to interfere with Hitler's *Putsch* of November 8, 1923. At the trial, he declared: 'For five years I did nothing but high treason.' He was convicted and sentenced to some months of imprisonment, but was prevented from serving his term by his election as a *Völkisch* deputy in the Bavarian Diet. After the trial he became disillusioned with the Nazis, mainly because of personalities, but did not openly break with them. He joined the German Nationalists. Some months later he was killed in an automobile accident.

Wilhelm Frick was one of the few Nazi leaders who did not fight in the War. Though in good health, he remained an official in the Palatinate. After the Republic of Councils he became chief of the political police in Munich. He spent some months in prison for his part in neutralizing the police during the *Putsch*, but was released when elected to the Reichstag. He was the first National Socialist appointed to a provincial cabinet — as minister of the interior in Thuringia. In 1930 he tried to obtain German citizenship for Hitler by making him a gendarme in the small Thuringian town of Hildburghausen. Hitler himself dropped the scheme as a result of public ridicule. When Hitler became Chancellor, in January, 1933, Frick became Reich Minister of the Interior. He helped the National Socialist revolution of March, 1933, by ordering the authorities not to resist.

venal officials the law of whose actions was prescribed, not by the interest of their people and the necessary uprising for its freedom, but by the boss's orders, without regard for the welfare of the national trust confided in them.

And above all he was one of those natures who, contrasting with most of the guardians of our so-called state authority, do not fear the enmity of traitors to the people and the nation, but long for it as for a treasure which a decent man must take for granted. The hatred of Jews and Marxists, their whole campaign of lies and slander, were for him the sole happiness amid the misery of our people.

A man of granite honesty, of antique simplicity and German straightforwardness, for whom the words 'Sooner dead than a slave' were no phrase but the essence of his whole being.

He and his collaborator, Dr. Frick, are in my eyes the only men in a state position who possess the right to be called co-creators of a national Bavaria.

Before we proceeded to hold our first mass meeting, not only did the necessary propaganda material have to be made ready, but the main points of the program also had to be put into print.

In the second volume I shall thoroughly develop the guiding principles which we had in mind, particularly in framing the program. Here I shall only state that it was done, not only to give the young movement form and content, but to make its aims understandable to the broad masses.

Circles of the so-called intelligentsia have mocked and ridiculed this and attempted to criticize it. But the soundness of our point of view at that time has been shown by the effectiveness of this program.

In these years I have seen dozens of new movements arise and they have all vanished and evaporated without trace. A single one remains: The National Socialist German Workers' Party. And today more than ever I harbor the conviction that people can combat it, that they can attempt to paralyze it, that petty party ministers can forbid us to speak and write, but that they will never prevent the victory of our ideas.

When not even memory will reveal the names of the entire present-day state conception and its advocates, the fundamentals of the National Socialist program will be the foundations of a coming state.

Our four months' activities at meetings up to January, 1920, had slowly enabled us to save up the small means that we needed for printing our first leaflet, our first poster, and our program.

If I take the movement's first large mass meeting as the conclusion of this volume, it is because by it the party burst the narrow bonds of a small club and for the first time exerted a determining influence on the mightiest factor of our time, public opinion.

I myself at that time had but one concern: Will the hall be filled, or will we speak to a yawning hall? [1] I had the unshakable inner conviction that if the people came, the day was sure to be a great success for the young movement. And so I anxiously looked forward to that evening.

The meeting was to be opened at 7:30. At 7:15 I entered the Festsaal of the Hofbräuhaus on the Platzl in Munich, and my heart nearly burst for joy. The gigantic hall — for at that time it still seemed to me gigantic — was overcrowded with people, shoulder to shoulder, a mass numbering almost two thousand people. And above all — those people to whom we wanted to appeal had come. Far more than half the hall seemed to be occupied by Communists and Independents.[2] They had resolved that our first demonstration would come to a speedy end.

But it turned out differently. After the first speaker had finished, I took the floor. A few minutes later there was a hail of shouts, there were violent clashes in the hall, a handful of the most faithful war comrades and other supporters battled with the disturbers, and only little by little were able to restore order.

---

[1] '*gähnende Halle.*' Second edition for obvious reasons changes this to '*gähnende Leere,*' a gaping void.

[2] In April, 1917, a group of Socialists opposing the war left the party and founded the Independent Social Democratic Party under the leadership of Haase and Kautsky.

I was able to go on speaking. After half an hour the applause slowly began to drown out the screaming and shouting.

I now took up the program and began to explain it for the first time.

From minute to minute the interruptions were increasingly drowned out by shouts of applause. And when I finally submitted the twenty-five theses, point for point, to the masses and asked them personally to pronounce judgment on them, one after another was accepted with steadily mounting joy, unanimously and again unanimously, and when the last thesis had found its way to the heart of the masses, there stood before me a hall full of people united by a new conviction, a new faith, a new will.

When after nearly four hours the hall began to empty and the crowd, shoulder to shoulder, began to move, shove, press toward the exit like a slow stream, I knew that now the principles of a movement which could no longer be forgotten were moving out among the German people.

A fire was kindled from whose flame one day the sword must come which would regain freedom for the Germanic Siegfried and life for the German nation.

And side by side with the coming resurrection, I sensed that the goddess of inexorable vengeance for the perjured deed of November 9, 1919, was striding forth.

Thus slowly the hall emptied.

The movement took its course.

MEIN KAMPF
VOLUME
TWO

THE NATIONAL SOCIALIST MOVEMENT

Philosophy and Party

O_N F_{EBRUARY} 24, 1920, the first great public demonstration of our young movement took place. In the Festsaal of the Munich Hofbräuhaus the twenty-five theses of the new party's program were submitted to a crowd of almost two thousand and every single point was accepted amid jubilant approval.

With this the first guiding principles and directives were issued for a struggle which was to do away with a veritable mass of old traditional conceptions and opinions and with unclear, yes, harmful, aims. Into the rotten and cowardly bourgeois world and into the triumphant march of the Marxist wave of conquest a new power phenomenon was entering, which at the eleventh hour would halt the chariot of doom.

It was self-evident that the new movement could hope to achieve the necessary importance and the required strength for this gigantic struggle only if it succeeded from the very first day in arousing in the hearts of its supporters the holy conviction that with it political life was to be given, not to a new *election slogan*, but to a new *philosophy* of fundamental significance.

We must bear in mind from what wretched viewpoints so-called *'party programs'* are normally patched together and from time to time refurbished or remodeled. We must submit the driving motives particularly of these bourgeois *'program-commissions'* to our magnifying glass, in order to achieve the necessary understanding for the evaluation of these programmatical *monstrosities*.

It is always one sole concern which impels men to set up new programs or to change existing ones: concern for the next election. As soon as it dawns on these parliamentary 'jugglers' that the beloved people are again revolting and would like to slip out of the harness of the old party cart, they begin to repaint the shafts. Then come the stargazers and party astrologers, the so-called 'experienced,' 'shrewd' men, old parliamentarians as a rule, who in their 'rich period of apprenticeship' can recall analogous cases when the patience of the masses had burst, and who now sense that something similar is again menacingly close. And so they take up the old prescriptions, form a 'commission,' go about listening to the voice of the beloved people, sniff at the products of the press, and thus slowly scent what the dear broad masses would like to have, what they detest and what they hope for. Every professional group, even every class of employees, is studied with the greatest precision and their most secret wishes investigated. Even the 'bad slogans' of the dangerous opposition then suddenly become ripe for examination, and, not seldom to the greatest amazement of their original inventors and disseminators, turn up, quite innocently and naturally, in the old parties' treasury of knowledge.

And so the commissions come together and 'revise' the old program and frame a new one (and in so doing the gentlemen change their convictions as a soldier in the field changes his shirt, which is when the old one is full of lice!), in which everybody gets his share. The peasant gets protection for his agriculture, the industrialist protection for his product, the consumer protection for his purchase, the teachers' salaries are raised, the civil servants' pensions are improved, widows and orphans are to be taken care of most liberally by the state, trade is promoted, tariffs are to be reduced, and taxes are pretty much, if not altogether, done away with. Occasionally it transpires that some group has been forgotten after all, or that some demand circulating among the people has not been heard of. Then anything there is room for is patched in with the greatest haste, until the framers can hope with a clear conscience that the army of run-of-the-mill petty bourgeois with their women have been pacified and are simply delighted. Thus inwardly armed with confidence in God and the unshakable stupidity of the voting citizenry, the politicians can begin the fight for the 'remaking' of the Reich as they call it.

Then, when election day is past and the parliamentarians have held their last mass meeting in five years, to turn from the training of the plebs to their higher and more agreeable tasks, the program commission again dissolves and the fight for the remodeling of things again takes the form of a struggle for daily bread: which in parliament is known as attendance fees.

Every morning Mr. People's deputy betakes himself to the exalted House, and even if he doesn't go in all the way, he at least goes as far as the anteroom where the attendance lists are kept. Aggressively serving the people, he there enters his name and as well-deserved reward accepts a small remuneration for these continuous and exhausting exertions.

After four years, or otherwise during critical weeks when the dissolution of the parliamentary bodies begins to loom closer and closer, an unconquerable urge suddenly comes over the gentlemen. Just as a caterpillar cannot help turning into a butterfly, these parliamentary larvae leave their parliamentary cocoons and, endowed with wings, fly out among the beloved people. Again they talk to their voters, speak of the enormous work they have done and the malignant stubbornness of their opponents, but the incomprehensible masses, instead of gratefully applauding, sometimes hurl vulgar, even bitter, epithets at their heads. If this ingratitude on the part of the people rises to a certain degree, only a single means can help: the party's sheen must be brushed off, the program needs improvement, the commission comes back to life, and the swindle begins again from the beginning. In view of the granite stupidity of our humanity, we have no need to be surprised at the outcome. Led by their press and dazzled by a new and alluring program, the 'bourgeois' as well as the 'proletarian' voting cattle return to the common stable and again vote for their old misleaders.

Thus, the man of the people and the candidate of the working classes turns himself back into the parliamentary caterpillar and again fattens on the foliage of state life, and again after four years turns back into a gleaming butterfly.

There is scarcely anything more depressing than to observe this whole process in the light of sober reality, to be obliged to witness this constantly repeated deception.

From such spiritual soil the bourgeois camp, you may be assured, cannot draw the strength to carry on the struggle with the organized power of Marxism.

And of this the gentlemen never think seriously. In view of all the admitted narrow-mindedness and mental inferiority of these parliamentary medicine-men of the white race, they themselves cannot seriously imagine that by way of Western democracy they can fight against a doctrine for which democracy, along with everything connected with it, is at best a means used to paralyze the adversary and to create a free path for its own activity. Though at present a part of the Marxists shrewdly try to pretend that they are inseparably linked with the principles of democracy, do not forget if you please that in the critical hour these gentlemen didn't care a damn about a majority decision in the Western-democratic sense! This was in the days when the bourgeois parliamentarians saw the security of the Reich guaranteed by the monumental small-mindedness of a superior number, while the Marxists, with a band of bums, deserters, party bosses, and Jewish journalists, abruptly seized power, thus giving democracy a resounding slap in the face. So it really takes the credulous mind of one of these parliamentary medicine-men of bourgeois democracy to imagine that now or in the future the brutal determination of those interested in and supporting that world plague could be exorcised merely by the magic formulas of a Western parliamentarianism.

The Marxists will march with democracy until they succeed in indirectly obtaining for their criminal aims the support of even the national intellectual world, destined by them for extermination. If today they came to the conviction that from the witches' cauldron of our parliamentary democracy a majority could be brewed, which — and even if only on the basis of its legislating majority — would seriously attack Marxism, the parliamentary jugglery would come to an end at once. The banner-bearers of the Red International would then, instead of addressing an appeal to the democratic conscience, emit a fiery call to the proletarian masses, and their struggle at one stroke would be removed from the stuffy air of our parliamentary meeting halls to the factories and the streets. Democracy would be done for immediately; what the mental dexterity of those people's apostles in the parliaments had failed to do, the crowbar and sledgehammer of incited proletarian masses would instantly succeed in doing, as in the fall of 1918: they would drive it home to the bourgeois world how insane it is to imagine that they can oppose Jewish world domination with the methods of Western democracy.

As I have said, it requires a credulous mind to bind oneself, in facing such a player, by rules which for him are only good for bluff or his own profit, and are thrown overboard as soon as they cease to be to his advantage.

Since with all parties of a so-called bourgeois orientation in reality the whole political struggle actually consists in nothing but a mad rush for seats in parliament, in which convictions and principles are thrown overboard like sand ballast whenever it seems expedient, their programs are naturally tuned accordingly and — inversely, to be sure — their forces also measured by the same standard. They lack that great magnetic attraction which alone the masses always follow under the compelling impact of towering great ideas, the persuasive force of absolute belief in them, coupled with a fanatical courage to fight for them.

At a time when one side, armed with all the weapons of a philosophy, a thousand times criminal though it may be, sets out to storm an existing order, the other side, now and forever can offer resistance only if it clads itself in the forms of a new faith, in our case a political one, and for a weak-kneed, cowardly defensive substitutes the battle-cry of courageous and brutal attack. And so, if today our movement gets the witty reproach that it is working toward a

'*revolution*,' especially from the so-called national bourgeois ministers, say of the Bavarian Center, the only answer we can give
one of political twerps is this: Yes, indeed, we are trying to make
up for what you in your criminal stupidity failed to do. By
the principles of your parliamentary cattle-trading, you helped to
drag the nation into the abyss; but we, in the form of attack
and by setting up a new philosophy of life and by fanatically and
indomitably defending its principles, shall build for our people
the steps on which it will some day climb back into the temple of
freedom.

And so, in the founding period of our movement, our first concern had always to be directed toward preventing the host of
warriors for an exalted conviction from becoming a mere club for
the advancement of parliamentary interests.

The first precautionary measure was the creation of a program
which aimed at a development which by its very inner greatness
seemed apt to scare away the small and feeble spirits of our present party politicians.

How correct was our conception of the necessity of programmatic aims of the sharpest stamp could be seen most clearly from
those catastrophic weaknesses which finally led to the collapse
of Germany.

From the realization of these weaknesses a new state conception, which in itself in turn is an essential ingredient of a new
world conception, would inevitably take form.

*   *   *

In the first volume I have dealt with the word 'folkish,' in so
far as I was forced to establish that this term seems inadequately
defined to permit the formation of a solid fighting community.
All sorts of people, with a yawning gulf between everything essential in their opinions, are running around today under the
blanket term 'folkish.' Therefore, before I proceed to the tasks
and aims of the National Socialist German Workers' Party, I

should like to give a clarification of the concept 'folkish,' as well as its relation to the party movement.

The concept *'folkish'* seems as vaguely defined, open to as many interpretations and as unlimited in practical application as, for instance, the word 'religious,' and it is very hard to conceive of anything absolutely precise under this designation, either in the sense of intellectual comprehension or of practical effects. The designation 'religious' only becomes tangibly conceivable in the moment when it becomes connected with a definitely outlined form of its practice. It is a very lovely statement and usually apt, to describe a man's nature as 'profoundly religious.' Perhaps there are a few people who feel satisfied by such a very general description, to whom it can even convey a definite, more or less sharp, picture of that soul-state. But, since the great masses consist neither of philosophers nor of saints, such a very general religious idea will as a rule mean to the individual only the liberation of his individual thought and action, without, however, leading to that efficacy which arises from religious inner longing in the moment when, from the purely metaphysical infinite world of ideas, a clearly delimited faith forms. Assuredly, this is not the end in itself, but only a means to the end; yet it is the indispensably necessary means which alone makes possible the achievement of the end. This end, however, is not only ideal, but in the last analysis also eminently practical. And in general we must clearly acknowledge the fact that the highest ideals always correspond to a deep vital necessity, just as the nobility of the most exalted beauty lies in the last analysis only in what is logically most expedient.

By helping to raise man above the level of bestial vegetation, faith contributes in reality to the securing and safeguarding of his existence. Take away from present-day mankind its education-based, religious-dogmatic principles — or, practically speaking, ethical-moral principles — by abolishing this religious education, but without replacing it by an equivalent, and the result will be a grave shock to the foundations of their existence. We may therefore state that not only does man live in order to serve higher ideals, but that, conversely, these higher ideals also provide the premise for his existence. Thus the circle closes.

Of course, even the general designation 'religious' includes various basic ideas or convictions, for example, the indestructibility of the soul, the eternity of its existence, the existence of a higher being, etc. But all these ideas, regardless how convincing they may be for the individual, are submitted to the critical examination of this individual and hence to a fluctuating affirmation or negation until emotional divination or knowledge assumes the binding force of apodictic faith. This, above all, is the fighting factor which makes a breach and opens the way for the recognition of basic religious views.

Without clearly delimited faith, religiosity with its unclarity and multiplicity of form would not only be worthless for human life, but would probably contribute to general disintegration.

The situation with the term 'folkish' is similar to that with the term 'religious.' In it, too, there lie various basic realizations. Though of eminent importance, they are, however, so unclearly defined in form that they rise above the value of a more or less acceptable opinion only if they are fitted into the framework of a political party as basic elements. *For the realization of philosophical ideals and of the demands derived from them no more occurs through men's pure feeling or inner will in themselves than the achievement of freedom through the general longing for it. No, only when the ideal urge for independence gets a fighting organization in the form of military instruments of power can the pressing desire of a people be transformed into glorious reality.*

Every philosophy of life, even if it is a thousand times correct and of highest benefit to humanity, will remain without significance for the practical shaping of a people's life, as long as its principles have not become the banner of a fighting movement which for its part in turn will be a party as long as its activity has not found completion in the victory of its ideas and its party dogmas have not become the new state principles of a people's community.

But if a spiritual conception of a general nature is to serve as a foundation for a future development, the first presupposition is to obtain unconditional clarity with regard to the nature, essence, and scope of this conception, since only on such a basis can a movement be formed which by the inner homogeneity of its convictions can develop the necessary force for struggle. From general ideas a political program must be stamped, from a general philosophy of life a definite political faith. The latter, since its goal must be practically attainable, will not only have to serve the idea in itself, but will also have to take into consideration the means of struggle which are available and must be used for the achievement of this idea. The abstractly correct spiritual conception, which the theoretician has to proclaim, must be coupled with the practical knowledge of the politician. And so an eternal ideal, serving as the guiding star of mankind, must unfortunately resign itself to taking the weaknesses of this mankind into consideration, if it wants to avoid shipwreck at the very outset on the shoals of general human inadequacy. To draw from the realm of the eternally true and ideal that which is humanly possible for small mortals, and make it take form, the search after truth must be coupled with knowledge of the people's psyche.

This transformation of a general, philosophical, ideal conception of the highest truth into a definitely delimited, tightly organized political community of faith and struggle, unified in spirit and will, is the most significant achievement, since on its happy solution alone the possibility of the victory of an idea depends. From the army of often millions of men, who as individuals more or less clearly and definitely sense these truths, and in part perhaps comprehend them, *one* man must step forward who with apodictic force will form granite principles from the wavering idea-world of the broad masses and take up the struggle for their sole correctness, until from the shifting waves of a free thought-world there will arise a brazen cliff of solid unity in faith and will.

The general right for such an activity is based on necessity, the personal right on success.

*　　　*　　　*

If from the word 'folkish' we try to peel out the innermost kernel of meaning, we arrive at the following:

Our present political world view, current in Germany, is based in general on the idea that creative, culture-creating force must indeed be attributed to the state, but that it has nothing to do with racial considerations, but is rather a product of economic necessities, or, at best, the natural result of a political urge for power. This underlying view, if logically developed, leads not only to a mistaken conception of basic racial forces, but also to an underestimation of the individual. For a denial of the difference between the various races with regard to their general culture-creating forces must necessarily extend this greatest of all errors to the judgment of the individual. The assumption of the equality of the races then becomes a basis for a similar way of viewing peoples and finally individual men. And hence international Marxism itself is only the transference, by the Jew, Karl Marx, of a philosophical attitude and conception, which had actually long been in existence, into the form of a definite political creed. Without the subsoil of such generally existing poisoning, the amazing success of this doctrine would never have been possible. Actually Karl Marx was only the *one* among millions who, with the sure eye of the prophet, recognized in the morass of a slowly decomposing world the most essential poisons, extracted them, and, like a wizard, prepared them into a concentrated solution for the swifter annihilation of the independent existence of free nations on this earth. And all this in the service of his race.

His Marxist doctrine is a brief spiritual extract of the philosophy of life that is generally current today. And for this reason alone any struggle of our so-called bourgeois world against it is impossible, absurd in fact, since this bourgeois world is also essentially infected by these poisons, and worships a view of life which in general is distinguished from the Marxists only by degrees and personalities. The bourgeois world is Marxist, but believes in the possibility of the rule of certain groups of men (bourgeoisie), while Marxism itself systematically plans to hand the world over to the Jews.

In opposition to this, the folkish philosophy finds the importance of mankind in its basic racial elements. In the state it sees on principle only a means to an end and construes its end as the preservation of the racial existence of man. Thus, it by no means believes in an equality of the races, but along with their difference it recognizes their higher or lesser value and feels itself obligated, through this knowledge, to promote the victory of the better and stronger, and demand the subordination of the inferior and weaker in accordance with the eternal will that dominates this universe. Thus, in principle, it serves the basic aristocratic idea of Nature and believes in the validity of this law down to the last individual. It sees not only the different value of the races, but also the different value of individuals. From the mass it extracts the importance of the individual personality, and thus, in contrast to disorganizing Marxism, it has an organizing effect. It believes in the necessity of an idealization of humanity, in which alone it sees the premise for the existence of humanity. But it cannot grant the right to existence even to an ethical idea if this idea represents a danger for the racial life of the bearers of a higher ethics; for in a bastardized and niggerized world all the concepts of the humanly beautiful and sublime, as well as all ideas of an idealized future of our humanity, would be lost forever.

Human culture and civilization on this continent are inseparably bound up with the presence of the Aryan. If he dies out or declines, the dark veils of an age without culture will again descend on this globe.

The undermining of the existence of human culture by the destruction of its bearer seems in the eyes of a folkish philosophy the most execrable crime. Anyone who dares to lay hands on the highest image of the Lord commits sacrilege against the benevolent creator of this miracle and contributes to the expulsion from paradise.

And so the folkish philosophy of life corresponds to the innermost will of Nature, since it restores that free play of forces which must lead to a continuous mutual higher breeding, until at last the best of humanity, having achieved possession of this earth, will have a free path for activity in domains which will lie partly above it and partly outside it.

We all sense that in the distant future humanity must be faced by problems which only a highest race, become master people and supported by the means and possibilities of an entire globe, will be equipped to overcome.

*   *   *

It is self-evident that so general a statement of the meaningful content of a folkish philosophy can be interpreted in thousands of ways. And actually we find hardly a one of our newer political formations which does not base itself in one way or another on this world view. And, by its very existence in the face of the many others, it shows the difference of its conceptions. And so the Marxist world view, led by a unified top organization, is opposed by a hodge-podge of views which even as ideas are not very impressive in face of the solid, hostile front. Victories are not gained by such feeble weapons! Not until the international world view — politically led by organized Marxism — is confronted by a folkish world view, organized and led with equal unity, will success, supposing the fighting energy to be equal on both sides, fall to the side of eternal truth.

A philosophy can only be organizationally comprehended on the basis of a definite formulation of that philosophy, and what dogmas represent for religious faith, party principles are for a political party in the making.

Hence an instrument must be created for the folkish world view which enables it to fight, just as the Marxist party organization creates a free path for internationalism.

This is the goal pursued by the National Socialist German Workers' Party.

That such a party formulation of the folkish concept is the precondition for the victory of the folkish philosophy of life is proved most sharply by a fact which is admitted indirectly at least by the enemies of such a party tie. Those very people who never weary of emphasizing that the folkish philosophy is not the 'hereditary estate' of an individual, but that it slumbers or 'lives' in the hearts of God knows how many millions, thus demonstrate the fact that the general existence of such ideas was absolutely unable to prevent the victory of the hostile world view, classically represented by a political party. If this were not so, the German people by this time would have been bound to achieve a gigantic victory and not be standing at the edge of an abyss. What gave the international world view success was its representation by a political party organized into storm troops; what caused the defeat of the opposite world view was its lack up to now of a unified body to represent it. Not by unlimited freedom to interpret a general view, but only in the limited and hence integrating form of a political organization can a world view fight and conquer.

Therefore, I saw my own task especially in extracting those nuclear ideas from the extensive and unshaped substance of a general world view and remolding them into more or less dogmatic forms which in their clear delimitation are adapted for holding solidly together those men who swear allegiance to them. In other words: *From the basic ideas of a general folkish world conception the National Socialist German Workers' Party takes over the essential fundamental traits, and from them, with due consideration of practical reality, the times, and the available human material as well as its weaknesses, forms a political creed which, in turn, by the strict organizational integration of large human masses thus made possible, creates the precondition for the victorious struggle of this world view.*

The State

Bᵧ 1920 to 1921, time and again the circles of the present outlived bourgeois world held it up to our young movement that our attitude toward the present-day state was negative, which made the political crooks of all tendencies feel justified in undertaking to suppress the young prophet of a new world view with all possible means. Of course they purposely forgot that the present bourgeois world itself can no longer form any unified picture of the *state* concept, that there neither is nor can be any uniform definition of it. For the explainers usually sit in our state universities in the form of political law professors, whose highest task it must be to find explanations and interpretations for the more or less unfortunate existence of their momentary source of bread. The more impossible the nature of such a state is, the more opaque, artificial, and unintelligible are the definitions regarding the purpose of its existence. What, for example, could a royal and imperial university professor formerly write about the sense and purpose of the state in a country whose state existence embodied the greatest monstrosity of the twentieth century? A grave task if we consider that for the present-day teacher of political law there is less obligation to truth than bondage to a definite purpose. And the purpose is: preservation at any price of the current monstrosity of human mechanism,[1] now called state. We have no call to be surprised if in the dis-

[1] '*Ein Monstrum von menschlichem Mechanismus.*'

cussion of this problem practical criteria are avoided as much as possible, and instead the professors dig themselves into a hodge-podge of 'ethical,' 'moral,' and other ideal values, tasks, and aims.

In general three conceptions can be distinguished:

(a) The troop of those who regard the state simply as a *more or less voluntary grouping of people under a governmental power.*

This group is the most numerous. In its ranks are found particularly the worshipers of our present-day principle of legitimacy, in whose eyes the will of the people plays no rôle in this whole matter. According to these saints, a sacred inviolability is based on the mere fact of the state's existence. To protect this madness of human brains, a positively dog-like veneration of so-called *state authority* is needed. In the minds of such people a means becomes an ultimate end in the twinkling of an eye. The state no longer exists to serve men; men exist in order to worship a state authority which embraces even the most humble spirit, provided he is in any sense an official. Lest this condition of silent, ecstatic veneration turn into one of unrest, the state authority for its part exists only to maintain peace and order. It, too, is now an end and no longer a means.[1] State authority must provide for peace and order, and peace and order in turn must conversely make possible the existence of state authority. Within these two poles all life must now revolve.

In Bavaria, such a conception is primarily represented by the political artists of the Bavarian Center, known as the 'Bavarian People's Party'; in Austria, it was the Black-and-Yellow Legitimists; in the Reich itself, unfortunately, it is often so-called conservative elements whose conception of the state moves along these paths.

(b) The second group of people is somewhat smaller in number, since among it must be reckoned those who at least attach a few conditions to the existence of the state. They desire not only

[1] '*ein Zweck.*' Second edition changes this to '*kein Zweck,*' though this may be a misprint. The meaning then would be: 'no longer an end or a means.'

uniform administration, but also, if possible, *uniform language* —
if only for general technical reasons of administration. State
authority is no longer the sole and exclusive purpose of the state,
but to it is added the promotion of the subjects' welfare. Ideas of
'freedom,' mostly of a misunderstood nature, inject themselves
into the state conceptions of these circles. The form of government no longer seems inviolable by the mere fact of its existence,
but is examined as to its expediency. The sanctity of age offers
no protection against the criticism of the present. Furthermore,
it is a conception which expects that the state above all will
beneficially shape the economic life of the individual, and which
therefore judges on the basis of practical criteria and general
economic conceptions of the profitable. We find the main representatives of these views in the circles of our normal German
bourgeoisie, especially in those of our liberal democracy.

(c) The third group is numerically the weakest.

It regards the state as a means for the realization of usually
very unclearly conceived aims of a state-people linguistically
stamped and united. The will for a uniform state language is
here expressed, not only in the hope of giving this state a foundation capable of supporting an outward increase of power, but not
less in the opinion — basically erroneous, incidentally — that
this will make it possible to carry through a nationalization in a
definite direction.

In the last hundred years it has been a true misery to observe
how these circles, sometimes in the best good faith, played with
the word 'Germanize.' I myself still remember how in my youth
this very term led to incredibly false conceptions. Even in Pan-
German circles the opinion could then be heard that the Austrian-
Germans, with the promotion and aid of the government, might
well succeed in a *Germanization* of the Austrian Slavs; these circles never even began to realize that *Germanization* can only be
applied to *soil* and never to *people*. For what was generally understood under this word was only the forced outward acceptance
of the German language. But it is a scarcely conceivable fallacy
of thought to believe that a *Negro* or a Chinese, let us say, will

turn into a German because he learns German and is willing to speak the German language in the future and perhaps even give his vote to a German political party. That any such Germanization is in reality a de-Germanization never became clear to our bourgeois national world. For if today, by forcing a universal language on them, obvious differences between different peoples are bridged over and finally effaced, this means the beginning of a bastardization, and hence in our case not a Germanization but a destruction of the Germanic element. Only too frequently does it occur in history that conquering people's outward instruments of power succeed in forcing their language on oppressed peoples, but that after a thousand years their language is spoken by another people, and the victors thereby actually become the vanquished.

Since nationality or rather race does not happen to lie in language but in the blood, we would only be justified in speaking of a Germanization if by such a process we succeeded in transforming the blood of the subjected people. But this is impossible. Unless a blood mixture brings about a change, which, however, means the lowering of the level of the higher race. The final result of such a process would consequently be the destruction of precisely those qualities which had formerly made the conquering people capable of victory. Especially the cultural force would vanish through a mating with the lesser race, even if the resulting mongrels spoke the language of the earlier, higher race a thousand times over. For a time, a certain struggle will take place between the different mentalities, and it may be that the steadily sinking people, in a last quiver of life, so to speak, will bring to light surprising cultural values. But these are only individual elements belonging to the higher race, or perhaps bastards in whom, after the first crossing, the better blood still predominates and tries to struggle through; but never final products of a mixture. In them a culturally backward movement will always manifest itself.

Today it must be regarded as a good fortune that a Germanization as intended by Joseph II in Austria was not carried out. Its result would probably have been the preservation of the Austrian state, but also the lowering of the racial level of the German nation induced by a linguistic union. In the course of the centuries a certain herd instinct would doubtless have crystallized out, but the herd itself would have become inferior. A state-people would perhaps have been born, but a culture-people would have been lost.

For the German nation it was better that such a process of mixture did not take place, even if this was not due to a noble insight, but to the shortsighted narrowness of the Habsburgs. If it had turned out differently, the German people could scarcely be regarded as a cultural factor.

Not only in Austria, but in Germany as well, so-called national circles were moved by similar false ideas. The Polish policy, demanded by so many, involving a Germanization of the East, was unfortunately based on the same false inference. Here again it was thought that a Germanization of the Polish element could be brought about by a purely linguistic integration with the German element. Here again the result would have been catastrophic; a people of alien race expressing its alien ideas in the German language, compromising the lofty dignity of our own nationality by their own inferiority.

How terrible is the damage indirectly done to our Germanism today by the fact that, due to the ignorance of many Americans, the German-jabbering Jews, when they set foot on American soil, are booked to our German account. Surely no one will call the purely external fact that most of this lice-ridden migration from the East speaks German a proof of their German origin and nationality.

What has been profitably Germanized in history is the soil which our ancestors acquired by the sword and settled with German peasants. In so far as they directed foreign blood into our national body in this process, they contributed to that catastrophic splintering of our inner being which is expressed in German super-individualism — a phenomenon, I am sorry to say, which is praised in many quarters.

Also in this third group, the state in a certain sense still passes

as an end in itself, and the preservation of the state, consequently, as the highest task of human existence.

In summing up we can state the following: All these views have
their deepest root, not in the knowledge that the forces which
create culture and values are based essentially on racial elements
and that the state must, therefore, in the light of reason, regard
its highest task as the preservation and intensification of the race,
this fundamental condition of all human cultural development.

It was the Jew, Karl Marx, who was able to draw the extreme
inference from those false conceptions and views concerning the
nature and purpose of a state: by detaching the state concept from
racial obligations without being able to arrive at any other
equally acknowledged formulation, the bourgeois world even
paved the way for a doctrine which denies the state as such.

Even in this field, therefore, the struggle of the bourgeois
world against the Marxist international must fail completely. It
long since sacrificed the foundations which would have been indispensably necessary for the support of its own ideological
world. Their shrewd foe recognized the weaknesses of their own
structure and is now storming it with the weapons which they
themselves, even if involuntarily, provided.

It is, therefore, the first obligation of a new movement, standing on the ground of a folkish world view, to make sure that its
conception of the nature and purpose of the state attains a uniform and clear character.

Thus the basic realization is: *that the state represents no end, but
a means. It is, to be sure, the premise for the formation of a higher
human culture, but not its cause, which lies exclusively in the existence of a race capable of culture.* Hundreds of exemplary states
might exist on earth, but if the Aryan culture-bearer died out,
there would be no culture corresponding to the spiritual level of
the highest peoples of today. We can go even farther and say that
the fact of human state formation would not in the least exclude
the possibility of the destruction of the human race, provided that
superior intellectual ability and elasticity would be lost due to the
absence of their racial bearers.

If today, for example, the surface of the earth were upset by some tectonic event and a new Himalaya rose from the ocean floods, by one single cruel catastrophe the culture of humanity would be destroyed. No state would exist any longer, the bands of all order would be dissolved, the documents of millennial development would be shattered — a single great field of corpses covered by water and mud. But if from this chaos of horror even a few men of a certain race capable of culture had been preserved, the earth, upon settling, if only after thousands of years, would again get proofs of human creative power. Only the destruction of the last race capable of culture and its individual members would desolate the earth for good. Conversely, we can see even by examples from the present that state formations in their tribal beginnings can, if their racial supporters lack sufficient genius, not preserve them from destruction. Just as great animal species of prehistoric times had to give way to others and vanish without trace, man must also give way if he lacks a definite spiritual force which alone enables him to find the necessary weapons for his self-preservation.

The *state* in itself does not create a specific cultural level; it can only preserve the race which conditions this level. Otherwise the state as such may continue to exist unchanged for centuries while, in consequence of a racial mixture which it has not prevented, the cultural capacity of a people and the general aspect of its life conditioned by it have long since suffered a profound change. The present-day state, for example, may very well simulate its existence as a formal mechanism for a certain length of time, but the racial poisoning of our national body creates a cultural decline which even now is terrifyingly manifest.

Thus, the precondition for the existence of a higher humanity is not the state, but the nation possessing the necessary ability.

This ability will fundamentally always be present and must only be aroused to practical realization by certain outward conditions. Culturally and creatively gifted nations, or rather races, bear these useful qualities latent within them, even if at the moment unfavorable outward conditions do not permit a realiza-

tion of these latent tendencies. Hence it is an unbelievable offense to represent the Germanic peoples of the pre-Christian era
as 'cultureless,' as barbarians. That they never were. Only the
harshness of their northern homeland forced them into circumstances which thwarted the development of their creative forces.
If, without any ancient world, they had come to the more favorable regions of the south, and if the material provided by lower
peoples had given them their first technical implements, the culture-creating ability slumbering within them would have grown
into radiant bloom just as happened, for example, with the Greeks.
But this primeval culture-creating force itself arises in turn not
from the northern climate alone. The Laplander, brought to
the south, would be no more culture-creating than the Eskimo.
For this glorious creative ability was given only to the Aryan,
whether he bears it dormant within himself or gives it to awakening life, depending whether favorable circumstances permit this
or an inhospitable Nature prevents it.

From this the following realization results:

*The state is a means to an end. Its end lies in the preservation
and advancement of a community of physically and psychically
homogeneous creatures. This preservation itself comprises first of all
existence as a race and thereby permits the free development of all the
forces dormant in this race. Of them a part will always primarily
serve the preservation of physical life, and only the remaining part
the promotion of a further spiritual development. Actually the one
always creates the precondition for the other.*

*States which do not serve this purpose are misbegotten, monstrosi-
ties in fact. The fact of their existence changes this no more than the
success of a gang of bandits can justify robbery.*

We National Socialists as champions of a new philosophy of
life must never base ourselves on so-called 'accepted facts' —
and false ones at that. If we did, we would not be the champions of a new great idea, but the coolies of the present-day lie.
We must distinguish in the sharpest way between the state as a
vessel and the race as its content. This vessel has meaning only
if it can preserve and protect the content; otherwise it is useless.

Thus, the highest purpose of a folkish state is concern for the preservation of those original racial elements which bestow culture and create the beauty and dignity of a higher mankind. We, as Aryans, can conceive of the state only as the living organism of a nationality which not only assures the preservation of this nationality, but by the development of its spiritual and ideal abilities leads it to the highest freedom.

But what they try to palm off on us as a state today is usually nothing but a monstrosity born of deepest human error, with untold misery as a consequence.

We National Socialists know that with this conception we stand as revolutionaries in the world of today and are also branded as such. But our thoughts and actions must in no way be determined by the approval or disapproval of our time, but by the binding obligation to a truth which we have recognized. Then we may be convinced that the higher insight of posterity will not only understand our actions of today, but will also confirm their correctness and exalt them.

*   *   *

From this, we National Socialists derive a standard for the evaluation of a state. This value will be relative from the standpoint of the individual nationality, absolute from that of humanity as such. This means, in other words:

The quality of a state cannot be evaluated according to the cultural level or the power of this state in the frame of the outside world, but solely and exclusively by the degree of this institution's virtue for the nationality involved in each special case.

A state can be designated as exemplary if it is not only compatible with the living conditions of the nationality it is intended to represent, but if in practice it keeps this nationality alive by its own very existence — quite regardless of the importance of this state formation within the framework of the outside world. For the function of the state is not to create abilities, but only to open the road for those forces which are present. *Thus, conversely, a state can be designated as bad if, despite a high cultural level, it dooms the bearer of this culture in his racial composition.* For thus it destroys to all intents and purposes the premise for the survival of this culture which it did not create, but which is the fruit of a culture-creating nationality safeguarded by a living integration through the state. The state does not represent the content, but a form. *A people's cultural level at any time does not, therefore, provide a standard for measuring the quality of the state* in which it lives. It is easily understandable that a people highly endowed with culture offers a more valuable picture than a Negro tribe; nevertheless, the state organism of the former, viewed according to its fulfillment of purpose, can be inferior to that of the Negro. Though the best state and the best state form are not able to extract from a people abilities which are simply lacking and never did exist, a bad state is assuredly able to kill originally existing abilities by permitting or even promoting the destruction of the racial culture-bearer.

Hence our judgment concerning the quality of a state can primarily be determined only by the relative utility it possesses for a definite nationality, and in no event by the intrinsic importance attributable to it in the world.

This relative judgment can be passed quickly and easily, but the judgment concerning absolute value only with great difficulty, since this absolute judgment is no longer determined merely by the state, but by the quality and level of the nationality in question.

If, therefore, we speak of a higher mission of the state, we must not forget that the higher mission lies essentially in the nationality whose free development the state must merely make possible by the organic force of its being.

Hence, if we propound the question of how the state which we Germans need should be constituted, we must first clearly understand what kind of people it is to contain and what purpose it is to serve.

Our German nationality, unfortunately, is no longer based on a unified racial nucleus. The blending process of the various original components has advanced so far that we might speak of a new race. On the contrary, the poisonings of the blood which have befallen our people, especially since the Thirty Years' War, have led not only to a decomposition of our blood, but also of our soul. The open borders of our fatherland, the association with un-German foreign bodies along these frontier districts, but above all the strong and continuous influx of foreign blood into the interior of the Reich itself, due to its continuous renewal, leaves no time for an absolute blending. No new race is distilled out, the racial constituents remain side by side, with the result that, especially in critical moments in which otherwise a herd habitually gathers together, the German people scatters to all the four winds. Not only are the basic racial elements scattered territorially, but on a small scale within the same territory. Beside Nordic men Easterners, beside Easterners Dinarics, beside both of these Westerners, and mixtures in between. On the one hand, this is a great disadvantage: the German people lack that sure herd instinct which is based on unity of the blood and, especially in moments of threatening danger, preserves nations from destruction in so far as all petty inner differences in such peoples vanish at once on such occasions and the solid front of a unified herd confronts the common enemy. This co-existence of unblended basic racial elements of the most varying kind accounts for what is termed *hyper-individualism* in Germany. In peaceful periods it may sometimes do good services, but taking all things together, it has robbed us of world domination. If the German people in its historic development had possessed that herd unity which other peoples enjoyed, the German Reich today would doubtless be mistress of the globe. World history would have taken a different course, and no one can distinguish whether in this way we would not have obtained what so many blinded pacifists today hope to gain by begging, whining, and whimpering: *a peace, supported not by the palm branches of tearful, pacifist female mourners, but based on the victorious sword of a master people, putting the world into the service of a higher culture.*

The fact of the non-existence of a nationality of unified blood has brought us untold misery. It has given capital cities to many small German potentates, but deprived the German people of the master's right.

Today our people are still suffering from this inner division; but what brought us misfortune in the past and present can be our blessing for the future. For detrimental as it was on the one hand that a complete blending of our original racial components did not take place, and that the formation of a unified national body was thus prevented, it was equally fortunate on the other hand that in this way at least a part of our best blood was preserved pure and escaped racial degeneration.

Assuredly, if there had been a complete blending of our original racial elements, a unified national body would have arisen; however, as every racial cross-breeding proves, it would have been endowed with a smaller cultural capacity than the highest of the original components originally possessed. This is the blessing of the absence of complete blending: that today in our German national body we still possess great unmixed stocks of Nordic-Germanic people whom we may consider the most precious treasure for our future. In the confused period of ignorance of all racial laws, when a man appeared to be simply a man, with full equality — clarity may have been lacking with regard to the different value of the various original elements. Today we know that a complete intermixture of the components of our people might, in consequence of the unity thus produced, have given us outward power, but that the highest goal of mankind would have been unattainable, since the sole bearer, whom Fate had clearly chosen for this completion, would have perished in the general racial porridge of the unified people.

But what, through none of our doing, a kind Fate prevented, we must today examine and evaluate from the standpoint of the knowledge we have now acquired.

Anyone who speaks of a mission of the German people on earth must know that it can exist only in the formation of a state which sees its highest task in the preservation and promotion of the most noble elements of our nationality, indeed of all mankind, which still remain intact.

Thus, for the first time the state achieves a lofty inner goal. Compared to the absurd catchword about safeguarding law and order, thus laying a peaceable groundwork for mutual swindles, the task of preserving and advancing the highest humanity, given to this earth by the benevolence of the Almighty, seems a truly high mission.

From a dead mechanism which only lays claim to existence for its own sake, there must be formed a living organism with the exclusive aim of serving a higher idea.

The German Reich as a state must embrace all Germans and has the task, not only of assembling and preserving the most valuable stocks of basic racial elements in this people, but slowly and surely of raising them to a dominant position.

*   *   *

Thus, a condition which is fundamentally one of paralysis is replaced by a period of struggle, but as everywhere and always in this world, here, too, the saying remains valid that 'he who rests — rusts,' and, furthermore, that victory lies eternally and exclusively in attack. The greater the goal we have in mind in our struggle, and the smaller the understanding of the broad masses for it may be at the moment, all the more gigantic, as the experience of world history shows, will be the success — and the significance of this success if the goal is correctly comprehended and the struggle is carried through with unswerving perseverance.

Of course it may be more soothing for many of our present official helmsmen of the state to work for the preservation of an *existing* condition than having to fight for a new one. They will find it much easier to regard the state as a mechanism which exists simply in order to keep itself alive, since in turn their lives *'belong to the state'* — as they are accustomed to put it. As though something which sprang from the nationality could logically serve anything else than the nationality or man could work for anything else than man. *Of course, as I have said before, it is easier to see in state authority the mere formal mechanism of an organization than the sovereign embodiment of a nationality's instinct of self-preservation on earth.* For in the one case the state, as well as state authority, is for these weak minds a purpose in itself, while in the other, it is only a mighty weapon in the service of the great, eternal life struggle for existence, a weapon to which everyone must submit because it is not formal and mechanical, but the expression of a common will for preserving life.

Hence, in the struggle for our new conception, which is entirely in keeping with the primal meaning of things, we shall find few fellow warriors in a society which not only is physically senile but, sad to say, usually, mentally as well. Only exceptions, old men with young hearts and fresh minds, will come to us from those classes, never those who see the ultimate meaning of their life task in the preservation of an existing condition.

We are confronted by the endless army, not so much of the deliberately bad as of the mentally lazy and indifferent, including those with a stake in the preservation of the present condition. But precisely in this apparent hopelessness of our gigantic struggle lies the greatness of our task and also the possibility of our success. The battle-cry which either scares away the small spirits at the very start, or soon makes them despair, will be the signal for the assemblage of real fighting natures. And this we must see clearly: *If in a people a certain amount of the highest energy and active force seems concentrated upon one goal and hence is definitively removed from the inertia of the broad masses, this small percentage has risen to be master over the entire number. World history is made by minorities when this minority of number embodies the majority of will and determination.*

What, therefore, may appear as a difficulty today is in reality the premise for our victory. Precisely in the greatness and the difficulties of our task lies the probability that only the best fighters will step forward to struggle for it. And in this selection lies the guaranty of success.

*   *   *

In general, Nature herself usually makes certain corrective decisions with regard to the racial purity of earthly creatures. She has little love for bastards. Especially the first products of such cross-breeding, say in the third, fourth, and fifth generation, suffer bitterly. Not only is the value of the originally highest element of the cross-breeding taken from them, but with their lack of blood unity they lack also unity of will-power and determination to live. In all critical moments in which the racially unified being makes correct, that is, unified decisions, the racially divided one will become uncertain; that is, he will arrive at half measures. Taken together, this means not only a certain inferiority of the racially divided being compared with the racially unified one, but in practice also the possibility of a more rapid decline. *In innumerable cases where race holds up, the bastard breaks down.* In this, we must see the correction of Nature. But often she goes even further. She limits the possibility of propagation. Thereby she prevents the fertility of continued crossings altogether and thus causes them to die out.

If, for example, an individual specimen of a certain race were to enter into a union with a racially lower specimen, the result would at first be a lowering of the standard in itself; but, in addition, there would be a weakening of the offspring as compared to the environment that had remained racially unmixed. If an influx of further blood from the highest race were prevented entirely, the bastards, if they continued mutually to cross, would either die out because their power of resistance had been wisely diminished by Nature, or in the course of many millenniums a new mixture would form in which the original individual elements would be completely blended by the thousandfold crossing and therefore no longer recognizable. Thus a new nationality would have formed with a certain herd resistance, but, compared to the highest race participating in the first crossing, seriously reduced in spiritual and cultural stature. But in this last case, moreover, the hybrid product would succumb in the mutual struggle for existence as long as a higher racial entity, which has remained unmixed, is still present as an opponent. All the herd solidarity of this new people, formed in the course of thousands of years, would, in consequence of the general lowering of the racial level and the resultant diminution of spiritual elasticity and creative ability, not suffice victoriously to withstand the struggle with an equally unified, but spiritually and culturally superior race.

Hence we can establish the following valid statement:

Every racial crossing leads inevitably sooner or later to the decline of the hybrid product as long as the higher element of this crossing is itself still existent in any kind of racial unity. The danger for the hybrid product is eliminated only at the moment when the last higher racial element is bastardized.

This is a basis for a natural, even though slow, process of regeneration, which gradually eliminates racial poisonings as long as a basic stock of racially pure elements is still present and a further bastardization does not take place.

Such a process can begin of its own accord in creatures with a strong racial instinct, who have only been thrown off the track of normal, racially pure reproduction by special circumstances or some special compulsion. As soon as this condition of compulsion is ended, the part which has still remained pure will at once strive again for mating among equals, thus calling a halt to further mixture. The results of bastardization spontaneously recede to the background, unless their number has increased so infinitely that serious resistance on the part of those who have remained racially pure is out of the question.

Man, once he has lost his instinct and fails to recognize the obligation imposed upon him by Nature, is on the whole not justified in hoping for such a correction on the part of Nature as long as he has not replaced his lost instinct by perceptive knowledge; this knowledge must then perform the required work of compensation. Yet the danger is very great that the man who has once grown blind will keep tearing down the racial barriers more and more, until at length even the last remnant of his best part is lost. Then in reality there remains nothing but a unified mash, such as the famous world reformers of our days idealize; but in a short time it would expel all ideals from this world. Indeed: *a great*

*herd could be formed in this way; a herd beast can be brewed from all
sorts of ingredients, but a man who will be a culture-bearer, or even
better, a culture-founder and culture-creator, never arises from such a
mixture.* The mission of humanity could then be looked upon as
finished.

Anyone who does not want the earth to move toward this condition must convert himself to the conception that it is the function above all of the Germanic states first and foremost to call a
fundamental halt to any further bastardization.

The generation of our present notorious weaklings will obviously cry out against this, and moan and complain about as-
saults on the holiest human rights. *No, there is only one holiest
human right, and this right is at the same time the holiest obligation,
to wit: to see to it that the blood is preserved pure and, by preserving
the best humanity, to create the possibility of a nobler development of
these beings.*

*A folkish state must therefore begin by raising marriage from the
level of a continuous defilement of the race, and give it the consecration of an institution which is called upon to produce images of the
Lord and not monstrosities halfway between man and ape.*

The protest against this on so-called *humane* grounds is particularly ill-suited to an era which on the one hand gives every depraved degenerate the possibility of propagating, but which burdens the products themselves, as well as their contemporaries,
with untold suffering, while on the other hand every drug store
and our very street peddlers offer the means for the prevention
of births for sale even to the healthiest parents. In this present-
day state of law and order in the eyes of its representatives, this
brave, bourgeois-national society, the prevention of the procreative faculty in sufferers from syphilis, tuberculosis, hereditary
diseases, cripples, and cretins is a crime, while the actual sup-
pression of the procreative faculty in millions of the very best
people is not regarded as anything bad and does not offend
against the morals of this hypocritical society, but is rather a
benefit to its short-sighted mental laziness. For otherwise these
people would at least be forced to rack their brains about pro-

viding a basis for the sustenance and preservation of those beings who, as healthy bearers of our nationality, should one day
serve the same function with regard to the coming generation.

How boundlessly unideal and ignoble is this whole system!
People no longer bother to breed the best for posterity, but let
things slide along as best they can. If our churches also sin
against the image of the Lord, whose importance they still so
highly emphasize, it is entirely because of the line of their present
activity which speaks always of the spirit and lets its bearer, the
man, degenerate into a depraved proletarian. Afterwards, of
course, they make foolish faces and are full of amazement at the
small effect of the Christian faith in their own country, at the
terrible 'godlessness,' at this physically botched and hence
spiritually degenerate rabble, and try with the Church's Blessing,
to make up for it by success with the Hottentots and Zulu Kaffirs.
While our European peoples, thank the Lord, fall into a condition
of physical and moral leprosy, the pious missionary wanders off
to Central Africa and sets up Negro missions until there, too, our
'higher culture' turns healthy, though primitive and inferior,
human beings into a rotten brood of bastards.

It would be more in keeping with the intention of the noblest
man in this world if our two Christian churches, instead of annoying Negroes with missions which they neither desire nor understand, would kindly, but in all seriousness, teach our European
humanity that where parents are not healthy it is a deed pleasing
to God to take pity on a poor little healthy orphan child and give
him father and mother, than themselves to give birth to a sick
child who will only bring unhappiness and suffering on himself
and the rest of the world.

The folkish state must make up for what everyone else today
has neglected in this field. *It must set race in the center of all life.
It must take care to keep it pure. It must declare the child to be the
most precious treasure of the people. It must see to it that only the
healthy beget children; that there is only one disgrace: despite one's
own sickness and deficiencies, to bring children into the world, and
one highest honor: to renounce doing so. And conversely it must be*

considered reprehensible: to withhold healthy children from the nation. Here the state must act as the guardian of a millennial future in the face of which the wishes and the selfishness of the individual must appear as nothing and submit. It must put the most modern medical means in the service of this knowledge. It must declare unfit for propagation all who are in any way visibly sick or who have inherited a disease and can therefore pass it on, and put this into actual practice. Conversely, it must take care that the fertility of the healthy woman is not limited by the financial irresponsibility of a state régime which turns the blessing of children into a curse for the parents. It must put an end to that lazy, nay criminal, indifference with which the social premises for a fecund family are treated today, and must instead feel itself to be the highest guardian of this most precious blessing of a people. Its concern belongs more to the child than to the adult.

Those who are physically and mentally unhealthy and unworthy must not perpetuate their suffering in the body of their children. In this the folkish state must perform the most gigantic educational task. And some day this will seem to be a greater deed than the most victorious wars of our present bourgeois era. By education it must teach the individual that it is no disgrace, but only a misfortune deserving of pity, to be sick and weakly, but that it is a crime and hence at the same time a disgrace to dishonor one's misfortune by one's own egotism in burdening innocent creatures with it; that by comparison it bespeaks a nobility of highest idealism and the most admirable humanity if the innocently sick, renouncing a child of his own, bestows his love and tenderness upon a poor, unknown young scion of his own nationality, who with his health promises to become some day a powerful member of a powerful community. And in this educational work the state must perform the purely intellectual complement of its practical activity. It must act in this sense without regard to understanding or lack of understanding, approval or disapproval.

A prevention of the faculty and opportunity to procreate on the part of the physically degenerate and mentally sick, over a period of only six hundred years, would not only free humanity from an immeasurable misfortune, but would lead to a recovery which today seems scarcely conceivable. If the fertility of the healthiest bearers of the nationality is thus consciously and systematically promoted, the result will be a race which at least will have eliminated the germs of our present physical and hence spiritual decay.

For once a people and a state have started on this path, attention will automatically be directed on increasing the racially most valuable nucleus of the people and its fertility, in order ultimately to let the entire nationality partake of the blessing of a highly bred racial stock.

The way to do this is above all for the state not to leave the settlement of newly acquired territories to chance, but to subject it to special norms. Specially constituted racial commissions must issue settlement certificates to individuals. For this, however, definite racial purity must be established. It will thus gradually become possible to found border colonies whose inhabitants are exclusively bearers of the highest racial purity and hence of the highest racial efficiency. This will make them a precious national treasure to the entire nation; their growth must fill every single national comrade with pride and confidence, for in them lies the germ for a final, great future development of our own people, nay — of humanity.

In the folkish state, finally, the folkish philosophy of life must succeed in bringing about that nobler age in which men no longer are concerned with breeding dogs, horses, and cats, but in elevating man himself, an age in which the one knowingly and silently renounces, the other joyfully sacrifices and gives.

That this is possible may not be denied in a world where hundreds and hundreds of thousands of people voluntarily submit to celibacy, obligated and bound by nothing except the injunction of the Church.

Should the same renunciation not be possible if this injunction is replaced by the admonition finally to put an end to the constant and continuous original sin of racial poisoning, and to give the Almighty Creator beings such as He Himself created?

Of course, the miserable army of our present-day shopkeepers will never understand this. They will laugh at it or shrug their crooked shoulders and moan forth their eternal excuse: 'That would be very nice in itself, but it can't be done!' True, it can no longer be done with you, your world isn't fit for it! You know but *one* concern: your personal life, and *one* God: your money! But we are not addressing ourselves to you, we are appealing to the great army of those who are so poor that their personal life cannot mean the highest happiness in the world; to those who do not see the ruling principle of their existence in gold, but in other gods. Above all we appeal to the mighty army of our German youth. They are growing up at a great turning point and the evils brought about by the inertia and indifference of their fathers will force them into struggle. Some day the German youth will either be the builder of a new folkish state, or they will be the last witness of total collapse, the end of the bourgeois world.

For if a generation suffers from faults which it recognizes, even admits, but nevertheless, as occurs today in our bourgeois world, contents itself with the cheap excuse that there is nothing to be done about it — such a society is doomed. The characteristic thing about our bourgeois world is precisely that it can no longer deny the ailments as such. It must admit that much is rotten and bad, but it no longer finds the determination to rebel against the evil, to muster the force of a people of sixty or seventy millions with embittered energy, and oppose it to the danger. On the contrary: if this is done elsewhere, silly comments are made about it, and they attempt from a distance at least to prove the theoretical impossibility of the method and declare success to be inconceivable. And no reason is too absurd to serve as a prop for their own dwarfishness and mental attitude. If, for example, a whole continent finally declares war on alcoholic poisoning, in order to redeem a people from the clutches of this devastating vice, our European bourgeois world has no other comment for it than a meaningless staring and head-shaking, a supercilious ridicule — which is particularly suited to this most ridiculous of all societies. But if all this is to no avail, and if somewhere in the world the sublime, inviolable old routine is opposed, and even with success, then, as said before, the success at least must be doubted and deprecated; and here they do not even shun to raise bourgeois-moral arguments against a struggle which strives to abolish the greatest immorality.

No, we must none of us make any mistake about all of this: our present bourgeoisie has become worthless for every exalted task of mankind, simply because it is without quality and no good; and what makes it no good is not so much in my opinion any *deliberate* malice as an incredible indolence and everything that springs from it. And therefore those political clubs which carry on under the collective concept of 'bourgeois parties' have long ceased to be anything else but associations representing the interests of certain professional groups and classes, and their highest task has ceased to be anything but the best possible selfish defense of their interests. It is obvious that such a political 'bourgeois' guild is good for anything sooner than struggle; especially if the opposing side does not consist of cautious pepper sacks [small tradesmen], but of proletarian masses, incited to extremes and determined to do their worst.

*   *   *

If as the first task of the state in the service and for the welfare of its nationality we recognize the preservation, care, and development of the best racial elements, it is natural that this care must not only extend to the birth of every little national and racial comrade, but that it must educate the young offspring to become a valuable link in the chain of future reproduction.

And as in general the precondition for spiritual achievement lies in the racial quality of the human material at hand, education in particular must first of all consider and promote physical health; for taken in the mass, a healthy, forceful spirit will be found only in a healthy and forceful body. The fact that geniuses are sometimes physically not very fit, or actually sick, is no argument against this. Here we have to do with exceptions which —

as everywhere — only confirm the rule. But if the mass of a people consists of physical degenerates, from this swamp a really great spirit will very seldom arise. In any case his activity will not meet with great success. The degenerate rabble will either not understand him at all, or it will be so weakened in will that it can no longer follow the lofty flight of such an eagle.

Realizing this, the folkish state must not adjust its entire educational work primarily to the inoculation of mere knowledge, but to the breeding of absolutely healthy bodies. The training of mental abilities is only secondary. And here again, first place must be taken by the development of character, especially the promotion of will-power and determination, combined with the training of joy in responsibility, and only in last place comes scientific schooling.

Here the folkish state must proceed from the assumption *that a man of little scientific education but physically healthy, with a good, firm character, imbued with the joy of determination and will-power, is more valuable for the national community than a clever weakling.* A people of scholars, if they are physically degenerate, weak-willed and cowardly pacifists, will not storm the heavens, indeed they will not even be able to safeguard their existence on this earth. In the hard struggle of destiny the man who knows least seldom succumbs, but always he who from his knowledge draws the weakest consequences and is most lamentable in transforming them into action. Here too, finally, a certain harmony must be present. *A decayed body is not made the least more aesthetic by a brilliant mind,* indeed the highest intellectual training could not be justified if its bearers were at the same time physically degenerate and crippled, weak-willed, wavering and cowardly individuals. What makes the Greek ideal of beauty a model is the wonderful combination of the most magnificent physical beauty with brilliant mind and noblest soul.

If Moltke's saying, 'In the long run only the able man has luck,' is anywhere applicable, it is surely to the relation between body and mind; the mind, too, if it is healthy, will as a rule and in the long run dwell only in the healthy body.

Physical training in the folkish state, therefore, is not an affair of the individual, and not even a matter which primarily regards the parents and only secondly or thirdly interests the community; it is a requirement for the self-preservation of the nationality, represented and protected by the state. Just as the state, as far as purely scientific education is concerned, even today interferes with the individual's right of self-determination and upholds the right of the totality toward him by subjecting the child to compulsory education without asking whether the parents want it or not — in far greater measure the folkish state must some day enforce its authority against the individual's ignorance or lack of understanding in questions regarding the preservation of the nationality. It must so organize its educational work that the young bodies are treated expediently in their earliest childhood and obtain the necessary steeling for later life. It must above all prevent the rearing of a generation of hothouse plants.

This work of care and education must begin with the young mother. Just as it became possible in the course of careful work over a period of decades to achieve antiseptic cleanliness in childbirth and reduce puerperal fever to a few cases, it must and will be possible, by a thorough training of nurses and mothers themselves, to achieve a treatment of the child in his first years that will serve as an excellent basis for future development.

The school as such in a folkish state must create infinitely more free time for physical training. It is not permissible to burden young brains with a ballast only a fraction of which they retain, as experience shows, not to mention the fact that as a rule it is unnecessary trifles that stick instead of essentials, since the young child cannot undertake a sensible sifting of the material that has been funneled into him. If today, even in the curriculum of the secondary schools, gymnastics gets barely two hours a week and participation in it is not even obligatory, but is left open to the individual, that is a gross incongruity compared to the purely mental training. Not a day should go by in which the young man does not receive one hour's physical training in the morning and one in the afternoon, covering every type of sport and gymnastics. And here one sport in particular must not be forgotten, which in the eyes of many 'folkish' minded people is considered vulgar and undignified: boxing. It is incredible what false opinions are widespread in 'educated' circles. It is regarded as natural and honorable that a young man should learn to fence and proceed to fight duels right and left, but if he boxes, it is supposed to be vulgar! Why? There is no sport that so much as this one promotes the spirit of attack, demands lightning decisions, and trains the body in steel dexterity. It is no more vulgar for two young men to fight out a difference of opinion with their fists than with a piece of whetted iron. It is not less noble if a man who has been attacked defends himself against his assailant with his fists, instead of running away and yelling for a policeman. But above all, the young, healthy body must also learn to suffer blows. Of course this may seem wild to the eyes of our present spiritual fighters. But it is not the function of the folkish state to breed a colony of peaceful aesthetes and physical degenerates. Not in the respectable shopkeeper or virtuous old maid does it see its ideal of humanity, but in the defiant embodiment of manly strength and in women who are able to bring men into the world.

And so sport does not exist only to make the individual strong, agile and bold; it should also toughen him and teach him to bear hardships.

If our entire intellectual upper crust had not been brought up so exclusively on upper-class etiquette; if instead they had learned boxing thoroughly, a German revolution of pimps, deserters, and such-like rabble would never have been possible; for what gave this revolution success was not the bold, courageous energy of the revolutionaries, but the cowardly, wretched indecision of those who led the state and were responsible for it. The fact is that our whole intellectual leadership had received only 'intellectual' education and hence could not help but be defenseless the moment not intellectual weapons but the crowbar went into action on the opposing side. All this was possible only because as a matter of principle especially our higher educational system did not train men, but officials, engineers, technicians, chemists, jurists, journalists, and to keep these intellectuals from dying out, professors.

Our intellectual leadership always performed brilliant feats, while our leadership in the matter of will-power usually remained beneath all criticism.

Certainly it will not be possible to turn a man of basically cowardly disposition into a courageous man by education, but just as certainly a man who in himself is not cowardly will be paralyzed in the development of his qualities if due to deficiencies in his education he is from the very start inferior to his neighbor in physical strength and dexterity. To what extent the conviction of physical ability promotes a man's sense of courage, even arouses his spirit of attack, can best be judged by the example of the army. Here, too, essentially, we have to deal not solely with heroes but with the broad average. But the superior training of the German soldier in peacetime inoculated the whole gigantic organism with that suggestive faith in its own superiority to an extent which even our foes had not considered possible. For the immortal offensive spirit and offensive courage achieved in the long months of midsummer and autumn 1914 by the forward-sweeping German armies was the result of that untiring training which in the long, long years of peace obtained the most incredible achievement often out of frail bodies, and thus cultivated that self-confidence which was not lost even in the terror of the greatest battles.

Particularly our German people which today lies broken and defenseless, exposed to the kicks of all the world, needs that suggestive force that lies in self-confidence. This self-confidence must be inculcated in the young national comrade from childhood on. His whole education and training must be so ordered as to give him the conviction that he is absolutely superior to others. Through his physical strength and dexterity, he must recover his faith in the invincibility of his whole people. For what formerly led the German army to victory was the sum of the confidence which each individual had in himself and all together in their leadership. What will raise the German people up again is confidence in the possibility of regain-

ing its freedom. And this conviction can only be the final product of the same feeling in millions of individuals.

Here, too, we must not deceive ourselves:

Immense was the collapse of our people, and the exertion needed to end this misery some day will have to be just as immense. Anyone who thinks that our present bourgeois education for peace and order will give our people the strength some day to smash the present world order, which means our doom, and to hurl the links of our slavery into the face of our enemies, is bitterly mistaken. Only by super-abundance of national will-power, thirst for freedom, and highest passion, will we compensate for what we formerly lacked.

*   *   *

The clothing of our youth should also be adapted to this purpose. It is truly miserable to behold how our youth even now is subjected to a fashion madness which helps to reverse the sense of the old saying: 'Clothes make the man' into something truly catastrophic.

Especially in the youth, dress must be put into the service of education. The boy who in summer runs around in long stovepipe trousers, and covered up to the neck, loses through his clothing alone a stimulus for his physical training. For we must exploit ambition and, we may as well calmly admit it, vanity as well. Not vanity about fine clothes which everyone cannot buy, but vanity about a beautiful, well-formed body which everyone can help to build.

This is also expedient for later life. The girl should get to know her beau. If physical beauty were today not forced entirely into the background by our foppish fashions, the seduction of hundreds of thousands of girls by bow-legged, repulsive Jewish bastards would not be possible. This, too, is in the interest of the nation: that the most beautiful bodies should find one another, and so help to give the nation new beauty.

Today, of course, all this is more necessary than ever, because there is no military training, and so the sole institution is excluded which in peacetime compensated at least in part for what was neglected by the rest of our educational system. And there, too, success was to be sought, not only in the training of the individual as such, but in the influence it exerted on the relations between the two sexes. The young girl preferred the soldier to the non-soldier.

The folkish state must not only carry through and supervise physical training in the official school years; in the post-school period as well it must make sure that, as long as a boy is in process of physical development, this development turns out to his benefit. It is an absurdity to believe that with the end of the school period the state's right to supervise its young citizens suddenly ceases, but returns at the military age. This right is a duty and as such is equally present at all times. Only the present-day state having no interest in healthy people has neglected this duty in a criminal fashion. It lets present-day youth go to the dogs on the streets and in brothels, instead of taking them in hand and continuing their physical education until the day when they grow up into a healthy man and a healthy woman.

In what form the state carries on this training is beside the point today; the important thing is that it should do so and seek the ways and means that serve this purpose. The folkish state will have to look on post-school physical training as well as intellectual education as a state function, and foster them through state institutions. This education in its broad outlines can serve as a preparation for future military service. The army will not have to teach the young men the fundamentals of the most elementary drill-book as hitherto, and it will not get recruits of the present type; no, it will only have to transform a young man who has already received flawless physical preparation into a soldier.

In the folkish state, therefore, the army will no longer have to teach the individual how to walk and to stand; it will be the last and highest school of patriotic education. In the army the young recruit will receive the necessary training in arms, and at the same time he will receive a further moulding for any other future career. But in the forefront of military training will stand what has to be regarded as the highest merit of the old army: in this school the boy must be transformed into a man; in this school he must not only learn to obey, but must thereby acquire a basis for commanding later. He must learn to be silent not only when he is *justly* blamed but must also learn, when necessary, to bear *injustice* in silence.

Furthermore, reinforced by faith in his own strength, filled with the force of a commonly experienced *esprit de corps,* he must become convinced of the invincibility of his nationality.

After the conclusion of his military service, two documents should be issued: *His citizen's diploma,* a legal document which admits him to public activity, and his health certificate, confirming his physical health for marriage.

*   *   *

Analogous to the education of the boy, the folkish state can conduct the education of the girl from the same viewpoint. There, too, the chief emphasis must be laid on physical training, and only subsequently on the promotion of spiritual and finally intellectual values. The goal of female education must invariably be the future mother.

*   *   *

Only secondarily must the folkish state promote the development of the *character* in every way.

Assuredly the most essential features of character are fundamentally preformed in the individual: the man of egotistic nature is and remains so forever, just as the idealist in the bottom of his heart will always be an idealist. But between the fully distinct characters there are millions that seem vague and unclear. The born criminal is and remains a criminal; but numerous people in whom there is only a certain tendency toward the criminal can by sound education still become valuable members of a national community; while conversely, through bad education, wavering characters can turn into really bad elements.

How often, during the War, did we hear the complaint that our people were so little able to be *silent*! How hard this made it to withhold even important secrets from the knowledge of our enemies! But ask yourself this question: What, before the War, did German education do to teach the individual silence? Even in school, sad to say, wasn't the little *informer* sometimes preferred to his more silent schoolmates? Was not and is not informing regarded as praiseworthy 'frankness,' discretion as reprehensible obstinacy? Was any effort whatever made to represent discretion as a manly and precious virtue? No, for in the eyes of our present school system these are trifles. But these trifles cost the state countless millions in court costs, for ninety per cent of all slander and similar suits have arisen only through lack of discretion. Irresponsibly dropped remarks are gossiped along just as frivolously, our national economy is constantly harmed by the frivolous revelation of important manufacturing processes, etc.; in fact, all our secret preparations for national defense are rendered illusory since the people simply have not learned how to be silent but pass everything on. This talkativeness can lead to the loss of battles and thus contribute materially to the unfavorable issue of the conflict. Here, again, we must realize that mature age cannot do what has not been practiced in youth. And this is the place to say that a teacher, for instance, must on principle not try to obtain knowledge of silly children's tricks by cultivating loathsome tattle-tales. Youth has its own state, it has a certain closed solidarity toward the grown-up, and this is perfectly natural. The ten year-old's bond with his playmate of the same age is more natural and greater than his bond with grown-ups. A boy who snitches on his comrade practices *treason* and thus betrays a mentality which, harshly expressed and enlarged, is the exact equivalent of treason to one's country.

Such a boy can by no means be regarded as a '*good, decent*' child; no, he is a boy of undesirable character. The teacher may find it convenient to make use of such vices for enhancing his authority, but in this way he sows in the youthful heart the germ of a mentality the later effect of which may be catastrophic. More than once, a little informer has grown up to be a big scoundrel!

This is only one example among many. Today the conscious development of good, noble traits of character in school is practically nil. In the future far greater emphasis must be laid on this. *Loyalty, spirit of sacrifice, discretion* are virtues that a great nation absolutely *needs,* and their cultivation and development in school are more important than some of the things which today fill out our curriculums. The discouragement of whining complaints, of bawling, etc., also belongs in this province. If a system of education forgets to teach the child in early years that sufferings and adversity must be borne in silence, it has no right to be surprised if later at a critical hour, when a man stands at the front, for example, the entire postal service is used for nothing but transporting whining letters of mutual complaint. If at the public schools a little less knowledge had been funneled into our youth and more self-control, this would have been richly rewarded in the years from 1915 to 1918.

And so the folkish state, in its educational work, must side by side with physical culture set the highest value precisely on the training of character. Numerous moral weaknesses in our present national body, if they cannot be entirely eliminated by this kind of education, can at least be very much attenuated.

*   *   *

Of the highest importance is the training of will-power and determination, plus the cultivation of joy in responsibility.

In the army the principle once held good that any command is better than none; related to youth this means primarily that any answer is better than none. The dread of giving no answer for fear of saying something wrong must be considered more humiliating than an incorrectly given answer. Starting from this most primitive basis, youth should be trained in such a way that it acquires courage for action.

People have often complained that in the days of November and December, 1918, every single authority failed, that from the monarchs down to the last divisional commander, no one was able to summon up the strength for an independent decision. This terrible fact is the handwriting on the wall for our educational system, for this cruel catastrophe expressed, hugely magnified, what was generally present on a small scale. It is this lack of will and not the lack of weapons which today makes us incapable of any serious resistance. It sits rooted in our whole people, prevents any decision with which a risk is connected, as though the greatness of a deed did not consist precisely in the risk. Without suspecting it, a German general succeeded in finding the classic formula for this miserable spinelessness: 'I act only if I can count on fifty-one per cent likelihood of success.' In these 'fifty-one per cent' lies the tragedy of the German collapse; anyone who demands of Fate a guaranty of success, automatically renounces all idea of a heroic deed. For this lies in undertaking a step which may lead to success, in the full awareness of the mortal danger inherent in a state of affairs. A cancer victim whose death is otherwise certain does not have to figure out fifty-one per cent in order to risk an operation. And if the operation promises only half a per cent likelihood of cure, a courageous man will risk it; otherwise he has no right to whimper for his life.

The plague of our present-day cowardly lack of will and determination is, all in all, mainly the result of our basically faulty education of youth, whose devastating effect extends to later life and finds its ultimate crowning conclusion in the lack of civil courage in our leading statesmen.

In the same line falls the present-day flagrant cowardice in the face of responsibility. Here, too, the error begins in the education of youth, goes on to permeate all public life, and finds its immortal completion in the parliamentary institution of government.

Even at school, unfortunately, more value is attached to 're-
pentant' confession and 'contrite abjuration' on the part of
the little sinner than to a frank admission. The latter seems to
many popular educators of today the surest mark of an incorrigible depravity and, incredible as it may seem, the gallows is
predicted for many a youth for qualities which would be of inestimable value if they constituted the common possession of a
whole people.

*Just as the folkish state must some day devote the highest attention
to the training of the will and force of decision, it must from an early
age implant joy in responsibility and courage for confession in the
hearts of youth.* Only if it recognizes this necessity in its full import will it finally, after an educational work enduring for centuries, obtain as a result a national body which will no longer
succumb to those weaknesses which today have contributed so
catastrophically to our decline.

*   *   *

The scientific school training which today is really the beginning and end of all state educational work can with only slight
changes be taken over by the folkish state. These changes lie in
three fields.

*In the first place the youthful brain should in general not be
burdened with things ninety-five per cent of which it cannot use and
hence forgets again.* Particularly, the curriculum of the elementary
and intermediate schools is today a mongrel; in many cases, the
material to be learned in the various subjects is so swollen that
only a fraction of it remains in the head of the individual pupil,
and only a fraction of this abundance can find application, while
on the other hand it is not adequate for the man working and
earning his living in a definite field. Take, for example, the
average government official, graduated from the *Gymnasium* or

the superior *Realschule*, at the age of thirty-five or forty, and examine him in the school learning that was once so painfully drummed into him. How little of all the stuff that was once funneled into him is still present! To be sure, you will get the answer: 'Well, the mass of material learned then was not intended only for the future possession of varied knowledge, but also for training mental receptivity, the power of thought and especially the memory. This is partly correct. Yet there is a danger in having the youthful brain flooded with so many impressions which only in the rarest cases it is able to master, and whose various elements it neither can sift nor evaluate according to their greater or lesser importance; and besides, as a rule, not the non-essential but the essential is forgotten and sacrificed. Thus the main purpose of learning so much is again lost; for it cannot consist after all in inducing learning power in the brain by an unmeasured heaping up of material, but must be to give the future man that store of knowledge which the individual needs and which through him in turn benefits the community. And this becomes illusory if the man, in consequence of the superabundance of the material forced on him in youth, later either possesses it not at all or has long since lost the very essentials. It is impossible to understand, for example, why millions of people in the course of the years must learn two or three foreign languages only a fraction of which they can make use of later and hence most of them forget entirely, for of a hundred thousand pupils who learn French for example, barely two thousand will have a serious use for this knowledge later, while ninety-eight thousand in the whole further course of their life will not find themselves in a position to make practical use of what they had once learned. They have in their youth, therefore, devoted thousands of hours to a subject which later is without value and meaning for them. And the objection that this material belongs to general education, is unsound, since it could only be upheld if people retained all through their life what they had learned. So in reality, because of the two thousand people for whom the knowledge of this language is profitable, ninety-eight thousand must be tormented for nothing and made to sacrifice valuable time.

And in this case we are dealing with a language of which it cannot even be said that it implies a training in sharp, logical thinking as applies, for example, to Latin. Hence it would be considerably more expedient if such a language were transmitted to the young student only in its general outlines or, better expressed, in its inner structure, thus giving him knowledge of the most salient essence of this language, introducing him perhaps to the fundamentals of its grammar and pronunciation, discussing syntax, etc., by model examples. This would suffice for general use and, because it is easier to visualize and remember, would be more valuable than the present-day manner of drumming in the whole language, which is not really mastered anyway and is later forgotten. In this way, moreover, the danger would be avoided that of all the overpowering abundance of material only a few unconnected crumbs would stick in the memory, as the young man would have to learn the most noteworthy aspects, and consequently the process of sifting according to value or the lack of it would have taken place in advance.

The general foundation thus imparted would suffice most people, even for later life, while it creates for those others who really need the language later the possibility of building further on it, and devoting themselves of their own free choice to learning it with the greatest thoroughness.

Thus the necessary time in the curriculum is gained for physical training as well as the increased demands in the abovementioned fields.

Particularly in the present method of teaching history a change must be made. Probably no people studies more history than the German; but probably there is no people that applies it worse than ours. If politics is history in the making, our historical education is directed by the nature of our political activity. Here, again, it is not permissible to complain about the wretched results of our political achievements unless we are determined to provide a better political education. The result of our present history instruction is wretched in ninety-nine cases out of a hundred. A few facts, dates, birthdays and names remain behind while a broad, clear line is totally lacking. The essentials which should really matter are not taught at all; it is left to the more or less gifted nature of the individual to find out the inner motives from the flood of dates and the sequence of events. We may argue as much as we like against this bitter statement; just read attentively the speeches on political problems, say questions of foreign policy, delivered during a single session by our parliamentary gentlemen; and bear in mind that these men — allegedly at least — are the cream of the German nation, and that at any rate a large part of them have even been at universities, and from this you will be able to see how totally inadequate the historical education of these people is. If they had not studied history at all, but only possessed a healthy instinct, it would be considerably better and more profitable for the nation.

Especially in historical instruction an abridgment of the material must be undertaken. The main value lies in recognizing the great lines of development. The more the instruction is limited to this, the more it is to be hoped that an advantage will later accrue to the individual from his knowledge, which summed up will also benefit the community. For we do not learn history just in order to know the past, we learn history in order to find an instructor for the future and for the continued existence of our own nationality. That is the *end*, and historical instruction is only a *means* to it. But today the means has become the end, and the end disappears completely. Let it not be said that thorough study of history requires attention to all these individual details, on the ground that only from them can a great line be developed. To lay down this line is the function of the special science. The normal, average man is no history professor. For him history exists primarily to give him that measure of historical insight which is necessary for him to take a position of his own on the political issues of his nation. Anyone who wants to become a history professor may later devote himself intensively to this study. It goes without saying that he will have to concern himself with all and even the smallest details. For this, however, even our present history instruction cannot suffice; for it is too extensive for the normal, average man, but much too limited for the specialized scholar.

Aside from this, it is the task of the folkish state to see to it that a world history is finally written in which the racial question is raised to a dominant position.

*   *   *

To sum up: the folkish state will have to put general, scientific instruction into an abbreviated form, embracing the essentials. In addition to this, the possibility of a thorough, specialized training must be offered. It suffices for the individual man to obtain a general knowledge in broad outlines as a foundation, and only in the field which will be that of his later life, to enjoy the most thorough specialized and detailed training. General education should be obligatory in all departments; the special training should remain free to the choice of the individual.

The shortening of the curriculum and the number of hours thus achieved will benefit the training of the body, of the character, of the will power and determination.

How irrevelant our present-day school training, especially in the high schools, is for a future profession is best demonstrated by the fact that today people from three schools of an entirely different nature can arrive at one and the same position. In reality only the general education is of decisive importance and not the specialized knowledge that is funneled into them. And where — as I have said before — a specialized knowledge is really necessary it can naturally not be obtained within the curriculums of our present high schools.

With such halfway methods, therefore, the folkish state must some day do away.

*   *   *

The second change of scientific curriculum in the folkish state must be the following:

It is the characteristic of our present materialized epoch that our scientific education is turning more and more toward practical subjects — in other words, mathematics, physics, chemistry, etc. Necessary as this is for a period in which technology and chemistry rule — embodying at least those of its characteristics which are most visible in daily life — it is equally dangerous when the general education of a nation is more and more exclusively directed toward them. This education on the contrary must always be ideal. It must be more in keeping with the humanistic subjects and offer only the foundations for a subsequent additional education in a special field. Otherwise we renounce the forces which are still more important for the preservation of the nation than all technical or other ability. Especially in historical instruction we must not be deterred from the study of antiquity. Roman history correctly conceived in extremely broad outlines is and remains the best mentor, not only for today, but probably for all time. The Hellenic ideal of culture should also remain preserved for us in its exemplary beauty. We must not allow the greater racial community to be torn asunder by the differences of the individual peoples. The struggle that rages today is for very great aims. A culture combining millenniums and embracing Hellenism and Germanism is fighting for its existence.

A sharp difference should exist between general education and specialized knowledge. As particularly today the latter threatens more and more to sink into the service of pure Mammon, general education, at least in its more ideal attitude, must be retained as a counterweight. Here, too, we must incessantly inculcate the principle *that industry, technology, and commerce can thrive only as long as an idealistic national community offers the necessary preconditions. And these do not lie in material egoism, but in a spirit of sacrifice and joyful renunciation.*

*   *   *

By and large the present education of youth has set itself the primary goal of pumping into the young person that knowledge which in his later career he needs for his own advancement. This is expressed in the words: 'The young man must some day become a useful member of society.' By this is meant his ability some day to earn his daily bread in a decent way. The superficial civic training carried on alongside rests on a weak base to begin with. Since the state in itself represents only a form, it is very hard to educate, let alone obligate people with regard to it. A form can too easily be shattered. But the concept 'state' — as we have seen — does not possess a clear content today. And so there remains nothing but the current 'patriotic' education. In old Germany its chief emphasis lay in a deification, often unintelligent and usually very insipid, of the small and smallest potentates, whose very quantity from the outset made it necessary to renounce any comprehensive appreciation of our nation's really great men. The result among our broad masses, consequently, was a very inadequate knowledge of German history. Here, too, the great line was lacking.

That a real national enthusiasm could not be achieved in this fashion is obvious. Our educational system lacked the art of picking a few names out of the historical development of our people and making them the common property of the whole German people, thus through like knowledge and like enthusiasm tying a uniform, uniting bond around the entire nation. They did not understand how to make the really significant men of our people appear as outstanding heroes in the eyes of the present, to concentrate the general attention upon them and thus create a unified mood. They were not able to raise what was glorious for the nation in the various subjects of instruction above the level of objective presentation, and fire the national pride by such gleaming examples. This would have seemed reprehensible chauvinism to that period, and in this form would not have met with much approval. Comfortable dynastic patriotism seemed more agreeable and easier to bear than the clamoring passion of higher national pride. The former was always ready to serve, the latter might some day become a master. Monarchistic patriotism ended in veterans' clubs, the national passion would have been hard to direct in its course. It is like a thoroughbred horse which does not carry everyone in the saddle. Is it any wonder that the powers of the time preferred to keep aloof from such a danger? No one seemed to consider it possible that some day there might come a war that would thoroughly test the inner steadfastness of our patriotic convictions in drumfire and clouds of gas. But when it came, the absence of the highest national passion brought the most frightful consequences. People had but little desire to die for their imperial and royal lords, and the 'nation' was unknown to most of them.

Since the revolution made its entry into Germany and monarchistic patriotism died out of its own accord, the purpose of instruction in history is really nothing more than the mere acquisition of knowledge. This state cannot use national enthusiasm; but what it would like to have it will never get. For no more than there could be a *dynastic patriotism* endowed with the ultimate *power of resistance* in an age governed by the *principle of nationalities*, much less can there be a *republican enthusiasm*. For there can be no doubt that under the motto, 'For the Republic,' the German people would not remain in the battlefield for any four and one-half years; least of all did those remain who have created this amazing structure.

Actually this Republic owes its unshorn existence only to its willingness, of which it gives assurance on all sides, voluntarily to assume all tribute payments and sign every renunciation of territory. It is liked by the rest of the world; just as every weakling is considered more agreeable by those who need him than a rough man. *True, this sympathy on the part of enemies is the most annihilating criticism for precisely this state form.* Our enemies love the German Republic and let it live because they could not find a better ally for their enslavement of our people. To this fact alone does this magnificent structure owe its present existence. That is why it can renounce any truly national education and content itself with cries of '*Hoch*' from Reichsbanner[1] heroes who, incidentally, if

---

[1] *Reichsbanner Schwarz-Rot-Gold*. Semi-military republican organization founded on February 22, 1924, by the Social Democrats Otto Horsing and Hölterman. In 1932, it had three and a half million members.

they had to protect this banner with their blood, would run away like rabbits.

The folkish state will have to fight for its existence. It will neither obtain it by Dawes signatures, nor be able to defend its existence by them. For its existence and for its protection, it will need the very things that people today think they can do without. The more incomparable and precious its form and content will be, the greater will be the envy and resistance of its enemies. Its best defense will lie not in its weapons, but in its citizens; no fortress walls will protect it, but a living wall of men and women filled with supreme love of their fatherland and fanatical national enthusiasm.

The third point to be considered in scientific education is the following:

Science, too, must be regarded by the folkish state as an instrument for the advancement of national pride. Not only world history but all cultural history must be taught from this standpoint. An inventor must not only seem great as an inventor, but must seem even greater as a national comrade. Our admiration of every great deed must be bathed in pride that its fortunate performer is a member of our own people. From all the innumerable great names of German history, the greatest must be picked out and introduced to the youth so persistently that they become pillars of an unshakable national sentiment.

The curriculum must be systematically built up along these lines so that when the young man leaves his school he is not a half pacifist, democrat, or something else, but a *whole German.*

In order that this national sentiment should be genuine from the outset and not consist in mere hollow pretense, beginning in youth one iron principle must be hammered into those heads which are still capable of education: *any man who loves his people proves it solely by the sacrifices which he is prepared to make for it. There is no such thing as national sentiment which is only out for gain. No more is there any nationalism which only embraces classes. Shouting hurrah proves nothing and gives no right to call oneself national if behind it there does not stand a great, loving concern for the preservation of a universal healthy nation. There is ground for pride in our people only if we no longer need be ashamed of any class. But a people, half of which is wretched and careworn, or even depraved, offers so sorry a picture that no one should feel any pride in it. Only when a nation is healthy in all its members, in body and soul, can every man's joy in belonging to it rightfully be magnified to that high sentiment which we designate as national pride. And this highest pride will only be felt by the man who knows the greatness of his nation.*

An intimate coupling of nationalism and a sense of social justice must be implanted in the young heart. Then a people of citizens will some day arise, bound to one another and forged together by a common love and a common pride, unshakable and invincible forever.

Our era's fear of chauvinism is the sign of its impotence. Not only lacking any exuberant force, but even finding it distasteful, it is no longer destined by Fate for a great deed. For the greatest revolutionary changes on this earth would not have been thinkable if their motive force, instead of fanatical, yes, hysterical passion, had been merely the bourgeois virtues of law and order.

And assuredly this world is moving toward a great revolution. The question can only be whether it will redound to the benefit of Aryan humanity or to the profit of the eternal Jew.

The folkish state will have to make certain that by a suitable education of youth it will some day obtain a race ripe for the last and greatest decisions on this earth.

And the people which first sets out on this path will be victorious.

* * *

The crown of the folkish state's entire work of education and training must be to burn the racial sense and racial feeling into the instinct and the intellect, the heart and brain of the youth entrusted to it. No boy and no girl must leave school without having been led to an ultimate realization of the necessity and essence of blood purity. Thus the groundwork is created by preserving the racial foundations of our nation and through them in turn securing the basis for its future cultural development.

For all physical and all intellectual training would in the last analysis remain worthless if it did not benefit a being which is ready and determined on principle to preserve himself and his special nature.

Otherwise that would occur which we Germans even now must greatly deplore, though perhaps the full extent of this tragic misfortune has hitherto not been realized: *that in the future we remain nothing but cultural fertilizer, not only in the limited conception of our present bourgeois view, which regards an individual national comrade lost as nothing more than a lost citizen, but with the painful realization that in this event, despite all our knowledge and ability, our blood is nevertheless doomed to decline. By mating again and again with other races, we may raise these races from their previous cultural level to a higher stage, but we will descend forever from our own high level.*

For the rest this education, too, from the racial viewpoint, must find its ultimate completion in military service. And in general, the period of military service must be regarded as the conclusion of the average German's normal education.

*   *   *

Important as the type of physical and mental education will be in the folkish state, equally important will be the human selection as such. Today this matter is taken lightly. In general it is the children of high-placed, at the time well-situated parents who are considered worthy of a higher education. Questions of talent play a subordinate rôle. Taken in itself, talent can only be evaluated relatively. A peasant boy can possess far more talents than the child of parents enjoying an elevated position in life for many generations, even if he is inferior to the bourgeois child in general knowledge. The latter's greater knowledge has in itself nothing to do with greater or lesser talent, but is rooted in the materially greater abundance of impressions which the child continuously receives as a result of his more varied education and rich environment. If the talented peasant boy from his early years had likewise grown up in such an environment, his intellectual ability would be quite different. Today, perhaps, there is a single field in which origin is really less decisive than the individual's native talent: the field of art. Here where a man cannot merely 'learn,' but everything has to be originally innate and is only later subject to a more or less favorable development in the sense of wise encouragement of existing gifts, the money and wealth of the parents are almost irrelevant. Hence it is here best shown that talent is not bound up with the higher walks of life, let alone with wealth. The greatest artists arise not seldom from the poorest houses. And many a poor village boy has later become a celebrated master.

It does not exactly argue great depth of thought in our time that this realization is not applied to our whole spiritual life. People imagine that what cannot be denied in art does not apply to the so-called exact sciences. Without doubt certain mechanical abilities can be taught a man, just as clever training can teach a docile poodle the most amazing tricks. But in animal training, the intelligence of the animal does not of itself lead to such exercises, and the same is the case with man. Without regard for any other talent, man too can be taught certain scientific tricks, but the process is just as lifeless and inwardly uninspired as with the animal. On the basis of a certain intellectual drill, knowledge above the average can be crammed into an average man; but it remains dead, and in the last analysis sterile knowledge. The result is a man who may be a living dictionary but nevertheless falls down miserably in all special situations and decisive moments in life; he will always have to be coached again for every situation, even the simplest, and by his own resources will not be able to make the slightest contribution to the development of humanity. Such a mechanically drilled knowledge suffices at most for assuming state positions in our present period.

It goes without saying that in the totality of a nation's population talents will be found for every possible domain of daily life. It is furthermore obvious that the value of knowledge will be the greater, the more the dead knowledge is animated by the relevant talent in the individual. *Creative achievements can only arise when ability and knowledge are wedded.*

The boundless sins of present-day humanity in this direction may be shown by one more example. From time to time illustrated papers bring it to the attention of the German petty-bourgeois that some place or other a Negro has for the first time become a lawyer, teacher, even a pastor, in fact a heroic tenor, or something of the sort. While the idiotic bourgeoisie looks with amazement at such miracles of education, full of respect for this marvelous result of modern educational skill, the Jew shrewdly draws from it a new proof for the soundness of his theory about the *equality of men* that he is trying to funnel into the minds of the nations. It doesn't dawn on this depraved bourgeois world that this is positively a sin against all reason; that it is criminal lunacy to keep on drilling a born half-ape until people think they have made a lawyer out of him, while millions of members of the highest culture-race must remain in entirely unworthy positions; that it is a sin against the will of the Eternal Creator if His most gifted beings by the hundreds and hundreds of thousands are allowed to degenerate in the present proletarian morass, while Hottentots and Zulu Kaffirs are trained for intellectual professions. For this is training exactly like that of the poodle, and not scientific 'education.' The same pains and care employed on intelligent races would a thousand times sooner make every single individual capable of the same achievements.

But intolerable as this state of affairs would be if it ever consisted of anything but exceptions, equally intolerable is it today in places where it is not talent and inborn gifts that decide who is chosen for higher education. Yes, indeed, it is an intolerable thought that every year hundreds of thousands of completely ungifted people are held worthy of a higher education, while other hundreds of thousands with great talent remain deprived of higher education. The loss which the nation thereby suffers is inestimable. If in the last decades the wealth of important inventions has increased amazingly, especially in North America, it is not least because there materially more talents from the lowest classes find opportunity for higher education than is the case in Europe.

For invention, drilled knowledge does not suffice, but only knowledge animated by talent. But in our country today no store is set on this; it is only good marks that matter.

Here, too, the folkish state will some day have to intervene by education. *Its task is not to preserve the decisive influence of an existing social class, but to pick the most capable kinds from the sum of all the national comrades and bring them to office and dignity.* It has not only the obligation of giving the average child a certain education in public school, but also the duty of putting talent on the track where it belongs. Above all, it must see its highest task in opening the gates of the higher state educational institutions to all talent, absolutely regardless from what circles it may originate. It must fulfill this task, since only in this way can representatives of a dead knowledge be transformed into brilliant leaders of a nation.

And for another reason the state must take measures in this direction: our intellectual classes, especially in Germany, are so segregated and so ossified that they lack a living connection with the people below them. We suffer from this in two ways: in the first place, they lack as a consequence any understanding and feeling for the broad masses. They have been torn out of this relation too long to possess the necessary psychological understanding for the people. They have become alien to the people. And in the second place, these intellectual strata lack the necessary will-power, which is always weaker in this secluded intellectual caste than in the mass of the primitive people. We Germans, by God, have never lacked scientific education; but we have been all the more lacking in any will power and determination. The more 'intellectual' our statesmen were, for example, the feebler, as a rule, was their actual accomplishment. The political preparations, as well as the technical armament for the

World War, was not inadequate because *insufficiently educated* minds ruled our people, but because the rulers were *overeducated* men, crammed full of knowledge and intellect, but bereft of any healthy instinct and devoid of all energy and boldness. It was a calamity that our people had to conduct its struggle for existence under the Chancellorship of a philosophizing weakling. If, instead of a Bethmann-Hollweg, we had had a robuster man of the people as a leader, the heroic blood of the common grenadier would not have flowed in vain. Likewise, the excessively rarefied pure intellect of our leader material was the best ally of the revolutionist November scoundrels. By disgracefully withholding the national treasure that had been entrusted to them, instead of staking it fully and wholly, these intellectuals themselves created the premise for the enemy's success.

In this the Catholic Church can be regarded as a model example. The celibacy of its priests is a force compelling it to draw the future generation again and again from the masses of the broad people instead of from their own ranks. But it is this very significance of celibacy that is not at all recognized by most people. It is the cause of the incredibly vigorous strength which resides in this age-old institution. For through the fact that this gigantic army of spiritual dignitaries is continuously complemented from the lowest strata of the nations, the Church not only obtains its instinctive bond with the emotional world of the people, but also assures itself a sum of energy and active force which in such a form will forever exist only in the broad masses of the people. From this arises the amazing youthfulness of this gigantic organism, its spiritual suppleness and iron will-power.

It will be the task of a folkish state to make certain through its educational system that a continuous renewal of the existing intellectual classes through an influx of fresh blood from below takes place. The state has the obligation to exercise extreme care and precision in picking from the total number of national comrades the human material visibly most gifted by Nature and to use it in the service of the community. For state and statesmen do not exist in order to provide individual classes with a living but to fulfill the tasks allotted to them. This will only be possible if as a matter of principle only capable and strong-willed personalities are trained to deal with these tasks. This applies not only to all official positions but to the intellectual leadership of the nation in all fields. Another factor for the greatness of the people is that it succeed in training the most capable minds for the field suited to them and placing them in the service of the national community. *If two peoples, equally well endowed, compete with one another, that one will achieve victory which has represented in its total intellectual leadership its best talents and that one will succumb whose leadership represents only a big common feeding crib for certain groups or classes, without regard to the innate abilities of the various members.*

To be sure, this looks impossible at first sight in our present world. The objection will at once be raised that the little son of a higher government official, for example, cannot be expected, let us say, to become an artisan because someone else whose parents were artisans seems more capable. This may be true in view of the present estimation of manual labor. For this reason the folkish state will have to arrive at a basically different attitude toward the concept of labor. *It will, if necessary, even by education extending over centuries, have to break with the mischief of despising physical activity. On principle it will have to evaluate the individual man not according to the type of work he does but according to the form and quality of his achievement.* This may appear positively monstrous to an era in which the most brainless columnist, just because he works with the pen, seems superior to the most intelligent precision mechanic. This false estimation, as has been said, does not lie in the nature of things, but is artificially cultivated and formerly did not exist. The present unnatural condition is based on the generally diseased condition of our present materialized epoch.

Fundamentally, the value of all work is twofold: *a purely material value and an ideal value.* The material value resides in the importance, that is to say, the material importance of a piece of work for the life of the totality. The more national comrades draw profit from a certain achievement performed, including direct and indirect profit, the greater the material value is to be estimated. This estimation, in turn, finds its plastic expression in the material reward which the individual obtains from his work. Contrasting with this purely material value, we now have the ideal value. It does not rest in the importance of the work performed measured materially, but in its necessity in itself. As surely as the material profit of an invention can be greater than that of an everyday handy-man's service, just as surely does the totality need the small service just as much as the great one. It may make a material distinction in evaluating the benefit of the individual piece of work for the totality, and can express this by a corresponding reward; in an ideal sense, however, it must recognize the equality of all as long as every individual endeavors to do his best in his field — whatever it may be. It is on this that the estimation of a man must be based, and not on his reward.

Since the concern of a sensible state must be to allot to the individual the activity which is in keeping with his ability or, otherwise expressed, to train the capable minds for the work that is suited to them, but since ability and principle are not taught but must be inborn, hence are a gift of Nature and not an achievement of man, general civic estimation cannot depend on the work that has, so to speak, been allotted to the individual. For this work falls to the account of his birth and to the training which he has consequently received through the community. The evaluation of the man must be based on the manner in which he fulfills the task entrusted him by the community. For the activity which the individual performs is not the end of his existence, but only the means to it. It is more important for him to develop and ennoble himself as a man, but he can do this only within the framework of his cultural community which must always rest on the fundament of a state. He must make his contribution to the preservation of this fundament. The form of this contribution is determined by Nature; his duty is only to return to the national community with honest industry what it has given him. Anyone who does this deserves the highest estimation and the highest respect. *Material reward may be granted to him whose achievement brings corresponding benefit to the community; his ideal reward, however, must lie in the esteem which everyone can claim who dedicates to the service of his nationality the forces which Nature gave him and which the national community has trained.* Then it is no longer a disgrace to be an honest manual worker, but it is a disgrace to be an incompetent official, stealing the daylight from his maker and daily bread from honest people. Then it will be taken for granted that a man will not be allotted tasks to which he is not equal to begin with.

Moreover, such activity provides the sole standard for right in universal, equal, juridical civic activity.[1]

The present era is liquidating itself: it introduces universal suffrage, shoots off its mouth about equal rights, but finds no basis for them. It sees in material reward the expression of a man's worth and thereby shatters the foundation for the noblest equality that there can be. For equality does not rest and never can rest on the achievements of individuals in themselves, but it is possible in the form in which everyone fulfills his special obligations. Thereby alone is the accident of Nature excluded in the judgment of the man's worth, and the individual himself becomes the smith of his own importance.

In the present period, when entire human groups can estimate one another only according to salary classes, there is — as said before — no understanding for this. But for us this cannot be a reason to renounce the fight for our ideas. On the contrary: *anyone who wants to cure this era, which is inwardly sick and rotten, must first of all summon up the courage to make clear the causes of this disease. And this should be the concern of the National Socialist movement: pushing aside all philistinism, to gather and to organize from the ranks of our nation those forces capable of becoming the vanguard fighters for a new philosophy of life.*

* * *

[1] *'Den einzigen Maszstab fur das Recht bei der allgemeinen bürgerlichen Betätigung.'*

Of course, the objection will be made that in general the ideal estimation is hard to separate from the material, indeed, that the diminishing estimation of physical labor is brought about precisely by its diminished reward. And that this diminished reward is in turn the cause for the limitation of the individual man's participation in the cultural treasures of his nation. And that precisely the ideal culture of man, which does not necessarily have anything to do with his activity as such, is impaired thereby. That the dread of physical labor is really based on the fact that, as a result of the inferior reward, the cultural level of the manual worker is necessarily lowered and that this provides the justification for a general diminished estimation.

In this there lies much truth. For this very reason we must in future guard ourselves against an excessive differentiation of wage rates. Let it not be said that this would destroy achievement. It would be the saddest sign of the decay of a period if the impetus to a higher spiritual achievement lay only in the increased wage. If this criterion had been the sole determinant in the world up to now, humanity would never have received its greatest scientific and cultural treasures. For the greatest inventions, the greatest discoveries, the most revolutionary scientific work, the most magnificent monuments of human culture, have not been given to the world through the urge for money. On the contrary, their birth not seldom meant positive renunciation of the earthly happiness of riches.

It may be that today gold has become the exclusive ruler of life, but the time will come when man will again bow down before a higher god. Many things today may owe their existence solely to the longing for money and wealth, but there is very little among them whose non-existence would leave humanity any the poorer.

This, too, is a task of our movement; even now it must herald a day which will give to the individual what he needs for living, but uphold the principle that man does not live exclusively for the sake of material pleasures. This must some day find its expression in a wisely limited gradation of earnings which in any event will give every decent working man an honest, regular existence as a national comrade and a man.

Let it not be said that this is an ideal condition which this world will not tolerate in practice and will actually never achieve.

We are not simple enough, either, to believe that it could ever be possible to bring about a perfect era. But this relieves no one of the obligation to combat recognized errors, to overcome weaknesses, and strive for the ideal. Harsh reality of its own accord will create only too many limitations. For that very reason, however, man must try to serve the ultimate goal, and failures must not deter him, any more than he can abandon a system of justice merely because mistakes creep into it, or any more than a medicament is discarded because there will always be sickness in spite of it.

Care must be taken not to underestimate the force of an idea. I should like to remind those who become faint-hearted in this connection — in case they were ever soldiers — of a time whose heroism represented the most overpowering proof of the force of idealistic motives. For what made men die then was not concern for their daily bread, but love of the fatherland, faith in its greatness, a general feeling for the honor of the nation. It was when the German people moved away from these ideals to follow the material premises of the revolution, and exchanged their arms for knapsacks,[1] that they arrived, not at the earthly paradise, but at the purgatory of general contempt and, no less, of general misery.

Therefore it is really necessary to confront the master bookkeepers of the present *material republic* by faith in an *ideal* Reich.

---

[1] In German there are two words for knapsack. A military knapsack is a *Tornister*; the word here used is *Rucksack*. This is the type worn by hikers. But Hitler is referring to the hungry proletarians after the War who went out foraging with knapsacks.

Subjects and Citizens

IN GENERAL the formation which today is erroneously designated as a state knows only two varieties of people: citizens and foreigners. Citizens are all those who either by their birth or subsequent naturalization possess the right of citizenship. Foreigners are all those who enjoy this same right in another state. In between, there are comet-like phenomena: the so-called stateless. These are people who have the honor of belonging to no present-day state; in other words, who nowhere possess the right of citizenship.

Today the right of citizenship, as mentioned above, is primarily achieved by birth *within* the borders of a state. In this, race or nationality play no rôle whatever. A Negro, who formerly lived in the German protectorates and now has his residence in Germany, gives birth to a 'German citizen' in the person of his child. Likewise every Jewish or Polish, African or Asiatic child can be declared a German citizen without further ado.

Aside from becoming a citizen through birth, there is the possibility of naturalization later. It is connected with certain requirements; for example, that the candidate in question is if possible no burglar or pimp; that he furthermore be politically unobjectionable, in other words, a harmless political idiot; that finally he should not fall a burden to the country which grants him citizenship. In this materialistic age this means, of course, a financial burden. Yes, it is even considered a desirable recommendation if you are presumably a good future taxpayer to hasten the acquisition of present-day citizenship.

Racial objections play no rôle whatsoever in this.

The whole process of acquiring citizenship takes place not far differently than admission into an automobile club. The man makes his application, it is examined and passed upon, and one day he receives a note informing him that he has become a citizen, and even the form of this is cute and kittenish. The former Zulu Kaffir in question is informed: 'You have hereby become a German!'

This magic trick is performed by a state president. What the heavens could not accomplish, such an official Theophrastus Paracelsus has accomplished in the twinkling of an eye. A simple dab of the pen and a Mongolian Wenceslaus has suddenly become a regular 'German.'

But not only do they not concern themselves about the race of such a new citizen; they do not even pay any attention to his physical health. Such a fellow may be as eaten by syphilis as he likes, for the present state he is nevertheless highly welcome as a citizen, provided that he does not, as above stated, represent a financial burden and a political danger.

And so every year these formations, called states, take into themselves poison elements which they can scarcely ever overcome.

The citizen himself then is only distinguished from the foreigner by the fact that the road to all public offices is open to him, that he may have to do military service, and that to make up for this he can actively and passively participate in elections. By and large this is all. For the protection of personal rights and of personal freedom is equally enjoyed by foreigners, not seldom more so; in any case, this applies in our present German Republic.

I know that people do not like to hear all this; but anything more thoughtless, more hare-brained than our present-day citizenship laws scarcely exists. There is today one state in which at least weak beginnings toward a better conception are noticeable. Of course, it is not our model German Republic, but the American

Union, in which an effort is made to consult reason at least partially. By refusing immigration on principle to elements in poor health, by simply excluding certain races from naturalization, it professes in slow beginnings a view which is peculiar to the folkish state concept.

The *folkish state* divides its inhabitants into three classes: citizens, subjects, and foreigners.

On principle only the status of *subject* is acquired by birth. The status of subject as such does not confer the right to hold public office, nor to carry on political activity in the sense of active or passive participation in elections. As a matter of principle, the race and nationality of every subject must be determined. The subject is free at any time to renounce his status of subject and become a citizen in the country whose nationality corresponds to his own. *The foreigner* is distinguished from the subject only by the fact that he is a subject of a foreign state.

The young subject of German nationality is obligated to undergo the schooling prescribed for every German. He thus submits to education to make him a racially conscious and patriotic national comrade. Later he must perform the supplementary physical exercises prescribed by the state, and finally he enters the army. The training in the army is general; it must embrace every individual German and train him in the field of military service made possible by his physical and intellectual ability. Thereupon, after completion of his military duty, the *right of citizenship* is most solemnly bestowed on the irreproachable, healthy young man. It is the most precious document for his whole life on earth. With it he enters upon all the rights of citizen and partakes of all his advantages. For the state must make a sharp distinction between those who, as national comrades, are the cause and bearer of its existence and its greatness, and those who only take up residence within a state, as 'earning' elements.

The bestowal of the *certificate of citizenship* must be associated with a solemn oath to the national community and the state. In this document there must lie a common bond which bridges all other gaps. *It must be a greater honor to be a street-cleaner and citizen of this Reich than a king in a foreign state.*

The citizen is privileged as against the foreigner. He is the lord of the Reich. But this higher dignity also obligates. The man without honor or character, the common criminal, the traitor to the fatherland, etc., can at any time be divested of this honor. He thus again becomes a subject.

The German girl is a subject and only becomes a citizen when she marries. But the right of citizenship can also be granted to female German subjects active in economic life.

Personality and the Conception of the Folkish State

T HE folkish National Socialist state sees its chief task in *educating and preserving the bearer of the state.* It is not sufficient to encourage the racial elements as such, to educate them and finally instruct them in the needs of practical life; the state must also adjust its own organization to this task.

It would be lunacy to try to estimate the value of man according to his race, thus declaring war on the Marxist idea that men are equal, unless we are determined to draw the ultimate consequences. And the ultimate consequence of recognizing the importance of blood — that is, of the racial foundation in general — is the transference of this estimation to the individual person. In general, I must evaluate peoples differently on the basis of the race they belong to, and the same applies to the individual men within a national community. The realization that peoples are not equal transfers itself to the individual man within a national community, in the sense that men's minds cannot be equal, since here, too, the blood components, though equal in their broad outlines, are, in particular cases, subject to thousands of the finest differentiations.

The first consequence of this realization might at the same time be called the cruder one: an attempt to promote in the most exemplary way those elements within the national community that have been recognized as especially valuable from the racial viewpoint and to provide for their special increase.

This task is cruder because it can be recognized and solved almost mechanically. It is more difficult to recognize among the whole people the minds that are most valuable in the intellectual and ideal sense, and to gain for them that influence which not only is the due of these superior minds, but which above all is beneficial to the nation. This sifting according to capacity and ability cannot be undertaken mechanically; it is a task which the struggle of daily life unceasingly performs.

A philosophy of life which endeavors to reject the democratic mass idea and give this earth to the best people — that is, the highest humanity — must logically obey the same aristocratic principle within this people and make sure that the leadership and the highest influence in this people fall to the best minds. Thus, it builds, not upon the idea of the majority, but upon the idea of personality.

Anyone who believes today that a folkish National Socialist state must distinguish itself from other states only in a purely mechanical sense, by a superior construction of its economic life — that is, by a better balance between rich and poor, or giving broad sections of the population more right to influence the economic process, or by fairer wages by elimination of excessive wage differentials — has not gone beyond the most superficial aspect of the matter and has not the faintest idea of what we call a philosophy. All the things we have just mentioned offer not the slightest guaranty of continued existence, far less of any claim to greatness. A people which did not go beyond these really superficial reforms would not obtain the least guaranty of victory in the general struggle of nations. A movement which finds the content of its mission only in such a general leveling, assuredly just as it may be, will truly bring about no great and profound, hence real, reform of existing conditions, since its entire activity does not, in the last analysis, go beyond externals, and does not give the people that inner armament which enables it, with almost inevitable certainty I might say, to overcome in the end those weaknesses from which we suffer today.

To understand this more easily, it may be expedient to cast

one more glance at the real origins and causes of human cultural development.

The first step which outwardly and visibly removed man from the animal was that of invention. Invention itself is originally based on the finding of stratagems and ruses, the use of which facilitates the life struggle with other beings, and is sometimes the actual prerequisite for its favorable course. These most primitive inventions do not yet cause the personality to appear with sufficient distinctness, because, of course, they enter the consciousness of the future, or rather the present, human observer, only as a mass phenomenon. Certain dodges and crafty measures which man, for example, can observe in the animal catch his eye only as a summary fact, and he is no longer in a position to establish or investigate their origin, but must simply content himself with designating such phenomena as 'instinctive.'

But in our case this last word means nothing at all. For anyone who believes in a higher development of living creatures must admit that every expression of their life urge and life struggle must have had a beginning; that *one* subject must have started it, and that subsequently such a phenomenon repeated itself more and more frequently and spread more and more, until at last it virtually entered the subconscious of all members of a given species, thus manifesting itself as an instinct.

This will be understood and believed more readily in the case of man. His first intelligent measures in the struggle with other beasts assuredly originate in the actions of individual, particularly able subjects. Here, too, the personality was once unquestionably the cause of decisions and acts which later were taken over by all humanity and regarded as perfectly self-evident. Just as any obvious military principle, which today has become, as it were, the basis of all strategy, originally owed its appearance to one absolutely distinct mind, and only in the course of many, perhaps even thousands of years, achieved universal validity and was taken entirely for granted.

Man complements this first invention by a second: he learns to place other objects and also living creatures in the service of his own struggle for self-preservation; and thus begins man's real inventive activity which today is generally visible. These material inventions, starting with the use of stone as a weapon and leading to the domestication of beasts, giving man artificial fire, and so on up to the manifold and amazing inventions of our day, show the individual creator the more clearly, the closer the various inventions lie to the present day, or the more significant and incisive they are. At all events, we know that all the material inventions we see about us are the result of the creative power and ability of the individual personality. And all these inventions in the last analysis help to raise man more and more above the level of the animal world and finally to remove him from it. Thus, fundamentally, they serve the continuous process of higher human development. But the very same thing which once, in the form of the simplest ruse, facilitated the struggle for existence of the man hunting in the primeval forest, again contributes, in the shape of the most brilliant scientific knowledge of the present era, to alleviate mankind's struggle for existence and to forge its weapons for the struggles of the future. All human thought and invention, in their ultimate effects, primarily serve man's struggle for existence on this planet, even when the so-called practical use of an invention or a discovery or a profound scientific insight into the essence of things is not visible at the moment. All these things together, by contributing to raise man above the living creatures surrounding him, strengthen him and secure his position, so that in every respect he develops into the dominant being on this earth.

Thus, all inventions are the result of an individual's work. All these individuals, whether intentionally or unintentionally, are more or less great benefactors of all men. Their work subsequently gives millions, nay, billions of human creatures, instruments with which to facilitate and carry out their life struggle.

If in the origin of our present material culture we always find individuals in the form of inventors, complementing one another and one building upon another, we find the same in the practice and execution of the things devised and discovered by the inventors. For all productive processes in turn must in their origin be considered equivalent to inventions, hence dependent on the individual. Even purely theoretical intellectual work, which in particular cases is not measurable, yet is the premise for all further material inventions, appears as the exclusive product of the individual person. It is not the mass that invents and not the majority that organizes or thinks, but in all things only and always the individual man, the person.

A human community appears well organized only if it facilitates the labors of these creative forces in the most helpful way and applies them in a manner beneficial to all. The most valuable thing about the invention itself, whether it lie in the material field or in the world of ideas, is primarily the inventor as a personality. Therefore, to employ him in a way benefiting the totality is the first and highest task in the organization of a national community. Indeed, the organization itself must be a realization of this principle. Thus, also, it is redeemed from the curse of mechanism and becomes a living thing. *It must itself be an embodiment of the endeavor to place thinking individuals above the masses, thus subordinating the latter to the former.*

Consequently, the organization must not only not prevent the emergence of thinking individuals from the mass; on the contrary, it must in the highest degree make this possible and easy by the nature of its own being. In this it must proceed from the principle that the salvation of mankind has never lain in the masses, but in its creative minds, which must therefore really be regarded as benefactors of the human race. To assure them of the most decisive influence and facilitate their work is in the interest of the totality. Assuredly this interest is not satisfied, and is not served by the domination of the unintelligent or incompetent, in any case uninspired masses, but solely by the leadership of those to whom Nature has given special gifts for this purpose.

The selection of these minds, as said before, is primarily accomplished by the hard struggle for existence. Many break and perish, thus showing that they are not destined for the ultimate, and in the end only a few appear to be chosen. In the fields of thought, artistic creation, even, in fact, of economic life, this selective process is still going on today, though, especially in the latter field, it faces a grave obstacle. The administration of the state and likewise the power embodied in the organized military might of the nation are also dominated by these ideas. Here, too, the idea of personality is everywhere dominant — its authority downward and its responsibility toward the higher personality above. Only political life has today completely turned away from this most natural principle. While all human culture is solely the result of the individual's creative activity, everywhere, and particularly in the highest *leadership* of the national community, the *principle of the value of the majority* appears decisive, and from that high place begins to gradually poison all life; that is, in reality to dissolve it. The destructive effect of the Jew's activity in other national bodies is basically attributable only to his eternal efforts to undermine the position of the personality in the host-peoples and to replace it by the mass. Thus, the organizing principle of Aryan humanity is replaced by the destructive principle of the Jew. He becomes 'a ferment of decomposition' among peoples and races, and in the broader sense a dissolver of human culture.

Marxism presents itself as the perfection of the Jew's attempt to exclude the pre-eminence of personality in all fields of human life and replace it by the numbers of the mass. To this, in the political sphere, corresponds the parliamentary form of government, which, from the smallest germ cells of the municipality up to the supreme leadership of the Reich, we see in such disastrous operation, and in the economic sphere, the system of a trade-union movement which does not serve the real interests of the workers, but exclusively the destructive purposes of the international world Jew. In precisely the measure in which the economy is withdrawn from the influence of the personality principle and instead exposed to the influences and effects of the masses, it must lose its efficacy in serving all and benefiting all, and gradually succumb to a sure retrogression. All the shop organizations which, instead of taking into account the interests of their employees, strive to gain influence on production, serve the same purpose. They injure collective achievement, and thus in reality injure individual achievement. For the satisfaction of the members of a national body does not in the long run occur exclusively through mere theoretical phrases, but by the goods of daily life that fall to the individual and the ultimate resultant conviction that a national community in the sum of its achievement guards the interests of individuals.

It is of no importance whether Marxism, on the basis of its mass theory, seems capable of taking over and carrying on the economy existing at the moment. Criticism with regard to the soundness or unsoundness of this principle is not settled by the proof of its capacity to *administer* the existing order for the future, but exclusively by the proof that it can itself *create* a higher culture. Marxism might a thousand times take over the existing economy and make it continue to work under its leadership, but even success in this activity would prove nothing in the face of the fact that it would not be in a position, by applying its principle *itself*, to create the same thing which today it takes over in a finished state.

Of this Marxism has furnished practical proof. Not only that it has nowhere been able to found and create a culture by itself; actually it has not been able to continue the existing ones in accordance with its principles, but after a brief time has been forced to return to the ideas embodied in the personality principle, in the form of *concessions;* — even in its own organization it cannot dispense with these principles.

The folkish philosophy is basically distinguished from the Marxist philosophy by the fact that it not only recognizes the value of race, but with it the importance of the personality, which it therefore makes one of the pillars of its entire edifice. These are the factors which sustain its view of life.

If the National Socialist movement did not understand the fundamental importance of this basic realization, but instead were merely to perform superficial patchwork on the present-day state, or even adopt the mass standpoint as its own — then it would really constitute nothing but a party in competition with the

Marxists; in that case, it would not possess the right to call itself a philosophy of life. If the social program of the movement consisted only in pushing aside the personality and replacing it by the masses, National Socialism itself would be corroded by the poison of Marxism, as is the case with our bourgeois parties.

The folkish state must care for the welfare of its citizens by recognizing in all and everything the importance of the value of personality, thus in all fields preparing the way for that highest measure of productive performance which grants to the individual the highest measure of participation.

And accordingly, the folkish state must free all leadership and especially the highest — that is, the political leadership — entirely from the parliamentary principle of majority rule — in other words, mass rule — and instead absolutely guarantee the right of the personality.

From this the following realization results:

The best state constitution and state form is that which, with the most unquestioned certainty, raises the best minds in the national community to leading position and leading influence.

But as, in economic life, the able men cannot be appointed from above, but must struggle through for themselves, and just as here the endless schooling, ranging from the smallest business to the largest enterprise, occurs spontaneously, with life alone giving the examinations, obviously political minds cannot be 'discovered.' Extraordinary geniuses permit of no consideration for normal mankind.

From the smallest community cell to the highest leadership of the entire Reich, the state must have the personality principle anchored in its organization.

There must be no majority decisions, but only responsible persons, and the word 'council' must be restored to its original meaning. Surely every man will have advisers by his side, but *the decision will be made by one man.*

The principle which made the Prussian army in its time into the most wonderful instrument of the German people must some day, in a transferred sense, become the principle of the construction of our whole state conception: *authority of every leader downward and responsibility upward.*

Even then it will not be possible to dispense with those corporations which today we designate as parliaments. But their councillors will then actually give counsel; responsibility, however, can and may be borne only by *one* man, and therefore only he alone may possess the authority and right to command.

Parliaments as such are necessary, because in them, above all, personalities to which special responsible tasks can later be entrusted have an opportunity gradually to rise up.

This gives the following picture:

The folkish state, from the township up to the Reich leadership, has no representative body which decides anything by the majority, but only *advisory bodies* which stand at the side of the elected leader, receiving their share of work from him, and in turn if necessary assuming unlimited responsibility in certain fields, just as on a larger scale the leader or chairman of the various corporations himself possesses.

As a matter of principle, the folkish state does not tolerate asking advice or opinions in special matters — say, of an economic nature — of men who, on the basis of their education and activity, can understand nothing of the subject. It, therefore, divides its representative bodies from the start into *political and professional chambers*.

In order to guarantee a profitable cooperation between the two, a special *senate* of the élite always stands over them.

In no chamber and in no senate does a vote ever take place. They are working institutions and not voting machines. The individual member has an advisory, but never a determining, voice. The latter is the exclusive privilege of the responsible chairman.

This principle — absolute responsibility unconditionally combined with absolute authority — will gradually breed an élite of leaders such as today, in this era of irresponsible parliamentarianism, is utterly inconceivable.

Thus, the political form of the nation will be brought into

agreement with that law to which it owes its greatness in the cultural and economic field.

*   *   *

As regards the possibility of putting these ideas into practice, I beg you not to forget that the parliamentary principle of democratic majority rule has by no means always dominated mankind, but on the contrary is to be found only in brief periods of history, which are always epochs of the decay of peoples and states.

But it should not be believed that such a transformation can be accomplished by purely theoretical measures from above, since logically it may not even stop at the state constitution, but must permeate all other legislation, and indeed all civil life. Such a fundamental change can and will only take place through a movement which is itself constructed in the spirit of these ideas and hence bears the future state within itself.

Hence the National Socialist movement should today adapt itself entirely to these ideas and carry them to practical fruition within its own organization, so that some day it may not only show the state these same guiding principles, but can also place the completed body of its own state at its disposal.

Philosophy and Organization

THE folkish state, a general picture of which I have attempted to draw in broad outlines, will not be realized by the mere knowledge of what is necessary to this state. It is not enough to know how a folkish state should look. Far more important is the program for its creation. We may not expect the present parties, which after all are primarily beneficiaries of the present state, to arrive of their own accord at a change of orientation and of their own free will to modify their present attitude. What makes this all the more impossible is that their real leading elements are always Jews and only Jews. And the development we are going through today, if continued unobstructed, would fulfill the Jewish prophecy — the Jew would really devour the peoples of the earth, would become their master.

Thus, confronting the millions of German 'bourgeois' and 'proletarians,' who for the most part, from cowardice coupled with stupidity, trot toward their ruin, he pursues his way inexorably, in the highest consciousness of his future goal. A party which is led by him can, therefore, stand for no other interests beside his interests; and with the concerns of Aryan nations, these have nothing in common.

And so, if we wish to transform the ideal image of a folkish state into practical reality, we must, independent of the powers that have thus far ruled public life, seek a new force that is willing and able to take up the struggle for such an idea. For it will take a struggle, in view of the fact that the first task is not creation of a folkish state conception, but above all elimination of the existing Jewish one. As so frequently in history, the main difficulty lies, not in the form of the new state of things, but in making place for it. Prejudices and interests unite in a solid phalanx and attempt with all possible means to prevent the victory of an idea that is displeasing to them or that menaces them.

And so, unfortunately, the fighter for such a new idea, important as it may be to put positive emphasis on it, is forced to carry through first of all the negative part of the fight, that part which should lead to the elimination of the present state of affairs.

A young doctrine of great and new fundamental significance will, displeasing as this may be to the individual, be forced to employ as its first weapon the probe of *criticism* in all its sharpness.

It indicates a lack of deep insight into historical developments when today people who call themselves folkish make a great point of assuring us over and over that they do not plan to engage in *negative criticism*, but only in *constructive work;* this absurd childish stammering is 'folkish' in the worst sense and shows how little trace the history even of their own times has left in these minds. *Marxism* also had a goal, and it, too, has a *constructive activity* (even if it is only to erect a despotism of international world Jewish finance); but previously, nevertheless, it *practiced criticism* for *seventy years*, annihilating, disintegrating criticism, and again criticism, which continued until the old state was undermined by this persistent corrosive acid and brought to collapse. Only then did its actual 'construction' begin. And that was self-evident, correct and logical. An existing condition is not eliminated just by emphasizing and arguing for a future one. For it is not to be presumed that the adherents, let alone the beneficiaries of the condition now existing, could all be converted and won over to the new one merely by demonstrating its necessity. On the contrary, it is only too possible that in this case two conditions will remain in existence side by side, and that the so-called *philosophy* will become a *party*, unable to raise itself above its limitations. For the philosophy is intolerant; it cannot content itself with the rôle of one 'party beside others,' but imperiously demands, not only its own exclusive and unlimited recognition, but the complete transformation of all public life in accordance with its views. It can, therefore, not tolerate the simultaneous continuance of a body representing the former condition.

This is equally true of religions.

Christianity could not content itself with building up its own altar; it was absolutely forced to undertake the destruction of the heathen altars. Only from this fanatical intolerance could its apodictic faith take form; this intolerance is, in fact, its absolute presupposition.

The objection may very well be raised that such phenomena in world history arise for the most part from specifically Jewish modes of thought, in fact, that this type of intolerance and fanaticism positively embodies the Jewish nature. This may be a thousand times true; we may deeply regret this fact and establish with justifiable loathing that its appearance in the history of mankind is something that was previously alien to history — yet this does not alter the fact that this condition *is* with us today. The men who want to redeem our German people from its present condition have no need to worry their heads thinking how lovely it would be if this and that did not exist; they must try to ascertain how the given condition can be eliminated. A philosophy filled with infernal intolerance will only be broken by a new idea, driven forward by the same spirit, championed by the same mighty will, and at the same time pure and absolutely genuine in itself.

The individual may establish with pain today that with the appearance of Christianity the first spiritual terror entered into the far freer ancient world, but he will not be able to contest the fact that since then the world has been afflicted and dominated by this coercion, and that coercion is broken only by coercion, and terror only by terror. Only then can a new state of affairs be constructively created.

Political parties are inclined to compromises; philosophies never. Political parties even reckon with opponents; philosophies proclaim their infallibility.

Political parties, too, almost always have the original purpose of attaining exclusive despotic domination; a slight impulse toward a philosophy is almost always inherent in them. Yet the very narrowness of their program robs them of the heroism which a philosophy demands. The conciliatory nature of their will attracts small and weakly spirits with which no crusades can be fought. And so, for the most part, they soon bog down in their own pitiful pettiness: They abandon the struggle for a philosophy and attempt instead, by so-called '*positive collaboration,*' to conquer as quickly as possible a little place at the feeding trough of existing institutions and to keep it as long as possible. That is their entire endeavor. And if they should be pushed away from the general feeding crib by a somewhat brutal competing boarder, their thoughts and actions are directed solely, whether by force or trickery, toward pushing their way back to the front of the hungry herd and finally, even at the cost of their holy conviction, toward refreshing themselves at the beloved swill pail. Jackals of politics!

Since a philosophy of life is never willing to share with another, it cannot be willing either to collaborate in an existing régime which it condemns, but feels obligated to combat this régime and the whole hostile world of ideas with all possible means; that is, to prepare its downfall.

This purely destructive fight — the danger of which is at once recognized by all others and which consequently encounters general resistance — as well as the positive struggle, attacking to make way for its own world of ideas, requires determined fighters. And so a philosophy will lead its idea to victory only if it unites the most courageous and energetic elements of its epoch and people in its ranks, and puts them into the solid forms of a fighting organization. For this, however, taking these elements into consideration, it must pick out certain ideas from its general world picture and clad them in a form which, in its precise, slogan-like brevity, seems suited to serve as a creed for a new community of men. While the program of a solely political *party* is the formula for a healthy outcome of the next elections, the program of a *philosophy* is the formulation of a declaration of war against the existing order, against a definite state of affairs; in short, against an existing view of life in general.

It is not necessary that every individual fighting for this philosophy should obtain a full insight and precise knowledge of the ultimate ideas and thought processes of the leaders of the movement. What is necessary is that some few, really great ideas be made clear to him, and that the essential fundamental lines be burned inextinguishably into him, so that he is entirely permeated by the necessity of the victory of his movement and its doctrine. The individual soldier is not initiated into the thought processes of higher strategy either. He is, on the contrary, trained in rigid discipline and fanatical faith in the justice and power of his cause, and taught to stake his life for it without reservation; the same must occur with the individual adherent of a movement of great scope, great future, and the greatest will.

Useless as an army would be, whose individual soldiers were all generals, even if it were only by virtue of their education and their insight, equally useless is a political movement, fighting for a *philosophy*, if it is only a reservoir of 'bright' people. No, it also needs the primitive soldier, since otherwise an inner discipline is unobtainable.

It lies in the nature of an *organization* that it can only exist if a broad mass, with a more emotional attitude, serves a high intellectual leadership. A company of two hundred men of equal intellectual ability would in the long run be harder to discipline than a company of a hundred and ninety intellectually less capable men and ten with higher education.

Social Democracy in its day drew the greatest profit from this fact. It took members of the broad masses, discharged from military service where they had been trained in discipline, and

drew them into its equally rigid party discipline. And its organization represented an army of officers and soldiers. The German
manual worker became the soldier, the *Jewish intellectual* the
officer; and the German trade-union officials can be regarded as
the corps of noncommissioned officers. The thing which our
bourgeoisie always viewed with headshaking, the fact that only
the so-called uneducated masses belonged to the Marxist movement, was in reality the basis for its success. For while the
bourgeois parties with their one-sided intellectualism constituted
a worthless undisciplined band, the Marxists with their unin-
tellectual human material formed an army of party soldiers,
who obeyed their Jewish leader as blindly as formerly their
German officer. The German bourgeoisie, which as a matter of
principle never concerned itself with psychological problems because it stood so high above them, found it, here too, unnecessary
to reflect, and recognize the deeper meaning, as well as the secret
danger, of this fact. They thought, on the contrary, that a
political movement, formed only from the circles of the 'intelligentsia,' is for this very reason more valuable and possesses a
greater claim, in fact a greater likelihood, of taking over the
government than the uneducated masses. *They never understood
that the strength of a political party lies by no means in the greatest
possible independent intellect of the individual members, but rather
in the disciplined obedience with which its members follow the intellectual leadership.* The decisive factor is the leadership itself.
If two bodies of troops battle each other, the one to conquer
will not be the one in which every individual has received the
highest strategic training, but *that one* which has the most
superior leadership and at the same time the most disciplined,
blindly obedient, best-drilled troop.

This is the basic insight which we must constantly bear in
mind in examining the possibility of transforming a philosophy
into action.

And so, if, in order to carry a philosophy to victory, we must
transform it into a fighting movement, logically the program of
the movement must take into consideration the human material

that stands at its disposal. As immutable as the ultimate aims
and the leading ideas must be, with equal wisdom and psychological soundness the recruiting program must be adapted to the
minds of those without whose aid the most beautiful idea would
remain eternally an idea.

*If the folkish idea wants to arrive at a clear success from the
unclear will of today, it must pick out from the broad world of its
ideas certain guiding principles, suited in their essence and content
to binding a broad mass of men, that mass which alone guarantees
the struggle for this idea as laid down in our philosophy.*

Therefore, the program of the new movement was summed up
in a few *guiding principles, twenty-five* in all. They were devised
to give, primarily to the man of the people, a rough picture of
the movement's aims. They are in a sense a *political creed*, which
on the one hand recruits for the movement and on the other is
suited to unite and weld together by a commonly recognized
obligation those who have been recruited.

Here the following insight must never leave us: Since the
so-called *program of the movement* is absolutely correct in its
ultimate aims, but in its formulation had to take psychological
forces into account, in the course of time the conviction may well
arise that in individual instances certain of the guiding principles
ought perhaps to be framed differently, given a better formulation. Every attempt to do this, however, usually works out
catastrophically. For in this way something which should be
unshakable is submitted to discussion, which, as soon as a single
point is deprived of its dogmatic, creedlike formulation, will not
automatically yield a new, better, and above all unified, formulation, but will far sooner lead to endless debates and a general
confusion. In such a case, it always remains to be considered
which is better: a new, happier formulation which causes an
argument within the movement, or a form which at the moment
may not be the very best, but which represents a solid, unshakable, inwardly unified organism. And any examination will show
that the latter is preferable. For, since in changes it is always
merely the outward formulation that is involved, such correc-

tions will again and again seem possible or desirable. *Finally, in view of the superficial character of men, there is the great danger that they will see the essential task of a movement in this purely outward formulation of a program.* Then the will and the power to fight for an idea recede, and the activity which should turn outward will wear itself out in inner programmatic squabbles.

With a doctrine that is really sound in its broad outlines, it is less harmful to retain a formulation, even if it should not entirely correspond to reality, than by improving it to expose what hitherto seemed a granite principle of the movement to general discussion with all its evil consequences. Above all, it is impossible as long as a movement is still fighting for victory. For how shall we fill people with blind faith in the correctness of a doctrine, if we ourselves spread uncertainty and doubt by constant changes in its outward structure?

The truth is that the most essential substance must never be sought in the outward formulation, but only and always in the inner sense. This is immutable; and in the interest of this immutable inner sense, we can only wish that the movement preserve the necessary strength to fight for it by avoiding all actions that splinter and create uncertainty.

Here, too, we can learn by the example of the Catholic Church. Though its doctrinal edifice, and in part quite superfluously, comes into collision with exact science and research, it is none the less unwilling to sacrifice so much as one little syllable of its dogmas. It has recognized quite correctly that its power of resistance does not lie in its lesser or greater adaptation to the scientific findings of the moment, which in reality are always fluctuating, but rather in rigidly holding to dogmas once established, for it is only such dogmas which lend to the whole body the character of a faith. And so today it stands more firmly than ever. It can be prophesied that in exactly the same measure in which appearances evade us, it will gain more and more blind support as a static pole amid the flight of appearances.[1]

---

[1] '*Der ruhende Pol in der Erscheinungen Flucht*' (the static pole in the flight of appearances). A familiar quotation. From Schiller's *Der Spaziergang*. line 134.

And so, anyone who really and seriously desires the victory of a folkish philosophy must not only recognize that, for the achievement of such a success in the first place, only a movement capable of struggle is suitable, but that, in the second place, such a movement itself will stand firm only if based on unshakable certainty and firmness in its program. It must not run the risk of making concessions in its formulation to the momentary spirit of the times, but must retain forever a form that has once been found favorable, in any case until crowned by victory. Before that, *any attempt* to bring about arguments as to the expediency of this or that point in the program splinters the solidity and the fighting force of the movement, proportionately as its adherents participate in such an inner discussion. This does not mean that an 'improvement' carried out today might not tomorrow be subjected to renewed critical tests, only to find a better substitute the day after tomorrow. Once you tear down barriers in this connection, you open a road, the beginning of which is known, but whose end is lost in the infinite.

This important realization had to be applied in the young National Socialist movement. The *National Socialist German Workers' Party obtained with its program of twenty-five theses a foundation which must remain unshakable.* The task of the present and future members of our movement must not consist in a critical revision of these theses, but rather in being bound by them. For otherwise the next generation in turn could, with the same right, squander its strength on such purely formal work within the party, instead of recruiting new adherents and thereby new forces for the movement. For the great number of the adherents, the essence of our movement will consist less in the letter of our theses than in the meaning which we are able to give them.

It was to these realizations that the young movement owed its name; the program was later framed according to them and, furthermore, the manner of their dissemination is based on them. In order to help the folkish ideas to victory, a party of the people had to be created, a party which consists not only of intellectual leaders, but also of manual workers!

Any attempt to realize folkish ideas without such a militant organization would today, just as in the past and in the eternal future, remain without success. And so the movement has not only the *right*, but also the *duty*, of regarding itself as a pioneer and representative of these ideas. To the same degree as the basic ideas of the National Socialist movement are *folkish*, the *folkish ideas* are *National Socialist*. And if National Socialism wants to conquer, it must unconditionally and exclusively espouse this truth. Here, too, it has not only the *right*, but also the duty, of sharply emphasizing the fact that any attempt to put forward the folkish idea outside the framework of the National Socialist German Workers' Party is impossible, and in most cases based on a positive swindle.

If today anyone reproaches the movement for acting as if the folkish idea were their monopoly, there is but one answer: *Not only a monopoly, but a working monopoly.*

For what previously existed under this concept was not suited to influence the destiny of our people even in the slightest, since all these ideas lacked a clear and coherent formulation. For the most part there were *single*, disconnected ideas of greater or lesser soundness, not seldom standing in mutual contradiction, in no case having any inner tie between them. And even had such a tie been present, in its weakness it would never have sufficed to orientate and build a movement on.

Only the National Socialist movement accomplished this.

*   *   *

If today all sorts of clubs and clublets, groups and grouplets, and, if you will, 'big parties' lay claim to the word 'folkish,' this in itself is a consequence of the influence of the *National Socialist movement. Without its work, it would never have occurred to all these organizations even to pronounce the word 'folkish.'* This word would have meant nothing to them, and especially their leading minds would have stood in no relation of any sort to this concept. Only the work of the NSDAP for the first time made this concept a word full of content, which is now taken up by every conceivable kind of people; above all, in its own success-

ful campaigning activity, it snowed and demonstrated the force
of these folkish ideas, so that mere desire to get ahead forces the
others, ostensibly at least, to desire similar ends.

Just as hitherto they used everything for their petty election
speculation, the folkish concept has today remained for them
only an external empty slogan with which they attempt to
counterbalance, among their own members, the attractive force
of the National Socialist movement. For it is only concern for
their own existence as well as fear of the rise of our new phi-
losophy-borne movement, whose universal importance as well as
its dangerous exclusiveness they sense, that puts into their
mouth words which eight years ago they did not know, seven
years ago ridiculed, six years ago branded as absurd, five years
ago combated, four years ago hated, three years ago persecuted,
only at length to annex them two years ago, and, combined with
the rest of their vocabulary, to use them as a battle-cry in the
fight.

And even today we must point out again and again that all
these parties lack the slightest idea of *what the German people
needs*. The most striking proof of this is the superficiality with
which they mouth the word 'folkish.'

And no less dangerous are all those who horse around pretend-
ing to be folkish, forge fantastic plans, for the most part based
on nothing but some *idée fixe*, which in itself might be sound, but
in its isolation remains none the less without any importance for
the formation of a great unified fighting community, and in no
case is suited to building one. These people, who partly from
their own thinking, partly from what they have read, brew up a
program, are frequently more dangerous than the open enemies
of the folkish idea. In the best case they are sterile theoreticians,
but for the most part disastrous braggarts, and not seldom they
believe that with flowing beards and primeval Teutonic gestures
they can mask the intellectual and mental hollowness of their
activities and abilities.

In contrast to all these useless attempts, it is therefore good
if we recall to mind the time in which the young National Socialist
movement began its struggle.

The Struggle of the Early Period — the Significance of the Spoken Word

T HE first great meeting on February 24, 1920, in the Festsaal [Banquet Hall] of the Hofbräuhaus, had not died down in our ears when the preparations for the next were made. While up till then it had been considered risky to hold a little meeting once a month or even once every two weeks in a city like Munich, a large mass meeting was now to take place every seven days; in other words, once a week. I do not need to assure you that there was but one fear that constantly tormented us: would the people come and would they listen to us? — though I personally, even then, had the unshakable conviction that once they were there, the people would stay and follow the speech.

In this period the Festsaal of the Munich Hofbräuhaus assumed an almost sacred significance for us National Socialists. Every week a meeting, almost always in this room, and each time the hall better filled and the people more devoted. Beginning with the 'War Guilt,' which at that time nobody bothered about, and the 'Peace Treaties,' nearly everything was taken up that seemed agitationally expedient or ideologically necessary. Especially to the peace treaties themselves the greatest attention was given. What prophecies the young movement kept making to the great masses! And nearly all of which have now been realized! Today it is easy to speak or write about these things. But in those days a public mass meeting, attended, not by bourgeois shopkeepers, but by incited proletarians, and dealing with the topic, 'The Peace Treaty of Versailles,' was taken as an attack on the Republic and a sign of a reactionary if not monarchistic attitude. At the very first sentence containing a criticism of Versailles, you had the stereotyped cry flung at you: 'What about Brest-Litovsk?' 'And Brest-Litovsk?' The masses roared this again and again, until gradually they grew hoarse or the speaker finally gave up his attempt to convince them. You felt like dashing your head against the wall in despair over such people! They did not want to hear or understand that Versailles was a shame and a disgrace, and not even that this dictated peace was an unprecedented pillaging of our people. The destructive work of the Marxists and the poison of enemy propaganda had deprived the people of any sense. And yet we had not even the right to complain! For how immeasurably great was the blame on another side! What had the bourgeoisie done to put a halt to this frightful disintegration, to oppose it and open the way to truth by a better and more thorough enlightenment? Nothing, and again nothing. In those days I saw them nowhere, all the great folkish apostles of today. Perhaps they spoke in little clubs, at teatables, or in circles of like-minded people, but where they should have been, among the wolves, they did not venture; except if there was a chance to howl with the pack.

But to me it was clear in those days that for the small basic nucleus which for the present constituted the movement, the question of war guilt had to be cleared up, and cleared up in the sense of historic truth. That our movement should transmit to the broadest masses knowledge of the peace treaty was the premise for the future success of the movement. At that time, when they all still regarded this peace as a success of democracy, we had to form a front against it and engrave ourselves forever in the minds of men as an enemy of this treaty, so that later, when the harsh reality of this treacherous frippery would be revealed in its naked hate, the recollection of our position at that time would win us confidence.

Even then I always came out in favor of taking a position in important questions of principle against all public opinion when it assumed a false attitude — disregarding all considerations of popularity, hatred, or struggle. The NSDAP should not become a constable of public opinion, but must dominate it. It must not become a servant of the masses, but their master!

There exists, of course, and especially for every movement that is still weak, a great temptation, in moments when a more powerful enemy has succeeded in driving the people to a mad decision or to a false attitude through his arts of seduction, to go along and join the shouting, particularly when there are a few reasons — even if they are merely illusory — which, from the standpoint of the young movement itself, might argue for this course. Human cowardice will seek such reasons so vigorously that it almost always finds something which would give a semblance of justification, even from one's 'own standpoint,' for participating in such a crime.

I have several times experienced such cases, in which supreme energy was necessary to keep the ship of the movement from drifting with the artificially aroused general current or rather from being driven by it. The last time was when our infernal press, to which the existence of the German people is Hecuba, succeeded in puffing up the South Tyrol question to an importance which will be catastrophic for the German people. Without considering whom they were serving thereby, many so-called 'national' men and parties and organizations, solely from cowardice in the face of Jew-incited public opinion, joined the general outcry and senselessly helped to support the fight against a system which we Germans, precisely in this present-day situation, must feel to be the sole ray of light in this degenerating world. While the international world Jew slowly but surely strangles us, our so-called patriots shouted against a man and a system which dared, in one corner of the earth at least, to free themselves from the Jewish-Masonic embrace and oppose a nationalistic resistance to this international world poisoning. It was, however, too alluring for weak characters simply to set their sails by the wind and capitulate to the clamor of public opinion. And a capitulation it was! Men are such base liars that they may not admit it, even to themselves, but it remains the truth that only cowardice and fear of the popular sentiment stirred up by the Jews impelled them to join in. All other explanations are miserable evasions devised by the petty sinner conscious of his guilt.

And so it was necessary to shake the movement with an iron fist to preserve it from ruin by this tendency. To attempt such a shift at a moment when public opinion, fanned by every driving force, was burning only in one direction is indeed not very popular at the moment and sometimes puts the venturesome leader in almost mortal peril. But not a few men in history have at such moments been stoned for an action for which posterity, at a later date, had every cause to thank them on its knees.

It is with this that a movement must reckon and not with the momentary approval of the present. It may be that in such hours the individual feels afraid; but he must not forget that after every such hour salvation comes at length, and that a movement that wants to renew a world must serve, not the moment, but the future.

In this connection it can be established that the greatest and most enduring successes in history tend for the most part to be those which in their beginnings found the least understanding because they stood in the sharpest conflict with general public opinion, with its ideas and its will.

Even then, on the first day of our public appearance, we had a chance to experience this. Truly we did not 'curry favor with the masses,' but everywhere opposed the lunacy of these people. Nearly always it came about that in these years I faced an assemblage of people who believed the opposite of what I wanted to say, and wanted the opposite of what I believed. Then it was the work of two hours to lift two or three thousand people out of a previous conviction, blow by blow to shatter the foundation of their previous opinions, and finally to lead them across to our convictions and our philosophy of life.

In those days I learned something important in a short time, *to strike the weapon of reply out of the enemy's hand myself.* We soon noticed that our opponents, especially their discussion speakers, stepped forward with a definite 'repertory' in which constantly recurring objections to our assertions were raised, so that the uniformity of this procedure pointed to a conscious, unified schooling. And that was indeed the case. Here we had an opportunity to become acquainted with the incredible discipline of our adversaries' propaganda, and it is still my pride today to have found the means, not only to render this propaganda ineffective, but in the end to strike its makers with their own weapon Two years later I was a master of this art.

In every single speech it was important to realize clearly in advance the presumable content and form of the objections to be expected in the discussion, and to pull every one of them apart in the speech itself. Here it was expedient to cite the possible objections ourselves at the outset and demonstrate their untenability; thus, the listener, even if he had come stuffed full of the objections he had been taught, but otherwise with an honest heart, was more easily won over when we disposed of the doubts that had been imprinted on his memory. The stuff that had been drummed into him was automatically refuted and his attention drawn more and more to the speech.

This is the reason why, right after my first lecture on the 'Peace Treaty of Versailles,' which I had delivered to the troops while still a so-called 'educator,' I changed the lecture and now spoke of the 'Peace Treaties of Brest-Litovsk and Versailles.' For after a short time, in fact, in the course of the discussion about this first speech of mine, I was able to ascertain that the people really knew nothing at all about the peace treaty of Brest-Litovsk, but that the adroit propaganda of their parties had succeeded in representing this very treaty as one of the most shameful acts of rape in the world. The persistence with which this lie was presented over and over to the great masses accounted for the fact that millions of Germans regarded the peace treaty of Versailles as nothing more than just retribution for the crime committed by us at Brest-Litovsk, thus viewing any real struggle against Versailles as an injustice and sometimes remaining in the sincerest moral indignation. And this among other things was why the shameless and monstrous word *'reparations'* was able to make itself at home in Germany. This vile hypocrisy really seemed to millions of our incited national comrades an accomplishment of higher justice. Dreadful, but it was so. The best proof of this was offered by the propaganda I initiated against the peace treaty of Versailles, which I introduced by some enlightenment regarding the treaty of Brest-Litovsk. I contrasted the two peace treaties, compared them point for point, showed the actual boundless humanity of the one treaty compared to the inhuman cruelty of the second, and the result was telling. At that time I spoke on this theme at meetings of two thousand people, and often I was struck by the glances of three thousand six hundred hostile eyes. And three hours later I had before me a surging mass full of the holiest indignation and boundless wrath. Again a great lie had been torn out of the hearts and brains of a crowd numbering thousands, and a truth implanted in its place.

I considered these two lectures on 'The True Causes of the World War' and on 'The Peace Treaties of Brest-Litovsk and Versailles,' the most important of all, and so I repeated and repeated them dozens of times, always renewing the form, until, on this point at least, a certain clear and unified conception became current among the people from among whom the movement gathered its first members.

For myself, moreoever, the meetings had the advantage that I gradually transformed myself into a speaker for mass meetings, that I became practiced in the pathos and the gestures which a great hall, with its thousands of people, demands.

At that time, except — as already emphasized — in small circles, I saw no enlightenment in this direction from the parties which today have their mouths so full of words and act as if *they* had brought about the change in public opinion. When a so-called *'national politician'* somewhere delivered a speech along these lines, it was only to circles who for the most part already shared his conviction, and for whom his utterances represented at most an intensification of their own opinions. This was not the important thing at that time; the important thing was to win by enlightenment and propaganda those people who, by virtue of their education and opinions, still stood on hostile ground.

The leaflet, too, was put into the service of this enlightenment. While still in the army, I had written a leaflet comparing the peace treaties of *Brest-Litovsk* and *Versailles*, and it was distributed in large editions. Later I took over stocks of it for the party, and here again the effect was good. The first meetings, in general, were distinguished by the fact that the tables were covered with all sorts of leaflets, newspapers, pamphlets, etc. But the chief emphasis was laid on the spoken word. And actually it alone — for general psychological reasons — is able to bring about really great changes.

I have already stated in the first volume that all great, world-shaking events have been brought about, not by written matter, but by the spoken word. This led to a lengthy discussion in a part of the press, where, of course, such an assertion was sharply attacked, particularly by our bourgeois wiseacres. But the very reason why this occurred confutes the doubters. For the bourgeois intelligentsia protest against such a view only because they themselves obviously lack the power and ability to influence the masses by the spoken word, since they have thrown themselves more and more into purely literary activity and renounced the real agitational activity of the spoken word. Such habits necessarily lead in time to what distinguishes our bourgeoisie today; that is, to the loss of the psychological instinct for *mass effect* and *mass influence*.

While the speaker gets a continuous correction of his speech from the crowd he is addressing, since he can always see in the faces of his listeners to what extent they can follow his arguments with understanding and whether the impression and the effect of his words lead to the desired goal — the writer does not know his readers at all. Therefore, to begin with, he will not aim at a definite mass before his eyes, but will keep his arguments entirely general. By this to a certain degree he loses psychological subtlety and in consequence suppleness. And so, by and large, a brilliant speaker will be able to write better than a brilliant writer can speak, unless he continuously practices this art. On top of this there is the fact that the mass of people as such is lazy; that they remain inertly in the spirit of their old habits and, left to themselves, will take up a piece of written matter only reluctantly if it is not in agreement with what they themselves believe and does not bring them what they had hoped for. Therefore, an article with a definite tendency is for the most part read only by people who can already be reckoned to this tendency. At most a leaflet or a poster can, by its brevity, count on getting a moment's attention from someone who thinks differently. The picture in all its forms up to the film has greater possibilities. Here a man needs to use his brains even less; it suffices to look, or at most to read extremely brief texts, and thus many will more readily accept a *pictorial presentation* than *read* an *article* of any *length*. The picture brings them in a much briefer time, I might almost say at one stroke, the enlightenment which they obtain from written matter only after arduous reading.

The essential point, however, is that a piece of literature never knows into what hands it will fall, and yet must retain its definite form. In general the effect will be the greater, the more this form corresponds to the intellectual level and nature of those very people who will be its readers. A book that is destined for the broad masses must, therefore, attempt from the very beginning to have an effect, both in style and elevation, different from a work intended for higher intellectual classes.

Only by this kind of adaptability does written matter approach the spoken word. To my mind, the speaker can treat the same theme as the book; he will, if he is a brilliant popular orator, not be likely to repeat the same reproach and the same substance twice in the same form. He will always let himself be borne by the great masses in such a way that instinctively the very words come to his lips that he needs to speak to the hearts of his audience. And if he errs, even in the slightest, he has the living correction before him. As I have said, he can read from the facial expression of his audience whether, firstly, they *understand* what he is saying, whether, secondly, they can *follow the speech as a whole*, and to what extent, thirdly, he has *convinced* them of the *soundness* of what he has said. If — firstly — he sees that they do not understand him, he will become so primitive and clear in his explanations that even the last member of his audience has to understand him; if he feels — secondly — that they cannot follow him, he will construct his ideas so cautiously and slowly that even the weakest member of the audience is not left behind, and he will — thirdly — if he suspects that they do not seem convinced of the soundness of his argument, repeat it over and over in constantly new examples. He himself will utter their objections, which he senses though unspoken, and go on confuting them and exploding them, until at length even the last group of an opposition, by its very bearing and facial expression, enables him to recognize its capitulation to his arguments.

Here again it is not seldom a question of overcoming prejudices which are not based on reason, but, for the most part unconsciously, are supported only by sentiment. To overcome this barrier of instinctive aversion, of emotional hatred, of prejudiced rejection, is a thousand times harder than to correct a faulty or erroneous scientific opinion. False concepts and poor knowledge can be eliminated by instruction, the resistance of the emotions never. Here only an appeal to these mysterious powers themselves can be effective; and the writer can hardly ever accomplish this, but almost exclusively the orator.

The most striking proof of this is furnished by the fact that, despite a bourgeois press that is often very skillfully gotten up, flooding our people with editions running into millions, this press could not prevent the masses from becoming the sharpest enemy of its own bourgeois world. The whole newspaper flood and all the books that are turned out year after year by the intellectuals slide off the millions of the lower classes like water from oiled leather. This can prove only two things: either the unsoundness of the content of this whole literary production of our bourgeois world or the impossibility of reaching the heart of the broad masses solely by written matter. Especially, indeed, when this

written matter demonstrates so unpsychological an attitude as
is here the case.

Let no one reply (as a big German national newspaper in
Berlin tried to do) that *Marxism* itself, by its writings, especially
by the effect of the great basic work of Karl Marx, provides proof
counter to this assertion. Seldom has anyone made a more superficial attempt to support an erroneous view. What gave *Marxism*
its astonishing power over the great masses is by no means the
formal written work of the Jewish intellectual world, but rather
the enormous oratorical propaganda wave which took possession
of the great masses in the course of the years. Of a hundred
thousand German workers, not a hundred on the average know
this work, which has always been studied by a thousand times
more intellectuals and especially Jews than by real adherents of
this movement from the great lower classes. And this work was
not written for the great masses, but exclusively for the intellectual leadership of that Jewish machine for world conquest; it
was stoked subsequently with an entirely different fuel: the press.
For that is what distinguishes the Marxist press from our bourgeois press. *The Marxist press is written by agitators, and the
bourgeois press would like to carry on agitation by means of writers.*
The Social Democratic yellow journalist, who almost always
goes from the meeting hall to the newspaper office, knows his
public like no one else. But the bourgeois scribbler who comes
out of his study to confront the great masses is nauseated by
their very fumes and faces them helplessly with the written word.

What has won the millions of workers for Marxism is less the
literary style of the Marxist church fathers than the indefatigable and truly enormous propaganda work of tens of thousands
of untiring agitators, from the great agitator down to the small
trade-union official and the shop steward and discussion speaker;
this work consisted of the hundreds of thousands of meetings at
which, standing on the table in smoky taverns, these people's
orators hammered at the masses and thus were able to acquire
a marvelous knowledge of this human material which really put
them in a position to choose the best weapons for attacking the

fortress of public opinion. And it consisted, furthermore, in the gigantic mass demonstrations, these parades of hundreds of thousands of men, which burned into the small, wretched individual the proud conviction that, paltry worm as he was, he was nevertheless a part of a great dragon,[1] beneath whose burning breath the hated bourgeois world would some day go up in fire and flame and the proletarian dictatorship would celebrate its ultimate final victory.

Such propaganda produced the people who were ready and prepared to read a Social Democratic press, however, a press which itself in turn is not written, but which is spoken. For, while in the bourgeois camp professors and scholars, theoreticians and writers of all sorts, occasionally attempt to speak, in the Marxist movement the speakers occasionally try to write. And precisely the Jew, who is especially to be considered in this connection, will, in general, thanks to his lying dialectical skill and suppleness, even as a writer be more of an agitational orator than a literary creator.

That is the reason why the bourgeois newspaper world (quite aside from the fact that it, too, is mostly Jewified and therefore has no interest in really instructing the great masses) cannot exert the slightest influence on the opinion of the broadest sections of our people.

How hard it is to upset emotional prejudices, moods, sentiments, etc., and to replace them by others, on how many scarcely calculable influences and conditions success depends, the sensitive speaker can judge by the fact that even the time of day in which the lecture takes place can have a decisive influence on the effect. The same lecture, the same speaker, the same theme, have an entirely different effect at ten o'clock in the morning, at three o'clock in the afternoon, or at night. I myself as a beginner organized meetings for the morning, and especially remember a rally which we held in the Munich Kindl Keller as a protest 'against the oppression of German territories.' At that time it was Munich's largest hall and it seemed a very great

[1] '*Als kleiner Wurm dennoch Glied eines grossen Drachens zu sein.*'

venture. In order to make attendance particularly easy for the
adherents of the movement and all the others who came, I set
the meeting for a Sunday morning at ten o'clock. The result
was depressing, yet at the same time extremely instructive:
the hall was full, the impression really overpowering, but the
mood ice cold; no one became warm, and I myself as a speaker
felt profoundly unhappy at being unable to create any bond,
not even the slightest contact, between myself and my audience.
I thought I had not spoken worse than usual; but the effect
seemed to be practically nil. Utterly dissatisfied, though richer
by one experience, I left the meeting. Tests of the same sort
that I later undertook led to the same result.

   This should surprise no one. Go to a theater performance and
witness a play at three o'clock in the afternoon and the same
play with the same actors at eight at night, and you will be
amazed at the difference in effect and impression. A man with
fine feelings and the power to achieve clarity with regard to this
mood will be able to establish at once that the impression made
by the performance at three in the afternoon is not as great as
that made in the evening. The same applies even to a movie.
This is important because in the theater it might be said that
perhaps the actor does not take as much pains in the afternoon
as at night. But a film is no different in the afternoon than at
nine in the evening. No, the *time* itself exerts a definite effect,
just as the hall does on me. There are halls which leave people
cold for reasons that are hard to discern, but which somehow
oppose the most violent resistance to any creation of mood.
Traditional memories and ideas that are present in a man can
also decisively determine an impression. Thus, a performance of
Parsifal in Bayreuth will always have a different effect than
anywhere else in the world. The mysterious magic of the house
on the Festspielhügel in the old city of the margraves cannot be
replaced or even compensated for by *externals*.

   In all these cases we have to do with an encroachment upon
man's freedom of will. This applies most, of course, to meetings
attended by people with a contrary attitude of will, who must

now be won over to a new will. In the morning and even during the day people's will power seems to struggle with the greatest energy against an attempt to force upon them a strange will and a strange opinion. At night, however, they succumb more easily to the dominating force of a stronger will. For, in truth, every such meeting represents a wrestling bout between two opposing forces. The superior oratorical art of a dominating preacher will succeed more easily in winning to the new will people who have themselves experienced a weakening of their force of resistance in the most natural way than those who are still in full possession of their mental tension and will.

The same purpose, after all, is served by the artificially made and yet mysterious twilight in Catholic churches, the burning lamps, incense, censers, etc.

In this wrestling bout of the speaker with the adversaries he wants to convert, he will gradually achieve that wonderful sensitivity to the psychological requirements of propaganda, which the writer almost always lacks. Hence the written word in its limited effect will in general serve more to retain, to reinforce, to deepen, a point of view or opinion that is already present. Really great historical changes are not induced by the *written* word, but at most *accompanied* by it.

Let no one believe that the French Revolution would ever have come about through philosophical theories if it had not found an army of agitators led by demagogues in the grand style, who whipped up the passions of the people tormented to begin with, until at last there occurred that terrible volcanic eruption which held all Europe rigid with fear. And likewise the greatest revolutionary upheaval of the most recent period, the Bolshevist Revolution in Russia, was brought about, not by Lenin's writings, but by the hate-fomenting oratorical activity of countless of the greatest and the smallest apostles of agitation.

The illiterate common people were not, forsooth, fired with enthusiasm for the Communist Revolution by the theoretical reading of Karl Marx, but solely by the glittering heaven which thousands of agitators, themselves, to be sure, all in the service of an idea, talked into the people.

And that has always been so and will eternally remain so.

It is entirely in keeping with the stubborn unworldliness of our German intelligentsia to believe that the writer must necessarily be mentally superior to the speaker. This conception is illustrated in the most precious way by a criticism appearing in the above-mentioned national newspaper, in which it is stated that one is so often disappointed to see the speech of a recognized great orator suddenly in print. This reminds me of another criticism which came into my hands in the course of the War; it painfully subjected the speeches of Lloyd George, who at that time was still munitions minister, to the magnifying glass, only to arrive at the brilliant discovery that these speeches were scientifically inferior products and hackneyed to boot. Later, in the form of a little volume, these speeches came into my own hands, and I had to laugh aloud that an average German knight of the ink-pot should possess no understanding for these psychological masterpieces in the art of mass propaganda. This man judged these speeches solely according to the impression they left on his own blasé nature, while the great English demagogue had set out solely to exert the greatest possible effect on the mass of his listeners, and in the broadest sense on the entire English lower class. Regarded from this standpoint, the speeches of this Englishman were the most wonderful performances, for they testified to a positively amazing knowledge of the soul of the broad masses of the people. And their effect was truly powerful.

Compare to it the helpless stammering of a Bethmann-Hollweg. These speeches, to be sure, were apparently wittier, but in reality they only showed this man's inability to speak to his people, which he simply did not know. Nevertheless, the average sparrow brain of a German scribbler, equipped, it goes without saying, with a high scientific education, manages to judge the intelligence of the English minister by the impression which a speech aimed at mass effect makes on his own brain, calcified with sheer science, and to compare it with that of a German statesman whose brilliant chatter naturally finds more receptive soil in him. Lloyd George proved that he was not only the equal in genius of a Bethmann-Hollweg, but was a thousand times his superior, precisely by the fact that in his speeches he found that form and that expression which opened to him the heart of his people and in the end made this people serve his will completely. Precisely in the primitiveness of his language, the primordiality of its forms of expression, and the use of easily intelligible examples of the simplest sort lies the proof of the towering political ability of this Englishman. *For I must not measure the speech of a statesman to his people by the impression which it leaves in a university professor, but by the effect it exerts on the people.* And this alone gives the standard for the speaker's genius.

*   *   *

The amazing development of our movement, which only a few years ago was founded out of the void and today is considered worthy to be sharply persecuted by all the inner and outer enemies of our people, must be attributed to the constant consideration and application of these realizations.

Important as the movement's literature may be, it will in our present position be more important for the equal and uniform training of the upper and lower leaders than for the winning of the hostile masses. Only in the rarest cases will a convinced Social Democrat or a fanatical Communist condescend to acquire a National Socialist pamphlet, let alone a book, to read it and from it gain an insight into our conception of life or to study the critique of his own. Even a newspaper will be read but very seldom if it does not bear the party stamp. Besides, this would be of little use; for the general aspect of a single copy of a newspaper is so chopped up and so divided in its effect that looking at it once cannot be expected to have any influence on the reader. We may and must expect no one, for whom pennies count, to subscribe steadily to an opposing newspaper merely from the urge for objective enlightenment. Scarcely one out of ten thousand will do this. Only a man who has already been won to

the movement will steadily read the party organ, and he will
read it as a running news service of his movement.

The case is quite different with the 'spoken' leaflet! The man
in the street will far sooner take it into his hands, especially if he
gets it for nothing, and all the more if the headlines plastically
treat a topic which at the moment is in everyone's mouth. By a
more or less thorough perusal, it may be possible by such a
leaflet to call his attention to new viewpoints and attitudes, even
in fact to a new movement. But even this, in the most favorable
case, will provide only a slight impetus, never an accomplished
fact. For the leaflet, too, can only suggest or point to something,
and its effect will only appear in combination with a subsequent
more thoroughgoing instruction and enlightenment of its readers.
And this is and remains the *mass meeting*.

*The mass meeting is also necessary for the reason that in it the
individual, who at first, while becoming a supporter of a young
movement, feels lonely and easily succumbs to the fear of being alone,
for the first time gets the picture of a larger community, which in
most people has a strengthening, encouraging effect.* The same man,
within a company or a battalion, surrounded by all his comrades,
would set out on an attack with a lighter heart than if left entirely on his own. In the crowd he always feels somewhat shel-
tered, even if a thousand reasons actually argue against it.

But the community of the great demonstration not only
strengthens the individual, it also unites and helps to create an
esprit de corps. The man who is exposed to grave tribulations,
as the first advocate of a new doctrine in his factory or workshop,
absolutely needs that strengthening which lies in the conviction
of being a member and fighter in a great comprehensive body.
And he obtains an impression of this body for the first time in the
mass demonstration. When from his little workshop or big factory, in which he feels very small, he steps for the first time into
a mass meeting and has thousands and thousands of people of
the same opinions around him, when, as a seeker,[1] he is swept

---

[1] *'Als Suchender.'* A Wagnerian phrase, which Hitler was apparently
determined to use at all costs.

away by three or four thousand others into the mighty effect of suggestive intoxication and enthusiasm, when the visible success and agreement of thousands confirm to him the rightness of the new doctrine and for the first time arouse doubt in the truth of his previous conviction — then he himself has succumbed to the magic influence of what we designate as 'mass suggestion.' The will, the longing, and also the power of thousands are accumulated in every individual. The man who enters such a meeting doubting and wavering leaves it inwardly reinforced: he has become a link in the community.

The National Socialist movement must never forget this and in particular it must never let itself be influenced by those bourgeois simpletons who know everything better, but who nevertheless have gambled away a great state including their own existence and the rule of their class. Oh, yes, they are very, very clever, they know everything, understand everything — only one thing they did not understand, how to prevent the German people from falling into the arms of Marxism. In this they miserably and wretchedly failed, so that their present conceit is only arrogance,[1] which in the form of pride, as everyone knows, always thrives on the same tree as stupidity.

If today these people attribute no special value to the spoken word, they do so, it must be added, only because, thank the Lord, they have become thoroughly convinced by now of the ineffectualness of their own speechmaking.

---

[1] '*so dass ihre jetzige Eingebildetheit nur Dünkel ist.*'

CHAPTER
VII

The Struggle with the Red Front

I~N~ 1919–20 and also in 1921 I personally attended bourgeois meetings. They always made the same impression on me as in my youth the prescribed spoonful of cod-liver oil. You've got to take it, and it's supposed to be very good, but it tastes terrible. If the German people were tied together with cords and pulled forcibly into these bourgeois '*demonstrations*,' and the doors were locked till the end of the performance and no one allowed to leave, it might lead to success in a few centuries. Of course, I must frankly admit that in this case I should probably lose all interest in life and would rather not be a German at all. But since, thank the Lord, this cannot be done, we have no need to be surprised that the healthy, unspoiled people avoid 'bourgeois mass meetings' as the devil holy water.

I came to know them, these prophets of a bourgeois philosophy, and I am really not surprised; I understand why they attribute no importance to the spoken word. In those days I attended meetings of the Democrats, the German Nationalists, the German People's Party, and also the Bavarian People's Party (Bavarian Center). What struck you at once was the homogeneous solidity of the audience. It was almost always solely party members that took part in one of these rallies. The whole thing was without any discipline, more like a yawning bridge club than a meeting of the people which had just been through their greatest revolution.

The speakers did everything they could to preserve this peaceful mood. They spoke, or rather, as a rule, they read speeches in the style of a witty newspaper article or of a scientific treatise, avoided all strong words, and here and there threw in some feeble professorial joke, at which the honorable committee dutifully began to laugh; though not loudly, provocatively, but in a dignified, subdued, reserved fashion.

And what a committee!

Once I saw a meeting in the Wagner-Saal in Munich; it was a demonstration on the occasion of the anniversary of the Battle of Nations at Leipzig. The speech was delivered or read by a dignified old gentleman, a professor at some university. On the platform sat the committee. To the left a monocle, to the right a monocle, and in between one without a monocle. All three in frock coats, so that you got the impression either of a court of justice planning an execution or of a solemn baptism, in any case more of a religious solemnity. The so-called speech, which might have cut a perfectly good figure in print, was simply terrible in its effect. After only three quarters of an hour the whole meeting was dozing along in a state of trance, which was interrupted only by the departure of individual men and women, the clattering of the waitresses, and the yawning of more and more numerous listeners. Three workers, who, either from curiosity or because they had been commissioned to attend, were present at the meeting, and behind whom I posted myself, looked at each other from time to time with ill-concealed grins, and finally nudged one another, whereupon they very quietly left the hall. You could see that they did not want to disturb the meeting at any price. And in this company it was really not necessary. Finally the meeting seemed to be drawing to its end. After the professor, whose voice had meanwhile grown steadily softer and softer, had finished his lecture, the chairman of the meeting, sitting between the two monocle-bearers, arose and roared at the 'German sisters' and 'brothers' present how great his gratitude was and how great their feelings on this order must be for the unique lecture, as enjoyable as it was thorough and deeply penetrating, which Professor X had given them, and which in the truest sense of the word was an 'inner experience,' in fact, an 'achievement.' It would be a profanation of this solemn hour to add a discussion to these lucid remarks; therefore, speaking for all those present, he would dispense with any such discussion and instead bid them all rise from their seats and join in the cry: 'We are a united people of brothers,' etc. Finally, to conclude the meeting he asked us all to sing the *Deutschland* song.

And then they sang, and it seemed to me that even at the second verse the voices were becoming somewhat fewer and only swelled mightily at the refrain, and at the third verse this impression grew stronger, and I believed that not all of them could have been quite sure of the text.

But what does this matter if such a song rings to the heavens in all fervor from the heart of a German National soul.

Thereupon the meeting scattered; that is, everyone rushed to get out quickly, some to their beer, others to a café, and still others into the fresh air.

Yes, indeed, out into the fresh air, at all costs out. That was my own one feeling, too. And this was supposed to serve for the glorification of a heroic struggle on the part of hundreds of thousands of Prussians and Germans? Phooey, I say, and again phooey!

The government, of course, may like this kind of thing. Naturally this is a 'peaceful' meeting. The minister for law and order really has no need to fear that the waves of enthusiasm will suddenly burst the legal measure of bourgeois propriety; that suddenly in a frenzy of enthusiasm, the people will pour forth from the hall, not to hurry to a café or tavern, but to march through the streets of the city in rows of four with measured tread, singing 'Deutschland hoch in Ehren,' thus creating unpleasantness for a police force in need of rest.

No, with such citizens they can be well pleased.

*   *   *

By contrast, it must be admitted, the National Socialist meetings were not '*peaceful.*' There the waves of two outlooks clashed, and they did not end with the insipid rattling off of some patriotic song, but with a fanatical outburst of folkish and national passion.

From the very beginning it was important to introduce blind discipline in our meetings and absolutely to guarantee the authority of the committee in charge. For what we said in our speeches was not the feeble bilge of a bourgeois 'speaker,' but in content and form was always suited to provoke a reply from our opponents. And opponents there were in our meetings! How often they came in dense crowds, individual agitators among them, and all their faces reflecting the conviction: Today we'll make an end of you!

How often, indeed, they were led in, literally in columns, our Red friends, with exact orders, poured into them in advance, to smash up the whole show tonight and put an end to the whole business. And how often it was touch and go, and only the ruthless energy of our people in charge and the brutal activism of our guards was able again and again to thwart the enemy's purpose.

And they had every reason to feel provoked.

The red color of our posters in itself drew them to our meeting halls. The run-of-the-mill bourgeoisie were horrified that we had seized upon the red of the Bolsheviks, and they regarded this as all very ambiguous. The German national souls kept privately whispering to each other the suspicion that basically we were nothing but a species of Marxism, perhaps Marxists, or rather, socialists in disguise. For to this very day these scatterbrains have not understood the difference between socialism and Marxism. Especially when they discovered that, as a matter of principle, we greeted in our meetings no '*ladies and gentlemen*' but only '*national comrades*,' and among ourselves spoke only of *party comrades*, the Marxist spook seemed demonstrated for many of our enemies. How often we shook with laughter at these simple bourgeois scare-cats, at the sight of their ingenious witty guessing games about our origin, our intentions, and our goal.

We chose the red color of our posters after careful and thorough reflection, in order to provoke the Left, to drive them to indignation and lead them to attend our meetings if only to break them up, in order to have some chance to speak to the people.

It was really a treat in those years to follow the perplexity and helplessness of our adversaries in their perpetually vacillating tactics. First they called on their adherents to take no notice of us and to avoid our meetings.

And on the whole this advice was followed.

But since in the course of time individuals came notwithstanding, and this number slowly but steadily increased and the impression made by our doctrine was obvious, the leaders gradually became nervous and uneasy and became obsessed with the conviction that they must not forever stand idly by and watch this development, but must put an end to it by terror.

Thereupon came appeals to the '*class-conscious proletarians*' to attend our meetings in masses and strike the representatives of '*monarchistic, reactionary agitation*' with the fists of the proletariat.

All at once our meetings were filled with workers, three quarters of an hour in advance. They were like a powder barrel that could blow up at any moment, with a burning fuse already under it. But it always turned out differently. The people came in as our enemies, and when they left, if they were not our supporters, at least they had grown thoughtful, indeed critical; they had begun to examine the soundness of their own doctrine. But gradually it transpired that after my speech lasting three hours adherents and adversaries fused into a single enthusiastic mass. Then any signal to smash up the meeting was in vain. Then the leaders really began to be afraid, and they turned back to those who had previously come out against this tactic and who now, with a certain semblance of justification, emphasized their opinion that the only correct method was to forbid the workers to attend our meeting on principle.

Then they stopped coming, or at least there were fewer of

them. But after a short while the whole game began again from the beginning.

The prohibition was not observed; more and more of the comrades came, and again the adherents of the radical tactic were victorious. Our meetings must be broken up, they decided.

Then, after two, three, or often eight and ten meetings it turned out that to break up the meetings was easier said than done, and the result of every single meeting was a crumbling away of the Red fighting troops. Suddenly the other watchword was back again: *'Proletarians, comrades! Avoid the meetings of the National Socialist agitators!'*

And the same, eternally vacillating tactic was found in the Red press. Sometimes they tried to kill us by silence, then becoming convinced of the uselessness of this effort and again trying the contrary. Every day we were 'mentioned' somewhere, usually with the intent of making the absolute absurdity of our whole existence clear to the workers. But after a certain time the gentlemen could not help but feel that not only did this do us no harm, but on the contrary benefited us, since naturally many individuals could not help but ask themselves why so many words were devoted to this phenomenon if it was absurd. The people became curious. Then there was a sudden shift, and they began for a time to treat us as humanity's biggest criminals. Article upon article, in which our criminality was explained and proved again and again, and scandalous stories, even if pulled out of the air from A to Z, were expected to do the rest. But after a short time they seem to have convinced themselves of the inefficacy of these attacks; essentially all this only helped really to concentrate the general attention upon us.

At that time I adopted the standpoint: It makes no difference whatever whether they laugh at us or revile us, whether they represent us as clowns or criminals; the main thing is that they mention us, that they concern themselves with us again and again, and that we gradually in the eyes of the workers themselves appear to be the only power that anyone reckons with at the moment. What we really are and what we really want, we will show the wolves of the Jewish press when the time comes.

One more reason why, as a rule, our meetings were not directly broken up in those days was the absolutely incredible cowardice of the leaders of our adversaries. In all critical cases they sent little rank-and-filers ahead, at most waiting outside for the results of the disturbances.

We were almost always very well informed with regard to the intentions of these gentry. Not only because, for reasons of expediency, we had left many party comrades within the Red formations, but because the Red wirepullers themselves were afflicted with a talkativeness which in this case was very useful to us, and which, unfortunately, is very frequently found among the German people in general. They couldn't keep it to themselves when they had hatched out such a plan, and as a rule they began to cackle even before the egg was laid. And so, many and many a time, we had made the most comprehensive preparations and the Red shock troops hadn't so much as a suspicion how close they were to being thrown out.

The times compelled us to take the defense of our meetings into our own hands; one can never count on protection on the part of the authorities; on the contrary, experience shows that it always and exclusively benefits the disturbers. For the sole actual result of intervention by the authorities — that is, the police — was at best to dissolve, in other words, to close the meeting. And that was the sole aim and purpose of the hostile disturbers.

In this connection the police has developed a practice which represents the most monstrous form of injustice that can be conceived of. If through some sort of threats it becomes known to the authorities that there is danger of a meeting being broken up, they do not arrest the threateners, but forbid the others, the innocent, to hold the meeting, and what is more, the run-of-the-mill police mind is mighty proud of such wisdom. They call this a 'precautionary measure for the prevention of an illegal act.'

Thus, the determined gangster is always in a position to make political activity and efforts impossible for decent people. In the name of law and order, the state authority gives in to the gangster and requests the others please not to provoke him. And so if National Socialists wanted to hold meetings in certain places and the unions declared that this would lead to resistance on the part of their members, the police, you may rest assured, did not put these blackmailing scoundrels behind the bars, but forbade our meeting. Yes, these organs of the law even had the incredible shamelessness to inform us of this innumerable times in writing.

If we wanted to defend ourselves against such eventualities, we had, therefore, to make sure that any attempt at a disturbance was forestalled [1] in the bud.

In this connection the following had also to be considered: *Any meeting which is protected exclusively by the police discredits its organizers in the eyes of the broad masses.* Meetings which are guaranteed only by the presence of a large police force do not attract support, since the presupposition for winning the lower strata of a people is always a strength that is visibly present.

Just as a courageous man can more easily conquer women's hearts than a coward, a heroic movement will sooner win the heart of a people than a cowardly one which is kept alive only by police protection.

Especially for this last reason, the young party had to make sure of defending its own existence, of protecting itself and of breaking the enemy terror with its own hands.

The protection of meetings was based:

(1) *On an energetic and psychologically sound conduct of the meeting.*[2]

If we National Socialists held a meeting in those days, *we* were its masters and no one else. And every minute, uninterruptedly, we sharply emphasized this master right. Our opponents knew perfectly well that anyone creating a provocation would be mercilessly thrown out, even if we were only a dozen among half

---

[1] '*schon in Keim unmöglich gemacht wurde.*'

[2] In the first edition this series concludes abortively with No. 1. The second edition inserts: '(2) *On an organized monitor troop.*'

a thousand. In the meetings of those days, especially outside of Munich, there would be five, six, seven, and eight hundred adversaries to fifteen or sixteen National Socialists. But nevertheless we tolerated no provocation, and those who attended our meetings knew full well that we would rather have let ourselves be beaten to death than capitulate. And it happened more than once that a handful of party comrades heroically fought their way to victory against a roaring, flailing Red majority.

In such cases these fifteen or twenty men would in the end have assuredly been overcome. But the others knew that previously at least twice or three times as many of them would have had their skulls bashed in, and this they did not gladly risk.

Here we tried to learn from the study of Marxist and bourgeois meeting technique, and learn we did.

The Marxists had always had a blind discipline, so that the idea of breaking up a Marxist meeting, by the bourgeoisie at least, could not even arise. But the Reds busied themselves all the more with such intentions. Gradually they had not only achieved a certain virtuosity in this field, but ultimately in large sections of the Reich they went so far as to designate a non-Marxist meeting as such as a provocation of the proletariat; especially when the wirepullers sensed that the meeting might draw up the catalogue of their own sins and unmask the treachery with which they deceived and lied to the people. Then, as soon as such a meeting was announced, the whole Red press raised a furious outcry, and these men who in principle despised the law were not seldom the first to turn to the authorities, with the urgent and threatening request that this 'provocation of the proletariat' be prohibited at once, 'in order to prevent worse things from happening.' They chose their language and achieved their success according to the dimensions of the official bonehead. But if, in an exceptional case, there was a real German official in such a post, not an official toady, and he rejected the shameless imposition, there followed the well-known summons not to suffer such a *provocation of the proletariat,* but on such and such a date to attend the meeting *en masse,* and 'put a stop to the disgraceful activity of the bourgeois creatures, with the horny fist of the proletariat.'

You need to have seen such a bourgeois meeting, you need to have seen its leaders in all their miserable fear! Often, upon such threats, a meeting was simply called off. And always the fear was so great that instead of eight o'clock the meeting was seldom opened before a quarter to nine or nine o'clock. The chairman then endeavored, with twenty-nine compliments, to make it clear to the 'gentlemen of the opposition' present, how pleased he and all the others present were at heart (a plain lie!) with the visit of men who did not yet stand on the same ground, because after all only mutual discussion (to which he thereby most solemnly consented in advance) could bring them closer, arouse mutual understanding, and throw a bridge between them. And in passing he gave assurance that it was by no means the purpose of the meeting to turn people away from their previous views. No, indeed, let each man be happy in his own fashion, but let him not interfere with the happiness of others; and so he requested the audience to let the speaker complete his remarks, which would not be very long anyway, so that this meeting should not present to the world the shameful spectacle of German brothers quarreling among themselves . . . Brrr!

But the brethren on the Left usually had no understanding for this; no, before the speaker had even begun, he had to pack up his things amid the wildest abuse; and not seldom you got the impression that he was thankful to Fate for quickly cutting off the painful procedure. Amid a monstrous tumult such bourgeois meeting-hall toreadors left the arena, except when they flew down the steps with gashed heads, which was actually often the case.

And so, you may be sure, it was something new to the Marxists when we National Socialists organized our first meetings, and especially how we organized them. They came in convinced that, of course, they would be able to repeat on us the little game they had so often played. 'Today we'll finish you off!' How many a one boastfully shouted this sentence to another on entering our meeting, only to find himself outside the hall in the twinkling of an eye, even before he could shout his second interruption.

In the first place, the committee in charge was different with us. No one begged the audience graciously to permit our speech, nor was everyone guaranteed unlimited time for discussion; it was simply stated that we were the masters of the meeting, that in consequence we had the privilege of the house, and that any-one who should dare to utter so much as a single cry of interrup-tion would be mercilessly thrown out where he came from. That, furthermore, we must reject any responsibility for such a fellow; if there was time left and it suited us, we would permit a dis-cussion to take place, if not, there would be none, and the speaker, Party Comrade So-and-So, had the floor.

This in itself filled them with amazement.

In the second place, we disposed of a rigidly organized house guard. In the bourgeois parties this house guard, or rather monitor service, usually consisted of gentlemen who believed that the dignity of their years gave them a certain claim to authority and respect. But since the Marxist-incited masses did not have the least regard for age, authority, and respect, the existence of this bourgeois house guard was for practical purposes nullified, so to speak.

At the very beginning of our big meetings, I began the organ-ization of a house guard in the form of a *monitor service*, which as a matter of principle included only young fellows. These were in part comrades whom I knew from military service; others were newly won party comrades who from the very outset were in-structed and trained in the viewpoint that terror can only be broken by terror; that on this earth success has always gone to the courageous, determined man; that we are fighting for a mighty idea, so great and noble that it well deserves to be guarded and protected with the last drop of blood. They were imbued with the doctrine that, as long as reason was silent and violence had the last word, the best weapon of defense lay in attack; and that our monitor troop must be preceded by the reputation of not being a debating club, but a combat group determined to go to any length.

And how this youth had longed for such a slogan!

How disillusioned and outraged was this front-line generation, how full of disgust and revulsion at bourgeois cowardice and shilly-shallying!

Thus, it became fully clear that the revolution had been possible thanks only to the disastrous bourgeois leadership of our people. The fists to protect the German people would have been available even then, but the heads to play the game were lacking. How many a time the eyes of my lads glittered when I explained to them the necessity of their mission and assured them over and over again that all the wisdom on this earth remains without success if force does not enter into its service, guarding it and protecting it; that the gentle Goddess of Peace can walk only by the side of the God of War; and that every great deed of this peace requires the protection and aid of force. How much more vividly the idea of military service now dawned on them! Not in the calcified sense of old, ossified officials serving the *dead authority* of a *dead state*, but in the living consciousness of the duty to fight for the existence of our people as a whole by sacrificing the life of the individual, always and forever, at all times and places.

And how these lads did fight!

Like a swarm of hornets they swooped down on the disturbers of our meetings, without regard for their superior power, no matter how great it might be, without regard for wounds and bloody victims, filled entirely with the one great thought of creating a free path for the holy mission of our movement.

As early as midsummer, 1920, the organization of the monitor troop gradually assumed definite forms, and in the spring of 1921 little by little divided into hundreds, which themselves in turn were split up into groups.

And this was urgently necessary, for in the meanwhile our public meeting activity had steadily increased. Even now, to be sure, we still often met in the Festsaal of the Munich Hofbräuhaus, but even more often in the larger halls of the city. The Festsaal of the Bürgerbräu and the Münchener Kindl-Keller saw mightier and mightier mass meetings in the fall and winter of 1920–21, and the picture was always the same: *rallies of the NSDAP even then usually had to be closed by the police even before beginning, because of overcrowding.*

*   *   *

The organization of our monitor troop clarified a very important question. Up till then the movement possessed no party insignia and no party flag. The absence of such symbols not only had momentary disadvantages, but was intolerable for the future. The disadvantages consisted above all in the fact that the party comrades lacked any outward sign of their common bond, while it was unbearable for the future to dispense with a sign which possessed the character of a symbol of the movement and could as such be opposed to the International.

What importance must be attributed to such a symbol from the psychological point of view I had even in my youth more than one occasion to recognize and also emotionally to understand. Then, after the War, I experienced a mass demonstration of the Marxists in front of the Royal Palace and the Lustgarten. A sea of red flags, red scarves, and red flowers gave to this demonstration, in which an estimated hundred and twenty thousand persons took part, an aspect that was gigantic from the purely external point of view. I myself could feel and understand how easily the man of the people succumbs to the suggestive magic of a spectacle so grandiose in effect.

The bourgeoisie, which in its party politics neither represents nor advocates any outlook at all, had therefore no flag of its own. They consisted of '*patriots*' and therefore ran around in the colors of the Reich. If these had been the symbol of a definite philosophy, it would have been understandable that the owners of the state viewed its flag as the representative of its philosophy, since the symbol of their philosophy had become the flag of the state and the Reich through their own activity.

But this was not the case.

The Reich had been formed without any move on the part of the German bourgeoisie, and the flag itself had been born from the womb of war. Hence it was really nothing but a state flag and possessed no meaning of any sort in the sense of a special philosophical mission.

Only in one spot of the German language area was anything like a bourgeois party flag in existence — in German Austria. By choosing the colors of 1848, black, red, and gold, for its party symbol, a part of the national bourgeoisie in that country had created a symbol, which, though without any meaning in a philosophical sense, nevertheless had a revolutionary character, politically speaking. *The sharpest enemies of this black, red, and gold flag were then — and today this should not be forgotten — the Social Democrats and the Christian Social Party, or Clericals.* It was precisely they who in those days reviled, befouled, and soiled these colors, just as later, in 1918, they dragged the black, white, and red into the gutter. At all events, the black, red, and gold of the German parties of old Austria were the colors of 1848; that is, of a time which may have been fantastic, but which was represented by the most honorable individual German souls, though the Jew stood in the background as the invisible wire-puller. Therefore, it was high treason and the shameless selling-out of the German people and German treasure which made these flags so agreeable to the Marxists and the Center that today they honor them as their most sacred possession and create organizations of their own for the protection of the flag they once spat upon.

And so, up to 1920, Marxism was actually confronted by no flag which philosophically would have represented its polar opposite. For even if the best parties of the German bourgeoisie after 1918 would no longer consent to take over the suddenly discovered black, red, and gold flag as their own symbol, they themselves had no program of their own for the future to oppose to the new development; at best they had the idea of a reconstruction of the past Reich

And it is to this idea that the black, white, and red banner of the old Reich owes its resurrection as the flag of our so-called national bourgeois parties.

It is obvious that the symbol of a state of affairs, which could be overcome by Marxism under conditions and attendant circumstances that were anything but glorious, is ill-suited for a symbol under which to annihilate this same Marxism. Sacred and beloved as these old and uniquely beautiful colors, in their fresh, youthful combination, must be to every decent German who has fought under them and beheld the sacrifice of so many, the flag is worthless as a symbol for a struggle for the future.

Unlike the bourgeois politicians, I have, in our movement, always upheld the standpoint that it is a true good fortune for the German nation to have lost the old flag. What the Republic does beneath its flag, can remain indifferent to us. But from the bottom of our hearts we should thank Fate for having been gracious enough to preserve the most glorious war flag of all times from being used as a bedsheet for the most shameful prostitution. The present-day Reich, which sells itself and its citizens, must never be permitted to fly the black, white, and red flag of honor and heroes.

As long as the November disgrace endures, let it bear its own outer covering and not try to steal this like everything else from a more honorable past. Let our bourgeois politicians remind their conscience that anyone who desires the black, white, and red flag for this state is burglarizing our past. Truly, the former flag was suited only to the former Reich, just as, God be praised and thanked, the Republic chose the one suited to it.

This was also the reason why we National Socialists could have seen no expressive symbol of our own activity in hoisting the old flag. For we do not desire to awaken from death the old Reich that perished through its own errors, but to build a new state.

The movement which today fights Marxism with this aim must therefore bear the symbol of the new state in its very flag.

The question of the new flag — that is, its appearance — occupied us intensely in those days. From all sides came suggestions, which for the most part it must be admitted were more well-intended than successful. For the new flag had to be equally a symbol of our own struggle, since on the other hand it was expected also to be highly effective as a poster. Anyone who has to concern himself much with the masses will recognize these apparent trifles to be very important matters. An effective insignia can in hundreds of thousands of cases give the first impetus toward interest in a movement.

For this reason we had to reject all suggestions of identifying our movement through a white flag with the old state, or, more correctly, with those feeble parties whose sole political aim was the restoration of past conditions, as was proposed by many quarters. Besides, white is not a stirring color. It is suitable for chaste virgins' clubs, but not for world-changing movements in a revolutionary epoch.

Black was also suggested: in itself suitable for the present period, it contained nothing, however, that could in any way be interpreted as a picture of the will of our movement. Finally, this color has not a stirring enough effect either.

White and blue were out of the question despite their wonderful esthetic effect, for these were the colors of an individual German state, and of an orientation toward particularistic narrow-mindedness which unfortunately did not enjoy the best reputation. Here, too, moreover, it would have been hard to find any reference to our movement. The same applied to black and white.

Black, red, and gold were in themselves out of the question.

So were black, white, and red, for reasons already mentioned, at least in their previous composition. In effect, to be sure, this color combination stands high above all others. It is the most brilliant harmony in existence.

I myself always came out for the retention of the old colors, not only because as a soldier they are to me the holiest thing I know, but because also in their esthetic effect they are by far the most compatible with my feeling. Nevertheless, I was obliged to reject without exception the numerous designs which poured in from the circles of the young movement, and which for the most part had drawn the swastika into the old flag. I myself — as Leader — did not want to come out publicly at once with my own design, since after all it was possible that another should produce one just as good or perhaps even better. Actually, a dentist from Starnberg did deliver a design that was not bad at all, and, incidentally, was quite close to my own, having only the one fault that a swastika with curved legs was composed into a white disk.

I myself, meanwhile, after innumerable attempts, had laid down a final form; a flag with a red background, a white disk, and a black swastika in the middle. After long trials I also found a definite proportion between the size of the flag and the size of the white disk, as well as the shape and thickness of the swastika.

And this remained final.

Along the same lines arm-bands were immediately ordered for the monitor detachments, a red band, likewise with the white disk and black swastika.

The party insignia was also designed along the same lines: a white disk on a red field, with the swastika in the middle. A Munich goldsmith by the name of Füss furnished the first usable design, which was kept.

In midsummer of 1920 the new flag came before the public for the first time. It was excellently suited to our new movement. It was young and new, like the movement itself. No one had seen it before; it had the effect of a burning torch. We ourselves experienced an almost childlike joy when a faithful woman party comrade for the first time executed the design and delivered the flag. Only a few months later we had half a dozen of them in Munich, and the monitor troop, which was growing bigger and bigger, especially contributed to spreading the new symbol of the movement.

And a symbol it really is! Not only that the unique colors, which all of us so passionately love and which once won so much honor for the German people, attest our veneration for the past; they were also the best embodiment of the movement's will. As National Socialists, we see our program in our flag. In *red* we see the social idea of the movement, in *white* the nationalistic idea, in the *swastika* the mission of the struggle for the victory of the Aryan man, and, by the same token, the victory of the idea of creative work, which as such always has been and always will be anti-Semitic.

Two years later, when the monitor troop had long since become a *Sturm-Abteilung* (storm section), embracing many thousands of men, it seemed necessary to give this armed organization a special symbol of victory: the *standard*. This, too, I designed myself and then gave it to a loyal old party comrade, master goldsmith Gahr, for execution. Since then the standard is among the symbols and battle signs of the National Socialist struggle.

*   *   *

Our public meeting activity, which increased more and more in 1920, finally led to the point where we held as many as two meetings in some weeks. People crowded in front of our posters, the largest halls of the city were always filled, and tens of thousands of misled Marxists found the way back to their national community to become warriors for a free German Reich to come. The Munich public had come to know us. People spoke of us, the word 'National Socialist' became familiar to many and already meant a program. The host of adherents, and even of members, began to grow uninterruptedly, so that in the winter of 1920–21 we could already be regarded as a strong party in Munich.

Aside from the Marxist parties there was in those days no party, above all no *national* party, which could boast of such mass demonstrations as ours. The Münchener-Kindl-Keller, holding five thousand people, had more than once been filled to the bursting point, and there was only a single hall into which we had not yet ventured, and this was the Zirkus Krone.

At the end of January, 1921, grave cares arose once more for Germany. The Paris Agreement, according to which Germany obligated herself to pay the insane sum of a hundred billion gold marks, was to be realized in the form of the London dictate.[1]

A working federation of so-called *folkish leagues*, long existing in Munich, wanted to call a large common protest meeting on this occasion. Time was pressing, and I myself was nervous in view of the eternal hesitation and delay in carrying out decisions that had been taken. First there was talk of a demonstration on the Königsplatz, but this was abandoned for fear of being broken up by the Reds and a protest demonstration in front of the Feldherrn-halle was projected. But this too was abandoned and finally a common demonstration in the Münchener-Kindl-Keller was suggested. Meanwhile, day after day had passed, the big parties had taken no notice whatever of the great event, and the action committee could not make up its mind to set a definite date for the intended demonstration.

On Tuesday, February 1, 1921, I most urgently demanded a final decision. I was put off till Wednesday. So on Wednesday I absolutely insisted on clear information when and whether the demonstration should take place. The answer was again indefinite and evasive; I was told that they 'intended' to call a demonstration for Wednesday a week.

With this the cord of my patience snapped and I decided to carry through the protest demonstration alone. On Wednesday noon I dictated the poster into the typewriter in ten minutes and at the same time had the Zirkus Krone rented for the following day, Thursday, February 3.

At that time this was a tremendous venture. Not only that it seemed questionable whether we could fill the gigantic hall, but we also ran the danger of being broken up.

Our monitor troop was far from being adequate for this colossal

---

[1] The Supreme Allied Council met in Paris from January 24 to 30, and elaborated a plan of reparations payments. Annual payments were to begin at two billion gold marks a year and gradually increase to six billions at the end of eleven years.

The London Conference on Reparations, held from April 29 to May 5 of the same year, sent an ultimatum to Germany demanding one billion gold marks on penalty of occupying the Ruhr. The Germans accepted the terms and paid the sum by borrowing in London.

hall. And I had no proper idea about the kind of procedure possible in case of an attempt to break the meeting up. At that time I thought this would be much harder for us in the Circus building than in a normal hall. Yet, as it later turned out, the truth was exactly the opposite. Actually, in this gigantic hall, it was easier to master a troop of disturbers than in small halls where you were penned in.

Only one thing was certain: any failure could throw us back for a long time to come. For if we were once successfully broken up, it would have destroyed our nimbus at one stroke and encouraged our opponents to attempt again what had once succeeded. This could have led to a sabotage of our whole further meeting activity, which would have taken many months and the hardest struggles to overcome.

We had only one day's time to put up posters, that was Thursday itself. Unfortunately, it was raining in the morning, and the fear seemed founded that under such circumstances many people would prefer to stay home, instead of hurrying through the rain and snow to a meeting at which there might possibly be murder and homicide.

Altogether, I suddenly became afraid on Thursday morning that the hall would not be filled after all (and in this case I would have been discredited in the eyes of the working federation), so now I hastily dictated a few leaflets and had them printed for circulation in the afternoon. They naturally contained an appeal to attend the meeting.

Two trucks that I had hired were swathed in as much *red* as possible, a few of our flags were planted on top of them and each one was manned with fifteen to twenty party comrades; they received the command to drive conscientiously through the streets of the city and throw off leaflets; in short, to make propaganda for the mass demonstration in the evening. It was the first time that trucks had driven through the city with banners and no Marxists on them. Consequently the bourgeoisie stared open-mouthed after the red car decked out with fluttering swastika flags, while in the outer sections numerous clenched fists arose whose owners seemed obviously burned up with rage at this newest 'provocation of the proletariat.' For only the Marxists had the right to hold meetings or to drive around in trucks.

At seven that night the Circus was not yet well filled. Every ten minutes I was notified by phone, and even I was pretty worried; for at seven or a quarter after, the other halls had usually been half, in fact, often almost entirely, full. This, however, was soon explained. I had not reckoned with the gigantic dimensions of the new hall: a thousand persons made the Hofbräuhaus seem very well filled, while they were simply swallowed up by the Zirkus Krone. You could hardly see them. A short time later, however, more favorable reports came in, and at a quarter to eight word came that the hall was three-quarters full and that large crowds were standing outside the box office windows. Thereupon I set out.

At two minutes past eight I arrived in front of the Circus. There was still a crowd to be seen in front, partly just curious people, with many opponents among them who wanted to stay outside and see what would happen.

As I entered the mighty hall, the same joy seized me as a year previous in the first meeting at the Munich Hofbräuhaus Festsaal. But only after I had pressed my way through the human walls and reached the lofty platform did I see the success in all its magnitude. Like a giant shell this hall lay before me, filled with thousands and thousands of people. Even the ring was black with people. Over five thousand six hundred tickets had been sold, and if we included the total number of unemployed, of poor students and our monitor detachments, there must have been six and a half thousand persons.

'Future or Ruin' was the theme, and my heart rejoiced in the conviction that down there before me the future lay.

I began to speak, and spoke about two and a half hours; and my feeling told me after the first half hour that the meeting would be a great success. Contact with all these thousands of individuals had been established. After the first hour the applause began to interrupt me in greater and greater spontaneous outbursts, ebbing off after two hours into that solemn stillness which I have later experienced so very often in this hall, and which will remain unforgettable to every single member of the audience. Then you could hardly hear more than the breathing of this gigantic multitude, and only when the last word had been spoken did the applause suddenly roar forth to find its release and conclusion in the *Deutschland* song, sung with the highest fervor.

I stayed to watch as the giant hall slowly began to empty and for nearly twenty minutes an enormous sea of human beings forced its way through the mighty center exit. Only then did I myself, overjoyed, leave my place to go home.

Photographs were made of this first meeting in the Zirkus Krone. They show better than words the magnitude of the demonstration. Bourgeois papers ran pictures and notices, but they only mentioned that there had been a 'national' demonstration and with their usual modesty passed over the organizers in silence.

With this we had for the first time far overstepped the bounds of an ordinary party of the day. We could no longer be ignored. And now, lest the impression arise that this successful meeting was nothing more than fly-by-night, I immediately fixed a second meeting in the Circus for the coming week, and the success was the same. Again the gigantic hall was full to the bursting point with human masses, so that I decided to hold a meeting in the coming week in the same style for the third time. And for the third time the giant Circus was packed full of people from top to bottom.

After this introduction to the year 1921, I increased our public meeting activity in Munich even more. I now switched over to holding not only one meeting every week, but in some weeks two mass meetings; in fact, in midsummer and late fall, it was sometimes three. We still met in the Circus and to our satisfaction noted that all our evenings brought the same success.

The result was a steadily increasing number of adherents to the movement and a great increase in members.

*   *   *

Such successes naturally did not leave our enemies inactive. Always wavering in their tactics, they had alternated between a policy of terror and one of killing us by silence, and now, as they themselves were forced to recognize, they could in no way obstruct the development of the movement with either the one or the other. And so, with a last exertion, they decided upon an act of terror that would definitely bar any further public meeting activity on our part.

As outward occasion for this action they used a highly mysterious attack upon a deputy in the Bavarian Diet by the name of Erhard Auer.[1] The said Erhard Auer was said to have been shot at one night by someone. That is, he had not actually been shot, but an attempt had been made to shoot him. Amazing presence of mind, as well as the proverbial courage of the Social Democratic Party leader, had ostensibly not only frustrated the insidious attack, but put the infamous assailants to ignominious flight. They had fled so hastily and so far that even later the police could not catch the slightest trace[2] of them. This mysterious occurrence was now used by the organ of the Social Democratic Party in Munich to agitate against the movement in the most unrestrained fashion, and among other things to hint with their customary loose tongue at what must soon follow. Measures had been taken, they hinted, to keep us from getting out of hand; proletarian fists would intervene before it was too late.

And a few days later the day of intervention was at hand.

A meeting in the Munich Hofbräuhaus Festsaal, at which I myself was to speak, had been chosen for the final reckoning.

On November 4, 1921, between six and seven in the evening, I received the first positive news that the meeting would definitely

---

[1] Erhard Auer was leader of the Munich Social Democrats and a member of the Bavarian Diet. Hitler's story is accurate to the extent that Auer himself reported the attempt to murder him and that nothing was ever learned of his assailants.

[2] 'nicht die leiseste Spur erwischen.'

be broken up, and that for this purpose they intended to send in great masses of workers, especially from a few Red factories.

It must be laid to an unfortunate accident that we did not get this information earlier. On the same day we had given up our venerable old business office in the Sterneckergasse in Munich and had moved to a new one; that is, we were out of the old one, but could not yet move into the new one because work was still going on inside. Since the telephone had already been taken out of the old one and not yet installed in the new one, a number of attempts to inform us by telephone of the intended invasion had been in vain.

The consequence of this was that the meeting itself was protected only by extremely weak monitor groups. Only a numerically weak company, comprising about forty-six heads,[1] was present, and the alarm apparatus was not yet sufficiently developed to bring ample reinforcement in the space of an hour in the evening. Added to this was the fact that such alarmist rumors had come to our ears innumerable times without anything special happening. The old saying that announced revolutions usually fail to take place had up to this time always proved correct in our experience.

And so, for this reason, too, perhaps everything was not done which could have been done that day, to counter any attempt to break up the meeting with the most brutal determination.

Finally, we regarded the Festsaal of the Munich Hofbräuhaus as most unsuited for an attempt to break up a meeting. We had been more afraid for the largest halls, especially the Circus. In this connection this day gave us a valuable lesson. Later we studied all these questions with a method which I should call truly scientific and came to results which in part were as incredible as they were interesting and in the ensuing period were of

[1] The German word I have translated as 'company' is '*Hundertschaft*,' literally a company of one hundred. The effect in German is somewhat ludicrous, but not impossible, as it would be in English to say 'a hundred comprising about forty-six heads.'

basic importance for the organizational and tactical leadership of our storm troops.

When I entered the vestibule of the Hofbräuhaus at a quarter of eight, there could indeed be no doubt with regard to the existing intention. The room was overcrowded and had therefore been closed by the police. Our enemies who had appeared very early were for the most part in the hall, and our supporters for the most part outside. The small S.A. awaited me in the vestibule. I had the doors to the large hall closed and then ordered the forty-five or forty-six men to line up. I made it clear to the lads that today probably for the first time they would have to show themselves loyal to the movement through thick and thin, and that not a man of us must leave the hall unless we were carried out dead; I myself would remain in the hall, and I did not believe that a single one of them would desert me; but if I should see anyone playing the coward, I myself would personally tear off his arm-band and take away his insignia. Then I called upon them to advance immediately at the slightest attempt to break up the meeting, and to bear in mind that the best defense lies in your own offensive.

The answer was a threefold *Heil* that sounded rougher and hoarser than usual.

Then I went into the hall and surveyed the situation with my own eyes. They were sitting in there, tight-packed, and tried to stab me with their very eyes. Innumerable faces were turned toward me with sullen hatred, while again others, with mocking grimaces, let out cries capable of no two interpretations. Today they would 'make an end of us,' we should look out for our guts, they would stop our mouths for good, and all the rest of these lovely phrases. They were conscious of their superior power and felt accordingly.

Nevertheless, the meeting could be opened and I began to speak. In the Festsaal of the Hofbräuhaus I always stood on one of the long sides of the hall and my platform was a beer table. And so I was actually in the midst of the people. Perhaps this circumstance contributed to creating in this hall a mood such as I have never found anywhere else.

In front of me, especially to the left of me, only enemies were sitting and standing. They were all robust men and young fellows, in large part from the Maffei factory, from Kustermann's, from the Isaria Meter Works, etc. Along the left wall they had pushed ahead close to my table and were beginning to collect beer mugs; that is, they kept ordering beer and putting the empty mugs under the table. In this way, whole batteries grew up and it would have surprised me if all had ended well this time.

After about an hour and a half — I was able to talk that long despite interruptions — it seemed almost as if I was going to be master of the situation. The leaders of the invading troops seemed to feel this themselves; for they were becoming more and more restless, they often went out, came in again, and talked to their men with visible nervousness.

A small psychological mistake I committed in warding off an interruption, and which I myself realized no sooner had I let the word out of my mouth, gave the signal for them to start in.

A few angry shouts and a man suddenly jumped on a chair and roared into the hall: '*Freiheit!*' [1] (Freedom.) At which signal the fighters for freedom began their work.

In a few seconds the whole hall was filled with a roaring, screaming crowd, over which, like howitzer shells, flew innumerable beer mugs, and in between the cracking of chair-legs, the crashing of the mugs, bawling, howling, and screaming.

It was an idiotic spectacle.

I remained standing in my place and was able to observe how thoroughly my boys fulfilled their duty.

I should have liked to see a bourgeois meeting under such circumstances.

The dance had not yet begun when my storm troopers — for so they were called from this day on — attacked. Like wolves they flung themselves in packs of eight or ten again and again on their enemies, and little by little actually began to thrash them out of the hall. After only five minutes I hardly saw a one of them who

[1] Greeting and slogan of the German Social Democrats.

was not covered with blood. How many of them I only came really to know on that day; at the head my good Maurice,[1] my present private secretary Hess, and many others, who, even though gravely injured themselves, attacked again and again as long as their legs would hold them. For twenty minutes the hellish tumult lasted, but then our enemies, who must have numbered seven and eight hundred men, had for the most part been beaten out of the hall and chased down the stairs by my men numbering not even fifty. Only in the left rear corner of the hall a big group stood its ground and offered embittered resistance. Then suddenly two shots were fired from the hall entrance toward the platform, and wild shooting started. Your heart almost rejoiced at such a revival of old war experiences.

Who was shooting could not be distinguished from that point on; only one thing could be definitely established, that from this point on the fury of my bleeding boys exceeded all bounds and finally the last disturbers were overcome and driven out of the hall.

About twenty-five minutes had passed; the hall looked almost as if a shell had struck it. Many of my supporters were being bandaged; others had to be driven away, but we had remained masters of the situation. Hermann Esser, who had assumed the chair this evening, declared: '*The meeting goes on. The speaker has the floor.*' And then I spoke again.

After we ourselves had closed the meeting, an excited police lieutenant came dashing in, and, wildly swinging his arms, he cackled into the hall: 'The meeting is dismissed.'

Involuntarily I had to laugh at this late-comer; real police pompousness. The smaller they are, the bigger they have to try and look at least.[2]

---

[1] Emil Maurice. By trade a watchmaker. An early associate of Hitler and first leader of the storm troops. He was in prison with Hitler in Landsberg after the *Putsch*, and Hitler first dictated *Mein Kampf* to him. After the National Socialists seized power, he became a municipal councilor in Munich. He was active in the blood purge killings of 1934.

[2] Persons residing in Munich at the time report that the Social Democrats were expelled from the meeting by the police.

That night we had really learned a good deal and our enemies never again forgot the lesson they for their part had received.

After that the *Münchener Post* threatened us with no more fists of the proletariat up to the autumn of 1923.

The Strong Man Is Mightiest Alone [1]

IN THE ABOVE I have already mentioned the existence of a *Working Federation of German Folkish Associations,* and in this place would like to discuss very briefly the problem of these working federations.

In general we understand by a *working federation* a group of associations which for the facilitation of their work enter into a certain mutual relationship, choose a common leadership of greater or lesser competence, and proceed to carry out common actions. From this alone it results that we must be dealing with clubs, associations, or parties whose aims and methods do not lie too far apart. It is claimed that this is always the case. For the usual average citizen it is equally pleasant and comforting to hear that such associations, by combining in such a '*working federation,*' have discovered a '*common bond*' and '*set aside all dividing factors.*' Here the general conviction prevails that such a unification brings an enormous increase in strength, and that the otherwise weak little groups have thereby suddenly become a power.

This, however, is usually false.

It is interesting and in my eyes important for the better understanding of this question to attain clarity as to how associations, clubs, and the like can arise which all claim to pursue the same goal. In the nature of things, it would after all be logical that *one* goal should be advocated by only *one* association, and that, rea-

[1] Familiar quotation from Schiller's *Wilhelm Tell,* Act I, Scene III.

sonably speaking, several associations should not pursue the same goal. Without doubt that goal had first been envisaged by *one* association. One man somewhere proclaims a truth and forms a movement which is intended to serve the realization of his purpose.

Thus, an association or a party is founded which, according to its program, should either bring about the elimination of existing evils or the achievement of a particular state of affairs in the future.

Once such a movement has been called to life, it possesses a certain practical *right of priority*. It should really be obvious that all men who mean to fight for the same goal should join into such a movement and thereby add to its strength, thus better to serve the common purpose. Especially every active mind must feel that the premise for any real success in the common struggle lies in such a coordination. Therefore, reasonably, and presupposing a certain honesty (much depends on this, as I shall later demonstrate), there should be only one movement for one goal.

That this is not the case can be attributed to two causes. One of these I might designate as almost tragic, while the second is miserable and to be sought in human weakness itself. But most fundamentally, I see in both only facts which are suited to enhancing the will as such, its energy and intensity, and, through this higher cultivation of human energy, ultimately to make possible a solution of the problem in question.

The tragic reason why in the solution of a single task we usually do not content ourselves with a single association is the following: Every deed in the grand manner on this earth will in general be the fulfillment of a desire which had long since been present in millions of people, a longing silently harbored by many. Yes, it can come about that centuries wish and yearn for the solution of a certain question, because they are sighing beneath the intolerable burden of an existing condition and the fulfillment of this general longing does not materialize. Nations which no longer find any heroic solution for such distress can be designated as *impotent*, while we see the vitality of a people, and the predestination for life guaranteed by this vitality, most strikingly demonstrated when, for a people's liberation from a great oppression, or for the elimination of a bitter distress, or for the satisfaction of its soul, restless because it has grown insecure — Fate some day bestows upon it the man endowed for this purpose, who finally brings the long yearned-for fulfillment.

Now it lies entirely in the essence of so-called great questions of the day that thousands are active in their solution, that many feel called, indeed, that Fate itself puts forward many for selection, and then ultimately, in the free play of forces, gives victory to the stronger and more competent, entrusting him with the solution of the problem.

Thus, it may be that centuries, dissatisfied with the form of their religious life, yearn for a renewal, and that from this psychic urge dozens and more men arise who on the basis of their insight and their knowledge believe themselves chosen to solve this religious distress, to manifest themselves as prophets of a new doctrine, or at least as warriors against an existing one.

Here, too, assuredly, by virtue of a natural order, the strongest man is destined to fulfill the great mission; yet the realization that this *one* is the exclusively elect usually comes to the others very late. On the contrary, they *all* see themselves as *chosen* and *having equal rights* for the solution of the task, and their fellow men are usually able least of all to distinguish which among them — being solely endowed with the highest ability — deserves their sole support.

Thus, in the course of centuries, often indeed within the same period, different men appear and found movements to fight for goals which, allegedly at least, are the same or at least are felt to be the same by the great masses. The common people themselves harbor indefinite desires and have general convictions, but cannot obtain precise clarity regarding the actual nature of their aim or of their own desire, let alone the possibility of its fulfillment.

The tragedy lies in the fact that these men strive for the same goal in entirely different ways, without knowing one another, and hence, with the highest faith in their own mission, consider themselves obligated to go their own ways without consideration for others.

The fact that such movements, parties, religious groups, arise entirely independent of one another, solely from the general will of the times to act in the same direction, is what, at least at first sight, seems tragic, because people incline too much to the opinion that the forces scattered among the different ways, could, if concentrated upon a single one, lead more quickly and surely to success. This, however, is not the case. For Nature itself in its inexorable logic makes the decision, by causing the different groups to enter into competition with one another and struggle for the palm of victory, and leads that movement to the goal which has chosen the clearest, shortest, and surest way.

But how should the correctness or incorrectness of a road be determined from outside unless free course is given to the play of forces, unless the ultimate decision is withdrawn from the doctrinaire opinion of human know-it-alls and entrusted to the infallible logic of visible success, which in the end will always render the ultimate confirmation of an action's correctness!

And so if different groups march toward the same goal on separate paths, once they have become aware of the existence of similar efforts, they will more thoroughly examine the nature of their own way; where possible they will shorten it, and by stretching their energy to the utmost will strive to reach the goal more quickly.

This competition helps to cultivate the individual fighter, and mankind often owes its successes in part to the doctrines that have been derived from the ill fate of previous unsuccessful efforts.

And so, in the fact of an incipient scattering of forces, which arose through no conscious fault of individuals and at first sight seemed tragic, we can recognize the means through which in the end the best method was achieved.

We see in history that in the opinion of most people the two roads which it was once possible to take for the solution of the German question and whose chief representatives and champions were Austria and Prussia, Habsburg and Hohenzollern, should have been joined together from the start; in their view, people should have entrusted themselves with united strength to the one or the other road. And then the road of the representative who in the end proved more significant would have been taken; the Austrian intention, however, would never have led to a German Reich.

And then the Reich of strongest German unity arose from the very thing which millions of Germans with bleeding heart felt to be the ultimate and most terrible sign of our fratricidal quarrel: the German imperial throne was in truth won on the field of Königgrätz and not in the battles outside Paris as people afterwards came to think.

And thus the founding of the German Reich as such was not the result of any common will along common paths, but the result of a conscious and sometimes unconscious struggle for hegemony, from which struggle Prussia ultimately issued victorious. And anyone who is not blinded by party politics into renouncing the truth, will have to confirm that so-called human wisdom would never have made the same wise decision which the wisdom of life, that is, the free play of forces, finally turned into reality. For who in German territories two hundred years ago would seriously have believed that the Prussia of the Hohenzollerns would some day become the germ cell, founder, and mentor of the new German Reich, and not the Habsburgs? And who, on the other hand, would deny today that Destiny acted more wisely in this respect; in fact, who today could even conceive of a German Reich based on the principles of a rotten and degenerate dynasty?

No, the natural development, though after a struggle enduring centuries, finally brought the best man to *the* place where he belonged.

This will always be so and will eternally remain so, as it always has been so.

Therefore, it must not be lamented if so many men set out on the road to arrive at the same goal: the most powerful and swiftest will in this way be recognized, and will be the victor.

Now there is a second reason why often in the life of nations movements of apparently the same nature nevertheless try to reach the same goal in different ways. *This* cause not only is not tragic, but is positively miserable. It lies in the sorry mixture of envy, jealousy, ambition, and thievish mentality which unfortunately we sometimes find combined in individual specimens of mankind.

For as soon as a man appears who profoundly recognizes the distress of his people and then, after he has attained the ultimate clarity with regard to the nature of the disease, seriously tries to cure it, when he has set a goal and chosen the road that can lead to this goal — immediately small and petty minds take notice and begin to follow eagerly the activity of this man who has attracted the public eye. These people are just like sparrows who, apparently uninterested, but in reality most attentive, keep watching a more fortunate comrade who has found a piece of bread, in hopes of suddenly robbing him in an unguarded moment. A man need only embark upon a new road and all sorts of lazy loiterers prick up their ears and sniff some worth-while morsel which might lie at the end of this road. Then, as soon as they have found out where it may be, they eagerly start out in order to reach the goal by some other road, if possible a shorter one.

So if a new movement has been founded and has received its definite program, those people come and claim to be fighting for the same goal; but, rest assured, not by honestly joining the ranks of such a movement and thus recognizing its priority; no, they steal the program and base a new party of their own upon it. With all this, they are shameless enough to assure their thoughtless fellow men that they had desired the same as the other movement long before, and not seldom they thus succeed in placing themselves in a favorable light, instead of winning universal contempt as they deserve. For is it not a tremendous gall to aspire to write on their own banner the task that another has written on his, to borrow his programmatical principles, and then, as though *he* had created all this, to go his own ways? And the gall is especially manifested in the fact that the same elements who have caused the split by founding their new movements ao the most talking, as experience shows, about the need of unification and unity as soon as they think they have observed that the opponent has too much of a headstart to be overtaken.

The so-called 'folkish splintering' is due to such a process.

To be sure, the foundation in 1918–19 of a considerable number of groups, parties, etc., designated as folkish, occurred through the natural development of things through no fault of the founders. From all these the NSDAP had slowly crystallized out as the victor by 1920. The basic honesty of those individual founders could be proved by nothing more splendidly than by the truly admirable decision taken by many to sacrifice their own obviously less successful movements to the stronger one; that is, to disband them or fuse them unconditionally.

This applies especially to the chief fighter of the German-Socialist Party (*Deutsch-Sozialistische Partei*) of those days in Nuremberg, Julius Streicher.[1] The NSDAP and the DSP had arisen with the same ultimate aims, yet absolutely independently of one another. The main fighter for the DSP, as I have said, was Julius Streicher, then a teacher in Nuremberg. At first he, too, had a holy conviction of the mission and the future of his movement. But as soon as he could recognize the greater power and superior growth of the NSDAP clearly and beyond all doubt, he ceased his activity for the DSP and the Working Federation, and called on his adherents to join the NSDAP, which had issued victoriously from the mutual struggle, and to fight on in its ranks for the common goal. A decision as grave from the personal point of view as it was profoundly decent.

And no form of split has remained from this first period of the movement; the honorable intention of the men of those days led almost entirely to an honorable, straight, and correct conclusion.

---

[1] Julius Streicher, *Gauleiter* of Nuremberg and publisher of the anti-Semitic paper *Der Stürmer*, devoted chiefly to pornographic exposures of sexual relations between Jews and Aryans, retained until the Second War a local independence enjoyed by no other party leader. His fusion with Hitler was not as peaceable as Hitler makes it.

In 1921, Streicher tried to wrest the party leadership from Hitler, but failed because of a revolt in the ranks of his own Nuremberg supporters.

What we designate today as 'folkish splintering' owes its existence, as we have already emphasized, exclusively to the second of the two causes I have cited: ambitious men who previously had no ideas, much less goals of their own, felt themselves 'called' at the very moment in which they saw the success of the *NSDAP* undeniably maturing.

Suddenly programs arose which from start to finish were copied from ours, ideas were put forward which had been borrowed from us, aims set up for which we had fought for years, roads chosen which the NSDAP had long traveled. By every possible means they sought to explain why they had been forced to found these movements despite the NSDAP which had long been in existence; but the nobler the alleged motives, the falser were their phrases.

In truth a single reason had been determining: the personal ambition of the founders to play a rôle to which their own dwarfish figure really brought nothing except a great boldness in taking over the ideas of others, a boldness which elsewhere in civil life is ordinarily designated as crooked.

There was no conception or idea belonging to other people, which one of these political kleptomaniacs did not rapidly collect for his own business. And those who did this were the same people who later with tears in their eyes profoundly bemoaned the 'folkish splintering' and spoke incessantly of the 'need for unity,' in the secret hope that in the end they would so outwit the others that, weary of the eternal accusing clamor, they would, in addition to the stolen ideas, toss the movements created for their execution to the thieves.

But if this proved unsuccessful, and if, thanks to the small intellectual dimensions of their owners, the new enterprises did not prove as profitable as they had hoped, they usually reduced their prices and considered themselves happy if they could land in one of the so-called *working federations*.

Everyone who at that time could not stand on his own feet joined in such working federations; no doubt proceeding from the belief that eight cripples joining arms are sure to produce one gladiator.

And if there were really one healthy man among the cripples, he used up all his strength just to keep the others on their feet, and in this way was himself crippled.

We have always regarded fusion in so-called working federations as a question of tactics; but in this we must never depart from the following basic realization:

By the formation of a working federation weak organizations are never transformed into strong ones, but a strong organization can and will not seldom be weakened. The opinion that a power factor must result from an association of weak groups is incorrect, since the majority in any form whatsoever and under all presuppositions will, as experience shows, be the representative of stupidity and cowardice, and therefore any multiplicity of organizations, as soon as it is directed by a self-chosen multiple leadership, is sacrificed to cowardice and weakness. Also, by such a fusion, the free play of forces is thwarted, the struggle for the selection of the best is stopped, and hence the necessary and ultimate victory of the healthier and stronger prevented forever. Therefore, such fusions are enemies of natural development, for usually they hinder the solution of the problem being fought for, far more than they advance it.

It can occur that from purely tactical considerations the top leadership of a movement which looks into the future nevertheless enters into an agreement with such associations for a short time as regards the treatment of definite questions and perhaps undertakes steps in common. But this must never lead to the perpetuation of such a state of affairs, unless the movement itself wants to renounce its redeeming mission. For once it has become definitely involved in such a union, it loses the possibility and also the right of letting its own strength work itself out to the full and thus overcome its rivals and victoriously achieve the goal it has set itself.

It must never be forgotten that nothing that is really great in this world has ever been achieved by coalitions, but that it has always been the success of a single victor. Coalition successes bear by the very nature of their origin the germ of future crumbling, in fact of the loss of what has already been achieved. Great, truly world-shaking revolu-

tions of a spiritual nature are not even conceivable and realizable except as the titanic struggles of individual formations, never as enterprises of coalitions.

And thus the folkish state above all will never be created by the compromising will of a folkish working federation, but solely by the iron will of a single movement that has fought its way to the top against all.

Basic Ideas Regarding the Meaning and Organization of the SA

T HE STRENGTH of the old state rested on three pillars: the monarchistic state form, the civil service, and the army. The revolution of 1918 eliminated the state form, disintegrated the army, and delivered the civil service to party corruption. Thus the most essential pillars of a so-called state authority were shattered. State authority as such rests almost always on the three elements which lie at the basis of all authority.

The first foundation for the creation of authority is always provided by popularity. But an authority which rests solely on this foundation is still extremely weak, uncertain, and shaky. Every bearer of such an authority based purely on popularity must, therefore, endeavor to improve and secure the foundation of this authority by the creation of power. *In power, therefore, in force, we see the second foundation of all authority.* It is already considerably more stable and secure, but by no means always stronger than the first. *If popularity and force are combined, and if in common they are able to survive for a certain time, an authority on an even firmer basis can arise, the authority of tradition. If finally, popularity, force, and tradition combine, an authority may be regarded as unshakable.*

Through the revolution this last case was completely excluded. Indeed, there is no longer even an authority of tradition. With the collapse of the old Reich, the elimination of the old state form, the destruction of the former sovereign emblems and symbols of the Reich, tradition was abruptly broken off. The consequence of this was the gravest shaking of state authority.

Even the second pillar of state authority, *force*, was no longer present. In order to carry out the revolution in the first place, it was necessary to disintegrate the embodiment of the organized force and power of the state, the army; indeed, it was necessary to use the infected parts of the army itself as revolutionary fighting elements. Even though the front-line armies had not succumbed to this disintegration in a uniform degree, they, nevertheless, the more they felt the glorious sites of their four and a half years of heroic struggle behind them, were corroded more and more by the homeland's acid of disorganization, and, arrived in the demobilization organizations, likewise ended up in the confusion of so-called voluntary obedience belonging to the epoch of the soldiers' councils.

Naturally no authority could be based on these mutinous bands of soldiers, who conceived of military service in terms of the eight-hour day. And thus the second element, the element which guarantees the firmness of authority, was also eliminated and the revolution now possessed only the original element, *popularity*, on which to build its authority. But this particular basis was extremely uncertain. To be sure, the revolution succeeded in shattering the old state structure with one mighty blow, but at bottom only because the normal balance within the structure of our people had already been eliminated by the war.

Every national body can be divided into three great classes: into an extreme of the best humanity on the one hand, good in the sense of possessing all virtues, especially distinguished by courage and self-sacrifice; on the other hand, an extreme of the worst human scum, bad in the sense that all selfish urges and vices are present. Between the two extremes there lies a third class, the great, broad, middle stratum, in which neither brilliant heroism nor the basest criminal mentality is embodied.

Times when a nation is rising are distinguished, in fact exist only, by the absolute leadership of the extreme best part.

Times of a normal, even development or of a stable state of affairs are distinguished and exist by the obvious domination of the elements of the middle, in which the two extremes mutually balance one another, or cancel one another.

Times when a nation is collapsing are determined by the dominant activity of the worst elements.

In this connection it is noteworthy that the broad masses, the class of the middle as I shall designate them, only manifest themselves perceptibly when the two extremes are locked in mutual struggle, but that in case of the victory of one of the extremes, they complaisantly submit to the victor. In case the best people dominate, the broad masses will follow them; in case the worst element rises up, they will at least offer them no resistance; for the masses of the middle themselves will never fight.

Now the war, with its four and a half years of bloody events, disturbed the inner balance of these three classes, in so far as — though recognizing all the sacrifices and victims of the middle — we must nevertheless recognize that it drained the extreme of the best humanity almost entirely of its blood. For the amount of irreplaceable German heroes' blood that was shed in these four and a half years was really enormous. Just sum up all the hundreds of thousands of individual cases in which again and again the watchword was: *volunteers* to the front, *volunteer* patrols, *volunteer* dispatch carriers, *volunteers* for telephone squads, *volunteers* for bridge crossings, *volunteers* for U-boats, *volunteeers* for airplanes, *volunteers* for storm battalions, etc. — again and again through four and a half years, on thousands of occasions, volunteers and more volunteers — and always you see the same result: the beardless youth or the mature man, both filled with fervent love of their fatherland, with great personal courage or the highest consciousness of duty, *they* stepped forward. Tens of thousands, yes, hundreds of thousands of such cases occurred, and gradually this human element became sparser and sparser. Those who did not fall were either shot to pieces and crippled, or they gradually crumbled away as a result of their small remaining number. Consider above all that the year 1914 set up whole armies of so-called volunteers who, thanks to the criminal unscrupulousness of our parliamentary good-for-nothings, had received no adequate peacetime training, and thus became helpless cannon fodder at the mercy of the enemy. The four hundred thousand who then fell or were maimed in the battles of Flanders could not be replaced. Their loss was more than the loss of a mere number. By their loss the scale, too lightly weighted on the good side, shot upward, and the elements of baseness, treachery, cowardice, in short, the mass of the bad extreme, weighed more heavily than before.

For one more factor was added:

Not only that the extreme of the best had been most frightfully thinned on the battlefields in the course of the four and a half years, but the bad extreme had meanwhile preserved itself in the most miraculous way. For every hero who had volunteered and mounted the steps of Valhalla after a heroic death, you can be sure there was a slacker who had cautiously turned his back on death, in order to engage in more or less useful activity at home.

And so the end of the War gives us the following picture: The middle broad stratum of the nation has given its measure of blood sacrifices; the extreme of the best, with exemplary heroism, has sacrificed itself almost completely; the extreme of the bad, supported by the most senseless laws on the one hand and by the non-application of the Articles of War on the other hand, has unfortunately been preserved almost as completely.

This well-preserved scum of our people then made the revolution and was able to make it only because no longer opposed by the extreme of the best elements: — they were no longer among the living.

This, however, made the German revolution only a relatively popular affair from the start. It was not the German people as such that committed this act of Cain, but its deserters, pimps, and other rabble that shun the light.

The man at the front welcomed the end of the bloody struggle; he was glad to return home again, to see his wife and children. But with the revolution itself he had at heart nothing in common;

he did not love it, and even less did he love its instigators and organizers. In the four and a half years of hardest struggle he had forgotten the party hyenas, and all their quarrels had grown alien to him.

Only with a small part of the German people had the revolution really been popular: among that class of its helpers who had chosen the knapsack [1] as the badge of recognition of all honorable citizens of this new state. They did not love revolution for its own sake, as some people erroneously still believe today, but because of its consequences.

In truth, these Marxist gangsters could hardly base an authority on popularity for any length of time. And yet precisely the young Republic needed authority at any price, if after a brief chaos it did not want to be suddenly devoured by a force of retribution gathering from the last elements of the good part of our people.

There was nothing they more feared, those champions of the revolution, than to lose all foothold in the whirlpool of their own confusion, and suddenly to be seized by an iron fist, such as more than once in such periods has grown out of the life of peoples, and have the ground shifted under them. The Republic had to consolidate itself at any price.

And so it was compelled almost instantaneously to create, by the side of the tottering pillar of its weak popularity, an organization of force, in order to base a firmer authority upon it.

When in the days of December, January, February of 1918–19 the matadors of the revolution felt the ground trembling beneath their feet, they looked around for men who would be ready to strengthen the weak position which the love of their people offered them, by the force of arms. The 'anti-militaristic' Republic needed soldiers. But since the first and sole support of their state authority — popularity — rooted only in the society of pimps, thieves, burglars, deserters, slackers, etc., in other words, in that part of the people which we must designate as the bad extreme — every effort to recruit men who were prepared to sacrifice their

[1] See page 437, note.

own lives in the service of the new ideal in these circles, was love's labor lost. *The class supporting the revolutionary idea and carrying out the revolution was neither able nor willing to provide the soldiers for its protection. For this class by no means wanted the organization of a republican state body, but the disorganization of the existing state body for the better satisfaction of their instincts. Their watchword was not: order and building up of the German Republic, but: pillage it.*

And so the cry for help which the representatives of the people let out in their agony of fear inevitably went unheard; on the contrary, in fact, it aroused resistance and bitterness. For in such an undertaking people felt a breach of loyalty and faith; in the formation of an authority based no longer solely on their popularity but supported by force, they sensed the beginning of the struggle against the one aspect of the revolution that was essential for these elements: against the right to rob and the undisciplined rule of a horde of thieves and plunderers who had broken out of the prison walls and been freed of their chains, in short, of foul rabble.

The representatives of the people could cry as much as they liked; no one stepped forward from their ranks, and only the answering cry, 'traitor,' informed them of the state of mind of those supporters of their popularity.

Then for the first time numerous young Germans once again stood ready to button up their soldier's tunics, to shoulder carbine and rifle, and don their steel helmets in the service of 'law and order' as they thought, to oppose the destroyers of their homes. *As volunteer soldiers they banded into free corps and began, though grimly hating the revolution, to protect, and thus for practical purposes to secure, this same revolution.*

This they did in the best good faith.

The *real* organizer of the revolution and its actual wirepuller, the international Jew, had correctly estimated the situation. The German people was not yet ripe for being forced into the bloody Bolshevistic morass, as had happened in Russia. This was due in large part to the greater racial unity that still existed between the German intelligentsia and the German manual worker. Further in the great permeation of even the broadest strata of the people with educated elements, such as prevailed only in the other countries of Western Europe, but was totally lacking in Russia. There the intelligentsia itself was in large part not of Russian nationality or at least was of non-Slavic racial character. The thin intellectual upper stratum of the Russia of that time could at any time be removed, due to the total lack of connecting intermediary ingredients with the mass of the great people. And the intellectual and moral level of these last was horribly low.

Once it was possible in Russia to incite the uneducated hordes of the great masses, unable to read or write, against the thin intellectual upper crust that stood in no relation or connection to them, the fate of the country was decided, the revolution had succeeded; the Russian illiterate had thus become the defenseless slave of his Jewish dictators, who for their part, it must be admitted, were clever enough to let this dictatorship ride on the phrase of 'people's dictatorship.'

In Germany there was the following additional factor: As certainly as the revolution could succeed only in consequence of the gradual disintegration of the army, just as certainly the real maker of the revolution and disintegrator of the army was not the soldier at the front, but the more or less light-shy rabble which either hung around the home garrisons or, supposedly 'indispensable,' were in the economic service somewhere. This army was strengthened by tens of thousands of deserters, who were able to turn their backs on the front without special risk. The real coward at all times naturally shuns nothing so much as death. And at the front, day after day, he faced death in thousands of different forms. *If you want to hold weak, wavering or actually cowardly fellows to their duty, there has at all times been only one possibility: The deserter must know that his desertion brings with it the very thing that he wants to escape. At the front a man* **can** *die, as a deserter he* **must** *die.* Only by such a Draconic threat against any attempt at desertion can a deterring effect be obtained, not only for the individual, but for the whole army.

And here lay the meaning and purpose of the Articles of War.

It was lovely to believe that the great fight for the existence of a people could be fought on the sole basis of *voluntary* loyalty born out of and preserved by the realization of necessity. Voluntary fulfillment of duty has always determined the best men in their actions; but not the average. Therefore, such laws are necessary, as for example those against theft, which were not made for those who are basically the most honest, but for the pusillanimous, weak elements. Such laws, by frightening the bad, are intended to prevent the development of a condition in which ultimately the honest man is regarded as the stupider, and consequently people come more and more to the view that it is more expedient likewise to participate in theft than to look on with empty hands, or even to let themselves be robbed.

So it was false to believe that in a struggle, which by all human prognosis might rage for years to come, we could dispense with the instruments which the experience of many centuries, in fact millenniums, showed to be those which, in the gravest times and moments of the heaviest strain on the nerves, can compel weak and uncertain men to the fulfillment of their duty.

For the volunteer hero we obviously needed no Articles of War, but we did for the cowardly egotist, who in the hour of his people's distress sets his own life higher than that of the totality. Such a spineless weakling can only be deterred from giving in to his cowardice by the application of the hardest penalty. When men struggle ceaselessly with death and have to hold out for weeks without rest in mud-filled shell holes, sometimes with the worst possible food, the vacillating soldier cannot be held in line by threatening him with prison or even the workhouse, but only by ruthless application of the death penalty. For experience shows that at such a time he regards prison as a thousand times more attractive a place than the battlefield, considering that in prison at least his invaluable life is not menaced. And the fact that in the War the death penalty was excluded, that in reality the Articles of War were thus suspended, had terrible consequences. An army of deserters, especially in 1918, poured into the reserve posts and the home towns, and helped to form that great criminal organization which, after November 7, 1918, we suddenly beheld as the maker of the revolution.

The front itself really had nothing to do with it. All its members felt only a longing for peace. But in this very fact lay tremendous danger for the revolution. For when after the armistice the German armies began to near home, the anxious question of the revolutionaries was again and again: *What will the front-line troops do? Will the men in field gray stand for this?*

In these weeks the revolution in Germany had to appear at least *outwardly* moderate, if it did not want to run the risk of suddenly being smashed to bits by a few German divisions. *For if at that time even a single divisional commander had taken the decision to pull down the red rags with the help of his loyal and devoted division and to stand the 'councils' up against the wall, to break possible resistance with mine-throwers and hand-grenades, the division in less than four weeks would have swollen to an army of sixty divisions.* This made the Jewish wirepullers tremble more than anything else. And precisely to prevent this, they had to cover the revolution with a certain moderation; it could not take the form of Bolshevism, but, as things happened to stand, had to make a pretense of 'law and order.' Hence the innumerable great concessions, the appeal to the old civil service personnel, to the old army leaders. They were needed for a certain time at least, and only after the Moors had done their duty,[1] could the wirepuller venture to give them the kicks they had coming to them and take the Republic out of the hands of the old state servants and surrender it into the claws of the revolutionary vultures.

Only in this way could they hope to dupe old generals and old civil officials, to disarm in advance any possible resistance on their part by an apparent innocence and mildness in the new régime.

And practice showed to what an extent this succeeded.

However, the revolution had not been made by elements of law and order, but by elements of riot, theft, and plunder. And for

[1] This is a reference to Hitler's pet quotation from Schiller's *Fiesko*. See page 294, note.

them, the development of the revolution neither accorded with their will, nor for tactical reasons could the course of events be explained and made palatable to them.

With the gradual growth of the Social Democracy, it had lost more and more the character of a brutal revolutionary party. Not that its thoughts had ever served any other goal than that of the revolution, or that its leaders had ever had other intentions; by no means. But what finally remained was only the purpose and a body no longer suited to its execution. *With a party of ten millions it is no longer possible to make a revolution.* In such a movement you no longer have an extreme of activity, but the great mass of the middle, that is, of inertia.

Out of this realization, while the War was still going on, the famous split of the Social Democracy by the Jews took place; that is: while the Social Democratic Party, in keeping with the inertia of its mass, hung on national defense like a lead weight, the radical-activistic elements were drawn out of it and formed into forceful new assault columns. *The Independent Party and the Spartacus League were the storm battalions of revolutionary Marxism.* Their task was to create the accomplished fact, the groundwork of which could be taken over by the masses of the Social Democratic Party, which had been prepared for this over a period of decades. The cowardly bourgeoisie, however, was not rightly estimated by the Marxists, and were simply treated '*en canaille.*' Of them no notice was taken whatever, for it was realized that the doglike submissiveness of the political formations of an old outlived generation would never be capable of serious resistance.

As soon as the revolution had succeeded and the main pillars of the old state could be regarded as broken, but the front-line army, marching home, began to appear as a terrifying sphinx, a brake had to be applied to the natural development; the van of the Social Democratic army occupied the conquered position, and the Independent and Spartacist storm battalions were shoved aside.

This, however, did not take place without a struggle.

Not only that the activistic assault formations of the revolution were dissatisfied and felt cheated, and wanted to go on fight-

ing on their own hook, but their unruly rowdyism was only too
welcome to the wirepullers of the revolution. For no sooner was
the revolution over than there rose within it two apparent camps:
the party of law and order and the group of bloody terror. Now
what was more natural than that our bourgeoisie should at once,
with flying colors, move into the camp of law and order? Now,
all at once, these wretched political organizations had an opportunity for an activity, in which, without being obliged to say so,
they nevertheless quietly found some ground beneath their feet
and came into a certain solidarity with the power which they
hated but even more fervently feared. The political German
bourgeoisie had received the high honor of being permitted to sit
down at the table with the accursed Marxist leaders to combat
the Bolshevists.

Thus, as early as December, 1918, and January, 1919, the following condition took form:

With a minority of the worst elements a revolution has been
made, and immediately backed by all the Marxist parties. The
revolution itself has an apparently moderate stamp, which nets
it the hostility of the fanatical extremists. The latter begin to
shoot off machine guns and hand grenades, to occupy public
buildings, in short, to menace the moderate revolution. To
suppress the terror of such a further development, an armistice
is concluded between the supporters of the new state of affairs
and the adherents of the old one, for the purpose of carrying on
the struggle in common against the extremists. The result is that
the enemies of the Republic have given up their fight against the
Republic as such, and help to force down those who, though from
totally different angles, are likewise enemies of this Republic.
And the further result is that the danger of a struggle of the adherents of the old state against those of the new one seems definitely averted.

We cannot consider this fact often and closely enough. Only
those who understand it can realize how it was possible that a
people, nine tenths of whom did not make a revolution, seven
tenths of whom reject it, and six tenths of whom hated it,

nevertheless could have this revolution forced on them by one tenth.

Gradually the Spartacist barricade fighters on the one hand and the nationalist fanatics and idealists on the other were bled white, and in exact proportion as the two extremes wore each other out, as always, the mass of the middle was victorious. The bourgeoisie and Marxism met on a 'realistic basis,' and the Republic began to be 'consolidated.' Which for the present, to be sure, did not prevent the bourgeois parties, especially before elections, from citing the monarchist idea for a time, in order, by means of the spirits of the past, to be able to conjure the smaller spirits of their adherents and ensnare them once more.

Honorable this was not. At heart they had all broken with the monarchy long since, and the filth of the new condition had begun to spread its seductive influences to the bourgeois party camp. The usual bourgeois politician feels more at home today in the muck of republican corruption than in the clean hardness which he still remembers from the past state.

*   *   *

As already stated, the revolution, after the smashing of the old army, had been forced to create a new power factor for the reinforcement of its state authority. As things were, it could gain this only from supporters of an outlook that was really opposed to it. From them alone there could slowly arise a new army which, externally limited by the peace treaties, would, with regard to its mentality, have to be reshaped in the course of time into an instrument of the new state conception.

If we put to ourselves the question how — aside from all the real mistakes of the old state, which were among its causes — the revolution as an action could succeed, we come to the conclusion:

1. *In consequence of the paralysis of our concepts of duty and obedience,* and

2. *In consequence of the cowardly passivity of our so-called state-preserving parties.*

On these points the following may be said:

The paralysis of our concepts of duty and obedience has its ultimate ground in our totally unnational education, oriented solely toward the state. Here again this gives rise to a confusion between means and end. Consciousness of duty, fulfillment of duty, and obedience are not ends in themselves, any more than the state is an end in itself; they should all be the means for making possible and safeguarding on this earth the existence of a community of spiritually and physically homogeneous beings. *In an hour when a national body is visibly collapsing and to all appearances is exposed to the gravest oppression, thanks to the activity of a few scoundrels, obedience and fulfillment of duty toward them amount to doctrinaire formalism, in fact pure insanity, if the refusal of obedience and 'fulfillment of duty' would make possible the salvation of a people from its ruin.* According to our present-day bourgeois state conception, the divisional commander who at that time received from above the command not to shoot, acted dutifully and hence rightly in not shooting, since to bourgeois society, thoughtless formal obedience is more valuable than the life of their own people. According to the National Socialist conception, however, it is not obedience toward weak superiors that goes into force at such moments, but obedience toward the national community. In such an hour, the duty of personal responsibility toward a whole nation manifests itself.

The fact that a living conception of these terms had been lost in our people or rather in our governments, giving way to a purely doctrinaire and formal conception, was the cause of the revolution's success.

On the second point, the following must be remarked:

The deeper reason for the cowardice of the 'state-preserving' parties is above all the departure of the activistic, well-intentioned part of our people from their ranks — those who bled to death in the field. Aside from this, our bourgeois parties, which we can designate as the sole political formations which supported the old state, were convinced that they were entitled to defend their views exclusively in the spiritual way and with spiritual weapons, since the use of physical weapons was the sole prerogative of the state. Not only that in such a conception we must see a symptom of a gradually developing decadent weakness, but it was also senseless at a time when a political opponent had long since abandoned this standpoint and openly emphasized his intention of putting forward his political aims by force when possible. At the moment when Marxism appeared in the world of bourgeois democracy, as one of its results, the bourgeois-democratic appeal to carry on the struggle with 'spiritual weapons' was an absurdity, which would one day bring dire consequences. For the Marxists themselves from the very beginning came out for the conception that the use of a weapon must be considered only according to criteria of expediency, and that the right to use it resides solely in success.

How correct this conception is was shown in the days of November 7 to 11, 1918. In those days the Marxists did not concern themselves in the least about parliamentarianism and democracy, but gave both of them the death blow with yelling and shooting mobs of criminals. It goes without saying that in this same moment the bourgeois talking clubs were defenseless.

After the revolution, when the bourgeois parties suddenly reappeared, though with modified firm names, and their brave leaders crawled out of the concealment of dark cellars and airy storerooms, like all the representatives of such formations, they had not forgotten their mistakes and likewise they had learned nothing new. Their political program lay in the past, in so far as they had not reconciled themselves at heart with the new state of affairs; their aim, however, was to participate if possible in the new state of affairs, and their sole weapons remained, as they had always been, words.

Even after the revolution, the bourgeois parties at all times miserably capitulated to the streets.

When the Law for the Protection of the Republic [1] came up for

[1] See page 270, note.

consideration, there was at first no majority in favor of it. But in the face of the two hundred thousand demonstrating Marxists, the bourgeois 'statesmen' were seized with such a fear that contrary to their conviction they accepted the law, in the miserable fear that otherwise when they left the Reichstag they would be beaten to a pulp by the furious masses. Which unfortunately, in consequence of the law's acceptance, did not take place.

And so the development of the new state went its ways, as though there had not been any national opposition at all.

The sole organizations which at this time would have had the courage and strength to oppose the Marxists and their incited masses, were for the present the free corps, later the self-defense organizations, citizens' guards, etc., and finally the tradition leagues.[1]

But why their existence brought about no sort of shift that was in any way discernible was due to the following:

Just as the so-called national parties could exert no sort of influence for lack of any threatening power on the streets, likewise the so-called defense organizations, in turn, could exert no sort of influence for lack of any political idea, and above all of any real political goal.

What had given Marxism its success was its complete combination of political will and activistic brutality. What excluded national Germany from any practical activity in shaping the German development was the lack of a unified collaboration of brutal force with brilliant political will.

Whatever the will of the 'national' parties might be, they had not the least power to fight for this will, least of all on the streets.

The combat leagues had all the power, they were the masters of the streets and the state, and possessed no political idea and no political goal for which their strength was or even could be thrown in for the benefit of national Germany. In both cases it was the slyness of the Jew who, by clever persuasion and insistence, was able to bring about a positive perpetuation, in any case

[1] Veterans' organizations based on glorification of the old army.

an increasing intensification, of this calamitous state of affairs.

It was the Jew who through his press knew how to launch with infinite dexterity the idea of the 'unpolitical character' of the combat leagues, as, on the other hand, in political life he always praised and encouraged, with equal slyness, the 'purely spiritual nature' of the struggle. Millions of German blockheads babbled this nonsense after him, without having even the faintest idea that in this way they were for practical purposes disarming themselves and exposing themselves defenseless to the Jew.

But for this, too, indeed, there is again a natural explanation. *The lack of a great, creative, renewing idea means at all times a limitation of fighting force. Firm belief in the right to apply even the most brutal weapons is always bound up with the existence of a fanatical faith in the necessity of the victory of a revolutionary new order on this earth.*

A movement that is not fighting for such highest aims and ideals will, therefore, never seize upon the ultimate weapon.

The fact of having a new great idea to show was the secret of the success of the French Revolution; the Russian Revolution owes its victory to the idea, and only through the idea did fascism achieve the power to subject a people in the most beneficial way to the most comprehensive creative renewal.

Of this, bourgeois parties are not capable.

But it was not only the bourgeois parties that saw their political goal in a restoration of the past, but also the combat leagues, in so far as they concerned themselves with any political aims at all. Old veterans' club and *Kyffhäuser* [1] tendencies were alive within them and contributed to politically blunting the sharpest weapon that national Germany had in those days and making it languish in the mercenary service of the Republic. The fact that in this they acted in the best conviction, and above all in the best good

---

[1] *Kyffhäuserbund der deutschen Landeskriegerverbände,* a veterans' organization founded in 1898. Before the War of 1914 it had over two million members. It takes its name from the mountain in the Harz, within which, according to the legend, Emperor Friedrich Barbarossa lies sleeping, to awaken in the hour of Germany's greatest need.

faith, changes nothing in the catastrophic madness of these occurrences.

Gradually Marxism obtained the required power to support its
authority in the Reichswehr that was being consolidated, and
thereupon, consistently and logically, began to disband as
superfluous the nationalist combat leagues, which seemed dangerous. Individual leaders of especial boldness, who were looked
on with distrust, were haled before the bars of justice and put
behind Swedish curtains.[1] But with all of them the destiny for
which they themselves were responsible was fulfilled.

*   *   *

With the founding of the NSDAP, for the first time a movement
had appeared whose goal did not, like that of the bourgeois parties, consist in a mechanical restoration of the past, but in the
effort to erect an organic folkish state in place of the present senseless state mechanism.

*The young movement, from the first day, espoused the standpoint
that its idea must be put forward spiritually, but that the defense of
this spiritual platform must if necessary be secured by strong-arm
means.* Faithful to its belief in the enormous significance of the
new doctrine, it seems obvious to the movement that for the at-
tainment of its goal no sacrifice can be too great.

I have already pointed to the forces which obligate a movement, in so far as it wants to win the heart of a people, to assume
from its own ranks its defense against the terrorist attempts of
its adversaries. And it is an eternal experience of world history
that a terror represented by a philosophy of life can never be
broken by a formal state power, but at all times can be defeated
only by another, new philosophy of life, proceeding with the same
boldness and determination. This will at all times be displeasing
to the sentiment of the official guardians of the state, but that will
not banish the fact. State power can only guarantee law and

[1] Thieves' slang for prison bars.

order when the content of the state coincides with the philosophy dominant at that particular time, so that violent elements possess only the character of individual criminal natures, and are not regarded as proponents of an idea in extreme opposition to the state views. In such a case, the state can for centuries apply the greatest measures of violence against a terror oppressing it; in the end it will nevertheless be able to do nothing against it, but will go down in defeat.

The German state is gravely attacked by Marxism. In its struggle of seventy years it has not been able to prevent the victory of this philosophy of life, but, despite a sum total of thousands of years in prison and jail sentences and the bloodiest measures which in innumerable cases it applied to the warriors of the menacing Marxist philosophy, has nevertheless been forced to almost total capitulation. (This, too, the run-of-the-mill bourgeois political leader will want to deny, though obviously he will be unable to convince anyone.)

The state which on November 9, 1918, unconditionally crawled on its belly before Marxism will not suddenly arise tomorrow as its conqueror; on the contrary: even today feeble-minded bourgeois in ministerial chairs are beginning to rave about the necessity of not governing against the workers, and what they have in mind under the concept 'worker' is Marxism. But by identifying the German worker with Marxism, they not only commit a falsification as cowardly as it is untrue, but attempt by this motivation to conceal their own collapse in the face of the Marxist idea and organization.

But in view of this fact — that is, the complete subjection of the present state to Marxism — the National Socialist movement really acquires the duty, not only of preparing the victory of its idea, but of taking over its defense against the terror of an International drunk with victory.

I have already described how in our movement a body for the protection of meetings gradually developed out of practical life, how it gradually assumed the character of a definite monitor troop, and strove for an organizational form.

Much as this gradually arising body might outwardly resemble a so-called combat league, it was nevertheless not to be compared with one.

As already mentioned, the German combat organizations had no definite political idea. They were really nothing but self-defense leagues of more or less competent training and organization, with the result that they actually represented an illegal complement to the state's momentary instruments of power. Their character of free corps was based only on the way in which they were formed and on the condition of the state at that time, but they were by no means deserving of such a title as free formations of the struggle for a free conviction of their own. This, despite all the opposition of individual leaders and whole leagues toward the Republic, they did not possess. *For being convinced of the inferiority of an existing condition does not suffice to entitle one to speak of a conviction in the higher sense; no, the latter is rooted only in the knowledge of a new condition and in the inner vision of a condition the achievement of which one feels as a necessity, and to stand up for whose realization one regards as one's highest life task.*

What distinguishes the monitor troop of the National Socialist organization of that time essentially from all combat leagues is that it was not and did not want to be in any way a servant of the conditions created by the revolution, but that it fought exclusively for a new Germany.

In the beginning, it is true, this monitor troop possessed only the character of a meeting-hall guard. Its first task was a limited one: it consisted in making it possible to hold meetings which without it would have been simply prevented by the enemy. Even then, it had been trained to carry out an attack blindly, but not, as stupid German-folkish circles nonsensically claimed, because it honored the blackjack as the highest spirit, but because it understood that the greatest spirit can be eliminated when its bearer is struck down with a blackjack, as in actual fact the most significant heads in history have not seldom ended beneath the blows of the pettiest helots.[1] They did not want to set up vio-

[1] *'Heloten.'* No change in second edition.

lence as a goal, but to protect the prophets of the spiritual goal from being shoved aside by violence. And in this they understood that they were not obligated to undertake the protection of a state which offers the nation no protection, but that, on the contrary, they had to assume the protection of a nation against those who threatened to destroy the people and the state.

After the meeting-hall battle in the Munich Hofbräuhaus the monitor troop, once and for all, in eternal memory of the heroic storm attacks of the small number they were then, received the name of *Sturmabteilung* (storm section). As this very designation indicates, it represents only a *section* of the movement. It is a link in it, just as propaganda, the press, the scientific institutes and so forth, constitute mere links in the party.

How necessary its development was, we could see, not only by this memorable meeting, but also by our attempt gradually to spread our movement from Munich into the rest of Germany. Once we had appeared dangerous to the Marxists, they missed no opportunity to nip any attempt at a National Socialist meeting in the bud, or prevent it from being held by breaking it up. And it was absolutely a matter of course that the party organizations of all shadings of Marxism blindly supported any such intentions and any such occurrences in the representative bodies. But what was one to say of bourgeois parties which themselves had been so thrashed by the Marxists that in many places they could no longer venture to have their speakers appear in public and which, nevertheless, followed any struggles against Marxism that in any way turned out unfavorably for us with an absolutely incomprehensible, idiotic satisfaction. They were happy that the enemy which could not be bested by them, which on the contrary bested them, could not be broken by us either. What should be said of state officials, police presidents, nay, even ministers, who with a really disreputable lack of principle liked to represent themselves publicly as 'national men,' but who in all conflicts that we National Socialists had with the Marxists, acted as the most disgraceful stooges for them? What should be said of men who went so far in their self-abasement that for a

pitiful word of praise in the Jewish newpsapers they did not hesitate to persecute the men to whose heroism in risking their own
lives they in part owed the fact that a few years previous they
were not tattered corpses hung up on lamp-posts by the Red
mob?

These were such sad figures that they once moved the unforgettable late President Pöhner, who in his hard straightforwardness hated all crawlers as only a man with an honest heart can
hate, to the harsh utterance: 'All my life I wanted to be nothing
else than first a German and then an official, and I would never
like to be confused with those creatures who prostitute themselves like official whores to everyone who can play the master at
the moment.'

And in all this it was especially sad that this kind of men gradually gained power over tens of thousands of the most honorable
and best German civil servants, but even gradually infected them
with their own disloyalty, and persecuted the honest ones with
grim hatred and finally drove them out of their posts and positions, while they themselves, with lying hypocrisy, still represented themselves as 'national' men.

From such men we could never hope for any support, and we
obtained it only in the very rarest cases. Solely the development
of our own defense organization could safeguard the activity of
the movement and at the same time win for it that public attention and general respect which are accorded to the man who,
when attacked, takes up his own defense.

As the directing idea for the inner training of this storm section, the intention was always dominant, aside from all physical
education, to teach it to be an unshakable, convinced defender of
the National Socialist idea, and finally to strengthen its discipline in the highest degree. It should have nothing in common
with a combat organization of bourgeois conception, but likewise
nothing in common with a secret organization.

The reason why, even at that time, I sharply opposed having
the SA of the NSDAP organized as a so-called combat league,
was based on the following consideration:

From the purely practical point of view, the military training of a people cannot be carried out by private leagues, except with the help of the most enormous state means. Any other belief is based on great overestimation of their own ability. And so it is out of the question that organizations possessing military value can be built up beyond certain limits with so-called 'voluntary discipline.' The most important support of the power to command is lacking, to wit, the power to punish. To be sure, it was possible in the fall, or even better in the spring of 1919, to set up so-called 'free corps,' but not only did most of them possess front-line fighters who had gone through the school of the old army, but the type of obligation which they laid upon the individuals subjected them, for a limited time at least, just as unconditionally to military obedience.

This is totally lacking in a voluntary 'combat organization' of today. The larger the league, the weaker its discipline will be, the smaller the demands made on the individual men, and the more the whole will take on the character of the old non-political soldiers' and veterans' clubs.

It will never be possible to carry out a voluntary training for army service among the great masses without guaranteed unconditional power of command. Never will more than a few be willing to submit of their own accord to such forced obedience as was considered self-evident and natural in the army.

Furthermore, real training cannot be given in consequence of the absurdly small means at the disposal of a so-called combat league for such a purpose. But the best, most reliable training should be precisely the main task of such an institution. Since the War, eight years have gone by, and since that time not a single age class among our German youth has been systematically trained. But it cannot be the function of a combat league to include the old classes that have already been trained, since otherwise it can at once be reckoned mathematically when the last member will leave this corporation. Even the youngest soldier of 1918 will in twenty years be incapable of fighting, and we are approaching this moment with a disquieting speed. Thus every so-called combat league must necessarily assume more and more the character of an old soldiers' association. This, however, cannot be the purpose of an organization that designates itself not as an *old soldiers'* league, but as a *Wehrverband* (combat league), and which by its very name endeavors to express the fact that it sees its mission, not only in the preservation of the tradition and common bond of former soldiers, but in the development of the military (*wehr*) idea, and in the practical advocacy of this idea, that is, in the creation of a military body.

This task, however, absolutely demands the training of elements which had previously received no military drill, and this in practice is actually impossible. With one or two hours training a week, you really cannot make a soldier. With the present-day enormously increased demands that warfare makes on the individual, a two-year period service is perhaps just adequate to transform an untrained young man into an expert soldier. We have all of us in the field seen the terrible consequences that resulted for young soldiers not thoroughly trained in their trade. Volunteer formations, which for fifteen or twenty weeks had been drilled with iron determination and boundless devotion, nevertheless represented nothing but cannon fodder at the front. Only distributed among the ranks of experienced old soldiers could younger recruits, trained for from four to six months, furnish useful members of a regiment; even then they were directed by the 'old men' and thus gradually grew into their functions.

How thoughtless in contrast seems an attempt to try to create troops with a so-called training period of one or two hours a week, without clear power of command and without extensive means! It might be possible to freshen up old soldiers in this way, but never to turn young men into soldiers.

How indifferent and totally worthless such a procedure would be in its results can be demonstrated especially by the fact that, while a so-called volunteer league, with puffing and blowing, with trouble and grief, trains or tries to train a few thousand essentially well-intentioned men (it does not get to any others) in the military idea, the state itself, by the pacifistic-democratic nature of its education, consistently robs millions and millions of young people of their natural instincts, poisons their logical patriotic thinking, and thus gradually transforms them into a herd of sheep, patiently accepting every arbitrary tyranny.

How absurd, in comparison with this, are all the exertions of the combat leagues to transmit their ideas to the German youth.

But almost more important is the following consideration, which had always made me take a position counter to any attempt at a so-called military rearming on the basis of volunteer leagues.

Assuming that despite the above-mentioned difficulties a league nevertheless succeeded in training a definite number of Germans year after year into arms-bearing men — equally with respect to their convictions as with respect to their physical fitness and schooling in the use of arms — the result would nevertheless be practically nil in a state which, by its whole tendency, absolutely does not desire such military education, in fact positively hates it, since it stands in complete contradiction to the aim of its leaders — the destroyers of this state.

In any case such a result would be worthless under governments which have not only demonstrated by their deeds that they care nothing about the military strength of the nation, but which above all would never be willing to issue an appeal to this strength, except at best for the support of their own ruinous existence.

And today this is the case. Or is it not absurd to try to train some tens of thousands of men for a government in the dim light of dawn and evening, when the state a few years previous disgracefully sacrificed eight and a half millions of the best-trained soldiers, not only ceasing to use them, but as thanks for their sacrifices actually exposing them to general vilification? And so they want to train soldiers for a state régime which befouled and spat upon the most glorious soldiers of former days, tore their decorations from their chest, took away their cockades, trampled their banners and degraded their achievements? Or has this present state régime ever undertaken a single step to restore the honor of the old army, to call to account those who have corrupted and reviled it? Not in the slightest. On the contrary: we can see these creatures enthroned in the highest state posts. — Remember the words spoken at Leipzig: 'Right goes with power.' But since today in our Republic the power lies in the hands of the same men who engineered the revolution, and this revolution represents the vilest high treason, nay, the most wretched piece of villainy in all German history, really no reason can be found for enhancing the power of these very characters by the formation of a new young army. In any event, all the arguments of reason speak against it.

But what importance this state, even after the revolution of 1918, attributed to the military strengthening of its position could be seen clearly and unmistakably by its attitude toward the large self-defense organizations that then existed. As long as they had to intervene for the protection of personally cowardly creatures of the revolution, they were not unwelcome. But as soon as, thanks to the gradually increasing depravity of our people, the danger to these creatures seemed eliminated and the existence of the leagues meant a strengthening of the national-political forces, they were superfluous, and everything was done to disarm them, in fact, if possible to break them up.

Only in the rarest examples does history show gratitude in princes. But to count on the gratitude of revolutionary pyromaniac murderers, plunderers of the people and traitors to the nation, is something that only a neo-bourgeois patriot can manage. In any case, I, in examining the question of whether volunteer combat leagues should be created, could never refrain from the question: for whom am I training the young people? For what purpose are they used and when are they to be called up? The answer to this question provides at the same time the best directives for our own attitude.

If the present state were ever to train forces of this sort, it would never be for the defense of national interests against the outside world, but only for the protection of the rapers of the nation at home against the general rage that some day perhaps will flare up in the swindled, betrayed, and sold-out people.

For this reason alone, the SA of the NSDAP could have nothing in common with a military organization. It was an instrument for defense and education in the National Socialist movement, and its tasks lay in an entirely different province from that of the so-called combat leagues.

But it could also constitute no secret organization. The aim of secret organizations can only be illegal. In this way the scope of such an organization is automatically limited. It is not possible, especially in view of the talkativeness of the German people, to build up an organization of any size and at the same time to keep it outwardly secret or even to veil its aims. Any such intention will be thwarted a thousand times. Not only that our police authorities today have a staff of pimps and similar rabble at their disposal who will betray anything they can find for thirty pieces of silver, and even invent things to betray, but the supporters themselves can never be brought to the silence that is necessary in such a case. Only very small groups, by years of sifting, can assume the character of real secret organizations. But the very smallness of such organizations would remove their value for the National Socialist movement. *What we needed and still need were and are not a hundred or two hundred reckless conspirators, but a hundred thousand and a second hundred thousand fighters for our philosophy of life. We should not work in secret conventicles, but in mighty mass demonstrations, and it is not by dagger and poison or pistol that the road can be cleared for the movement, but by the conquest of the streets. We must teach the Marxists that the future master of the streets is National Socialism, just as it will some day be the master of the state.*

The danger of secret organizations today lies, furthermore, in the fact that the members often totally misunderstand the magnitude of the task, and the opinion arises that the fate of a people really might be suddenly decided in a favorable sense by a single act of murder. Such an opinion can have its historical justification especially when a people languishes under the tyranny of some oppressor genius, of whom it is known that his outstanding personality alone guarantees the inner solidity and frightfulness of the hostile pressure. In such a case, a self-sacrificing man may suddenly spring forth from a people, to plunge the steel of death into the breast of the hated individual. And only the republican sentiment of petty scoundrels with a bad conscience will regard such a deed as horrible, while our people's greatest poet of freedom has dared to give a glorification of such an action in his *Tell*.

In the years 1919 and 1920 there existed a danger that the member of secret organizations, filled with enthusiasm by the great models of history and horrified by the boundless misfortune of his fatherland, should attempt to avenge himself against the destroyers of his homeland, in the belief that in this way he could put an end to the distress of his people. Any such attempt, however, was an absurdity, because Marxism had not been victorious thanks to the superior genius and personal significance of an individual, but by the boundless contemptibleness, the cowardly failure of the bourgeois world. The most cruel criticism that can be made of our bourgeoisie lies in the fact that the revolution itself did not produce a single leader of any greatness and nevertheless subjected it. It is understandable to capitulate to a Robespierre, a Danton or a Marat, but it is devastating to have crawled before the scrawny Scheidemann, the fat Herr Erzberger [1] and a Friedrich Ebert and all the other innumerable political midgets. In reality there was not *one* leader who might have been regarded as the genius of the revolution and hence the misfortune of the fatherland; they were all revolutionary bedbugs, knapsack Spartacists, wholesale and retail. To put any one of these out of the way was completely irrelevant and the chief result was that a few other bloodsuckers, just as big and just as threadbare, came into a job that much sooner.

In those years it was not possible to attack sharply enough a conception which had its cause and explanation in the really great

---

[1] On Ebert and Scheidemann see page 199, note. Mattias Erzberger was vice-chancellor in the cabinet of Gustav Bauer which accepted the peace treaty on June 23, 1919. He was murdered on August 26, 1921, by two young soldiers of the Ehrhardt Brigade, a Rightist military organization.

figures of history, but was not in the least suited to the present era of dwarfs.

Likewise, in the question of *eliminating so-called traitors against the nation* the same consideration is in order. It is absurdly illogical to kill a scamp who has informed about a cannon,[1] while next door in the highest posts and dignities sit scoundrels who have sold a whole Reich, who have the vain sacrifice of two millions on their consciences, who bear the responsibility for millions of cripples, and with all this calmly carry on their republican business deals. It is senseless to eliminate petty traitors in a country whose government itself frees these traitors against the nation from any punishment. For then it is possible that some day the honest idealist, who puts a scoundrelly armaments stool-pigeon out of the way, for his people, is called to account by capital traitors against the nation. Therefore, it is an important question: Should we have such a traitorous petty creature eliminated by another creature or by an idealist? In one case the success is doubtful and the treason for later almost certain; in the other case, a small scoundrel is eliminated and the life of a perhaps unreplaceable idealist is risked.

Further, in this question, my position is that there is no use in hanging petty thieves in order to let big ones go free; but that some day a German national court must judge and execute some ten thousand of the organizing and hence responsible criminals of the November betrayal and everything that goes with it. Such an example will provide the small armaments stool-pigeon with the necessary lesson for all time.

All these are considerations which caused me again and again to forbid participation in secret organizations and to preserve the SA itself from the character of such organizations. In those years I kept the National Socialist movement away from experiments, whose performers for the most part were glorious, idealistic-

---

[1] At this time there were various nationalist military leagues with unofficial Reichswehr connections, specializing in the illegal concealment of arms. Numerous murders occurred of persons believed to have reported secret arms caches to the Allied Control Commission.

minded young Germans, whose acts, however, only made victims of themselves, but were powerless to improve the lot of the fatherland even in the slightest.

*   *   *

Now if the SA could be neither a military combat organization nor a secret league, the following consequences inevitably resulted:

1. *Its training must not proceed from military criteria, but from criteria of expediency for the party.*

In so far as the members require physical training, the main emphasis must be laid, not on military drilling, but on athletic activity. Boxing and jiu-jitsu have always seemed to me more important than any inferior, because incomplete, training in marksmanship. Give the German nation six million bodies with flawless athletic training, all glowing with fanatical love of their country and inculcated with the highest offensive spirit, and a national state will, in less than two years if necessary, have created an army, at least in so far as a certain basic core is present. This, as things are today, can rest only in the Reichswehr and not in any combat league that has always done things by halves. Physical culture must inoculate the individual with the conviction of his superiority and give him that self-confidence which lies forever and alone in the consciousness of his own strength; in addition, it must give him those athletic skills which serve as a weapon for the defense of the movement.

2. *In order, at the outset, to prevent the SA from assuming any secret character, in addition to its uniform immediately recognizable to all, the very size of its membership must point the way which benefits the movement and is known to the whole public.* It must not hold sessions in secret, but must march beneath the open sky, thus being put unmistakably into a type of activity which destroys all legends of 'secret organization' once and for all. And in order to remove it, spiritually as well, from all attempts to satisfy its activism by petty conspiracies, it had from the very beginning to

be initiated completely into the great idea of the movement and to be educated so thoroughly in the task of fighting for this idea that its horizon broadened from the outset, and the individual man saw his mission, not in the elimination of any greater or lesser scoundrel, but in fighting for the erection of a new National Socialist folkish state. Thereby the struggle against the present-day state was removed from the atmosphere of petty actions of revenge and conspiracy, to the greatness of a philosophical war of annihilation against Marxism and its organization.

3. *The organizational formation of the SA, as well as its uniform and equipment, can therefore not reasonably emulate the models of the old army, but must pursue an expediency determined by its function.*

These views, which directed me in 1920 and 1921 and which I gradually endeavored to inject into the young organization, had the result that, as early as midsummer, 1922, we disposed of an imposing number of companies, which in late autumn, 1922, little by little received their special distinguishing uniforms. Three events were of infinite importance for the further shaping of the SA.

1. The great general demonstration of *all* patriotic leagues against the Law for the Protection of the Republic in late summer 1922 on the *Königsplatz in Munich.*

The patriotic leagues of Munich had issued an appeal summoning a gigantic demonstration as a protest against the introduction of the Law for the Protection of the Republic. The National Socialist movement was also expected to participate in it. The solid procession of the party was headed by six Munich companies, followed by the sections of the political party. In the column itself marched two brass bands, and about fifteen flags were carried along. The arrival of the National Socialists in the half-filled square, which was otherwise void of flags, aroused immeasureable enthusiasm. I myself had the honor of being privileged to address the crowd, now numbering sixty thousand heads, as one of the orators.

The success of the rally was overpowering, particularly because, in defiance of all Red threats, it was proved for the first time that national Munich, too, could march in the streets. Red republican defense corps (*Schutzbund*), who attempted to proceed with terror against the approaching columns, were within a few minutes scattered with bloody skulls by SA detachments. The National Socialist movement then for the first time showed its determination to claim for itself the right to the streets in the future, thus wresting this monopoly from the hands of the international traitors to the people and enemies of the fatherland.

The result of this day was an incontestable proof of the psychological and also organizational soundness of our conceptions with regard to the structure of the SA.

On the foundation which had been so successfully proven, it was energetically broadened, so that only a few weeks later double the number of companies had been set up.

2. The march to Coburg in October, 1922.

'Folkish' associations planned to hold a so-called 'German Day' in Coburg. I myself received an invitation to it, remarking that it would be desirable for me to bring an escort. This request, which I received at eleven o'clock in the morning, came very opportunely. An hour later the arrangements for attending this 'German Day' had been issued. As an 'escort' I appointed eight hundred men of the SA; we arranged to transport them in approximately fourteen companies by special train to the little city that had become Bavarian.[1] Similar orders went out to National Socialist SA groups which had meanwhile been formed in other places.

It was the first time that such a special train was used in Germany. At all towns where new SA men got in, the transport aroused much attention. Many people had never seen our flags before; the impression they made was very great.

When we arrived at the Coburg station, we were received by a deputation of the organizers of the 'German Day,' which con-

[1] Coburg, former co-capital of the Duchy of Saxony-Coburg-Gotha, was a separate administrative entity up to 1920 when it became part of Bavaria.

veyed to us an order from the local trade unions — in other words, from the Independent [1] and Communist Party — to the effect that we were forbidden to enter the town with flags unfurled, or with music (we had taken along a forty-two-piece band of our own), or to march in a solid column.

I at once flatly rejected these disgraceful conditions, and did not fail to express to the gentlemen present, the organizers of this congress, my surprise that they had carried on negotiations with these people and entered into agreements; I declared that the SA would immediately line up in companies and march into the city with resounding music and flags flying.

And that is just what happened.

On the square in front of the railroad station we were received by a howling, shrieking mob numbering thousands. 'Murderers,' 'bandits,' 'robbers,' 'criminals,' were the pet names which the model founders of the German Republic affectionately showered on us. The young SA kept exemplary order, the companies formed on the square in front of the station, and at first took no notice of the vulgar abuse. In the city that was strange to all of us, frightened police officials led the marching column, not, as arranged, to our quarters, a shooting gallery situated on the periphery of Coburg, but to the Hofbräuhauskeller, near the center of the city. To left and right of the procession, the uproar of the masses of people accompanying us increased more and more. Hardly had the last company turned into the courtyard of the Keller than great masses, amid deafening cries, tried to crowd in after us. To prevent this, the police locked the Keller. Since this state of affairs was intolerable, I had the SA line up once again, gave them a brief speech of admonition, and demanded that the police open the gates immediately. After a long hesitation, they yielded.

To get to our quarters, we marched back the way we had come, and now at last a stand had to be taken. After they had been unable to disturb the poise of our companies by cries and insulting shouts, the representatives of true socialism, equality, and fra-

[1] See page 369, note.

ternity had recourse to stones. At this our patience was at an end, and so for ten whole minutes a devastating hail fell from left and right, and a quarter of an hour later, there was nothing red to be seen in the streets.

In the evening there were serious clashes again. Some National Socialists had been assaulted singly, and patrols of the SA found them in a terrible condition. Thereupon we made short shrift of our foes. By next morning the Red terror, under which Coburg had suffered for years, had been broken.

With real Marxist-Jewish lies they now attempted to harry the 'comrades of the international proletariat' back into the streets, by totally twisting the facts and maintaining that our 'bands of murderers' had begun a 'war of extermination against peaceful workers' in Coburg. The great 'demonstration of the people,' which, it was hoped, tens of thousands of workers from the whole vicinity would attend, was set for half-past one. Therefore, firmly resolved to dispose of the Red terror for good, I ordered the SA, which had meanwhile swollen to nearly one and a half thousand men, to line up, and set out with them on the march for the Fortress of Coburg, by way of the great square on which the Red demonstration was to take place. I wanted to see whether they would dare to molest us again. When we entered the square, only a few hundred were present instead of the announced ten thousand, and at our approach they kept generally quiet, and some ran away. Only at a few points did Red troops, who had meanwhile come from the outside and who did not yet know us, try to pester us again; but in the twinkling of an eye, all their enthusiasm was spoiled. And now it could be seen how the frightened and intimidated population slowly woke up and took courage, and ventured to shout greetings at us, and in the evening as we were marching off broke into spontaneous cheering in many places.

At the station the railroad men suddenly informed us that they would not run the train. Thereupon I notified a few of the ringleaders that in that case I planned to round up whatever Red bosses fell into my hands, and that we would run the train ourselves; however, we would take along a few dozen of the brothers of international solidarity on the locomotive and the tender and in every car. Nor did I fail to call it to the gentlemen's attention that the trip with our own forces would, of course, be an extremely risky undertaking and that it was not excluded that the whole lot of us should break our necks and bones. But, anyway, in that case, we should be delighted to leave for the Hereafter, not alone but in equality and fraternity with the Red gentlemen.

Thereupon the train departed with the utmost punctuality, and we were back in Munich safe and sound the following morning.

Thus, for the first time since 1914 the equality of citizens before the law was re-established in Coburg. For if today some simpleton of a higher official ventures the assertion that the state protects the lives of its citizens, this was certainly not the case at that time; for at that time the citizens had to defend themselves against the representatives of the present-day state.

At first the importance of this day could not be fully evaluated by its consequences. Not only that the victorious SA had been enormously enhanced in its self-confidence and its faith in the soundness of its leadership, but the outside world also began to follow our doings more closely, and many for the first time recognized in the National Socialist movement the institution which in all probability would some day be called upon to put a suitable end to the Marxist madness.

Only the democrats groaned that anyone could dare not peacefully to let his skull be bashed in, and that under a democratic republic we had had the audacity to oppose a brutal attack with fists and cudgels instead of pacifistic songs.

On the whole, the bourgeois press, as usual, was partly pitiful and partly contemptible, and only a few honest newspapers greeted the fact that in one place at least someone had dared to call a halt to the activity of the Marxist highwaymen.

In Coburg itself, at least a part of the Marxist working class, which incidentally could be regarded only as misled, had learned a lesson from the fists of National Socialist labor and been taught to realize that these workers also fight for ideals, since, as experience shows, men fight only for something that they believe in and love.

The greatest benefit, however, was derived by the SA itself. It now grew with great rapidity, and at the Party Day held on January 27, 1923, approximately six thousand men could take part in the dedication of the flag, and the first companies were fully equipped with their new uniforms.

For the experience in Coburg had shown how necessary it is, and not only in order to strengthen the *esprit de corps*, but also to avoid confusion and forestall mutual non-recognition, to introduce uniform dress among the SA. Until then it wore only the armband; now the canvas jacket and the well-known cap were added.

And, furthermore, the experience of Coburg had the significance that we now began systematically, in all places where for many years the Red terror had prevented any meeting of people with different ideas, to break this terror and restore freedom of assembly. From now on, National Socialist battalions were assembled again and again in such localities, and in Bavaria gradually one Red citadel after another fell a victim to National Socialist propaganda. The SA had grown more and more into its task, and so had moved further and further away from the character of a senseless and unimportant defense movement and risen to the level of a living organization of struggle for the erection of a new German state.

This logical development lasted until March, 1923. Then there occurred an event which compelled me to shift the movement from its previous course and subject it to a modification.

3. *The occupation of the Ruhr* by the French in the first months of 1923 had in the following period a great significance for the development of the SA.

Even today it is not yet possible, and particularly in the national interest not expedient, to speak or write of this with full publicity. I can only express myself in so far as this theme has already been touched upon in public proceedings and thus brought to the knowledge of the public.

The occupation of the Ruhr, which came as no surprise to us, gave rise to the justified hope that now at length there would be an end to the cowardly policy of retreat, and that with this a definite task would fall to the combat leagues. And the SA, which then embraced many thousands of young powerful men, could not fittingly be excluded from this national service. In the spring and midsummer of 1923 it was reshaped into a military fighting organization. To it the later development of 1923, in so far as it concerned our movement, was attributable.

Since I treat the development of 1923 in broad outlines elsewhere, I shall only state here that the reorientation of the SA was a harmful one from the viewpoint of the movement, if the presuppositions that had led to its reorientation — that is, the resumption of active resistance against France — did not materialize.

The close of the year 1923, terrible as it may seem at first sight, may, if viewed from a higher standpoint, be regarded as positively necessary, in so far as with one stroke it ended the reorientation of the SA, made pointless by the attitude of the German Reich government and hence harmful for the movement, and thus created the possibility of building some day at the point where we had once been forced to relinquish the correct road.

The NSDAP, newly founded in 1925, must again set up, train, and organize its SA according to the aforementioned principles. It must thus return to the original healthy views, and must now once more find its highest task in creating, in its SA, an instrument for the conduct and reinforcement of the movement's struggle for its philosophy of life.

It must neither suffer the SA to degenerate into a kind of combat league nor into a secret organization; it must, on the contrary, endeavor to train it as a guard, numbering hundreds of thousands of men, for the National Socialist and hence profoundly folkish idea.

CHAPTER
X

Federalism as a Mask

I<small>N THE WINTER</small> of 1919, and even more in the spring and summer of 1920, the young party was forced to take a position on a question which even during the War rose to an immense importance. In the first volume, in my brief account of the symptoms of the threatening German collapse that were visible to me personally, I have pointed to the special type of propaganda which was carried on by the English as well as the French for the purpose of tearing open the old cleft between North and South. In spring, 1915, there appeared the first systematic agitational leaflets attacking Prussia, as solely responsible for the War. By 1916, this system had been brought to full perfection, as adroit as it was treacherous. And after a short time the agitation of the South German against the North German, calculated on the lowest instincts, began to bear fruit. It is a reproach that must be raised against the authorities of that time, in the government as well as the army command — or rather the Bavarian staff offices — a reproach which these last cannot shake off, that in their damnable blindness and disregard of duty they did not proceed against this with the necessary determination. Nothing was done! On the contrary, various quarters did not seem to take it so much amiss, and they were small-minded enough to believe that such a propaganda would not only put a bar in the path of the development of the German people toward unity, but that it would inevitably and automatically bring a strengthening of the federative forces. Scarcely ever in history has a malicious omission brought more evil consequences. The weakening these men thought they were administering to Prussia struck the whole of Germany. And its consequence was the acceleration of the collapse, which, however, not only shattered Germany herself, but primarily, in point of fact, the individual states themselves.

In the city where the artificially fanned hatred against Prussia raged most violently, the revolution was first to break out against the hereditary royal house.

Yet it would be false to believe that the manufacture of this anti-Prussian mood is attributable solely to hostile war propaganda and that the people affected by it had no grounds of justification. The incredible way in which our war economy was organized, a positively insane centralization that held the whole German Reich territory in tutelage and pillaged it to the limit, was one of the main reasons for the rise of this anti-Prussian sentiment. *For the average little man, the war societies,[1] which happened to have their central offices in Berlin, and Berlin itself, were synonymous with Prussia.* It scarcely dawned on the individual at that time that the organizers of this institute of robbery, known as 'war societies,' were neither Berliners nor Prussians, in fact, were not Germans at all. He saw only the great faults and the constant encroachments of this hated institution in the capital and then naturally transferred his whole hatred to the capital and Prussia simultaneously, all the more so since in certain quarters not only nothing was done about this, but such an interpretation was even secretly and smirkingly welcomed.

The Jew was far too shrewd not to realize in those days that the infamous campaign of pillage that he was then organizing against the German people, under the cloak of the war societies, would, nay, must, arouse resistance. But as long as it did not spring at

---

[1] *Kriegsgesellschaften.* These were the corporations, with state participation and control, established in 1914 to carry on war production. For some months at the beginning of the War, Walter Rathenau, a Jew, headed the raw material department of the Prussian war ministry.

his own throat, he had no need to fear it. But in order to prevent
an explosion of the masses driven to despair and indignation in
this direction, there could be no better prescription than to cause
their rage to flare up elsewhere, thus using it up.

Let Bavaria fight against Prussia and Prussia against Bavaria
as much as they wanted, the more the better! The hottest struggle between the two meant the securest peace for the Jew. In this
way, the general attention was entirely diverted from the international maggot of nations. And if the danger seemed to arise
that thoughtful elements, of which there were many in Bavaria,
as elsewhere, called for understanding, reflection, and restraint
and so the embittered struggle threatened to die down, the Jew
in Berlin needed only to stage a new provocation and wait for the
result. Instantly all the beneficiaries of the conflict between
North and South flung themselves on every such occurrence, and
kept on blowing until the flame of indignation had again burst
into a roaring blaze.

It was an adroit, subtle game that the Jew then played, constantly occupying and distracting the individual German tribes,
and meanwhile pillaging them the more thoroughly.

Then came the revolution.

If up to the year 1918, or rather up to November of this year,
the average man, and especially the little-educated petit bourgeois and worker, especially in Bavaria, could not yet correctly
estimate the real course and the inevitable consequences of the
quarrel of German tribes among themselves, the section calling
themselves 'national' should at least have had to recognize it on
the day of the outbreak of the revolution. For no sooner had the
action succeeded than in Bavaria the leader and organizer of the
revolution became the defender of 'Bavarian' interests. *The
international Jew Kurt Eisner began to play Bavaria against
Prussia.* It goes without saying, however, that this Oriental
who spent his time as a newspaper *journaille* [1] running all over

---

[1] *Journaille* is a brunch word composed of '*journal*' and '*canaille*,' attributed to Karl Kraus, editor and writer of the Viennese magazine '*Die
Fackel.*' *Zeitungsjournaille* is a typical Hitlerian redundancy.

Germany, was unquestionably the last man fitted to defend Bavarian interests, and that to him, in particular, Bavaria was a matter of the utmost indifference possible on God's earth.

In giving the revolutionary uprising in Bavaria a thoroughly con-scious edge against the rest of the Reich, Kurt Eisner did not in the least act from Bavarian motives, but solely as the servant of the Jews. He used the existing instincts and dislikes of the Bavarian people, to help him break up Germany the more easily. The shattered Reich would have easily fallen a prey to Bolshevism.

After his death the tactics applied by him were at first con-tinued. The Marxists, who had always covered the individual states and their princes in Germany with the bloodiest scorn, now came out as an *'Independent Party'* and suddenly appealed to those feelings and instincts which had their strongest roots in princely houses and individual states.

The fight of the Bavarian Republic of Councils[1] against the approaching contingents of liberation was dressed up by propa-ganda as mainly a 'struggle of the Bavarian workers' against 'Prussian militarism.' Only from this can it be understood why in Munich, quite unlike the other German territories, the over-throw of the Republic of Councils did not bring the great masses to their senses, but rather led to an even greater bitterness and rancor against Prussia.

The skill with which the Bolshevistic agitators were able to represent the elimination of the Republic of Councils as a 'Prus-sian militaristic' victory against the 'anti-militaristic' and 'anti-Prussian' Bavarian people, bore rich fruit. While Kurt Eisner, on the occasion of the elections to the legislative Bavarian Provincial Diet in Munich, still could summon not even ten thousand sup-

---

[1] After the murder of Kurt Eisner there was at first a Majority Socialist government under Hoffmann. On April 7, the Workers' Councils proclaimed the Republic of Councils. A temporary council was formed of Independents and the Bavarian Peasant League (*Bauernbund*). On April 10, this was overthrown by the Communists, partly under Russian leadership. Free corps from surrounding states marched on Munich and occupied it on May 1, 1919.

porters, and the Communist Party actually remained under three thousand, after the collapse of the Republic both parties together had risen to nearly a hundred thousand voters.

As early as this period my personal fight against the insane incitement of German tribes against each other began.

I believe that in all my life I have undertaken no more unpopular cause than my resistance at that time to the *anti-Prussian agitation*. In Munich, even during the Soviet period, the first mass meetings had taken place, in which hatred against the rest of Germany and in particular against Prussia was lashed to such a white heat that it not only involved a risk of his life for a North German to attend such a meeting, but the conclusion of such rallies as a rule ended quite openly with mad cries of: 'Away from Prussia!' — 'Down with Prussia!' — 'War against Prussia!' a mood which a particularly brilliant representative of Bavarian sovereign interests in the German Reichstag summed up in the battle-cry: '*Rather die Bavarian than rot Prussian!*'

You need to have lived through the meetings of that time to understand what it meant for me when, surrounded by a handful of friends, I for the first time, at a meeting in the Löwenbräukeller in Munich, offered resistance against this madness. It was war comrades who then supported me, and perhaps you can imagine how we felt when a mob — by far the greatest part of which had been deserters and slackers, hanging around the reserve posts or at home while we were defending the fatherland — lost all reason and bellowed at us and threatened to strike us down. For me, to be sure, these incidents had the virtue that the squad of my loyal followers came to feel really attached to me, and was soon sworn to live or die by my side.

These struggles, which were repeated again and again and dragged out through the entire year of 1919, seemed to become even sharper right at the beginning of 1920. There were meetings — I particularly remember one in the Wagner-Saal in the Sonnen-Strasse in Munich — in which my group, which had meanwhile grown stronger, had to withstand grave clashes, which not seldom ended in dozens of my supporters being mishandled, struck down, trampled under foot, and finally, more dead than alive, thrown out of the halls.

The struggle which I had first taken up as an individual, supported only by my war comrades, was now continued by the young movement as, I might also say, a sacred mission.

Today I am still proud to be able to say that in those days — dependent almost entirely on our Bavarian adherents — we nevertheless slowly but surely put an end to this mixture of stupidity and treason. I say stupidity and treason because, though fully convinced that the mass of followers were really nothing but good-natured fools, such simplicity cannot be attributed to the organizers and instigators. I regarded them, and today still regard them, as traitors bought and paid for by France. In one case, the Dorten case,[1] history has already given its verdict.

What made the affair especially dangerous at that time was the skill with which the true tendencies were concealed, by shoving federalistic intentions into the foreground as the sole motive for this activity. But it is obvious that stirring up hatred against Prussia has nothing to do with federalism. And a 'federative activity' which attempts to dissolve or split up another federal state makes a weird impression. An honorable federalist, for whom quotations of Bismarck's conception of the Reich are more than lying phrases, would hardly in the same breath want to separate portions from the Prussian state created or rather completed by Bismarck, let alone publicly support such separatist endeavors. How they would have yelled in Munich if a conservative Prussian party had favored the separation of Franconia from Bavaria, or actually demanded and promoted it by public action. In all this one could only feel sorry for the honest, federalist-minded souls who had not seen through this foul swindle; for they, first and foremost, were the cheated parties. By thus compromising the federative idea, its own supporters were digging its grave. It is impossible to preach a *federalistic* form for the Reich,

---

[1] Dorten was state's attorney in Wiesbaden after the World War. In 1919 he tried to establish a Rhenish Republic in collaboration with the French. He was associated with Bavarian Right Wing Catholic politicians.

at the same time *deprecating, reviling, and befouling* the most essential section of such a state structure, namely, Prussia, and in
short, making it, if possible, *impossible* as a *federal state*.[1] What
made this all the more incredible was that the fight of these so-
called federalists was directed precisely against *that Prussia*
which can least be brought into connection with the November
democracy. For it was not against the *fathers of the Weimar Constitution*, who themselves incidentally were for the most part
South Germans or Jews, that the vilifications and attacks of these
so-called *'federalists'* were directed, but against the representatives of the old *conservative Prussia*, hence the antipodes of the
Weimar Constitution. It must not surprise us that they took
special care not to attack the Jew, and perhaps this furnishes the
key to the solution of the whole riddle.

Just as before the revolution the Jew knew how to divert attention from his war societies, or rather from himself, and was
able to turn the masses, especially of the Bavarian people, against
Prussia, after the revolution he had somehow to cover his new
and ten times bigger campaign of pillage. And again he succeeded, in this case in inciting the so-called *'national elements'* of
Germany against one another: *conservative-minded Bavaria against
equally conservative-thinking Prussia*. And again he managed it
in the shrewdest way; he, who alone held the strings of the
Reich's destinies, provoked such brutal and tactless excesses that
the blood of those affected could not but be brought to the boil-
ing point each time anew. Yet never against the Jew, but always
against the German brother. *It was not the Berlin of four million
hard-working, producing people that the Bavarian saw, but the
rotten decaying Berlin of the foulest West End! But it was not
against this West End that his hatred turned, but against the 'Prussian' city.*

Often it could really drive you to despair.

This aptitude of the Jew for diverting public attention from
himself and occupying it elsewhere, can be studied again today.

In 1918 there could be no question of a systematic anti-

[1] '... *kurz als Bundesstaat wenn möglich, unmöglich macht.*'

Semitism. I still remember the difficulties one encountered if one so much as uttered the word Jew. Either one was stupidly gaped at, or one experienced the most violent resistance. Our first attempts to show the public the real enemy then seemed almost hopeless, and only very slowly did things begin to take a better turn. Bad as the organizational set-up of the *Watch and Ward League (Schutz- und Trutzbund)* [1] was, it nonetheless had great merit in having *reopened* the *Jewish question* as such. At all events, in the winter of 1918–19, something like *anti-Semitism* began slowly to take root. Later, to be sure, the National Socialist movement drove the Jewish question to the fore in quite a different way.[2] Above all, it succeeded in lifting this problem out of the narrow, limited circle of bourgeois and petit bourgeois strata and transforming it into the driving impulse of a great people's movement. But scarcely had it succeeded in giving the German people its great, unifying idea of struggle in this question than the Jew commenced to counter-attack. He seized upon his old weapon. With miraculous speed he threw the torch of discord into the folkish movement and sowed dissension. As things then stood, the only possibility of occupying the public attention with other questions and withholding a concentrated attack from the Jews lay in raising the *Ultramontane question,* and in the resulting clash between *Catholicism* and *Protestantism.* These men can never atone for the wrong they did our people in hurling this question into their midst. In any case the Jew reached his desired goal: Catholics and Protestants wage a merry war with one another, and the mortal enemy of Aryan humanity and all Christendom laughs up his sleeve.

[1] *Deutsch-Völkischer Schutz- und Trutzbund,* a supposedly non-political anti-Semitic organization, founded in 1919, ostensibly to protect German culture against Jewish influence, though anti-Semitism was not officially mentioned in its program. Its propaganda was first directed against the Jewish immigrants from Russia and Poland after the war. It petered out in 1922, its members merging with the National Socialists. It introduced the swastika to German political life.

[2] ' ... *die Judenfrage ganz anders vorwärtsgetrieben!* '

Just as formerly they were able to busy public opinion for years with the struggle between federalism and centralization, thus wearing it down with exhaustion, while the Jew sold the freedom of the nation and betrayed our fatherland to international high finance, now again he succeeds in causing the two German denominations to assail one another, while the foundations of both are corroded and undermined by the poison of the international world Jew.

Bear in mind the devastations which Jewish bastardization visits on our nation each day, and consider that this blood poisoning can be removed from our national body only after centuries, if at all; consider further how racial disintegration drags down and often destroys the last Aryan values of our German people, so that our strength as a culture-bearing nation is visibly more and more involved in a regression and we run the risk, in our big cities at least, of reaching the point where southern Italy is today. This contamination of our blood, blindly ignored by hundreds of thousands of our people, is carried on systematically by the Jew today. Systematically these black parasites of the nation defile our inexperienced young blond girls and thereby destroy something which can no longer be replaced in this world. Both, yes, both Christian denominations look on indifferently at this desecration and destruction of a noble and unique living creature, given to the earth by God's grace. The significance of this for the future of the earth does not lie in whether the Protestants defeat the Catholics or the Catholics the Protestants, but in whether the Aryan man is preserved for the earth or dies out. Nevertheless, the two denominations do not fight today against the destroyer of this man, but strive mutually to annihilate one another. The folkish-minded man, in particular, has the sacred duty, each in his own denomination, of making *people stop just talking superficially of God's will, and actually fulfill God's will, and not let God's word be desecrated.*

For God's will gave men their form, their essence and their abilities. Anyone who destroys His work is declaring war on the Lord's creation, the divine will. Therefore, let every man be active, each in his own denomination if you please, and let every man take it as his first and most sacred duty to oppose anyone who in his activity by word or deed steps outside the confines of his religious community and tries to butt into the other. For in Germany to attack the special characteristics of a denomination, within the religious schism we already have with us, necessarily leads to a war of annihilation between the two denominations. Our conditions permit here of no comparison say with France or Spain, let alone Italy. In all three countries, for example, a fight can be preached against clericalism or Ultramontanism, without running the risk that in this endeavor the French, Spanish, or Italian people as such will fall apart. In Germany, however, this may not be done, for here it is certain that the Protestants would also participate in such a movement. And thus the resistance, which would elsewhere be carried on only by Catholics against encroachments of a political nature by their own high clergy, immediately assumes the character of an attack of *Protestantism* against *Catholicism*. What is tolerated from members of the same denomination, even when it is unjust, is at once sharply rejected at the outset, as soon as the assailant belongs to another creed. This goes so far that even men, who themselves would be perfectly willing to correct a visible abuse within their own religious community, at once change their minds and turn their resistance outward as soon as such a correction is recommended, let alone demanded, by a source not belonging to their community. They regard this as an unjustified and impermissible, nay, indecent attempt to mix in matters which do not concern the party in question. Such attempts are not pardoned even if they are justified by the higher right of the interests of the national community, since today religious sentiments still go deeper than all considerations of national and political expediency. And this is in no way changed by driving the two denominations into an embittered mutual war, but could only change if, through compatibility on both sides, the nation is given a future which by its greatness would gradually have a conciliatory effect in this province as elsewhere.

I do not hesitate to declare that I regard the men who today draw the folkish movement into the crisis of religious quarrels as worse enemies of my people than any international Communist. For to convert the latter is the mission of the National Socialist movement. But anyone in its own ranks who leads it away from its true mission is acting damnably. Whether consciously or unconsciously is immaterial, he is a fighter for Jewish interests. For it is to the Jewish interest today to make the folkish movement bleed to death in a religious struggle at the moment when it is beginning to become a danger for the Jew. And I expressly emphasize the words bleed to death; for only a man without historical education can imagine that with this movement today a problem can be solved which has defied centuries and great statesmen.

For the rest, the facts speak for themselves. The gentlemen who in 1924 suddenly discovered that the highest mission of the folkish movement was the struggle against '*Ultramontanism*' did not break Ultramontanism, but tore apart the folkish movement.[1] I must also lodge protest against any immature mind in the ranks of the folkish movement imagining that he can do what even a Bismarck could not do. It will always be the highest duty of the top leadership of the National Socialist movement to offer the sharpest opposition to any attempt to drive the National Socialist movement into such struggles, and immediately to remove the propagandists of such an intention from the ranks of the movement. And actually, by autumn, 1923, we succeeded entirely in this. In the ranks of the movement, *the most devout Protestant* could sit beside *the most devout Catholic*, without coming into the slightest conflict with his religious convictions. The mighty common struggle which both carried on against the destroyer of Aryan humanity had, on the contrary, taught them mutually to respect and esteem one another. And yet, in these very years, the movement carried on the bitterest fight against the Center, though never on religious, but exclusively on national, racial, and

[1] An attack on the bitterly anti-Catholic Ludendorf, who broke with National Socialism after the *putsch* of November, 1923.

economico-political grounds. The results spoke in our favor, just as today they testify against the know-it-alls.

In recent years things have sometimes gone so far that folkish circles in the God-forsaken blindness of their denominational squabbles did not even recognize the madness of their actions from the fact that atheistic Marxist newspapers suddenly, when convenient, became advocates of religious communities, and tried to compromise one or the other by bandying back and forth remarks that were sometimes really too stupid, and thus stir up the fire to the extreme.

Especially with a people like the Germans, who have so often demonstrated in their history that they were capable of waging wars down to the last drop of blood for phantoms, any such battle-cry will be mortally dangerous. In this way our people has always been diverted from the real practical questions of its existence. While we devoured each other in religious squabbles, the rest of the world was distributed. And while the folkish movement considers whether the Ultramontane peril is greater than the Jewish peril or vice versa, the Jew destroys the racial foundations of our existence and thus destroys our people for all time. As far as this variety of '*folkish*' *warriors* are concerned, I can only wish the National Socialist movement and the German people with all my heart: Lord, protect them from such friends and they will settle with their enemies by themselves.

*   *   *

The struggle between federalism and centralization so shrewdly propagated by the Jews in 1919–20–21 and afterward, forced the National Socialist movement, though absolutely rejecting it, to take a position on its essential problems.

Should Germany be a *federated or a unified state*, and what for practical purposes must be understood by the two? To me the second seems the more important question, because it is not only fundamental to the understanding of the whole problem, but also because it is clarifying and possesses a conciliatory character.

What is a federated state?

By a federated state we understand a league of sovereign states which band together of their own free will, on the strength of their sovereignty; ceding to the totality that share of their particular sovereign rights which makes possible and guarantees the existence of the common federation.

In practice this theoretical formulation does not apply entirely to any of the federated states existing on earth today. Least of all to the American Union, where, as far as the overwhelming part of the individual states are concerned, there can be no question of any original sovereignty, but, on the contrary, many of them were sketched into the total area of the Union in the course of time, so to speak. Hence in the individual states of the American Union we have mostly to do with smaller and larger territories, formed for technical, administrative reasons, and, often marked out with a ruler, states which previously had not and could not have possessed any state sovereignty of their own. For it was not these states that had formed the Union, on the contrary it was the Union which formed a great part of such so-called states. The very extensive special rights granted, or rather assigned, to the individual territories are not only in keeping with the whole character of this federation of states, but above all with the size of its area, its spatial dimensions which approach the scope of a continent. And so, as far as the states of the American Union are concerned, we cannot speak of their state sovereignty, but only of their constitutionally established and guaranteed rights, or better, perhaps, privileges.

The above formulation is not fully and entirely applicable to Germany either. Although in Germany without doubt the individual states did exist first and in the form of states, and the Reich was formed out of them. But the very formation of the Reich did not take place on the basis of the free will or equal participation of the single states, but through the workings of the hegemony of one state among them, Prussia. The great difference between the German states, from the purely territorial standpoint, permits no comparison with the formation of the American Union, for instance. The difference in size between the smallest of the former federated states and the larger ones, let alone the largest, shows the non-similarity of their achievements, and also the inequality of their share in the founding of the Reich, the forming of the federated state. Actually, in most of these states there could be no question of a real sovereignty, except if state sovereignty was taken only as an official phrase. In reality, not only the past, but the present as well, had put an end to any number of these so-called 'sovereign states' and thus clearly demonstrated the weakness of these 'sovereign' formations.

I shall not state here how each of these states was historically formed, but I do want to say that in practically no case do they coincide with tribal boundaries. They are purely political phenomena, and their roots for the most part go back to the gloomiest epoch of the German Reich's impotence and of the national fragmentation which conditioned it and itself in turn was conditioned by it.

All this, in part at least, was taken into account by the constitution of the old Reich, in so far as it did not grant the individual states the same representation in the Bundesrat,[1] but set up gradations corresponding to size and actual importance, as well as the achievement of the individual states in the formation of the Reich.

The sovereign rights waived by the single states to make possible the formation of the Reich were only in the smallest part surrendered of their own free will; in the greatest part they were either practically non-existent to begin with or were simply taken away under the pressure of superior Prussian power. At the same time, Bismarck did not act on the principle of giving to the Reich everything that could in any way be taken away from the individual states; his principle was to demand of the individual states only what the Reich absolutely needed. A principle as moderate

[1] The Bundesrat was the federal council of the North German Confederation formed in 1867, and later of the German Empire. Made up of delegates of the various member states, it had equal powers with the Reichstag and was a sort of upper house.

as it was wise, which on the one hand took the highest consideration of custom and tradition, and on the other hand thereby
assured the new Reich a great measure of love and joyful collaboration at the very outset. It is absolutely wrong, however, to attribute this decision of Bismarck to his conviction that the Reich
thus possessed sufficient sovereign rights for all time. Bismarck
had no such conviction; on the contrary, he only wanted to put
off till the future what at the moment would have been hard to
accomplish and to endure. He put his hope in the gradual compromises brought about by time and the pressure of development
as such, which in the long run he credited with more strength than
any attempt to break the momentary resistance of the individual
states at once. Thus, he best demonstrated and proved the greatness of his statesmanship. For in reality the sovereignty of the
Reich steadily grew at the expense of the sovereignty of the
individual states. Time fulfilled Bismarck's expectations.

With the German collapse and the destruction of the German
state form, this development was necessarily accelerated. For
since the existence of the individual German states was attributable less to tribal foundations than to purely political causes, the
significance of these individual states inevitably shriveled into
nothing once the most essential embodiment of the political
development of these states, *the monarchic state form and their
dynasties*, had been excluded. A considerable number of these
'state formations' lost all internal stability, to such an extent that
they voluntarily renounced any further existence and for reasons
of pure expediency fused with others or merged with larger ones of
their own free will: the most striking proof of the extraordinary
weakness of these little formations and the small respect they enjoyed even among their own citizens.

And so, if the elimination of the monarchic state form and its
representatives in itself administered a strong blow to the Reich's
character as a federated state, this was even more true of the
assumption of the obligations resulting from the 'peace' treaty.

It was natural and self-evident that the financial sovereignty
previously vested in the provinces should be lost to the Reich at

the moment when the Reich due to the lost War was subjected to a financial obligation which would never have been met by individual contributions of the provinces. Also the further steps, which led to the taking over of the postal service and railroads by the Reich, were necessary effects of the enslavement of our people, gradually initiated by the peace treaties. The Reich was forced to take firm possession of more and more capital, in order to be able to meet the obligations which arose in consequence of further extortions.

Insane as the *forms* frequently were, in which the centralization was accomplished, the process in itself was logical and natural. Those to blame were the parties and the men who had not done everything in their power to end the War victoriously. Those to blame, especially in Bavaria, were the parties which in pursuit of selfish aims of their own, had during the War wrung from the principle of the Reich concessions which after its loss they had to restore ten times over. Avenging history! Seldom, however, has the punishment of Heaven come so swiftly after the crime as in this case. The same parties, which only a few years previous had placed the interests of their individual states — and this especially in Bavaria — above the interest of the Reich, were now compelled to look on as, beneath the pressure of events, the interest of the Reich throttled the existence of the individual states. All through their own complicity.

It is an unequaled hypocrisy to bemoan to the masses of voters (for only toward them is the agitation of our present-day parties directed) the loss of the sovereign rights of the individual provinces, while all these parties without exception outbid one another in a policy of fulfillment which in its ultimate consequences could not but lead to deep-seated changes inside Germany. Bismarck's Germany was free and unbound on the outside. This Reich did not possess financial obligations of so burdensome, and at the same time unproductive, a nature as the present Dawes-Germany has to bear. But internally as well, its competence was limited to a few matters and those absolutely necessary. Thus, it could very well dispense with a financial sovereignty of its own,

and live from the contributions of the provinces; and it goes
without saying that, on the one hand, the continued possession
of their own sovereign rights, and, on the other hand, comparatively small financial contributions to the Reich, were very con-
ducive to satisfaction with the Reich [1] in the provinces. However,
it is incorrect, dishonest in fact, to make propaganda today with
the assertion that the present lack of satisfaction with the Reich
can be attributed solely to the *financial bondage* of the provinces to
the Reich. No, that is not the real state of affairs. *The diminished
satisfaction with the Reich idea is not attributable to the loss of
sovereign rights on the part of the provinces, but is rather the result of
the deplorable way in which the German people is at present represented by its state.* Despite all the *Reichsbanner* [2] rallies and cele-
brations in honor of the constitution, the present Reich has remained alien to the heart of the people in all strata, and republican protective laws may deter people from transgressing against
republican institutions, but can never win the love of so much as
a single German. *In this excessive concern with protecting the Republic against its own citizens by means of penal laws and imprisonment lies the most annihilating criticism and disparagement of the
whole institution itself.*

But for another reason as well, the assertion, made by certain
parties today, that the disappearance of satisfaction in the Reich
is attributable to the encroachments of the Reich against certain
sovereign rights of the provinces, is untrue. Assuming that the
Reich had not undertaken the extension of its competencies, let
no one suppose that the love of the individual provinces for the
Reich would be any greater, if nonetheless their total contributions had to be the same as now. On the contrary: If the individual provinces today had to bear taxes to the amount which the
Reich requires for the fulfillment of the slave dictates, hostility
toward the Reich would be infinitely greater. Not only would the
contributions of the provinces to the Reich be very hard to bring
in, but they would have to be raised by means of downright coer-

[1] *'Reichsfreudigkeit.'* Literally: joy in the Reich.
[2] See page 425, note.

cion. For since the Republic stands on the basis of the peace treaties, and possesses neither the courage nor any intention whatever of breaking them, it must reckon with its obligations. *Solely to blame for this are again the parties which incessantly harangue the patient mass of the voters about the necessary independence of the provinces, but at the same time promote and support a Reich policy which must lead inevitably to the elimination of the very last of these so-called 'sovereign rights.'*

I say *inevitably* because the present Reich retains no other possibility of meeting the burdens imposed by its notorious domestic and foreign policy. Here again one wedge drives the next, and every new debt that the Reich heaps upon itself by its criminal handling of German interests abroad, must be balanced at home by a stronger downward pressure which in turn requires the gradual elimination of all sovereign rights of the individual states, in order to prevent germ cells of resistance from arising or merely persisting in them.

Altogether, a characteristic difference must be noted between the present Reich policy and that of former days: *The old Reich gave internal freedom and demonstrated strength on the outside, while the Republic shows weakness outside and represses its citizens internally.* In both cases one conditions the other: *The powerful national state needs fewer laws within in consequence of the greater love and attachment of its citizens; the international slave state can only hold its subjects to their slave labor by force.* For it is one of the present régime's most shameless impertinences to speak of 'free citizens.' Only the old Germany possessed such citizens. *The Republic is a slave colony of foreign countries and has no citizens, but at best subjects.* It therefore possesses no *national flag*, but only a *trade-mark*, introduced and protected by official decrees and legal measures. This symbol, regarded by everyone as the Gessler's hat of German democracy, will therefore always remain inwardly alien to our people. The Republic which, without any feeling for tradition and without much respect for the greatness of the past, trod its symbols in the mire, will some day be amazed how superficially its subjects are attached to its own symbols. It has given to itself the character of an intermezzo in German history.

And so today this state, for the sake of its own existence, is obliged to curtail the sovereign rights of the individual provinces more and more, not only out of general material considerations, but from ideal considerations as well. For in draining its citizens of their last drop of blood by its policy of financial extortion, it must inevitably withdraw their last rights if it does not want the general discontent to break out into open rebellion some day.

By inverting the above proposition, the following rule, basic for us National Socialists, is derived. *A powerful national Reich, which takes into account and protects the outward interests of its citizens to the highest extent, can offer freedom within, without having to fear for the stability of the state. On the other hand, a powerful national government can undertake and accept responsibility for great limitations on the freedom of the individual as well as the provinces, without damage to the Reich idea if in such measures the individual citizen recognizes a means toward the greatness of his nation.*

Certainly all the states in the world are moving toward a certain unification in their inner organization. And in this Germany will be no exception. Today it is an absurdity to speak of a 'state sovereignty' of individual provinces, which in reality the absurd size of these formations in itself fails to provide. The techniques of communication as well as administration steadily diminish the importance of the individual states. Modern communications, modern technology, make distance and space shrink more and more. A state of former days today represents only a province, and the states of the present formerly seemed like continents. The difficulty, from the purely technical point of view, of administering a state like Germany is no greater than the difficulty in directing a province like Brandenburg a hundred and twenty years ago. It is today easier to span the distance from Munich to Berlin than that from Munich to Starnberg a hundred years ago. And the whole Reich territory of today is smaller in relation to current communications technique than any medium German federated state at the time of the Napoleonic Wars. Anyone who disregards consequences resulting from undeniable facts cannot help but remain behind the times. At all times there have been men who do this, and in the future there will be too. But they can scarcely impede the wheel of history, and never bring it to a standstill.

We National Socialists must never blindly disregard the consequences of these truths. Here again we must not let ourselves be taken in by the phrases of our so-called national bourgeois parties. I use the term phrases because these parties themselves do not believe seriously in the possibility of carrying out their intentions, and because in the second place they themselves are the accomplices mainly responsible for the present development. Especially in Bavaria, the cry for the elimination of centralization is really nothing more than a party machination without any serious thought behind it. Every time that these parties should really have made something serious out of their phrases, they all of them fell down miserably. Every so-called 'theft of sovereign rights' from the Bavarian state by the Reich was accepted practically without resistance except for a repulsive yelping. *Indeed, if anyone really dared to put up serious opposition to this insane system, he was outlawed and damned and persecuted for 'contempt of the existing state' by these very parties, and in the end was silenced either by imprisonment or illegally forbidden to speak.* This more than anything should show our supporters the inner hypocrisy of these so-called federalistic circles. The federative state idea, like religion in part, is only an instrument for their often unclean party interests.

*   *   *

Natural as a certain unification may seem, particularly in the field of communications, for us National Socialists there nevertheless remains an obligation to take an energetic position against such a development in the present-day state, at times when the measures only serve the purpose of masking and making possible a catastrophic foreign policy. Precisely because the present Reich did not undertake the nationalization of the railroads, postal service, finances, etc., out of higher considerations of national policy, but only in order to lay hands on the financial means and securities for such a policy of unlimited *fulfillment*,[1] we National Socialists must do everything that seems in any way calculated to impede and if possible prevent the execution of such a policy. And to this belongs the struggle against the present centralization of vitally important institutions of our people, which is undertaken only in order to raise the billions of marks and the collateral for our post-War foreign policy.

And for this reason the National Socialist movement has taken a position against such attempts.

The second reason which can induce us to offer resistance to such a centralization is that it might stabilize the power of a system of internal government which in all its effects has brought the gravest disaster upon the German people. *The present Jewish-Democratic Reich, which has become a true curse for the German nation, seeks to make the criticism of the individual states, all of which are not yet imbued with this new spirit, ineffectual, by reducing them to total insignificance.* In the face of this, we National Socialists have every reason to attempt to give to the opposition of these individual states, not only the foundation of a state power promising success, but in general to make their struggle against centralization into an expression of a higher, national and universal German interest. And so, while the *Bavarian People's Party endeavors to preserve special rights for the Bavarian State out of smallhearted, particularistic motives, we must use this special position in the service of a higher national interest in opposition to the present November democracy.*

The third reason that can also determine us to fight against the current centralization is the conviction that a great part of the so-called nationalization is in reality no unification, and in no event a simplification, but that in many cases it is only a matter of removing institutions from the sovereign rights of the provinces, in order to open their gates to the interests of the revolu-

[1] Fulfillment of the peace treaties, here the reparations provisions.

tionary parties. Never in German history has there been a more shameless policy of favoritism than under the democratic Republic. *A large part of the present frenzy for centralization falls to the account of those parties which once promised to clear the road to ability, but when it came to filling offices and posts solely considered party membership.* Since the founding of the Republic, particularly Jews in incredible numbers poured into the economic concerns and administrative apparatuses snatched up by the Reich, so that today both have become a domain of Jewish activity.

This third consideration above all must obligate us on tactical grounds to examine sharply any further measure on the road to centralization and if necessary to take a position against it. *But in this our motives must always be higher motives of national policy and never petty particularistic ones.*

This last remark is necessary lest the opinion arise among our supporters that we National Socialists would not grant the Reich as such the right to embody a higher sovereignty than that of the individual states. Concerning this right, there must and can be no doubt among us. *Since for us the state as such is only a form, but the essential is its content, the nation, the people, it is clear that everything else must be subordinated to its sovereign interests. In particular we cannot grant to any individual state within the nation and the state representing it state sovereignty and sovereignty in point of political power.* The mischief of individual federated states maintaining so-called missions abroad and among each other must cease and will some day cease. As long as such things are possible, we must not be surprised if foreign countries still doubt the stability of our Reich structure and act accordingly. The mischief of these missions is all the greater as not the least benefit can be attributed to them along with the harm. The interests of a German abroad, which cannot be protected by the ambassador of the Reich, can much less be looked after by the envoy of a petty state that looks ridiculous in the setting of the present world order. In these petty federated states we can really see nothing but points of attack for separatist endeavors inside and outside of the German Reich, endeavors such as *one* state still particularly

welcomes. And we National Socialists can have no sympathy
with some noble family, grown feeble with age, providing new soil
for a withered scion by clothing him in an ambassadorial post.
Our diplomatic missions abroad were so miserable even at the
time of the old Reich that further additions to the experience then
gained are highly superfluous.

In future the significance of the individual provinces will unquestionably be shifted more to the field of cultural policy. The
monarch who did the most for the importance of Bavaria was not
some stubborn anti-German particularist, but Ludwig I, a man of
pan-German mind and artistic sensibilities. By using the forces of
the state primarily for the development of Bavaria's cultural
position and not for the strengthening of her political power, he
built better and more enduringly than would otherwise have been
possible. By pushing Munich from the level of an insignificant
provincial capital into the format of a great German art metropolis,[1] he created a spiritual center which even today is strong
enough to bind the essentially different Franks to this state.
Supposing that Munich had remained what it formerly was, the
same process that took place in Saxony would have been repeated
in Bavaria, only with the difference that Nuremberg, the Bavarian
Leipzig, would have become not a Bavarian but a Frankish city.
It was not the 'Down with Prussia' shouters that made Munich
great; this city was given its importance by the King who in it
wished to bestow upon the German nation an art treasure which
would have to be seen and respected, and which was seen and
respected. And therein lies a lesson for the future. *The importance
of the individual states will in the future no longer lie in the fields of
state power and policy; I see it either in the tribal field or the field of
cultural policy.* But even here time will have a leveling effect.
The ease of modern transportation so scatters people around that
slowly and steadily the tribal boundaries are effaced and thus
even the cultural picture gradually begins to even out.

[1] *'indem er München damals aus dem Rahmen einer wenig bedeutenden
provinziellen Residenz in das Format einer grossen deutschen Kunstmetropole
hineinschob . . .'*

The army in particular must be sharply removed from all influence of the individual states. The coming National Socialist state must not fall into the error of the past and attribute to the army a function which it has not and must not have. *The German army does not exist to be a school for the preservation of tribal peculiarities, but should rather be a school for the mutual understanding and adaptation of all Germans.* Whatever source of division there may be in the life of the nation must be given a unifying effect by the army. Furthermore, it must raise the individual young man from the narrow horizon of his little province and put him into the German nation. He must learn to see, not the boundaries of his home province, but those of his fatherland; for it is these that he will one day have to protect. It is, therefore, senseless to leave the young German at home; the expedient thing is to show him Germany during his period of military service. Today this is all the more necessary, as the young German no longer has his years of wandering apprenticeship with their broadening effect on his horizon. In view of this, is it not preposterous to leave, when possible, the young Bavarian in Munich, the Frank in Nuremberg, the Badener in Karlsruhe, the Württemberger in Stuttgart, etc., and is it not more sensible to show the young Bavarian the Rhine and the North Sea, the Hamburger the Alps, the East Prussian the mountains of Central Germany, and so on? Regional character should remain in the detachment, but not in the garrison. Every attempt at centralization may encounter our disapproval, but that of the army never! On the contrary, even if we welcomed no such endeavor, we should have to take pleasure in this one. Quite aside from the fact that, in view of the size of the present Reich army, the preservation of individual state troop formations would be absurd, we regard the unification of the Reich army that has taken place as a step which, in the future, when a national army has been introduced, we must never again abandon.

Moreover, a young victorious idea will have to reject any fetter which might paralyze its activity in pushing forward its conceptions. National Socialism as a matter of principle, must lay claim to the

*right to force its principles on the whole German nation without con-
sideration of previous federated state boundaries, and to educate it in
its ideas and conceptions. Just as the churches do not feel bound and
limited by political boundaries, no more does the National Socialist
idea feel limited by the individual state territories of our fatherland.*

*The National Socialist doctrine is not the servant of individual
federated states, but shall some day become the master of the German
nation. It must determine and reorder the life of a people, and must,
therefore, imperiously claim the right to pass over boundaries drawn
by a development we have rejected.*

*The more complete the victory of its ideas will be, the greater may be
the particular liberties it offers internally.*

Propaganda and Organization

IN SEVERAL RESPECTS the year 1921 had assumed a special significance for me and the movement.

After my entrance into the German Workers' Party, I at once took over the management of propaganda. I regarded this department as by far the most important. For the present, it was less important to rack one's brains over organizational questions than to transmit the idea itself to a larger number of people. Propaganda had to run far in advance of organization and provide it with the human material to be worked on. Moreover, I am an enemy of too rapid and too pedantic organizing. It usually produces nothing but a dead mechanism, seldom a living organization. For organization is a thing that owes its existence to organic life, organic development. Ideas which have gripped a certain number of people will always strive for a greater order, and a great value must be attributed to this inner molding. Here, too, we must reckon with the weakness of men, which leads the individual, at first at least, instinctively to resist a superior mind. If an organization is mechanically ordered from above, there exists a great danger that a once appointed leader, not yet accurately evaluated and perhaps none too capable, will from jealousy strive to prevent the rise of abler elements within the movement. The harm that arises in such a case can, especially in a young movement, be of catastrophic significance.

For this reason it is more expedient for a time to disseminate an idea by propaganda from a central point and then carefully to search and examine the gradually gathering human material for leading minds. Sometimes it will turn out that men inconspicuous in themselves must nevertheless be regarded as born leaders.

But it would be absolutely mistaken to regard a wealth of theoretical knowledge as characteristic proof for the qualities and abilities of a leader.

The opposite is often the case.

The great *theoreticians* are only in the rarest cases great *organizers*, since the greatness of the *theoretician* and *program-maker* lies primarily in the *recognition* and *establishment of abstractly correct laws*, while the *organizer* must primarily be a *psychologist*. He must take people as they are and must therefore know them. He must not overestimate them, any more than he must underestimate them in the mass. On the contrary, he must endeavor to take weakness and bestiality equally into account, in order, considering all factors, to create a formation which will be a living organism, imbued with strong and stable power, and thus suited to upholding an idea and paving the way for its success.

Even more seldom, however, is a great theoretician a great leader. Much more readily will an *agitator* be one, something which many who only work scientifically on the question do not want to hear. And yet that is understandable. An agitator who demonstrates the ability to transmit an idea to the broad masses must always be a psychologist, even if he were only a demagogue. Then he will still be more suited for leadership than the unworldly theoretician, who is ignorant of people. *For leading means: being able to move masses.* The gift of shaping ideas has nothing to do with ability as a leader. And it is quite useless to argue which is of greater importance, to set up ideals and aims for mankind, or to realize them. Here, as so often in life: one would be utterly meaningless without the other. The finest theoretical insight remains without purpose and value if the leader does not set the masses in motion toward it. And conversely, of what avail would be all the genius and energy of a leader, if the brilliant theoretician did not set up aims for the human struggle? However, the combination of theoretician, organizer, and leader in one person is the rarest thing that can be found on this earth; this combination makes the great man.

As I have already remarked, I devoted myself to propaganda in the first period of my activity in the movement. What it had to do was gradually to fill a small nucleus of men with the new doctrine, and so prepare the material which could later furnish the first elements of an organization.

When a movement harbors the purpose of tearing down a world and building another in its place, complete clarity must reign in the ranks of its own leadership with regard to the following principles:

Every movement will first have to sift the human material it wins into two large groups: supporters and members.

The function of propaganda is to attract supporters, the function of organization to win members.

A supporter of a movement is one who declares himself to be in agreement with its aims, a member is one who fights for them.

The supporter is made amenable to the movement by propaganda. The member is induced by the organization to participate personally in the recruiting of new supporters, from whom in turn members can be developed.

Since being a supporter requires only a passive recognition of an idea, while membership demands active advocacy and defense, to ten supporters there will at most be one or two members.

Being a supporter is rooted only in understanding, membership in the courage personally to advocate and disseminate what has been understood.

Understanding in its passive form corresponds to the majority of mankind which is lazy and cowardly. Membership requires an activistic frame of mind and thus corresponds only to the minority of men.

Propaganda will consequently have to see that an idea wins supporters, while the organization must take the greatest care only to make the most valuable elements among the supporters into members. Propaganda does not, therefore, need to rack its brains with regard to the importance of every individual instructed by it, with regard to his ability, capacity, and understanding, or character, while the organization must carefully gather from the mass of these elements those which really make possible the victory of the movement.

*   *   *

Propaganda tries to force a doctrine on the whole people; the organization embraces within its scope only those who do not threaten on psychological grounds to become a brake on the further dissemination of the idea.

*   *   *

Propaganda works on the general public from the standpoint of an idea and makes them ripe for the victory of this idea, while the organization achieves victory by the persistent, organic, and militant union of those supporters who seem willing and able to carry on the fight for victory.

*   *   *

The victory of an idea will be possible the sooner, the more comprehensively propaganda has prepared people as a whole and the more exclusive, rigid, and firm the organization which carries out the fight in practice.

From this it results that the number of supporters cannot be too large, but that the number of members can more readily be too large than too small.

*   *   *

If propaganda has imbued a whole people with an idea, the organization can draw the consequences with a handful of men. Propaganda and organization, in other words, supporters and members, thus stand in a certain mutual relation. The better the propaganda has worked, the smaller the organization can be; and the larger the number of supporters, the more modest the number of members can be; and vice versa: the poorer the propaganda is, the larger the organization must be, and the smaller the host of followers of a movement remains, the more extensive the number of its members must be, if it still hopes to count on any success at all.

* * *

The first task of propaganda is to win people for subsequent organization; the first task of organization is to win men for the continuation of propaganda. The second task of propaganda is the disruption of the existing state of affairs and the permeation of this state of affairs with the new doctrine, while the second task of organization must be the struggle for power, thus to achieve the final success of the doctrine.

* * *

The most striking success of a revolution based on a philosophy of life will always have been achieved when the new philosophy of life as far as possible has been taught to all men, and, if necessary, later forced upon them, while the organization of the idea, in other words, the movement, should embrace only as many as are absolutely required for occupying the nerve centers of the state in question.

This, in other words, means the following:

In every really great world-shaking movement, propaganda will first have to spread the idea of this movement. Thus, it will indefatigably attempt to make the new thought processes clear to the others, and therefore to draw them over to their own ground, or to make them uncertain of their previous conviction. Now, since the dissemination of an idea, that is, propaganda, must have a firm backbone, the doctrine will have to give itself a solid organization. The organization obtains its members from the general body of supporters won by propaganda. The latter will grow the more rapidly, the more intensively the propaganda is carried on, and the latter in turn can work better, the stronger and more powerful the organization is that stands behind it.

Hence it is the highest task of the organization to make sure that no inner disunities within the membership of the movement lead to a split and hence a weakening of the movement's work; further, that the spirit of determined attack does not die out, but is continuously renewed and reinforced. The number of members need not grow infinitely; on the contrary: since only a small fraction of mankind is by nature energetic and bold, a movement which endlessly enlarges its organization would inevitably be weakened some day as a result. *Organizations, in other words, membership figures, which grow beyond a certain level gradually lose their fighting power and are no longer capable of supporting or utilizing the propaganda of an idea resolutely and aggressively.*

The greater and more essentially revolutionary an idea is, the more activistic its membership will become, since the revolutionary force of a doctrine involves a danger for its supporters, which seems calculated to keep cowardly little shopkeepers away from it. They will privately regard themselves as supporters, but decline to make a public avowal of this by membership. *By virtue of this fact, the organization of a really revolutionary idea obtains as members only the most active among the supporters won over by propaganda.* And precisely in this activity of a movement's membership, guaranteed by natural selection, lies the premise for equally active future propaganda as well as a successful struggle for the realization of the idea.

The greatest danger that can threaten a movement is a membership which has grown abnormally as a result of too rapid successes. For, just as a movement is shunned by all cowardly and egotistic individuals, as long as it has to fight bitterly, these same people rush with equal alacrity to acquire membership when a success of the party has been made probable or already realized by developments.

To this it must be ascribed why [1] many victorious movements, on the point of success, or, rather, the ultimate completion of their will, suddenly from inexplicable inner weakness, flag, stop fighting, and finally die out. In consequence of their first victory, so many inferior, unworthy, and worst of all cowardly, elements have entered their organization that these inferior people finally achieve predominance over the militants and then force the movement into the service of their own interests, lower it to the level of their own scanty heroism, and do nothing to complete the victory of the original idea. The fanatical zeal has been blurred, the fighting force paralyzed, or, as the bourgeois world correctly puts it in such cases: 'Water has been mixed with the wine.' And when that happens, the trees can no longer grow skyward.

It is, therefore, most necessary that a movement, for pure reasons of self-preservation, should, once it has begun to achieve success, immediately block enrollments and henceforth increase its organization only with extreme caution and after the most thorough scrutiny. Only in this way will it be able to preserve the core of the movement in unvitiated freshness and health. *It must see to it that, from this point on, this core alone shall exclusively lead the movement, that is, determine the propaganda which should lead to its universal recognition, and, in full possession of the power, undertake the actions which are necessary for the practical realization of its ideas.*

It must not only occupy all the important positions of the conquered territory with the basic core of the old movement, but also constitute the entire leadership. And this until the principles and doctrines of the party have become the foundation and content of the new state. Only then can the reins gradually be handed over to the special government of this state, born of its spirit. This, however, in turn occurs for the most part only in mutual struggle, since it is less a question of human insight than of the play and workings of forces which can perhaps be recognized from the first, but cannot forever be guided.

All great movements, whether of a religious or a political nature, must attribute their mighty successes only to the recognition and ap-

[1] *'Dem ist es zuzuschreiben warum . . .'*

*plication of these principles, and all lasting successes in particular
are not even thinkable without consideration of these laws.*

*   *   *

As director of the party's propaganda I took much pains, not
only to prepare the soil for the future greatness of the movement,
but by an extremely radical conception in this work I also strove
to bring it about that the party should obtain only the best material. For the more radical and inflammatory my propaganda
was, the more this frightened weaklings and hesitant characters,
and prevented them from penetrating the primary core of our
organization. They might continue as supporters, but certainly
not with loud emphasis; they timidly concealed the fact. How
many thousands assured me at that time that they were essentially in agreement with everything we said, but that under no
circumstances could they become members. The movement,
they said, was so radical that membership in it would expose
the individual to the gravest difficulties, nay, dangers, and we
shouldn't take it amiss if the honest, peaceable citizen should
stand aside for the present at least, even if at heart he was entirely
with the cause.

And this was good.

If these men, who at heart were not for the revolution, had all
come into our party at that time, and as members, we could regard ourselves today as a pious fraternal organization, but no
longer as a young militant movement.

The live and aggressive form that I then gave to our propaganda reinforced and guaranteed the radical tendency of our
movement, since now only radical people — with some exceptions
— were ready for membership.

At the same time, this propaganda had the effect that after a
short while hundreds of thousands not only believed us to be right
but desired our victory. even if personally they were too cowardly
to make sacrifices for it, let alone fight for it.

Up to the middle of 1921 this purely propagandist activity could still suffice and benefit the movement. But special events in the midsummer of this year made it seem indicated that now, after the slowly visible success of our propaganda, the organization should be adapted to it and put on a par with it.

The attempt of a group of folkish lunatics to obtain the leadership of the party, with the aid and support of the party chairman of the time, led to the collapse of this little intrigue and, at a general membership meeting, unanimously gave me the leadership over the whole movement. Immediately, a new by-law was passed, transferring full responsibility to the first chairman of the party, eliminating committee decisions as a matter of principle, and introducing instead a system of division of labor which has since proved its worth in the most beneficial way.

Beginning on August 1, 1921, I took over this inner reorganization of the movement and in so doing found the support of a number of excellent people whom I consider it necessary to mention in a special appendix.

In the attempt to organizationally exploit the results of propaganda and thereby establish them for all time, I had to do away with a number of previous habits and introduce principles which none of the existing parties possessed or would even have recognized.

In the years from 1919 to 1920 the movement had for leadership a committee which was chosen by membership meetings, which themselves in turn were prescribed by rule. The committee consisted of a first and second treasurer, a first and second secretary, and at the head, a first and second chairman. Added to these was a membership secretary, the propaganda chief, and various assisting committeemen.

Strange as it may seem, this committee actually embodied exactly what the party most wanted to combat, namely, *parliamentarianism*. For it was obvious that we were involved with a principle which from the smallest local group, through the later districts, counties, and provinces, up to the Reich leadership, embodied the very same system under which we all suffered and today still suffer.

It was urgently necessary to bring about a change in this some day, unless the movement, in consequence of the poor foundation of its inner organization, were to be forever ruined and hence incapable of ever fulfilling its high mission.

The committee sessions, of which minutes were kept, and in which votes were taken and decisions made by a majority, represented in reality a parliament on a small scale. Here, too, all personal responsibility was lacking. Here, too, the same irrationality and the same unreasonableness reigned as in our great state representative bodies. For this committee, secretaries, treasurers, membership secretaries, propaganda chiefs, and God knows what else were appointed, and then all of them together were made to deliberate on every single question and decide by vote. And so the man who was there for propaganda voted on a matter that regarded the finance man, and he in turn voted on a matter regarding organization, and the latter in turn on a matter which should only have concerned the secretary, etc.

Why they bothered to appoint a special man for propaganda, when treasurers, secretaries, membership secretaries, etc., had to decide on questions regarding it, seems just as incomprehensible to a healthy mind as it would be incomprehensible if in a big industrial enterprise the directors or engineers of other departments and other branches had to decide on questions having nothing to do with their affairs.

I did not submit to this lunacy, but after a short time stayed away from the sessions. I did my propaganda work and let it go at that, and I did not stand for any incompetent trying to tell me what to do in this field. Just as, conversely, I did not interfere in the business of the others.

When the acceptance of the new statutes and my appointment to the position of first chairman had meanwhile given me the necessary authority and the rights that went with it, this nonsense immediately stopped. In the place of committee decisions, the principle of absolute responsibility was introduced.

The first chairman is responsible for the total leadership of the movement. He apportions the work to be performed among the committeemen subordinated to him and among whatever other collaborators are needed. And each one of these gentlemen is absolutely responsible for the tasks transferred to him. He is subordinated only to the first chairman, who must procure the cooperation of all, or else must bring about this cooperation by the choice of persons and the issuance of general directives.

This law of fundamental responsibility was gradually taken for granted within the movement, at least in so far as the party leadership was concerned. In the little local groups and perhaps even in the counties and districts, it will take years before these principles will be forced through, since scare-cats and incompetents will of course always fight against it; to them sole responsibility for an undertaking will always be unpleasant; they always felt freer and better when in every grave decision they were covered by the majority of a so-called committee. But to me it seems necessary to express myself with the greatest sharpness against such an attitude, to make no concession to cowardice in the face of responsibility, and thereby, even if it takes a long time, to achieve a conception of leader's duty and leader's ability, which will bring to leadership exclusively those who are really called and chosen for it.

In any case a movement that wants to combat the parliamentary madness must itself be free of it. Only on such a basis can it win the strength for its struggle.

A movement which in a time of majority rule orients itself in all things on the principle of the leader idea and the responsibility conditioned by it will some day with mathematical certainty overcome the existing state of affairs and emerge victorious.

This idea led to a complete reorganization within the movement. And in its logical effects also to an extremely sharp division between the business activities of the movement and the general political leadership. As a matter of principle, the idea of responsibility was extended to all the party activities and led inevitably to their recovery, in exact proportion as they were freed from political influences and adjusted to purely economic considerations.

When in the fall of 1919, I joined the handful of men who then constituted the party, it had neither a business office nor a clerk, not even forms or rubber stamps; and no printed matter existed. The committee room was first a tavern in the Herrengasse, and later a café on the Gasteig. That was an impossible state of affairs. Soon afterward I started out and visited a number of Munich restaurants and taverns with the intention of renting a back room or some other space for the party. In the former Sterneckerbräu in the Tal, there was a small vault-like room which had once served the imperial councilors [1] of Bavaria as a sort of taproom. It was dark and gloomy and thus was just as well suited for its former purpose as it was ill-suited for its projected new use. The alley on which its single window opened was so narrow that even on the brightest summer day the room remained gloomy and dark. This became our first business office. But since the monthly rent was only fifty marks (then an exorbitant sum for us!), we could make no greater demands and were not even in a position to complain when, before we moved in, the wall paneling, formerly intended for the imperial councilors, was quickly torn out, so that now the room really gave more the impression of a funeral vault than of an office.

And yet this was an immense step forward. Slowly we obtained electric light, even more slowly a telephone; a table and a few borrowed chairs were brought in, finally an open book-stand, still somewhat later a cupboard; two sideboards belonging to the landlord served for keeping pamphlets, posters, etc.

The previous system — that is, having the movement run by a committee session taking place once a week — was impossible in the long run. Only an official paid by the movement could guarantee the day-to-day business organization.

At the time that was very difficult. The movement still had so few members that it took great skill to find among them a suitable man who, making the smallest demands for his own person, could satisfy the innumerable demands of the movement.

[1] '*die Reichsräte von Bayern.*' Until 1918 the upper house of the Bavarian Diet, consisting of nobility, high clergy, and other notables.

In the person of a soldier, named Schüssler, one of my former comrades, the first business manager of the party was found. At first he came to our new office only daily from six to eight o'-clock, later from five to eight, finally every afternoon, and shortly afterward he was taken on full time and served from morning until late into the night. He was a man as conscientious as he was upright and absolutely honest, who personally took the greatest pains and was devoted with especial loyalty to the movement itself. Schüssler brought with him a small Adler typewriter that belonged to him. It was the first such instrument in the service of our movement. Later the party acquired it by installment payments. A small safe seemed necessary to safeguard the card index and the membership books from thieves. We did not acquire it in order to deposit any large sums of money we might have had at the time. On the contrary, everything was extremely threadbare, and often I contributed from my own small savings.

A year and a half later, the business office was too small, and we moved into a new place in the Corneliusstrasse. Again it was a tavern we moved to, but now we no longer possessed only a single room, but three rooms and one large additional room with a wicket-window. At the time that seemed to us like a good deal. Here we remained until November, 1923.

In December, 1920, we acquired the *Völkischer Beobachter*. This paper, which, as its name indicates, stood on the whole for folkish interests even then, was now to be transformed into the organ of the NSDAP. At first it appeared twice a week, at the beginning of 1923 became a daily, and at the end of August, 1923, it received its large format which later became well known.

As a total novice in the field of journalism, I sometimes had to pay dearly for my experience in those days.

The mere fact that in comparison with the enormous Jewish press there was hardly a single really significant folkish paper gave food for thought. This, as I later ascertained any number of times in practice, was in large part due to the unbusinesslike management of so-called folkish enterprises in general. They were too much conducted from the angle that loyalty takes precedence over achievement. An absolutely false standpoint, in so far as loyalty must not be an outward thing, but find its most eminent expression in achievement. Anyone who creates something really valuable for his people thus gives evidence of an equally valuable loyalty, while another, who merely displays hypocritical loyalty, but in reality performs no useful services for his people, is an enemy to any true loyalty. And his loyalty is a burden to the community.

The *Völkischer Beobachter*, as its very name indicates, was also a folkish organ, with all the advantages, and even more faults and weaknesses, that were characteristic of folkish institutions. Honest as its content was, the management of the enterprise was impossible from the commercial viewpoint. It, too, was run on the assumption that folkish newspapers must be supported by folkish contributions, instead of the principle that they must make their way in competition with other papers and that it is indecent to cover the negligence or mistakes of their business management by the donations of well-situated patriots.

In any case I attempted to eliminate this state of affairs, the objectionableness of which I had soon recognized, and luck favored me by making me acquainted with the man who since then, not only as business manager of the paper, but also of the party, has performed services of the greatest value for the movement. In 1914 — at the front, that is — I met Max *Amann*, the present general business manager of the party (then still my superior in rank). During the four years of the War, I had an almost continuous opportunity to observe the extraordinary ability, the industry and scrupulous conscientiousness of my future collaborator.

In midsummer of 1921, when the movement was in a grave crisis and I could no longer be satisfied with a number of employees, and with one in fact had had the bitterest experience, I turned to my former regimental comrade, whom chance brought to me one day, with the request that he become business manager of the movement. After long hesitation — Amann was holding a position with good prospects — he finally consented, though on condition that he would never serve as a stooge for any incompetent committees, but would exclusively recognize a single master.

It is the inextinguishable merit of this first business manager of the movement, a man of really comprehensive business training, to have brought order and neatness into the party's business affairs. Since that time they have remained exemplary and could be equaled, let alone surpassed, by none of the subdivisions of the movement, but, as always in life, outstanding ability is not seldom the cause of envy and disfavor. This, of course, had to be expected in this case and to be taken patiently into account.

By 1922 there existed, by and large, firm directives for the business as well as the purely organizational development of the movement. There was already a complete central card index which embraced all members belonging to the movement. Likewise the financing of the movement had been brought into healthy channels. Current expenses had to be covered by current receipts; extraordinary receipts were used only for extraordinary expenses. Despite the hard times, the movement thereby remained, apart from small running accounts, almost free of debt, and even succeeded in steadily increasing its resources. We worked as in a private business: the employed personnel had to distinguish itself by achievement, and could not get by on the strength of any of your famous 'loyalty.' The loyalty of every National Socialist is demonstrated primarily by his readiness to work, his industry and ability in accomplishing the work entrusted to him by the community. Anyone who does not fulfill his duty in this should not boast of his loyalty, against which he is actually committing an offense. With the utmost energy the new business manager, in opposition to all possible influences, upheld the standpoint that party enterprises must not be a sinecure for supporters or members with no great enthusiasm for work. A movement which fights in so sharp a form against the party corruption of our present administrative apparatus must keep its own apparatus pure of such vices. There were cases where employees were taken into the administration of the newspaper, who in their previous allegiance belonged to the Bavarian People's Party, but, measured by their achievements, showed themselves excellently qualified. The result of this attempt was in general outstanding. By this honest and frank recognition of the individual's real achievement, the movement more quickly and more thoroughly won the hearts of its employees than would otherwise have been the case. They later became good National Socialists and remained so, and not only in words; they also demonstrated it by the conscientious, regular, and honest work which they performed in the service of the new movement. It goes without saying that the well-qualified party comrade was given preference over the equally qualified non-party member. But no one obtained a position on the basis of his party membership alone. The firmness with which the new business manager upheld these principles, and gradually enforced them despite all opposition, was later of the greatest benefit to the movement. Through this alone was it possible, in the difficult inflation period, when tens of thousands of businesses collapsed and thousands of newspapers had to close, for the business leadership of the movement, not only to remain above water and fulfill its tasks, but for the *Völkischer Beobachter* to be expanded more and more. It had entered the ranks of the great newspapers.

The year 1921 had, furthermore, the significance that I gradually succeeded, through my position as chairman of the party, in withdrawing the various party services from the criticism and interference of dozens of committee members. This was important, because it was impossible to obtain a really capable mind for a job if incompetents kept on babbling and interfering, knowing everything better than anyone else and actually creating a hopeless muddle. Whereupon, to be sure, these know-it-alls usually withdrew quite modestly, to seek a new field for their inspiring supervisory activity. There were men who were possessed by a positive disease for finding something behind anything and everything, and who were in a kind of continuous pregnancy with excellent plans, ideas, projects, methods. Their highest and most ideal aim was usually the formation of a committee or controlling organ to put its expert nose into other people's serious work. It never dawned on many of these committee people how insulting and how un-National Socialist it is, when men who do not understand a thing keep interfering with real specialists. In any case, I regarded it as my duty in these years, to take all real workers, charged with responsibility in the movement, under my protection against such elements, to cover them in the rear, as it were, so as to leave them free to work forward.

The best means for making harmless such committees, who did nothing and only cooked up decisions that could not be practically carried out, was to assign them to some real work. It was laughable how silently one of these clubs would then disappear, and suddenly was impossible to locate. It made me think of our greatest institution of the sort, the Reichstag. How all its members would suddenly evaporate if, instead of talk, some real work were assigned to them; and particularly a task which every single one of these braggarts would have to perform with personal responsibility.

Even then I always raised the demand that, in the movement as everywhere in private life, we keep looking until the obviously capable official, administrator, or director for the various business sections had been found. And this man was then to receive unconditional authority and freedom of action downward, but to be charged with unlimited responsibility upward, and no one obtains authority toward subordinates who does not know the work involved better than they. In the course of two years, I enforced my opinion more and more, and today it is taken for granted in the movement, at least in so far as the top leadership is concerned.

The visible success of this attitude was shown on November 9, 1923: when I came to the movement four years previous, not even a rubber stamp was available. On November 9, the party was dissolved, its property confiscated. This, including all properties and the newspaper, already amounted to over a hundred and seventy thousand gold marks.

The Trade-Union Question

THE RAPID GROWTH of the movement compelled us in 1922 to take a position on a question which even today is not entirely solved.

In our attempts to study those methods which could most easily open up to the movement the way to the hearts of the masses, we always encountered the objection that the worker could never be entirely with us because the defense of his interests in the purely occupational and economic field lay in the hands of our enemies and their organizations.

This objection, of course, had much to be said for it. It was a matter of general belief that the worker who was active in a factory could not even exist unless he became a member of a union. Not only that his occupational interests seemed protected by this alone, but his position in the factory for any length of time was conceivable only as a union member. The majority of the workers were organized in trade unions. These, on the whole, had fought out the wage struggles and concluded the agreements which assured the worker of a certain income. Without doubt the results of these struggles benefited all the workers in the factory, and inevitably conflicts of conscience arose, especially for the decent man, if he pocketed the wage which the unions had won him, but remained aloof from the struggle.

It was hard to speak of these problems with the average

bourgeois employer. They neither had (or perhaps wanted to have) any understanding for the material side of the question nor for the moral side. Finally, their own supposed economic interests argue from the start against any organizational grouping of the workers under them, and for this reason alone most of them can hardly form an unprejudiced judgment. Here, as so often, it is therefore necessary to turn to outsiders who do not succumb to the temptation of not seeing the forest for the trees. These, with good will, will much more easily achieve understanding for a matter which in any event is among the most important of our present and future life.

In the first volume I have expressed myself with regard to the nature and purpose, and the necessity, of trade unions. There I espoused the viewpoint that, as long as no change in the attitude of employer to worker is brought about either by state measures (which for the most part, however, are fruitless) or by a universal new education, there remains nothing for the worker to do but stand on his rights as an equal contracting party and defend his own interests in economic life. I further emphasized that safeguarding his interests in this way was entirely compatible with a whole national community if it can prevent social injustices which must subsequently bring about excessive damage to the entire community of a people. I further declared that this necessity must be considered to prevail as long as there exist among employers men who, left to themselves, not only have no feeling for social duties, but not even for the most primitive human rights; and from this I drew the inference that, once such a self-defense is regarded as necessary, its form can reasonably exist only in a grouping of workers on a trade-union basis.

And in the year 1922 nothing changed in this general conception of mine. But now it was necessary to seek a clear and definite formulation of our attitude toward these problems. It was not acceptable to content ourselves in future with mere knowledge; it was necessary to draw practical inferences from it.

We required the answer to the following questions:

1. *Are trade unions necessary?*

*2. Should the NSDAP itself engage in trade-union activity or
direct its members to such activity in any form?*

*3. What must be the nature of a National Socialist trade union?
What are our tasks and aims?*

4. How shall we arrive at such unions?

I believe that I have adequately answered the *first question*.
As things stand today, the trade unions in my opinion cannot be
dispensed with. On the contrary, they are among the most important institutions of the nation's economic life. Their significance lies not only in the social and political field, but even more
in the general field of national politics. A people whose broad
masses, through a sound trade-union movement, obtain the
satisfaction of their living requirements and at the same time an
education, will be tremendously strengthened in its power of
resistance in the struggle for existence.

Above all, the trade unions are necessary as foundation stones
of the future economic parliament or chambers of estates.

The *second question*, too, is easy to answer. If the trade-union
movement is important, it is clear that National Socialism must
take a position on it, not only from the purely theoretical, but
from the practical viewpoint as well. Yet, to be sure, the *how* of
it is harder to clarify.

The National Socialist movement, which envisions the National
Socialist folkish state as the aim of its activity, cannot doubt that
all future institutions of this state some day to be must grow out
of the movement itself. It is the greatest error to believe that
suddenly, once we have power, we can undertake a definite reorganization out of the void, unless we previously possess a certain basic stock of men who above all have been educated with
regard to loyalty. Here, too, the principle applies that more important than the outward form, which can be created mechanically and very quickly, remains the spirit which fills such a form.
For instance, it is quite possible dictatorially to graft the leader
principle on a state organism by command. But it will only be
alive if it has gradually taken shape from smallest beginnings
in a development of its own, and, by the constant selection which

life's hard reality incessantly performs, has obtained in the course of many years the leader material necessary for the execution of this principle.

And so we must not imagine that we can suddenly pull the plans for a new state form out of a briefcase into the light of day and 'introduce' them by decree from above. Such a thing can be attempted, but the result will surely be incapable of survival, in most cases a stillborn child. This reminds me of the beginning of the Weimar régime and the attempt to present the German people with not only a new régime, but a new flag which had no inner bond with the experience of our people in the last half century.

The National Socialist state must beware of such experiments. It can, when the time comes, only grow out of an organization that has long existed. This organization must possess National Socialist life innate within itself, in order to finally create a living National Socialist state.

As already emphasized, the germ cells for the economic chambers will have to reside in bodies representing the most varied occupations, hence above all in the trade unions. And if this future body representing the estates and the central economic parliament are to constitute a National Socialist institution, these important germ cells must also embody a National Socialist attitude and conception. The institutions of the movement are to be transferred to the state, but the state cannot suddenly conjure up the required institutions from the void, unless they are to remain utterly lifeless structures.

From this highest standpoint alone, the National Socialist movement must recognize the necessity of a trade-union activity of its own.

It must, furthermore, do so because a truly National Socialist education of employers as well as workers, in the sense of an integration of both into the common framework of the national community, does not come about through theoretical instruction, proclamations, or remonstrances, but through the struggle of daily life. In it and through it the movement must educate the various great economic groups and bring them closer to one another on the main issues. Without such preliminary work, all hope that a true national community will some day arise remains pure illusion. Only the great philosophical ideal for which the movement fights can slowly form that universal style which will some day make the new era seem really solidly founded within, and not just outwardly manufactured.

And so the movement must not only take an affirmative attitude toward the idea of the trade union as such, but it must by practical participation impart to the multitudes of its [1] members and supporters the necessary education for the coming National Socialist state.

The answer to the *third question* follows from what has previously been said.

The National Socialist trade union is no organ of class struggle, but an organ for representing occupational interests. The National Socialist state knows no 'classes,' but politically speaking only citizens with absolutely equal rights and accordingly equal general duties, and, alongside of these, state subjects who in the political sense are absolutely without rights.

The trade union in the National Socialist sense does not have the function of grouping certain people within a national body and thus gradually transforming them into a class, to take up the fight against other similarly organized formations. We can absolutely not impute this function to the trade union as such; it became so only in the moment when the trade union became the instrument of Marxist struggle. *Not the trade union is characterized by class struggle; Marxism has made it an instrument for the Marxist class struggle.* Marxism created the economic weapon which the international world Jew uses for shattering the economic base of the free, independent national states, for the destruction of their national industry and their national commerce, and, accordingly, the enslavement of free peoples in the service of supra-state world finance Jewry.

---

[1] This confusion of pronouns exists in the German. 'Its' refers, of course, to the trade unions.

In the face of this, the National Socialist trade union must, by organizationally embracing certain groups of participants in the national economic process, increase the security of the national economy itself and intensify its strength by the corrective elimination of all those abuses which in their ultimate consequences have a destructive effect on the national body, injure the vital force of the national community, and hence also of the state, and last but not least redound to the wrack and ruin of the economy itself.

Hence, for the National Socialist union the strike is not a means for shattering and shaking national production, but for enhancing it and making it run smoothly by combating all those abuses which, due to their unsocial character, interfere with the efficiency of the economy and hence the existence of the totality. For the efficiency of the individual always stands in a casual connection with the general legal and social position that he occupies in the economic process and with his understanding, resulting from this alone, of the necessity that this process thrive for his own advantage.

The National Socialist worker must know that the prosperity of the national economy means his own material happiness.

The National Socialist employer must know that the happiness and contentment of his workers is the premise for the existence and development of his own economic greatness.

National Socialist workers and National Socialist employers are both servants and guardians of the national community as a whole. The high degree of personal freedom that is granted them in their activity can be explained by the fact that, as experience shows, the efficiency of the individual is increased much more by far-reaching freedom than by compulsion from above, and, furthermore, it is calculated to prevent the natural process of selection, which advances the most efficient, capable, and industrious from being thwarted.

For the National Socialist union, therefore, the strike is an instrument which may and actually must be applied only so long as a National Socialist folkish state does not exist. This state, to be sure, must, in place of the mass struggle of the two great groups — employers and workers — (which in its consequences always injures the national community as a whole by diminishing production) assume the legal care and the legal protection of all. Upon the *economic chambers* themselves it will be incumbent to keep the national economy functioning and eliminate the deficiencies and errors which damage it. The things for which millions fight and struggle today must in time be settled in the *chambers of estates* and the *central economic parliament*. Then employers and workers will not rage against one another in struggles over pay and wage scales, damaging the economic existence of both, but solve these problems jointly in a higher instance, which must above all constantly envision the welfare of the people as a whole and of the state, in gleaming letters.

Here, too, as everywhere, the iron principle must prevail that first comes the fatherland and then the party.

The function of the National Socialist union is the education and preparation for this aim itself, which is: *All working together for the preservation and safeguarding of our people and our state, in accordance with the abilities and strength innate in the individual and trained by the national community.*

The *fourth question:* How do we arrive at such unions? seemed at the time by far the hardest to answer.

It is in general easier to found an institution on new soil than in an old territory that already possesses a similar institution. In a town where no store of a certain type is present, it is easy to establish such a store. It is harder when a similar enterprise already is present, and hardest of all when the conditions are such that only one alone can prosper. For here the founders face the task of not only introducing their own business, but they must, in order to exist, destroy the one that has previously existed in the town.

A National Socialist union side by side with other unions is senseless. For it, too, must feel itself permeated by its philosophical task and the resultant obligation to be intolerant of other similar, let alone hostile, formations and to emphasize the exclusive necessity of its own ego. Here, too, there is no understanding and no compromise with related efforts, but only the maintenance of our *absolute sole right*.

There were two ways of arriving at such a development:

(1) *We could found a trade union and then gradually take up the struggle against the international Marxist unions;* or we could

(2) *penetrate the Marxist unions and try to fill them with the new spirit;* in other words, transform them into instruments of the new ideology.

To the first method there were the following objections: Our financial difficulties at that time were still very considerable, the means that stood at our disposal were quite insignificant. The gradually and increasingly spreading inflation made the situation even more difficult, since in those years one could hardly have spoken of any tangible benefit to the member from the trade union. The individual worker, viewed from his own standpoint, had no ground at that time to pay dues to the union. Even the already existing Marxist unions were on the point of collapse until suddenly, through Herr Cuno's brilliant Ruhr action, the millions fell into their lap. This so-called 'national' chancellor may be designated as the savior of the Marxist unions.

At that time we could not count on such financial possibilities; and it could allure no one to enter a new union which, owing to its financial impotence, could not have offered him the least benefit. On the other hand, I must sharply oppose creating such an organization as a soft spot for more or less great minds to take refuge in.

All in all, the question of personalities played one of the most important parts. At that time I had not a single personality whom I would have held capable of solving this gigantic task. *Anyone who at that time would really have shattered the Marxist unions, and in place of this institution of destructive class struggle, helped the National Socialist trade-union idea to victory, was among the very great men of our people, and his bust would some day have had to be dedicated to posterity in the Valhalla at Regensburg.*

But I did not know of any head that would have fitted such a pedestal.

It is absolutely wrong to be diverted from this view by the fact that the international trade unions themselves have only average minds at their disposal. This in reality means nothing at all; for at the time when they were founded, there was nothing else. Today the National Socialist movement must combat a colossal gigantic organization which has long been in existence, and which is developed down to the slightest detail. The conqueror must always be more astute than the defender if he wants to subdue him. The Marxist trade-union fortress can today be administered by ordinary bosses; but it will only be stormed by the wild energy and shining ability of an outstanding great man on the other side. If such a man is not found, it is useless to argue with Fate and even more useless to attempt forcing the matter with inadequate substitutes.

Here we must apply the maxim that in life it is sometimes better to let a thing lie for the present than to begin it badly or by halves for want of suitable forces.

There was also another consideration which should not be designated as demagogic. I had at that time and still possess today the unshakable conviction that it is dangerous to tie up a great politico-philosophical struggle with economic matters at too early a time. This is particularly true with our German people. For here, in such a case, the economic struggle will at once withdraw the energy from the political struggle. Once people have won the conviction that by thrift they can acquire a little house, they will dedicate themselves only to this task and will have no more time to spare for the political struggle against those who are planning to take away their saved-up pennies some day in one way or another. Instead of fighting in the political struggle for the insight and conviction they have won, they give themselves up entirely to their idea of 'settlement,' and in the end as a rule find themselves holding the bag.

The National Socialist movement today stands at the beginning of its struggle. In large part it has still to form and complete its philosophical picture. It must fight with all the fiber of its energy for the accomplishment of its great ideas, and success is thinkable only if all its strength goes completely into the service of this fight.

To what an extent concern with purely economic problems can paralyze active fighting strength, we can see at this very moment in a classical example:

The revolution of November, 1918, was not made by trade unions, but was accomplished against them. And the German bourgeoisie is carrying on no political struggle for the German future because it believes this future to be sufficiently guaranteed by the constructive work in the economic sphere.

We should learn from such experiences; for it would be no different with us. The more we muster the entire strength of our movement for the political struggle, the sooner may we count on success all along the line; but the more we *prematurely* burden ourselves with trade-union, settlement, and similar problems, the smaller will be the benefit for our cause taken as a whole. For important as these matters may be, their fulfillment will only occur on a large scale, when we are in a position to put the state power into the service of these ideas. Until then, these problems would paralyze the movement all the more, the sooner it concerned itself with them and the more its *philosophical will* was limited by them. *Then it might easily come about that trade-union motives would guide the movement instead of the philosophy forcing the trade union into its channels.*

Real benefit for the movement as well as our people can only arise from a trade-union movement, if philosophically this movement is already so strongly filled with our National Socialist ideas that it no longer runs the risk of falling into Marxist tracks. For a trade-union movement which sees its mission only in competition with the Marxist unions would be worse than none at all. It must declare war on the Marxist union, not only as an *organization*, but above all as an *idea*. In the Marxist union it must strike down the herald of the class struggle and the class idea and in its stead must become the protector of the occupational interests of German citizens.

All these criteria then argued and still argue *against* the foundation of *our own* trade unions, unless suddenly a *man* should appear who is obviously chosen by Fate for the solution of this very question.

And so there were only two other possibilities: either to recommend that our own party comrades leave the unions, or that they remain in them and work as destructively as possible.

In general I recommended this latter way.

Especially in the year 1922–23 this could be done without difficulty; for the financial benefit which during the inflation period accrued to the trade union from our members in their ranks, who due to the youth of our movement were not yet very numerous, was practically nil. But the damage to it was very great, for the National Socialist supporters were its sharpest critics and thus its inner disrupters.

At that time I totally rejected all experiments which contained the seeds of failure to begin with. I would have viewed it as a crime to take so and so much of a worker's meager earnings for an institution of whose benefit to its members I was at heart not convinced.

If one fine day a new political party disappears, it is scarcely ever a loss but almost always a benefit, and no one has any right to moan about it; for what the individual gives to a political movement, he gives *à fonds perdu*. But anyone who pays money into a union has a right to the fulfillment of the promised return services. If this is not taken into account, the leaders of such a union are swindlers, or at least frivolous characters who must be called to account.

And in 1922 we acted according to this view. Others thought they knew better and founded trade unions. They attacked our lack of unions as the most visible sign of our mistaken and limited views. But it was not long before these organizations themselves vanished, so that the final result was the same as with us. Only with the one difference, that we had deceived neither ourselves nor others.

German Alliance Policy After the War

T HE HEEDLESSNESS of the leaders of the
Reich's foreign policy when it came to establishing basic principles for an expedient alliance policy was not only continued
after the revolution but was even exceeded. For if before the
War general confusion of political concepts could be regarded as
the cause of our faulty leadership in foreign policy, after the
War it was a lack of honorable intentions. It was natural that
the circles who saw their destructive aims finally achieved by
the revolution could possess no interest in an alliance policy
whose final result would inevitably be the re-establishment of a
free German state. Not only that such a development would
have run counter to the inner sense of the November crime, not
only that it would have interrupted or actually ended the internationalization of the German economy: but also the domestic
political effects resulting from a victorious fight for freedom in
the field of foreign policy would in the future have meant doom
for the present holders of power in the Reich. For the resurrection of a nation is not conceivable without its preceding nationalization, as, conversely, every great success in the sphere of
foreign affairs inevitably produces reactions in the same direction. Every fight for freedom, as experience shows, leads to an
intensification of national sentiment, of self-reliance, and hence
also to a sharper sensibility toward anti-national elements and

tendencies. Conditions and persons who are tolerated in peaceable times, who often, in fact, pass unnoticed, are not only rebuffed in times of seething national enthusiasm, but encounter a resistance that is not seldom fatal to them. Just recall, for example, the general fear of spies which at the outbreak of wars suddenly bursts forth in the fever heat of human passions and leads to the most brutal, sometimes even unjust persecutions, though everyone might tell himself that the danger of spies will be greater in the long years of a peaceful period, even though, for obvious reasons, it does not receive general attention to the same extent.

For this reason alone the subtle instinct of the government parasites washed to the surface by the November events senses that an uprising of our people for freedom, supported by an intelligent alliance policy and the resultant outburst of national passions, would mean the possible end of their own criminal existence.

Thus, it becomes understandable why the government authorities in power since 1918 have failed us in the field of foreign affairs and why the leaders of the state have almost always worked systematically against the real interests of the German nation. For what at first sight might appear planless is revealed on closer examination as merely the logical continuation of the road which the November revolution for the first time openly trod.

Here, to be sure, we must distinguish between the responsible or rather 'should-be-responsible' leaders of our state affairs, the average parliamentary politicasters, and the great stupid sheep's herd of patient lamblike people.

The first know what they want. The others play along, either because they know it or are too cowardly to ruthlessly oppose what they have recognized and felt to be harmful. And the others submit from incomprehension and stupidity.

As long as the National Socialist German Workers' Party possessed only the scope of a small and little known club, problems of foreign policy could possess only a subordinate importance in

the eyes of many adherents. This especially because our movement in particular has always upheld and must always uphold
the conception that external freedom comes neither as a gift
from heaven nor from earthly powers, but can only be the fruit
of development of inner strength. *Only the elimination of the
causes of our collapse, as well as the destruction of its beneficiaries,
can create the premise for our outward fight for freedom.*

And so it is understandable if, due to such considerations, in
the first period of the young movement the value of questions of
foreign policy was set below the importance of its domestic reform
plans.

But once the limits of the small, insignificant club were broadened and finally broken, and the young formation obtained the
importance of a big organization, the necessity arose of taking a
position on the questions pertaining to the developments in
foreign affairs. It became necessary to lay down guiding principles which would not only not contradict the fundamental views
of our world concept, but actually represent an emanation of
this line of thought.

Precisely from our people's lack of schooling in foreign affairs,
there results for the young movement an obligation to transmit
to the individual leaders as well as the great masses through broad
guiding principles a line of thought in matters of foreign policy,
which is the premise for any practical execution in the future
of the preparations in the field of foreign policy for the work of
recovering the freedom of our people as well as a real sovereignty
of the Reich.

The essential fundamental and guiding principle, which we
must always bear in mind in judging this question, is that foreign
policy is only a means to an end, and that the end is solely the
promotion of our own nationality. No consideration of foreign
policy can proceed from any other criterion than this: *Does it
benefit our nationality now or in the future, or will it be injurious
to it?*

This is the sole preconceived opinion [1] permissible in dealing

[1] *'die einzige vorgefasste Meinung...'*

with this question. Partisan, religious, humanitarian, and all other criteria in general, are completely irrelevant.

*   *   *

If before the War the task of a German foreign policy was to safeguard the sustenance of our people and its children on this globe by the preparation of the roads that can lead to this goal, as well as the winning of the necessary helpers in the form of expedient allies, today it is the same, with the single difference: *Before the War, it was a question of helping to preserve the German nationality, taking into account the existing strength of the independent power state, today it is necessary first to restore to the nation its strength in the form of a free power state, which is the premise for the subsequent implementation of a practical foreign policy which will preserve, promote, and sustain our people for the future.*

In other words: **The aim of a German foreign policy of today must be the preparation for the reconquest of freedom for tomorrow.**

And here a fundamental principle must always be kept in mind: *The possibility of regaining independence for a nationality is not absolutely bound up with the integrity of a state territory, but rather with the existence of a remnant, even though small, of this people and state, which, in possession of the necessary freedom, not only can embody the spiritual community of the whole nationality, but also can prepare the military fight for freedom.*

When a nation of a hundred million people, in order to preserve its state integrity, suffers the yoke of slavery in common, it is worse than if such a state and such a people had been shattered and only a part of them remained in possession of full freedom. On condition, to be sure, that this last remnant were filled with the holy mission of not only proclaiming its spiritual and cultural inseparability, but also of accomplishing the military preparation for the final liberation and reunion of the unfortunate oppressed portions.

It should further be borne in mind that the question of regaining lost sections of a people's and state's territory is always primarily a question of regaining the political power and independence of the mother country; that, therefore, in such a case the interests of lost territories must be ruthlessly subordinated to the interest of regaining the freedom of the main territory. For the liberation of oppressed, separated splinters of a nationality or of provinces of a country does not take place on the basis of a desire on the part of the oppressed people or of a protest on the part of those left behind, but through the implements of power of those remnants of the former common fatherland that are still more or less sovereign.

Therefore, the presupposition for the gaining of lost territories is the intensive promotion and strengthening of the remaining remnant state and the unshakable decision slumbering in the heart to dedicate the new force thus arising to the freedom and unification of the entire nationality in the proper hour: therefore, *subordination* of the interests of the separated territories to the single interest of winning for the remaining remnant that measure of political power and strength which is the precondition for a correction of the will of hostile victors. *For oppressed territories are led back to the bosom of a common Reich, not by flaming protests, but by a mighty sword.*

To forge this sword is the task of a country's internal political leadership; to safeguard the work of forging and seek comrades in arms is the function of diplomatic leadership.

*   *   *

In the first volume of this work I have discussed the half-heartedness of our alliance policy before the War. Of the four roads to a future preservation of our nationality and its sustenance, the fourth and least favorable was chosen. In place of a healthy European land policy, a colonial and commercial policy was chosen. This was all the more fallacious as it was thought that an armed settlement could in this way be avoided. The result of this attempt to sit on several chairs was the proverbial fall between them, and the World War was only the last reckoning submitted the Reich for its faulty conduct of foreign affairs.

The correct road would even then have been the third: *a strengthening of our continental power by gaining new soil in Europe*, and precisely this seemed to place a completion by later acquisitions of colonial territory within the realm of the naturally possible. This policy, to be sure, could only have been carried out in alliance with England or with so abnormal an emphasis on the military implements of power that for forty or fifty years cultural tasks would have been forced into the background. This would have been quite justifiable. The cultural importance of a nation is almost always bound up with its political freedom and independence; therefore, the latter is the presupposition for the existence, or, better, the establishment, of the former. Therefore, no sacrifice can be too great for the securing of political freedom. What general cultural matters lose through an excessive promotion of the state's implements of military power, it will later be possible to restore most abundantly. Yes, it may be said that, after such a concentrated exertion in the sole direction of preserving state independence, a certain relaxation or compensation customarily ensues in the form of a really amazing golden age of the hitherto neglected cultural forces of a nation. From the hardships of the Persian Wars arose the Age of Pericles, and through the cares of the Punic Wars the Roman state began to dedicate itself to the service of a higher culture.

To be sure, such a complete subordination of all a nation's other interests to the sole task of preparing a coming contest of arms for the future security of the state cannot be entrusted to the decision of a majority of parliamentary idiots or good-for-nothings. The father of a Frederick the Great was able to prepare for a contest of arms, disregarding all other concerns, but the fathers of our democratic parliamentary nonsense of the Jewish variety cannot do so.

For this very reason the military preparation for an acquisition of land and soil in Europe could be only a limited one, and the support of suitable allies could hardly be dispensed with.

Since, however, our leaders wanted to know nothing of a systematic preparation for war, they renounced the acquisition of land in Europe and, by turning instead to a colonial and commercial policy, sacrificed the alliance with England which would otherwise have been possible, but did not, as would have been logical, seek the support of Russia, and finally, forsaken by all except the Habsburg hereditary evil, stumbled into the World War.

*   *   *

In characterizing our present foreign policy, it must be said that there exists no visible or even intelligible line. Before the War the fourth road was erroneously taken, and that pursued only by halves, while since the revolution no road at all has been discernible, even to the sharpest eye. Even more than before the War, any systematic thought is lacking, except perhaps an attempt to smash the last possibility of a resurrection of our people.

A cool appraisal of the present European relations of power leads to the following conclusion:

For three hundred years the history of our continent has been basically determined by the attempt of England to obtain the necessary protection in the rear for great British aims in world politics, indirectly through balanced, mutually interlocking relations of power.

The traditional tendency of British diplomacy, which in Germany can only be compared with the tradition of the Prussian army, was, since the efforts of Queen Elizabeth, directed solely toward preventing by all possible means the rise of any European great power above its place in the general hierarchy, and, if possible, to break it by military intervention. The instruments of power which England was accustomed to apply in this case varied according to the existing or presented task; [1] but the determination and will power for using them were always the same. Indeed, the more difficult England's situation became in the course of time, the more necessary it seemed to the leaders of the British Empire to keep the individual state powers of Europe in a state of general paralysis resulting from mutual rivalries. The political separation of the former North American colonial territory led, in the ensuing period, to the greatest exertions to keep the European rear absolutely covered. And so — after the destruction of Spain and the Netherlands as great sea powers — the strength of the English was concentrated against aspiring France until finally, with the fall of Napoleon, the danger to England of this most dangerous military power's hegemony could be regarded as broken.

The shift of British policy against Germany was undertaken only slowly, not only because, due to the lack of a national unification of the German nation, a visible danger for England did not exist, but also because public opinion, prepared by propaganda for a particular political goal, is slow in following new aims. The sober knowledge of the statesman seems transposed into emotional values which are not only more fruitful in their momentary efficacy, but also more stable with regard to duration. Therefore the statesman, after achieving one purpose, can without further ado turn his thought processes toward new goals, but it will be possible to transform the masses emotionally into an instrument of their leader's new view [2] only by slow propagandist efforts.

As early as 1870–71, England had meanwhile formulated her

---

[1] '*nach der vorhandenen oder gestellten Aufgabe.*' Second edition reads: '*nach der vorhandenen Lage oder*, etc.' (according to the existing situation or the task presented).

[2] '*zum Instrument der neuen Ansicht ihres Leiters.*' Second edition puts it more modestly. '*Leiters*' has been changed to '*Lebens*': the new view of their life. This doesn't quite make sense, but like many of the other corrections was apparently made with a view to avoid resetting the paragraph. As it happens, the same plates were not used in the second edition, but perhaps this was not the original intention.

new position. Fluctuations which occurred at times, due to the importance of America in world economy as well as Russia's development as a political power, were unfortunately not utilized by Germany, so that a steady intensification of the original tendency in British statesmanship was bound to result.

England saw in Germany the power whose importance in commercial and hence in world politics, not least as a result of her enormous industrialization, was increasing to such a menacing extent that the strength of the two states in identical fields could already be balanced. The *'peaceful, economic'* conquest of the world which to the helmsmen of our state seemed the highest emanation of the ultimate wisdom, became for the English politicians the ground for the organization of resistance against us. That this resistance assumed the form of a comprehensively organized attack was fully in keeping with the essence of a diplomacy whose aims did not lie in the *preservation* of a questionable *world peace*, but in the reinforcement of British *world domination*. That England used as allies all states which were in any way possible in the military sense was equally in keeping with her traditional caution in the estimation of the adversary's strength as with the appreciation of her own momentary weakness. This can, therefore, not be characterized as 'unscrupulousness,' because such a comprehensive organization of a war is to be judged by criteria, not of heroism, but of expediency. *Diplomacy must see to it that a people does not heroically perish, but is practically preserved. Every road that leads to this is then expedient, and not taking it must be characterized as criminal neglect of duty.*

With the revolutionization of Germany, the British concern over a threatening Germanic world hegemony found an end, to the relief of British statesmen.

Since then England has had no further interest in the complete effacement of Germany from the map of Europe. On the contrary, the terrible collapse which occurred in the November days of 1918 placed British diplomacy in a new situation which at first was not even considered possible.

For four and a half years the British world empire had fought in order to break the supposed preponderance of a continental power. Now suddenly a crash occurred which seemed to remove this power entirely from the picture. There was manifested such an absence of even the most primitive instinct of self-preservation that the European balance seemed thrown off its hinges by an action of scarcely forty-eight hours: *Germany destroyed and France the first continental power of Europe.*

The enormous propaganda which had made the British people persevere and hold out in this war, which recklessly incited them and stirred up all their deepest instincts and passions, now inevitably weighed like lead on the decisions of British statesmen. With the colonial, economic, and commercial destruction of Germany, the British war aim was achieved; anything beyond this was a curtailment of English interests. Through wiping out a German power state in continental Europe, only the enemies of England could gain. Nevertheless, in the November days of 1918 and up to midsummer of 1919, a reorientation of English diplomacy, which in this long war more than ever before had used up the emotional powers of the great masses, was no longer possible. It was not possible from the viewpoint of the existing attitude of their own people, and was not possible in view of the disposition of the military relation of forces. France had seized upon the law of action [1] and could dictate to the others. The single power, however, which in these months of haggling and bargaining might have brought about a change, Germany herself, lay in the convulsions of inner civil war and only kept on proclaiming, through the mouth of her so-called statesmen, her readiness to accept any dictate whatsoever.

Now, if in the life of peoples, a nation, in consequence of its total lack of an instinct of self-preservation, ceases to be a possible 'active' ally, she customarily sinks to the level of a slave nation and her land succumbs to the fate of a colony.

Precisely to prevent France's power from becoming excessive, a participation of England in her predatory lusts was the sole possible form for England herself.

‘‘*das Gesetz des Handelns an sich gerissen.*’

Actually England did not achieve her war aim. The rise of a European power above the relations of forces of the continental state system of Europe was not only not prevented but was given increased support.

In 1914, Germany as a military state was wedged in between two countries one of which disposed of an equal power and the other of a superior power. On top of this came the superior sea power of England; France and Russia alone offered obstacles and resistance to every disproportionate development of German greatness. The extremely unfavorable situation of the Reich from the viewpoint of military geography could be considered a further coefficient of security against an excessive increase in the power of this country. The coastline especially was unfavorable from the military standpoint for a fight with England; it was short and cramped, and the land front, on the other hand, disproportionately long and open.

The situation of France today is different: the first military power, without a serious rival on the continent; on her southern borders, as good as guaranteed against Spain and Italy; secured against Germany by the feebleness of the fatherland; her coastline on a long front poised directly opposite the vital nerves of the British Empire. Not only for airplanes and long-distance batteries do the English vital centers constitute worth-while targets, but also her trade lanes are exposed to the effects of submarine warfare. A submarine campaign, based on the long Atlantic coast and the equally long stretches of the French border territories of the Mediterranean in Europe and North Africa, would be devastating in effect.

Thus, the fruit of the struggle against the development of Germany's power was politically to bring about French hegemony on the continent. The military result: the reinforcement of France as the first prime power on land and the recognition of the Union as an equal sea power. Economically: the surrender of immense spheres of British interest to former allies.

Just as England's traditional political aims desire and necessitate a certain Balkanization of Europe, those of France necessitate a Balkanization of Germany.

*England's desire is and remains the prevention of the rise of a
continental power to world-political importance; that is, the maintenance of a certain balance of power between the European states;
for this seems the presupposition of a British world hegemony.*

*France's desire is and remains to prevent the formation of a
unified power in Germany, the maintenance of a system of German
petty states with balanced power relations and without unified
leadership, and occupation of the left bank of the Rhine as the presupposition for creating and safeguarding her position of hegemony
in Europe.*

*The ultimate aim of French diplomacy will always stand in con-
flict with the ultimate tendency of British statesmanship.*

*   *   *

Anyone who undertakes an examination of the present *alliance
possibilities* for Germany from the above standpoint must arrive
at the conclusion that the last practicable tie remains with
England. Terrible as the consequences of the English war policy
were and are for Germany, we must not close our eyes to the
fact that a necessary interest on the part of England in the
annihilation of Germany no longer exists *today*; that, on the
contrary, England's policy from year to year must be directed
more and more to an obstruction of France's unlimited drive for
hegemony. An alliance policy is not conducted from the standpoint of retrospective grudges, but is fructified by the knowledge
of retrospective experience. And experience should have taught
us that alliances for the achievement of *negative* aims languish
from inner weakness. *National destinies are firmly forged together
only by the prospect of a common success in the sense of common
gains, conquests; in short, of a mutual extension of power.*

How feebly our people think in terms of foreign policy can be
seen most clearly from the current press reports with regard to
the greater or lesser *'friendliness to Germany'* of this or that
foreign statesman; such reports see a special guaranty of a

benevolent policy toward our nationality in this supposed attitude on the part of such personalities. This is an utterly incredible absurdity, a speculation on the unparalleled simplicity of the average German shopkeeper dabbling in politics. No English or American or Italian statesman was ever '*pro-German.*' As a statesman, every Englishman will naturally be even more of an *Englishman*, every American an *American*, and no Italian will be found ready to pursue any other policy than a *pro-Italian* one. Therefore, anyone who thinks he can base alliances with foreign nations on a *pro-German* orientation of their leading statesmen is either an ass or a hypocrite. The premise for the linking of national destinies is never based on mutual *respect*, let alone *affection*, but on the prospect of *expediency* for both contracting parties. In other words: true as it is that an English statesman will always pursue a pro-English policy and never a pro-German one, certain definite interests of this *pro-English* policy may for the most varying reasons coincide with *pro-German* interests. This, of course, need only be the case up to a certain degree and can some day shift to the exact opposite; *but the skill of a leading statesman is manifested precisely in always finding at specified periods those partners for the achievement of their own needs, who must go the same road in pursuit of their own interests.*

The practical moral of all this for the present can result only from the answer to the following questions: *What states at the present time have no vital interest in having the French economic and military power achieve a position of dominant hegemony in Europe by the total exclusion of a German Central Europe? Yes, which states on the basis of their own requirements for existence and their previous political tradition see a threat to their own future in such a development?*

For on this point we must at length achieve full clarity: The inexorable mortal enemy of the German people is and remains France. It matters not at all who ruled or will rule in France, whether Bourbons or Jacobins, Bonapartists or bourgeois democrats, Clerical republicans or Red Bolshevists: the final goal of their activity in foreign affairs will always be an attempt to seize possession of the Rhine border and to secure this watercourse for France by means of a dismembered and shattered Germany.

England desires no Germany as a world power, but France wishes no power at all called Germany: quite an essential difference, after all! Today we are not fighting for a position as a world power; today we must struggle for the existence of our fatherland, for the unity of our nation and the daily bread of our children. If we look about us for European allies from this standpoint, there remain only two states: *England and Italy.*

England does not want a France whose military fist, unobstructed by the rest of Europe, can undertake a policy which, one way or another, must one day cross English interests. England can never desire a France which, in possession of the immense Western European iron and coal deposits, obtains the foundations of a menacing economic world position. And England, furthermore, cannot desire a France whose continental political situation, thanks to the shattering of the rest of Europe, seems so assured that the resumption of a French world policy along broader lines is not only made possible but positively forced. The Zeppelin bombs of former times might multiply a thousandfold every night; the military preponderance of France presses heavy on the heart of Great Britain's world empire.

And Italy, too, cannot and will not desire a further reinforcement of the French position of superior power in Europe. Italy's future will always be conditioned by a development which is geographically grouped around the Mediterranean basin.[1] What drove Italy into the war was really not the desire to aggrandize France, but the desire to give the hated Adriatic rival the death blow. Any further continental strengthening of France, however, is an obstacle to Italy in the future, and we must not delude ourselves that relations of parentage among nations can in any way exclude rivalries.

On soberest and coldest reflection, it is today primarily these two states, *England* and *Italy*, whose most natural selfish interests

---

[1] *'eine Entwicklung die sich gebietsmässig um das Mittellandische Meerbecken gruppiert.'*

are not, in the most essential points at least, opposed to the German nation's requirements for existence, and are, indeed, to a certain extent, identified with them.

* * *

We must, to be sure, in judging such a possibility of alliance, not overlook three factors. The first depends on us, the two others on the states in question.

Can any nation ally itself with the present-day Germany? Can a power which seeks in an alliance an aid for carrying out *offensive* aims of its own, ally itself with a state whose leaders for years have offered a picture of the most wretched incompetence, of pacifistic cowardice, and the greater part of whose population, in democratic-Marxist blindness, betray the interests of their own nation and country in a way that cries to high Heaven? Can any power hope today to create a valuable relation with a state, in the hope of some day fighting in common for common interests, when this country obviously possesses neither the courage nor the desire to stir so much as a finger in defense of its own bare existence? Will any power, for which an alliance is and should be more than a treaty for the guaranty and maintenance of a state of slow putrefaction like the old Triple Alliance, obligate itself for weal or woe to a state whose characteristic way of life consists only in cringing submissiveness without and disgraceful oppression of national virtues within; with a state that no longer possesses any greatness, since on the basis of its whole behavior it no longer deserves it; with governments which can boast of no respect whatsoever on the part of their citizens, so that foreign countries cannot possibly harbor any greater admiration for them?

No, a power which itself wants to be respected and which hopes to gain more from alliances than fees for hungry parliamentarians will not ally itself with present-day Germany; indeed, it cannot. *And in our present unfitness for alliance lies the*

deepest and ultimate ground for the solidarity of the enemy bandits.
Since Germany never defends herself, except by a few flaming
protests on the part of our parliamentary élite, and the rest of
the world has no reason for fighting in our defense, and as a
matter of principle God does not make cowardly nations free —
notwithstanding the whimpering of our patriotic leagues to that
effect — there remains nothing else even for the states which
possess no *direct* interest in our total annihilation but to take
part in France's campaigns of pillage, if only, by such cooperation
and participation in the pillage, at least to prevent the exclusive
strengthening of France alone.

Secondly, we must not overlook the difficulty in undertaking
a reorientation of the great popular masses of the countries previously hostile to us, who have been influenced in a certain direction by mass propaganda. For it is not possible to represent a
nationality as 'Huns,' 'robbers,' 'Vandals,' etc., over a period
of years, only to discover the opposite *suddenly overnight*, and
recommend the former enemy as the ally of tomorrow.

Yet even more attention must be given to a third fact which
will be of essential importance for the shaping of the coming
European alliance relations:

Little interest as England, from a British state viewpoint,
may have in a further annihilation of Germany, that of the international stock exchange Jews in such a development is great.
The cleavage between the official, or, better expressed, the traditional, British statesmanship and the controlling Jewish stock
exchange powers is nowhere better shown than in their different
position on the questions of British foreign policy. *Jewish finance
in opposition to the interests of the British state welfare desires not
only the complete economic annihilation of Germany, but also her
complete political enslavement.* The internationalization of our
German economy — that is, the appropriation of the German
labor power by Jewish world finance — can be completely carried out only in a politically Bolshevist state. But if the Marxist
shock troops of international Jewish stock exchange capital are
to break the back of the German national state for good and all,

this can only be done with friendly aid from outside. The armies of France must, therefore, besiege the German state structure until the Reich, inwardly exhausted, succumbs to the Bolshevistic shock troop of international Jewish world finance.

And so the Jew today is the great agitator for the complete destruction of Germany. Wherever in the world we read of attacks against Germany, Jews are their fabricators, just as in peacetime and during the War the press of the Jewish stock exchange and Marxists systematically stirred up hatred against Germany until state after state abandoned neutrality and, renouncing the true interests of the peoples, entered the service of the World War coalition.

The Jewish train of thought in all this is clear. The Bolshevization of Germany — that is, the extermination of the national folkish Jewish intelligentsia to make possible the sweating of the German working class under the yoke of Jewish world finance — is conceived only as a preliminary to the further extension of this Jewish tendency of world conquest. As often in history, Germany is the great pivot in the mighty struggle. If our people and our state become the victim of these bloodthirsty and avaricious Jewish tyrants of nations, the whole earth will sink into the snares of this octopus; if Germany frees herself from this embrace, this greatest of dangers to nations may be regarded as broken for the whole world.

Therefore, as surely as the Jews will bring their entire agitational efforts to bear, not only to maintain the hostility of the nations to Germany, but if possible to increase it even more, just as surely only a fraction of this activity coincides with the real interests of the peoples poisoned by it. *In general, the Jews will always fight within the various national bodies with those weapons which on the basis of the recognized mentality of these nations seem most effective and promise the greatest success.* In our national body, so torn with regard to blood, it is therefore the more or less 'cosmopolitan,' pacifistic-ideological ideas, arising from this fact; in short, the international tendencies which they utilize in their struggle for power: in France they work with the well-known and correctly estimated chauvinism; in

England with economic and world-political considerations; in short, they always utilize the most essential qualities that characterize the mentality of a people. Only when in such a way they have achieved a certain profusion of economic and political influence and predominance do they strip off the fetters of these borrowed weapons, and display in exactly the same measure the true inner purposes of their will and their struggle. They now begin to destroy with ever-greater rapidity, until they have turned one state after another into a heap of rubble on which they can then establish the sovereignty of the eternal Jewish empire.

In England as well as Italy the cleavage between the views of the better indigenous statesmanship and the will of the world stock exchange Jews is clear; sometimes, indeed, it is crassly obvious.

Only in France does there exist today more than ever an inner *unanimity* between the intentions of the Jew-controlled stock exchange and the desire of the *chauvinist-minded national statesmen.* But in this very identity there lies an immense danger for Germany. For this very reason, France is and remains by far the most terrible enemy. *This people, which is basically becoming more and more negrified, constitutes in its tie with the aims of Jewish world domination an enduring danger for the existence of the white race in Europe.* For the contamination by Negro blood on the Rhine in the heart of Europe is just as much in keeping with the perverted sadistic thirst for vengeance of this hereditary enemy of our people as is the ice-cold calculation of the Jew thus to begin bastardizing the European continent at its core and to deprive the white race of the foundations for a sovereign existence through infection with lower humanity.

What France, spurred on by her own thirst for vengeance and systematically led by the Jew, is doing in Europe today is a sin against the existence of white humanity and some day will incite against this people all the avenging spirits of a race which has recognized racial pollution as the original sin of humanity.

For Germany, however, the French menace constitutes an obligation to subordinate all considerations of sentiment and hold out a hand to those who, threatened as much as we are, will neither suffer nor tolerate France's desires for domination.

In the predictable future there can be only two allies for Germany in Europe: England and Italy.

*   *   *

Anyone who takes the trouble to glance back and follow Germany's leadership in foreign policy since the revolution will, in view of the constant and incomprehensible failure of our governments, be unable to do otherwise than take his head in his hands, and either simply despair or, in flaming indignation, declare war on such a régime. These actions no longer have anything in common with lack of understanding: for what would have seemed unthinkable to any thinking brain [1] has been done by these intellectual Cyclopses of our November parties: *they have courted France's favor.* Yes, indeed, in all these years, with the touching simplicity of incorrigible dreamers, they have tried again and again to make friends with France; over and over again they have bowed and scraped before the 'great nation'; in every shrewd trick of the French hangman they have felt justified in seeing the first sign of a visible change of attitude. *Our real political wirepullers, of course, never harbored this insane belief. For them currying favor with France was only the obvious means of sabotaging every practical alliance policy.* They were never in doubt as to the aims of France and her men behind the scenes. What compelled them to act as if they nevertheless honestly believed in the possibility of a change in the fate of Germany was the sober realization that otherwise our people would take things into their own hands.

Even for us, of course, it is hard to represent England as a possible future ally in the ranks of our own movement. Again and again our Jewish press has known how to concentrate special hatred on England, and many a good German simpleton has

---

[1] *'...was jedem denkenden Gehirn eben als undenkbar erschienen wäre...'*

fallen into the Jewish snare with the greatest willingness, drooled
about 'strengthening' German sea power, protested against the
rape of our colonies, recommended their reconquest, and thus
helped furnish the material which the Jewish scoundrel could
pass on to his fellow Jews in England for practical propagandist
use. For it should gradually dawn even on our political bourgeois
simpletons that what we have to fight for today is not '*sea power*,'
etc. The orientation of the German national strength toward
this aim, without the most thoroughgoing previous securing of
our position in Europe, was an absurdity even before the War.
Today such a hope must be counted among those *stupidities*
which in the field of politics are characterized as *crimes*.

Sometimes it was really maddening to be compelled to look on
as the Jewish wirepullers succeeded in occupying our people with
things that are today of the most secondary nature, inciting them
to demonstrations and protests, while at the same time France
was tearing piece after piece out of the flesh of our national body,
and the foundations of our independence were systematically
taken away from us.

Here I must recall a special hobby which in these years the Jew
rode with amazing adroitness: the **South Tyrol**.

Yes, the *South Tyrol*. If I here concern myself with this particular question, it is not least to settle accounts with that hypocritical rabble which, counting on the forgetfulness and stupidity of our broad strata, has the insolence to mimic on this
point a national indignation, which is more alien especially to the
parliamentary swindlers than honest conceptions of property to
a magpie.

I would like to emphasize that I personally am among the men
who, when the fate of the South Tyrol was being decided — that
is, beginning in August, 1914, up to November, 1918 — went
where this territory was being actively defended — I mean the
army. In those years I did my part of the fighting, not in order
that the South Tyrol should be lost, but in order that it should be
preserved for the fatherland just like every other German
province.

The ones who did not do their bit at that time were the parliamentary sneak-thieves, all the politics-playing party rabble. On the contrary, while we fought in the conviction that only a victorious issue to the War would preserve this South Tyrol for the German nationality, the big-mouths of these Ephialteses agitated and plotted against victory until at last the battling Siegfried succumbed to the treacherous dagger thrust. *For the preservation of the South Tyrol in German possession was naturally not guaranteed by the lying inflammatory speeches of parliamentary sharpers on the Vienna Rathausplatz or in front of the Munich Feldherrnhalle, but only by the battalions at the fighting front. Those who broke this front betrayed the South Tyrol, just as they betrayed all other German territories.*

And anyone who believes today that he can solve the South Tyrol question by *protests, declarations, clubby parades*, is either a very special scoundrel or a German petit bourgeois.

We must clearly recognize the fact that the recovery of the lost territories is not won through solemn appeals to the Lord or through pious hopes in a League of Nations, but only by **force of arms.**

And so the only question is, Who is ready to attempt the reconquest of these lost territories by defiant armed force?

As far as my person is concerned, I can here assure you with a clear conscience that I could still muster up enough courage to take part in the victorious conquest of the South Tyrol at the head of a parliamentary storm battalion that ought to be formed, consisting of parliamentary big-mouths and other party leaders plus various privy councilors. God knows it would give me pleasure if suddenly a few shrapnel would burst over the heads of such a 'flaming' protest demonstration. I think if a fox were to break into a chicken-coop the cackling could hardly be worse, or the rush of the feathered fowl for safety any quicker, than the flight of such a splendid 'protest rally.'

But the vile thing about the whole business is that the gentlemen themselves do not believe they can achieve anything in this way. They personally know, better than anyone else, the impossibility and innocuousness of all the fuss they are making.

But they carry on as they do, because it is naturally somewhat easier to shoot off their mouths for the recovery of the South Tyrol today than it once was to fight for *keeping* it. Everyone does his own part; then we sacrificed our blood, and today this company sharpen their beaks.

It is especially delightful, moreover, to see how Viennese legitimist circles literally bristle with their present activity for regaining the South Tyrol. Seven years ago, to be sure, their noble and exalted ruling house helped by a scoundrelly deed of treacherous perjury to make it possible for the victorious world coalition to win among other things the South Tyrol. At that time these circles supported the policy of their treacherous dynasty, and didn't care a damn about the South Tyrol or anything else. Today, of course, it is easier to take up the struggle for these territories, for today, after all, it is fought only with 'spiritual weapons,' and it is always easier to talk your throat hoarse in some 'protest meeting' — from noble, heartfelt indignation — and wear your fingers to the bone writing a newspaper article than, say, to blow up bridges during the occupation of the Ruhr.

The reason why in the last few years certain definite circles have made the 'South Tyrol' question the pivotal point of German-Italian relations is obvious. *Jews and Habsburg legitimists have the greatest interest in preventing a German alliance policy which might lead some day to the resurrection of a free German fatherland. All this fuss today is not made for love of the South Tyrol — which it does not help but only harms — but for fear of a possible German-Italian understanding.*

It is quite in keeping with the general hypocrisy and slanderous tendencies of these circles when they attempt with cold and brazen gall to make things look as if *we* had 'betrayed' the South Tyrol.

To these gentlemen let it be said with all plainness: the South Tyrol was 'betrayed' first and foremost by every German with sound limbs who in the years 1914–1918 did not stand somewhere at the front, putting his services at the disposal of the fatherland;

secondly, by every man who in those years did not help to strengthen our national body's power of resistance for the pursuit of the War and to fortify the endurance of our people for carrying through this fight to the end;

thirdly, the South Tyrol was betrayed by every man who cooperated in the outbreak of the November revolution — whether directly by deed or indirectly by the cowardly toleration of the deed — and thereby smashed the weapon which alone could have saved the South Tyrol.

Yes, my brave *lip-service protesters,* that is how things stand!

Today I am guided only by the sober realization that lost territories are not won back by sharp parliamentary big-mouths and their glibness of tongue, but by a sharp sword; in other words, by a bloody fight.

But I do not hesitate to declare that, now the dice have fallen, I not only regard a reconquest of the South Tyrol by war as impossible, but that I personally would reject it in the conviction that for this question the flame of national enthusiasm of the whole German people could not be achieved to a degree which would offer the premise for victory. I believe, on the contrary, that, if this blood some day were staked, it would be a crime to stake it for two hundred thousand Germans while next door more than seven millions languish under foreign domination and the vital artery of the German people runs through the hunting ground of African Negro hordes.

If the German nation wants to end a state of affairs that threatens its extermination in Europe, it must not fall into the error of the pre-War period and make enemies of God and the world; it must recognize the most dangerous enemy and strike at him with all its concentrated power. And if this victory is obtained through sacrifices elsewhere, the coming generations of our people will not condemn us. The more brilliant the resultant successes, the better they will appreciate the dire distress and profound cares, and the bitter decision born of them.

What must guide us today is again and again the basic insight that the reconquest of a Reich's lost territories is primarily the question of regaining the political independence and power of the motherland.

To make this possible and sure by an astute alliance policy is the first task of a powerful German leadership in the field of foreign affairs.

Especially we National Socialists must guard against being taken in tow by the Jewish-led bourgeois patriots of the word. *Heaven help us if our movement, instead of preparing for the struggle, were to spend its time in protests!*

The fantastic conception of the Nibelungen [1] alliance with the Habsburg state cadaver has been the ruin of Germany. Fantastic sentimentality in the treatment of today's diplomatic possibilities is the best means of preventing our resurrection forever.

*   *   *

Here I must briefly take up those objections which apply to the three questions raised above, to wit, the questions whether anyone will

first, make an alliance with the present-day Germany in her visible weakness that is clear for all to see;

secondly, whether the enemy nations seem capable of such a reorientation; and

thirdly, whether the existing influence of the Jews is not stronger than any understanding or good intentions and will thus frustrate and nullify all plans.

I think I have sufficiently discussed one half of the first question. Of course, no one will make an alliance with present-day Germany. No power in the world will venture to link its destiny to a state whose government is bound to destroy all confidence. And as regards the attempt of many of our national comrades to condone the government's actions because of the wretched mentality of our people at the time, and even accept this as an excuse, we must take the sharpest position against this.

---

[1] Although the *Niebelungenlied*, the German national epic, is a tale of treachery and deceit from beginning to end, in the German popular consciousness, the Nibelungs were conspicuous for their loyal, trusting natures. By Niebelungen alliance, Hitler means an alliance entered into by the loyal, naïvely trusting Germans.

It is true, the absence of character in our people for the last six years has been profoundly sad, their indifference toward the most important concerns of our nation has been truly crushing, their cowardice has sometimes cried out to high Heaven. But it must not be forgotten that we are nevertheless dealing with a people which a few years previous offered the world the most admirable example of the highest human virtues. From the August days of 1914 up to the end of the mighty conflict of nations, no people on earth revealed more manly courage, tenacious endurance, and patience in suffering than our German people which has today grown so wretched. No one will maintain that the disgrace of our present period is the characteristic expression of our nation's being. What we are compelled to experience around us and in us today is only the horrible, maddening, and infuriating influence of the perjuring deed of November 9, 1918. Here more than ever the poet's saying applies that evil begets evil. But even at the present time, our people has not entirely lost its good basic elements; they only are slumbering unawakened in the depths; and from time to time it has been possible to see, gleaming like summer lightning in an overcast firmament, virtues which the future Germany will some day remember as the first signs of an incipient recovery. More than once, thousands and thousands of young Germans have stepped forward with the self-sacrificing resolve to sacrifice their young lives freely and joyfully on the altar of the beloved fatherland, just as in 1914. Again, millions of men are diligently and industriously at work, as though the ravages of the revolution had never been. The blacksmith stands again at his anvil, the peasant guides his plow, and the scholar sits in his study, all with the same painstaking devotion to duty.

The repressions on the part of our enemies no longer meet the same condoning laughter as formerly, but grieved, embittered faces. Undoubtedly a great change in sentiment has taken place.

If today all this is not yet expressed in a rebirth of our people's concept of political power and instinct of self-preservation, it is the fault of those who, less by the grace of Heaven than by self-appointment, have governed our people to death since 1918.

Yes, if we bemoan the state of the nation today, we may ask: What has been done to improve it? Is the feeble support given by the people to the decisions of our governments — decisions which scarcely existed — only a sign of our nation's small vitality or is it not even more a sign of total failure in the handling of this precious treasure? *What have our governments done to reimplant the spirit of proud self-reliance, manly defiance, and wrathful hatred in this people?*

When in the year 1919 the German people was burdened with the peace treaty, we should have been justified in hoping that precisely through this instrument of boundless repression the cry for German freedom would have been immensely promoted. *Peace treaties whose demands are a scourge to nations not seldom strike the first roll of drums for the uprising to come.*

What could have been done with this peace treaty of Versailles?!

This instrument of boundless extortion and abject humiliation might, in the hands of a willing government, have become an instrument for whipping up the national passions to fever heat. With a brilliant propagandist exploitation of these sadistic cruelties, the indifference of a people might have been raised to indignation, and indignation to blazing fury!

How could every single one of these points have been burned into the brain and emotion of this people, until finally in sixty million heads, in men and women, a common sense of shame and a common hatred would have become a single fiery sea of flame, from whose heat a will as hard as steel would have risen and a cry burst forth:

Give us arms again!

Yes, my friends, that is what such a peace treaty would do. In the boundlessness of its oppression, the shamelessness of its demands, lies the greatest propaganda weapon for the reawakening of a nation's dormant spirits of life.

For this, to be sure, from the child's primer down to the last newspaper, every theater and every movie house, every advertising pillar and every billboard, must be pressed into the service of this one great mission, until the timorous prayer of our present parlor patriots: 'Lord, make us free!' is transformed in the brain of the smallest boy into the burning plea: *'Almighty God, bless our arms when the time comes; be just as thou hast always been; judge now whether we be deserving of freedom; Lord, bless our battle!'*

All this was neglected and nothing was done.

Who, then, will be surprised that our people is not as it should be and could be? If the rest of the world sees in us only a stooge, an obsequious dog, who gratefully licks the hands that have just beaten him?

Certainly our capacity for alliances today is injured by our people, but most of all by its governments. They in their corruption are to blame if after eight years of the most unlimited oppression so little will for freedom is present.

Much, therefore, as an active alliance policy is linked with the necessary evaluation of our people, the latter is equally dependent on the existence of a governmental power which does not want to be a handyman for foreign countries, not a taskmaster over its own strength, but a herald of the national conscience.

If our people has a state leadership which sees its mission in this light, six years will not pass before a bold Reich leadership in the field of foreign affairs will dispose of an equally bold will on the part of a people thirsting for freedom.

*   *   *

The second objection, the great difficulty of transforming hostile peoples into friendly allies, can be answered as follows:

The general anti-German psychosis cultivated in other countries by war propaganda will inevitably continue to exist until the German Reich, through the resurrection visible to all of a German will for self-preservation, achieves the character of a state which plays on the general European chessboard and with which it is possible to play. Only when government and people seem to provide absolute guaranty of a possible fitness for alliance can one or another power, out of parallel interest, think of reshaping public opinion by the effects of propaganda. This, too, naturally requires years of shrewd continuous work. The very need of this long period for altering the sentiments of a people necessitates caution in undertaking it; that is, no one will enter upon such an activity unless he is absolutely convinced of the value of such a labor and its fruits for the future. No one will want to change the spiritual orientation of a nation on the strength of the empty bragging of some more or less witty foreign minister, without possessing a tangible guaranty of the value of a new orientation. Otherwise this would lead to a complete shattering of public opinion. The most reliable certainty for the possibility of a future alliance with a state does not lie in the bombastic phrases of individual members of the government, but in the visible stability of a definite and seemingly expedient governmental tendency, and in a public opinion with an analogous orientation. The faith in this will be the firmer, the greater the visible activity of a governing power in the field of propagandist preparation and foundation of its work, and, conversely, the more unmistakably the will of public opinion is reflected in the governmental tendency.

A nation, then, will — in our situation — be regarded as fit for alliance, if government and public opinion with equal fanaticism proclaim and uphold the will to fight for freedom. This is the premise for beginning a reorientation in the public opinion of other nations, which on the basis of their knowledge are willing, in defense of their very own interests, to go a stretch of the way by the side of a partner who seems suitable to them — in other words, to conclude an alliance.

But there is one thing more to be said in this connection: *Since the transformation of a certain spiritual attitude in a people requires hard work in itself, and at first will not be understood by many, it is a crime and a stupidity at once, to furnish these opposing elements with weapons for their counter-efforts by mistakes of one's own.*

It must be realized that it will necessarily take a certain time before a people has completely comprehended the inner purposes of a government, since explanations cannot be given regarding the final ultimate aims of certain preliminary political work, and one can only reckon either with the blind faith of the masses or the intuitive insight of the intellectually superior leader strata. But since in many people this clairvoyant political sixth sense is not present, and for political reasons explanations cannot be given, a part of the intellectual leader class will always turn against new tendencies which due to their incomprehensibility can easily be interpreted as mere experiments. Thus, the resistance of the anxious conservative elements is aroused.

For this reason more than any other, it becomes our highest duty to make sure that all serviceable weapons are wrested from the hands of such disturbers of mutual understanding, especially when, as in our case, we are dealing with nothing but the totally impracticable, purely fantastic babble of inflated parlor patriots and petit bourgeois café politicians. For on calm reflection no one will seriously deny that screaming for a new battle fleet, for recovery of our colonies, etc., is in reality nothing but silly gossip, without so much as a thought of practical application. The way in which the senseless outpourings of these knights of the protest meeting, some of them innocent, some of them insane, but all of them in the silent service of our mortal enemies, are exploited in England, cannot be characterized as favorable to Germany. And so we wear ourselves out in harmful little demonstrations against God and the whole world and forget the first principle which is the premise for every success, to wit: *Whatever you do, do it completely. By beefing against five or ten states, we neglect the concentration of all our will power and physical force for the thrust to the heart of our infamous enemy, and sacrifice the possibility of strengthening ourselves by an alliance for this conflict.*

Here, too, lies a mission for the National Socialist movement. It must teach our people to look beyond trifles and see the biggest things, not to split up over irrelevant things, and never to forget that the aim for which we must fight today is the bare existence of our people, and the sole enemy which we must strike is and remains the power which is robbing us of this existence.

Some things may be profoundly painful to us. But this is far from being a ground for renouncing reason and bickering loudly and senselessly with the whole world instead of attacking the most mortal enemy in concentrated force.

Furthermore, the German people has no moral right to blame the rest of the world for its conduct as long as it has not called to account the criminals who sold and betrayed their whole country. Really, it is not serious for us to curse and protest against England, Italy, etc., from a distance, and leave the scoundrels at large who, in the pay of enemy war propaganda, took away our arms, broke our moral backbone, and auctioned off the crippled Reich for thirty pieces of silver.

The enemy does only what was to be predicted. We should learn from his conduct and his acts.

Anyone who is really unwilling to rise to the heights of such a conception should finally bear in mind that the only thing remaining in that case is renunciation, because then any alliance policy is impossible for the future. For if we cannot ally ourselves with England because she stole our colonies, or with Italy because she has the South Tyrol, with Poland and Czechoslovakia on their merits, then, aside from France — who incidentally did steal Alsace-Lorraine from us — there would remain no one else in Europe.

Whether this serves the German people is scarcely subject to doubt. The only thing that can remain in doubt is whether such an opinion is put forward by a simple dunce or by a shrewd adversary.

When it comes to leaders, I always believe the latter.

And so, in all human probability, a transformation of the psyche of individual peoples, who have hitherto been hostile but whose true future interests lie close to our own, may very well be possible if the inner strength of our state as well as our visible will for the preservation of our existence again make us seem worthy as an ally, and, further, if awkward movements of our own, or even criminal acts, do not furnish grist for the mill of the enemies of such a future tie with nations previously hostile to us.

*   *   *

The hardest to answer is the third objection.

Is it conceivable that the representatives of the real interests of the nations possible for alliance can put through their views in opposition to the will of the Jewish mortal enemy of free national states?

Can the forces of traditional British statesmanship, for example, break the devastating Jewish influence or not?

This question, as already stated, is very hard to answer. It depends on too many factors to permit of a conclusive judgment. One thing is certain in any case: *In one country the present state power can be regarded as so stabilized and serves the interests of the country so absolutely that we can no longer speak of a really effective obstruction of political necessities by international Jewish forces.*

The struggle that Fascist Italy is waging, though perhaps in the last analysis unconsciously (which I personally do not believe), against the three main weapons of the Jews is the best indication that, even though indirectly, the poison fangs of this supra-state power are being torn out. The prohibition of Masonic secret societies, the persecution of the supra-national press, as well as the continuous demolition of international Marxism, and, conversely, the steady reinforcement of the Fascist state conception, will in the course of the years cause the Italian government to serve the interests of the Italian people more and more, without regard for the hissing of the Jewish world hydra.

Things are more difficult in England. In this country of the 'freest democracy,' the Jew exerts an almost unlimited dictatorship indirectly through public opinion. And yet, even there an incessant struggle is taking place between the advocates of British state interests and the proponents of a Jewish world dictatorship.

How sharply these opposites often clash could be seen most clearly for the first time after the War in the different attitude toward the Japanese problem of the British government leaders on the one hand and of the press on the other.

Immediately after the end of the War, the old strain in the relations of America and Japan began to reappear. Of course, the great European powers could not remain indifferent to this new war danger. No ties of kinship can prevent a certain feeling of envious concern in England toward the growth of the American Union in all fields of international economic and power politics. The former colonial country, child of the great mother, seems to be growing into a new master of the world. It is understandable that England today re-examines her old alliances with anxious concern and British statesmen gaze with trepidation toward a period in which it will no longer be said:

'*Britannia rules the waves!*' But instead: '*The seas for the Union!*'

It is harder to attack the gigantic American colossus of states with the enormous wealth of its virgin soil than the wedged-in German Reich. If the dice and the ultimate decision should ever roll,[1] England, if left to her own resources would be doomed. And so they snatch eagerly at the yellow fist and cling to an alliance which, from the racial viewpoint, is perhaps unjustifiable, but from the viewpoint of state politics nevertheless represents the sole possibility of strengthening the British world position in the face of the upsurging American continent.

While the English state leadership, despite the common struggle on the European battlefields, could not resolve to relax its alliance with the Asiatic partner, the whole Jewish press fell on this alliance from behind.

How is it possible that the organs of a Northcliffe,[2] until 1918 the faithful armor-bearer of the British struggle against the German Reich, should now break their loyalty and go their own ways?

The annihilation of Germany was not an English interest, but primarily a Jewish one, just as today a destruction of Japan serves British state interests less than it does the widespread desires of the leaders of the projected Jewish world empire. While England sweats to maintain her position in this world, the Jew organizes his attack for its conquest.

[1] '*wenn jemals ... die Würfel und die letzte Entscheidung rollen würden ...*'
[2] Second edition omits Northcliffe's name, speaking merely of 'Jewish organs.'

He already sees the present-day European states as will-less tools in his fist, whether indirectly through a so-called Western democracy, or in the form of direct domination by Jewish Bolshevism. But it is not only the Old World that he holds thus enmeshed, the same fate menaces the New. It is Jews who govern the stock exchange forces of the American Union. Every year makes them more and more the controlling masters of the producers in a nation of one hundred and twenty millions; only a single great man, Ford,[1] to their fury, still maintains full independence.

With astute shrewdness they knead public opinion and make it into an instrument for their own future.

Already the greatest heads of Jewry see the approaching fulfillment of their testamentary prophecy about the great devouring of nations.

Within this great herd of denationalized colonial territories, a single independent state might still wreck the whole work at the eleventh hour. For a Bolshevistic world can exist only if it embraces everything.

If only a single state is preserved in its national strength and greatness, the world empire of Jewish satrapies, like every tyranny in this world, must succumb to the force of the national idea.

Now the Jew knows only too well that in his thousand years of adaptation he may have been able to undermine European peoples and train them to be raceless bastards, but that he would scarcely be in a position to subject an Asiatic national state like Japan to this fate. Today he may mimic the German and the Englishman, the American and Frenchman, but he lacks the bridges to the yellow Asiatic. And so he strives to break the Japanese national state with the strength of similar existing formations, in order to rid himself of the dangerous adversary before the last state power is transformed in his hand into a despotism over defenseless beings.

In his millennial Jewish empire he dreads a Japanese national

[1] Second edition substitutes 'only very few' for 'a single great man, Ford.'

state, and, therefore, desires its annihilation even before establishing his own dictatorship.

And so he incites the nations against Japan as he once did against Germany, and this is what brings it about that, while British statesmen are still striving to build on the alliance with Japan, the British-Jewish press already demands struggle against the ally, and prepares the war of annihilation under the proclamation of democracy and under the battle-cry: Down with Japanese militarism and imperialism!

That is how insubordinate the Jew has become in England today.

And for this reason it is there that the struggle against the Jewish world menace will begin.

And again the National Socialist movement has the mightiest task to fulfill.

It must open the eyes of the people on the subject of foreign nations and must remind them again and again of the true enemy of our present-day world. In place of hatred against Aryans, from whom almost everything may separate us, but with whom we are bound by common blood or the great line of a kindred culture, it must call eternal wrath upon the head of the foul enemy of mankind as the real originator of our sufferings.

It must make certain that in our country, at least, the mortal enemy is recognized and that the fight against him becomes a gleaming symbol of brighter days, to show other nations the way to the salvation of an embattled Aryan humanity.

For the rest, may reason be our guide, may our will be our strength. May the sacred duty to act in this way give us determination, and above all may our faith protect us.

Eastern Orientation or Eastern Policy

THERE are two reasons which induce me to submit to a special examination the relation of Germany to Russia:

1. Here perhaps we are dealing with the most decisive concern of all German foreign affairs; and

2. This question is also the touchstone for the political capacity of the young National Socialist movements to think clearly and to act correctly.

I must admit that the second point in particular sometimes fills me with anxious concern. Since our young movement does not obtain membership material from the camp of the indifferent, but chiefly from very extreme outlooks, it is only too natural if these people, in the field of understanding foreign affairs as in other fields, are burdened with the preconceived ideas or feeble understanding of the circles to which they previously belonged, both politically and philosophically. And this by no means applies only to the man who comes to us from the *Left*. On the contrary. Harmful as his previous instruction with regard to such problems might be, in part at least it was not infrequently balanced by an existing remnant of natural and healthy instinct. Then it was only necessary to substitute a better attitude for the influence that was previously forced upon him, and often the essentially healthy instinct and impulse of self-preservation that still survived in him could be regarded as our best ally.

It is much harder, on the other hand, to induce clear political thinking in a man whose previous education in this field was no less devoid of any reason and logic, but on top of all this had also sacrified his last remnant of natural instinct on the altar of objectivity. Precisely the members of our so-called intelligentsia are the hardest to move to a really clear and logical defense of their interests and the interests of their nation. They are not only burdened with a dead weight of the most senseless conceptions and prejudices, but what makes matters completely intolerable is that they have lost and abandoned all healthy instinct of self-preservation. The National Socialist movement is compelled to endure hard struggles with these people, hard because, despite total incompetence, they often unfortunately are afflicted with an amazing conceit, which causes them to look down without the slightest inner justification upon other people, for the most part healthier than they. Supercilious, arrogant know-it-alls, without any capacity for cool testing and weighing, which, in turn, must be recognized as the pre-condition for any will and action in the field of foreign affairs.

Since these very circles are beginning today to divert the tendency of our foreign policy in the most catastrophic way from any real defense of the folkish interests of our people, placing it instead in the service of their fantastic ideology, I feel it incumbent upon me to discuss for my supporters the most important question in the field of foreign affairs, our relation to Russia, in particular, and as thoroughly as is necessary for the general understanding and possible in the scope of such a work.

But first I would like to make the following introductory remarks:

If under foreign policy we must understand the regulation of a nation's relations with the rest of the world, the manner of this regulation will be determined by certain definite facts. As National Socialists we can, furthermore, establish the following principle concerning the nature of the foreign policy of a folkish state:

The foreign policy of the folkish state must safeguard the exist-

*ence on this planet of the race embodied in the state, by creating a
healthy, viable natural relation between the nation's population and
growth on the one hand and the quantity and quality of its soil on
the other hand.*

As a healthy relation we may regard only that condition which
assures the sustenance of a people on its own soil. Every other
condition, even if it endures for hundreds, nay, thousands of
years, is nevertheless unhealthy and will sooner or later lead to
the injury if not annihilation of the people in question.

*Only an adequately large space on this earth assures a nation of
freedom of existence.*

Moreover, the necessary size of the territory to be settled cannot be judged exclusively on the basis of present requirements,
not even in fact on the basis of the yield of the soil compared to
the population. For, as I explained in the first volume, under
'German Alliance Policy Before the War,' *in addition to its importance as a direct source of a people's food, another significance,
that is, a military and political one, must be attributed to the area of
a state.* If a nation's sustenance as such is assured by the amount
of its soil, the safeguarding of the existing soil itself must also be
borne in mind. This lies in the general power-political strength
of the state, which in turn to no small extent is determined by geo-
military considerations.

Hence, the German nation can defend its future only as a world
power. For more than two thousand years the defense of our
people's interests, as we should designate our more or less fortunate activity in the field of foreign affairs, was *world history*.
We ourselves were witnesses to this fact: for the gigantic struggle
of the nations in the years 1914–1918 was only the struggle of the
German people for its existence on the globe, but we designated
the type of event itself as a *World War*.

The German people entered this struggle as a *supposed* world
power. I say here 'supposed,' for in reality it was none. If the
German nation in 1914 had had a different relation between area
and population, Germany would really have been a world power,
and the War, aside from all other factors, could have been ter-
minated favorably.

Germany today is no world power. Even if our momentary military impotence were overcome, we should no longer have any claim to this title. What can a formation, as miserable in its relation of population to area as the German Reich today, mean on this planet? In an era when the earth is gradually being divided up among states, some of which embrace almost entire continents, we cannot speak of a world power in connection with a formation whose political mother country is limited to the absurd area of five hundred thousand square kilometers.

From the purely territorial point of view, the area of the German Reich vanishes completely as compared with that of the so-called world powers. Let no one cite England as a proof to the contrary, for England in reality is merely the great capital of the British world empire which calls nearly a quarter of the earth's surface its own. In addition, we must regard as giant states, first of all the American Union, then Russia and China. All are spatial formations having in part an area more than ten times greater than the present German Reich. And even France must be counted among these states. Not only that she complements her army to an ever-increasing degree from her enormous empire's reservoir of colored humanity, but racially as well, she is making such great progress in negrification that we can actually speak of an African state arising on European soil. The colonial policy of present-day France cannot be compared with that of Germany in the past. If the development of France in the present style were to be continued for three hundred years, the last remnants of Frankish blood would be submerged in the developing European-African mulatto state. An immense self-contained area of settlement from the Rhine to the Congo, filled with a lower race gradually produced from continuous bastardization.

This distinguishes French colonial policy from the old German one.

The former German colonial policy, like everything we did, was carried out by halves. It neither increased the settlement area of the German Reich, nor did it undertake any attempt — criminal though it would have been — to strengthen the Reich by the use of black blood. The Askaris in German East Africa were a short, hesitant step in this direction. Actually they served only for the defense of the colonies themselves. The idea of bringing black troops into a European battlefield, quite aside from its practical impossibility in the World War, never existed even as a design to be realized under more favorable circumstances, while, on the contrary, it was always regarded and felt by the French as the basic reason for their colonial activity.

Thus, in the world today we see a number of power states, some of which not only far surpass the strength of our German nation in population, but whose area above all is the chief support of their political power. Never has the relation of the German Reich to other existing world states been as unfavorable as at the beginning of our history two thousand years ago and again today. Then we were a young people, rushing headlong into a world of great crumbling state formations, whose last giant, Rome, we ourselves helped to fell. Today we find ourselves in a world of great power states in process of formation, with our own Reich sinking more and more into insignificance.

We must bear this bitter truth coolly and soberly in mind. We must follow and compare the German Reich through the centuries in its relation to other states with regard to population and area. I know that everyone will then come to the dismayed conclusion which I have stated at the beginning of this discussion: *Germany is no longer a world power, regardless whether she is strong or weak from the military point of view.*

We have lost all proportion to the other great states of the earth, and this thanks only to the positively catastrophic leadership of our nation in the field of foreign affairs, thanks to our total failure to be guided by what I should almost call a testamentary aim in foreign policy, and thanks to the loss of any healthy instinct and impulse of self-preservation.

If the National Socialist movement really wants to be consecrated by history with a great mission for our nation, it must be permeated by knowledge and filled with pain at our true situation in this world; boldly and conscious of its goal, it must take up the struggle against the aimlessness and incompetence which have hitherto guided our German nation in the line of foreign affairs. Then, without consideration of 'traditions' and prejudices, it must find the courage to gather our people and their strength for an advance along the road that will lead this people from its present restricted living space to new land and soil, and hence also free it from the danger of vanishing from the earth or of serving others as a slave nation.

The National Socialist movement must strive to eliminate the disproportion between our population and our area — viewing this latter as a source of food as well as a basis for power politics — between our historical past and the hopelessness of our present impotence. And in this it must remain aware that we, as guardians of the highest humanity on this earth, are bound by the highest obligation, and the more it strives to bring the German people to racial awareness so that, in addition to breeding dogs, horses, and cats, they will have mercy on their *own* blood, the more it will be able to meet this obligation.

*   *   *

If I characterize German policy up to now as aimless and incompetent, the proof of my assertion lies in the actual failure of this policy. If our people had been intellectually inferior or cowardly, the results of its struggle on the earth could not be worse than what we see before us today. Neither must the development of the last decades before the War deceive us on this score; for we cannot measure the strength of an empire by itself, but only by comparison with other states. And just such a comparison furnishes proof that the increase in strength of the other states was not only more even, but also greater in its ultimate effect; that consequently, despite its apparent rise, Germany's road actually diverged more and more from that of the other states and fell far behind; in short, the difference in magnitudes increased to our disfavor. Yes, as time went on, we fell behind more and more even in population. But since our people is certainly excelled by none on earth in heroism, in fact, all in all has certainly given the most blood of all the nations on earth for the preservation of its existence, the failure can reside only in the *mistaken way* in which it was given.

If we examine the political experiences of our people for more than a thousand years in this connection, passing all the innumerable wars and struggles in review and examining the present end result they created, we shall be forced to admit that this sea of blood has given rise to only three phenomena which we are justified in claiming as enduring fruits of clearly defined actions in the field of foreign and general politics:

(1) The colonization of the *Ostmark*, carried out mostly by Bavarians;

(2) the acquisition and penetration of the territory east of the Elbe; and

(3) the organization by the Hohenzollerns of the Brandenburg-Prussian state as a model and nucleus for crystallization of a new Reich.

An instructive warning for the future!

The first two great successes of our foreign policy have remained the most enduring. Without them our nation today would no longer have any importance at all. They were the first, but unfortunately the only successful attempt to bring the rising population into harmony with the quantity of our soil. And it must be regarded as truly catastrophic that our German historians have never been able to estimate correctly these two achievements which are by far the greatest and most significant for the future, but by contrast have glorified everything conceivable, praised and admired fantastic heroism, innumerable adventurous wars and struggles, instead of finally recognizing how unimportant most of these events have been for the nation's great line of development.

The third great success of our political activity lies in the formation of the Prussian state and the resultant cultivation of a special state idea, as also of the German army's instinct of self-preservation and self-defense, adapted to the modern world and put into organized form. The development of the idea of individual militancy into the duty of national militancy [conscription] has grown out of every state formation and every state conception. The significance of this development cannot be overestimated. Through the discipline of the Prussian army organism, the German people, shot through with hyperindividualism by their racial divisions, won back at least a part of the capacity for organization which they had long since lost. What other peoples still primitively possess in their herd community instinct, we, partially at least, regained artificially for our national community through the process of military training. Hence the elimination of universal conscription — which for dozens of other peoples might be a matter of no importance — is for us fraught with the gravest consequences. Ten German generations without corrective and educational military training, left to the evil effects of their racial and hence philosophical division — and our nation would really have lost the last remnant of an independent existence on this planet. Only through individual men, in the bosom of foreign nations, could the German spirit make its contribution to culture, and its origin would not even be recognized. Cultural fertilizer, until the last remnant of Aryan-Nordic blood in us would be corrupted or extinguished.

It is noteworthy that the significance of these real political successes won by our nation in its struggles, enduring more than a thousand years, were far better understood and appreciated by our adversaries than by ourselves. Even today we still rave about a heroism which robbed our people of millions of its noblest blood-bearers, but in its ultimate result remained totally fruitless.

The distinction between the real political successes of our people and the national blood spent for fruitless aims is of the greatest importance for our conduct in the present and the future.

We National Socialists must never under any circumstances join in the foul [1] *hurrah patriotism of our present bourgeois world. In particular it is mortally dangerous to regard the last pre-War de-*

---

[1] '*übel*.' In preparing the second edition, the frugal copy-reader substitutes '*üblich*,' usual.

velopments as binding even in the slightest degree for our own course.
From the whole historical development of the nineteenth century, not a single obligation can be derived which was grounded
in this period itself. In contrast to the conduct of the representatives of this period, we must again profess the highest aim of
all foreign policy, to wit: *to bring the soil into harmony with the
population.* Yes, from the past we can only learn that, in setting
an objective for our political activity, we must proceed in two
directions: *Land and soil as the goal of our foreign policy, and a
new philosophically established, uniform foundation as the aim of
political activity at home.*

* * *

I still wish briefly to take a position on the question as to what
extent the demand for soil and territory seems ethically and morally justified. This is necessary, since unfortunately, even in so-
called folkish circles, all sorts of unctuous big-mouths step forward, endeavoring to set the rectification of the injustice of 1918
as the aim of the German nation's endeavors in the field of foreign
affairs, but at the same time find it necessary to assure the whole
world of folkish brotherhood and sympathy.

I should like to make the following preliminary remarks: *The
demand for restoration of the frontiers of 1914 is a political absurdity
of such proportions and consequences as to make it seem a crime.
Quite aside from the fact that the Reich's frontiers in 1914 were anything but logical. For in reality they were neither complete in the
sense of embracing the people of German nationality, nor sensible
with regard to geo-military expediency. They were not the result of a
considered political action, but momentary frontiers in a political
struggle that was by no means concluded; partly, in fact, they were
the results of chance.* With equal right and in many cases with
more right, some other sample year of German history could be
picked out, and the restoration of the conditions at that time
declared to be the aim of an activity in foreign affairs. The above

demand is entirely suited to our bourgeois society, which here as
elsewhere does not possess a single creative political idea for the
future, but lives only in the past, in fact, in the most immediate
past; for even their backward gaze does not extend beyond their
own times. The law of inertia binds them to a given situation
and causes them to resist any change in it, but without ever increasing the activity of this opposition beyond the mere power of
perseverance. So it is obvious that the political horizon of these
people does not extend beyond the year 1914. By proclaiming the
restoration of those borders as the political aim of their activity,
they keep mending the crumbling league of our adversaries.
Only in this way can it be explained that eight years after a world
struggle in which states, some of which had the most heterogeneous desires, took part, the coalition of the victors of those days
can still maintain itself in a more or less unbroken form.

All these states were at one time beneficiaries of the German
collapse. Fear of our strength caused the greed and envy of the
individual great powers among themselves to recede. By grab-
bing as much of the Reich as they could, they found the best
guard against a future uprising. A bad conscience and fear of our
people's strength is still the most enduring cement to hold together the various members of this alliance.

And we do not disappoint them. By setting up the restoration
of the borders of 1914 as a political program for Germany, our
bourgeoisie frighten away every partner who might desire to leave
the league of our enemies, since he must inevitably fear to be attacked singly and thereby lose the protection of his individual
fellow allies. Each single state feels concerned and threatened by
this slogan.

Moreover, it is senseless in two respects:

(1) because the instruments of power are lacking to remove it
from the vapors of club evenings into reality; and

(2) because, if it could actually be realized, the outcome would
again be so pitiful that, by God, it would not be worth while to
risk the blood of our people for *this*.

For it should scarcely seem questionable to anyone that even the restoration of the frontiers of 1914 could be achieved only by blood. Only childish and naïve minds can lull themselves in the idea that they can bring about a correction of Versailles by wheedling and begging. Quite aside from the fact that such an attempt would presuppose a man of Talleyrand's talents, which we do not possess. One half of our political figures consist of extremely sly, but equally spineless elements which are hostile toward our nation to begin with, while the other is composed of good-natured, harmless, and easy-going soft-heads. Moreover, the times have changed since the Congress of Vienna: *Today it is not princes and princes' mistresses who haggle and bargain over state borders; it is the inexorable Jew who struggles for his domination over the nations.* No nation can remove this hand from its throat except by the sword. Only the assembled and concentrated might of a national passion rearing up in its strength can defy the international enslavement of peoples. Such a process is and remains a bloody one.

If, however, we harbor the conviction that the German future, regardless what happens, demands the supreme sacrifice, quite aside from all considerations of political expediency as such, we must set up an aim worthy of this sacrifice and fight for it.

The boundaries of the year 1914 mean nothing at all for the German future. Neither did they provide a defense of the past, nor would they contain any strength for the future. Through them the German nation will neither achieve its inner integrity, nor will its sustenance be safeguarded by them, nor do these boundaries, viewed from the military standpoint, seem expedient or even satisfactory, nor finally can they improve the relation in which we at present find ourselves toward the other world powers, or, better expressed, the real world powers. The lag behind England will not be caught up, the magnitude of the Union will not be achieved; not even France would experience a material diminution of her world-political importance.

Only one thing would be certain: even with a favorable outcome, such an attempt to restore the borders of 1914 would lead to a further bleeding of our national body, so much so that there would be no worth-while blood left to stake for the decisions and actions really to secure the nation's future. On the contrary, drunk with such a shallow success, we should renounce any further goals, all the more readily as 'national honor' would be repaired and, for the moment at least, a few doors would have been reopened to commercial development.

As opposed to this, we National Socialists must hold unflinchingly to our aim in foreign policy, namely, *to secure for the German people the land and soil to which they are entitled on this earth.* And this action is the only one which, before God and our German posterity, would make any sacrifice of blood seem justified: before God, since we have been put on this earth with the mission of eternal struggle for our daily bread, beings who receive nothing as a gift, and who owe their position as lords of the earth only to the genius and the courage with which they can conquer and defend it; and before our German posterity in so far as we have shed no citizen's blood out of which a thousand others are not bequeathed to posterity. The soil on which some day German generations of peasants can beget powerful sons will sanction the investment of the sons of today, and will some day acquit the responsible statesmen of blood-guilt and sacrifice of the people, even if they are persecuted by their contemporaries.

And I must sharply attack those folkish pen-pushers who claim to regard such an acquisition of soil as a 'breach of sacred human rights' and attack it as such in their scribblings. One never knows who stands behind these fellows. But one thing is certain, that the confusion they can create is desirable and convenient to our national enemies. By such an attitude they help to weaken and destroy from within our people's will for the only correct way of defending their vital needs. For no people on this earth possesses so much as a square yard of territory on the strength of a higher will or superior right. Just as Germany's frontiers are fortuitous frontiers, momentary frontiers in the current political struggle of any period, so are the boundaries of other nations' living space. And just as the shape of our earth's surface can seem immutable as granite only to the thoughtless soft-head, but in reality only represents at each period an apparent pause in a continuous development, created by the mighty forces of Nature in a process of continuous growth, only to be transformed or destroyed tomorrow by greater forces, likewise the boundaries of living spaces in the life of nations.

State boundaries are made by man and changed by man.

The fact that a nation has succeeded in acquiring an undue amount of soil constitutes no higher obligation that it should be recognized eternally. At most it proves the strength of the conquerors and the weakness of the nations. And in this case, right lies in this strength alone. If the German nation today, penned into an impossible area, faces a lamentable future, this is no more a commandment of Fate than revolt against this state of affairs constitutes an affront to Fate. No more than any higher power has promised another nation more territory than the German nation, or is offended by the fact of this unjust distribution of the soil. Just as our ancestors did not receive the soil on which we live today as a gift from Heaven, but had to fight for it at the risk of their lives, in the future no folkish grace will win soil for us and hence life for our people, but only the might of a victorious sword.

Much as all of us today recognize the necessity of a reckoning with France, it would remain ineffectual in the long run if it represented the whole of our aim in foreign policy. It can and will achieve meaning only if it offers the rear cover for an enlargement of our people's living space in Europe. For it is not in colonial acquisitions that we must see the solution of this problem, but exclusively in the acquisition of a territory for settlement, which will enhance the area of the mother country, and hence not only keep the new settlers in the most intimate community with the land of their origin, but secure for the total area those advantages which lie in its unified magnitude.

The folkish movement must not be the champion of other peoples, but the vanguard fighter of its own. Otherwise it is superfluous and above all has no right to sulk about the past. For in that case it is behaving in exactly the same way. The old

German policy was wrongly determined by dynastic considerations, and the future policy must not be directed by cosmopolitan folkish drivel. In particular, we are not constables guarding the well-known 'poor little nations,' but soldiers of our own nation.

But we National Socialists must go further. *The right to possess soil can become a duty if without extension of its soil a great nation seems doomed to destruction.* And most especially when not some little nigger nation or other is involved, but the Germanic mother of life, which has given the present-day world its cultural picture. *Germany will either be a world power or there will be no Germany.* And for world power she needs that magnitude which will give her the position she needs in the present period, and life to her citizens.

*   *   *

And so we National Socialists consciously draw a line beneath the foreign policy tendency of our pre-War period. We take up where we broke off six hundred years ago. We stop the endless German movement to the south and west, and turn our gaze toward the land in the east. At long last we break off the colonial and commercial policy of the pre-War period and shift to the soil policy of the future.

If we speak of soil in Europe today, we can primarily have in mind only *Russia* and her vassal border states.

Here Fate itself seems desirous of giving us a sign. By handing Russia to Bolshevism, it robbed the Russian nation of that intelligentsia which previously brought about and guaranteed its existence as a state. For the organization of a Russian state formation was not the result of the political abilities of the Slavs in Russia, but only a wonderful example of the state-forming efficacity of the German element in an inferior race. Numerous mighty empires on earth have been created in this way. Lower nations led by Germanic organizers and overlords have more than once grown to be mighty state formations and have endured as long as the racial nucleus of the creative state race maintained itself. For centuries Russia drew nourishment from this Germanic nucleus of its upper leading strata. Today it can be regarded as almost totally exterminated and extinguished. It has been replaced by the Jew. Impossible as it is for the Russian by himself to shake off the yoke of the Jew by his own resources, it is equally impossible for the Jew to maintain the mighty empire forever. He himself is no element of organization, but a ferment of decomposition. The Persian [1] empire in the east is ripe for collapse. And the end of Jewish rule in Russia will also be the end of Russia as a state. We have been chosen by Fate as witnesses of a catastrophe which will be the mightiest confirmation of the soundness of the folkish theory.

Our task, the mission of the National Socialist movement, is to bring our own people to such political insight that they will not see their goal for the future in the breath-taking sensation of a new Alexander's conquest, but in the industrious work of the German plow, to which the sword need only give soil.

* * *

It goes without saying that the Jews announce the sharpest resistance to such a policy. Better than anyone else they sense the significance of this action for their own future. This very fact should teach all really national-minded men the correctness of such a reorientation. Unfortunately, the opposite is the case. Not only in German-National, but even in 'folkish' circles, the idea of such an eastern policy is violently attacked, and, as almost always in such matters, they appeal to a higher authority. The spirit of Bismarck is cited to cover a policy which is as senseless as it is impossible and in the highest degree harmful to the German nation. Bismarck in his time, they say, always set store on good relations with Russia. This, to a certain extent, is true. But they forget to mention that he set just as great store on good

[1] Second edition has 'giant' instead of 'Persian.'

relations with Italy, for example; in fact, that the same Herr von Bismarck once made an alliance with Italy in order to finish off Austria the more easily. Why, then, don't they continue *this* policy? 'Because the Italy of today is not the Italy of those days,' they will say. Very well. But then, honored sirs, will you permit the objection that present-day Russia is not the Russia of those days either? It never entered Bismarck's head to lay down a political course tactically and theoretically for all time. In this respect he was too much master of the moment to tie his hands in such a way. *The question, therefore, must not be: What did Bismarck do in his time?* But rather: *What would he do today?* And this question is easier to answer. *With his political astuteness, he would never ally himself with a state that is doomed to destruction.*

Furthermore, Bismarck even then viewed the German colonial and commercial policy with mixed feelings, since for the moment he was concerned only with the surest method of internally consolidating the state formation he had created. And this was the only reason why at that time he welcomed the Russian rear cover, which gave him a free hand in the west. But what was profitable to Germany then would be detrimental today.

As early as 1920–21, when the young National Socialist movement began slowly to rise above the political horizon, and here and there was referred to as the movement for German freedom, the party was approached by various quarters with an attempt to create a certain bond between it and the *movements for freedom in other countries*. This was in the line of the '*League of Oppressed Nations*,' propagated by many. Chiefly involved were representatives of various Balkan states, and some from Egypt and India, who as individuals always impressed me as pompous big-mouths without any realistic background. But there were not a few Germans, especially in the nationalist camp, who let themselves be dazzled by such inflated Orientals and readily accepted any old Indian or Egyptian student from God knows where as a 'representative' of India or Egypt. These people never realized that they were usually dealing with persons who had absolutely nothing behind them, and above all were authorized by no one to conclude any pact with anyone, so that the practical result of any relations with such elements was nil, unless the time wasted were booked as a special loss. I always resisted such attempts. Not only that I had better things to do than twiddle away weeks in fruitless 'conferences,' but even if these men had been authorized representatives of such nations, I regarded the whole business as useless, in fact, harmful.

Even in peacetime it was bad enough that the German alliance policy, for want of any aggressive intentions of our own, ended in a defensive union of ancient states, pensioned by world history. The alliance with Austria as well as Turkey had little to be said for them. While the greatest military and industrial states on earth banded into an active aggressive union, we collected a few antique, impotent state formations and with this decaying rubbish attempted to face an active world coalition. Germany received a bitter accounting for this error in foreign policy. But this accounting does not seem to have been bitter enough to prevent our eternal dreamers from falling headlong into the same error. For the attempt to disarm the almighty victors through a 'League of Oppressed Nations' is not only ridiculous, but catastrophic as well. It is catastrophic because it distracts our people again and again from the practical possibilities, making them devote themselves to imaginative, yet fruitless hopes and illusions. The German of today really resembles the drowning man who grasps at every straw. And this can apply even to men who are otherwise exceedingly well educated. If any will-o'-the-wisp of hope, however unreal, turns up anywhere, these men are off at a trot, chasing after the phantom. Whether it is a League of Oppressed Nations, a League of Nations, or any other fantastic new invention, it will be sure to find thousands of credulous souls.

I still remember the hopes, as childish as they were incomprehensible, which suddenly arose in folkish circles in 1920–21, to the effect that British power was on the verge of collapse in India. Some Asiatic jugglers, for all I care they may have been real 'fighters for Indian freedom,' who at that time were wandering around Europe, had managed to sell otherwise perfectly reasonable people the *idée fixe* that the British Empire, which has its pivot in India, was on the verge of collapse at that very point. Of course, it never entered their heads that here again their own wish was the sole father of all their thoughts. No more did the inconsistency of their own hopes. For by expecting the end of the British Empire to follow from a collapse of British rule in India, they themselves admitted that India was of the most paramount importance to England.

It is most likely, however, that this vitally important question is not a profound secret known only to German-folkish prophets; presumably it is known also to the helmsmen of English destiny. It is really childish to suppose that the men in England cannot correctly estimate the importance of the Indian Empire for the British world union. And if anyone imagines that England would let India go without staking her last drop of blood, it is only a sorry sign of absolute failure to learn from the World War, and of total misapprehension and ignorance on the score of Anglo-Saxon determination. It is, furthermore, a proof of the German's total ignorance regarding the whole method of British penetration and administration of this empire. *England will lose India either if her own administrative machinery falls a prey to racial decomposition* (which at the moment is completely out of the question in India) *or if she is bested by the sword of a powerful enemy.* Indian agitators, however, will never achieve this. How hard it is to best England, we Germans have sufficiently learned. Quite aside from the fact that I, as a man of Germanic blood, would, in spite of everything, rather see India under English rule than under any other.

Just as lamentable are the hopes in any mythical uprising in Egypt. The '*Holy War*' can give our German *Schafkopf* players the pleasant thrill of thinking that now perhaps others are ready to shed their blood for us — for this cowardly speculation, to tell the truth, has always been the silent father of all hopes; in reality it would come to an infernal end under the fire of English machine-gun companies and the hail of fragmentation bombs.

It just happens to be impossible to overwhelm with a coalition of cripples a powerful state that is determined to stake, if necessary, its last drop of blood for its existence. As a folkish man, who appraises the value of men on a racial basis, I am prevented by mere knowledge of the racial inferiority of these so-called 'oppressed nations' from linking the destiny of my own people with theirs.

And today we must take exactly the same position toward Russia. Present-day Russia, divested of her Germanic upper stratum, is, quite aside from the private intentions of her new masters, no ally for the German nation's fight for freedom. *Considered from the purely military angle, the relations would be simply catastrophic in case of war between Germany and Russia and Western Europe, and probably against all the rest of the world. The struggle would take place, not on Russian, but on German soil,* and Germany would not be able to obtain the least effective support from Russia. The present German Reich's instruments of power are so lamentable and so useless for a foreign war, that no defense of our borders against Western Europe, including England, would be practicable, and particularly the German industrial region would lie defenselessly exposed to the concentrated aggressive arms of our foes. There is the additional fact that between Germany and Russia there lies the Polish state, completely in French hands. In case of a war between Germany and Russia and Western Europe, Russia would first have to subdue Poland before the first soldier could be sent to the western front. Yet it is not so much a question of soldiers as of technical armament. In this respect, the World War situation would repeat itself, only much more horribly. Just as German industry was then drained for our glorious allies, and, technically speaking, Germany had to fight the war almost single-handed, likewise in this struggle Russia would be entirely out of the picture as a technical factor. We could oppose practically nothing to the general motorization of the world, which in the next war will manifest itself overwhelmingly and decisively. For not only that Germany herself has remained shamefully backward in this all-important field, but from the little she possesses she would have to sustain Russia, which even today cannot claim possession of a single factory capable of producing a motor vehicle that really runs. Thus, such a war would assume the character of a plain massacre. Germany's youth would be bled even more than the last time, for as always the burden of the fighting would rest only upon us, and the result would be inevitable defeat.

But even supposing that a miracle should occur and that such a struggle did not end with the total annihilation of Germany, the ultimate outcome would only be that the German nation, bled white, would remain as before bounded by great military states and that her real situation would hence have changed in no way.

Let no one argue that in concluding an alliance with Russia we need not immediately think of war, or, if we did, that we could thoroughly prepare for it. *An alliance whose aim does not embrace a plan for war is senseless and worthless.* Alliances are concluded only for struggle. And even if the clash should be never so far away at the moment when the pact is concluded, the prospect of a military involvement is nevertheless its cause. And do not imagine that any power would ever interpret the meaning of such an alliance in any other way. Either a German-Russian coalition would remain on paper, or from the letter of the treaty it would be translated into visible reality — and the rest of the world would be warned. How naïve to suppose that in such a case England and France would wait a decade for the German-Russian alliance to complete its technical preparations. No, the storm would break over Germany with the speed of lightning.

And so the very fact of the conclusion of an alliance with Russia embodies a plan for the next war. Its outcome would be the end of Germany.

On top of this there is the following:

1. *The present rulers of Russia have no idea of honorably entering into an alliance, let alone observing one.*

Never forget that the rulers of present-day Russia are common blood-stained criminals; that they are the scum of humanity which, favored by circumstances, overran a great state in a tragic hour, slaughtered and wiped out thousands of her leading intelligentsia in wild blood lust, and now for almost ten years have been carrying on the most cruel and tyrannical régime of all time. Furthermore, do not forget that these rulers belong to a race which combines, in a rare mixture, bestial cruelty and an inconceivable gift for lying, and which today more than ever is conscious of a mission to impose its bloody oppression on the whole world. Do not forget that the international Jew who completely dominates Russia today regards Germany, not as an ally, but as a state destined to the same fate. *And you do not make pacts with anyone whose sole interest is the destruction of his partner.* Above all, you do not make them with elements to whom no pact would be sacred, since they do not live in this world as representatives of honor and sincerity, but as champions of deceit, lies, theft, plunder, and rapine. If a man believes that he can enter into profitable connections with parasites, he is like a tree trying to conclude for its own profit an agreement with a mistletoe.

2. *The danger to which Russia succumbed is always present for Germany.* Only a bourgeois simpleton is capable of imagining that Bolshevism has been exorcised. With his superficial thinking he has no idea that this is an instinctive process; that is, the striving of the Jewish people for world domination, a process which is just as natural as the urge of the Anglo-Saxon to seize domination of the earth. And just as the Anglo-Saxon pursues this course in his own way and carries on the fight with his own weapons, likewise the Jew. He goes his way, the way of sneaking in among the nations and boring from within, and he fights with his weapons, with lies and slander, poison and corruption, intensifying the struggle to the point of bloodily exterminating his hated foes. *In Russian Bolshevism we must see the attempt undertaken by the Jews in the twentieth century to achieve world domination.* Just as in other epochs they strove to reach the same goal by other, though inwardly related processes. Their endeavor lies profoundly rooted in their essential nature. No more than another nation renounces of its own accord the pursuit of its impulse for

the expansion of its power and way of life, but is compelled by outward circumstances or else succumbs to impotence due to the symptoms of old age, does the Jew break off his road to world dictatorship out of voluntary renunciation, or because he represses his eternal urge. He, too, will either be thrown back in his course by forces lying outside himself, or all his striving for world domination will be ended by his own dying out. But the impotence of nations, their own death from old age, arises from the abandonment of their blood purity. And this is a thing that the Jew preserves better than any other people on earth. And so he advances on his fatal road until another force comes forth to oppose him, and in a mighty struggle hurls the heaven-stormer back to Lucifer.

Germany is today the next great war aim of Bolshevism. It requires all the force of a young missionary idea to raise our people up again, to free them from the snares of this international serpent, and to stop the inner contamination of our blood, in order that the forces of the nation thus set free can be thrown in to safeguard our nationality, and thus can prevent a repetition of the recent catastrophes down to the most distant future. If we pursue this aim, it is sheer lunacy to ally ourselves with a power whose master is the mortal enemy of our future. How can we expect to free our own people from the fetters of this poisonous embrace if we walk right into it? How shall we explain Bolshevism to the German worker as an accursed crime against humanity if we ally ourselves with the organizations of this spawn of hell, thus recognizing it in the larger sense? By what right shall we condemn a member of the broad masses for his sympathy with an outlook if the very leaders of the state choose the representatives of this outlook for allies?

The fight against Jewish world Bolshevization requires a clear attitude toward Soviet Russia. You cannot drive out the Devil with Beelzebub.

If today even folkish circles rave about an alliance with Russia, they should just look around them in Germany and see whose support they find in their efforts. Or have folkish men lately begun to view an activity as beneficial to the German people which is recommended and promoted by the international Marxist press? Since when do folkish men fight with armor held out to them by a Jewish squire?

There is one main charge that could be raised against the old German Reich with regard to its alliance policy: not, however, that it failed to maintain good relations with Russia, but only that it ruined its relations with everyone by continuous shilly-shallying, in the pathological weakness of trying to preserve world peace at any price.

I openly confess that even in the pre-War period I would have thought it sounder if Germany, renouncing her senseless colonial policy and renouncing her merchant marine and war fleet, had concluded an alliance with England against Russia, thus passing from a feeble global policy to a determined European policy of territorial acquisition on the continent.

I have not forgotten the insolent threat which the pan-Slavic Russia of that time dared to address to Germany; I have not forgotten the constant practice mobilizations, whose sole purpose was an affront to Germany; I cannot forget the mood of public opinion in Russia, which outdid itself in hateful outbursts against our people and our Reich; I cannot forget the big Russian newspapers, which were always more enthusiastic about France than about us.

But in spite of all that, before the War there would still have been a second way: we could have propped ourselves on Russia and turned against England.

Today conditions are different. If before the War we could have choked down every possible sentiment and gone with Russia, today it is no longer possible. The hand of the world clock has moved forward since then, and is loudly striking the hour in which the destiny of our nation must be decided in one way or another. The process of consolidation in which the great states of the earth are involved at the moment is for us the last warning signal to stop and search our hearts, to lead our people out of the dream world back to hard reality, and show them the way to the future which alone will lead the old Reich to a new golden age.

If the National Socialist movement frees itself from all illusions with regard to this great and all-important task, and accepts reason as its sole guide, the catastrophe of 1918 can some day become an infinite blessing for the future of our nation. Out of this collapse our nation will arrive at a complete reorientation of its activity in foreign relations, and, furthermore, reinforced within by its new philosophy of life, will also achieve outwardly a final stabilization of its foreign policy. Then at last it will acquire what England possesses and even Russia possessed, and what again and again induced France to make the same decisions, essentially correct from the viewpoint of her own interests, to wit: *A political testament.*

The political testament of the German nation to govern its outward activity for all time should and must be:

Never suffer the rise of two continental powers in Europe. Regard any attempt to organize a second military power on the German frontiers, even if only in the form of creating a state capable of military strength, as an attack on Germany, and in it see not only the right, but also the duty, to employ all means up to armed force to prevent the rise of such a state, or, if one has already arisen, to smash it again. — See to it that the strength of our nation is founded, not on colonies, but on the soil of our European homeland. Never regard the Reich as secure unless for centuries to come it can give every scion of our people his own parcel of soil. Never forget that the most sacred right on this earth is a man's right to have earth to till with his own hands, and the most sacred sacrifice the blood that a man sheds for this earth.

*   *   *

I should not like to conclude these reflections without pointing once again to the sole alliance possibility which exists for us at the moment in Europe. In the previous chapter on the alliance problem I have already designated England and Italy as the only two states in Europe with which a closer relationship would be desirable and promising for us. Here I shall briefly touch on the *military* importance of such an alliance.

The military consequences of concluding this alliance would in every respect be the opposite of the consequences of an alliance with Russia. The most important consideration, first of all, is *the fact that in itself an approach to England and Italy in no way conjures up a war danger.* France, the sole power which could conceivably oppose the alliance, would not be in a position to do so. *And consequently the alliance would give Germany the possibility of peacefully making those preparations for a reckoning with France, which would have to be made in any event within the scope of such a coalition.* For the significant feature of such an alliance lies precisely in the fact that upon its conclusion Germany would not suddenly be exposed to a hostile invasion, but that the opposing alliance would break of its own accord; the Entente, to which we owe such infinite misfortune, would be dissolved, and hence *France, the mortal enemy of our nation, would be isolated.* Even if this success is limited at first to moral effect, it would suffice to give Germany freedom of movement to an extent which today is scarcely conceivable. *For the law of action would be in the hands of the new European Anglo-German-Italian alliance and no longer with France.*

The further result would be that at one stroke Germany would be freed from her unfavorable strategic position. The most powerful protection on our flank on the one hand, complete guaranty of our food and raw materials on the other, would be the beneficial effect of the new constellation of states.

But almost more important would be the fact that the new league would embrace states which in technical productivity almost complement one another in many respects. For the first time Germany would have allies who would not drain our own economy like leeches, but could and would contribute their share to the richest supplementation of our technical armament.

And do not overlook the final fact that in both cases we should be dealing with allies who cannot be compared with Turkey or present-day Russia. *The greatest world power on earth and a youth-*

ful national state would offer different premises for a struggle in Europe than the putrid state corpses with which Germany allied herself in the last war.

Assuredly, as I emphasized in the last chapter, the difficulties opposing such an alliance are great. But was the formation of the Entente, for instance, any less difficult? *What the genius of a King Edward VII achieved, in part almost counter to natural interests, we, too, must and will achieve, provided we are so inspired by our awareness of the necessity of such a development that with astute self-control we determine our actions accordingly.* And this will become possible in the moment when, imbued with admonishing distress,[1] we pursue, not the diplomatic aimlessness of the last decades, but a conscious and determined course, and stick to it. *Neither western nor eastern orientation must be the future goal of our foreign policy, but an eastern policy in the sense of acquiring the necessary soil for our German people. Since for this we require strength, and since France, the mortal enemy of our nation, inexorably strangles us and robs us of our strength, we must take upon ourselves every sacrifice whose consequences are calculated to contribute to the annihilation of French efforts toward hegemony in Europe. Today every power is our natural ally, which like us feels French domination on the continent to be intolerable. No path to such a power can be too hard for us, and no renunciation can seem unutterable if only the end result offers the possibility of downing our grimmest enemy.* Then, if we can cauterize and close the biggest wound, we can calmly leave the cure of our slighter wounds to the soothing effects of time.

Today, of course, we are subjected to the hateful yapping of the enemies of our people within. We National Socialists must never let this divert us from proclaiming what in our innermost conviction is absolutely necessary. Today, it is true, we must brace ourselves against the current of a public opinion confounded by Jewish guile exploiting German gullibility; sometimes, it is true, the waves break harshly and angrily about us, but he who swims with the stream is more easily overlooked than he who

[1] *'erfüllt von der Mahnenden Not,'* a Wagnerism.

bucks the waves. Today we are a reef; in a few years Fate may raise us up as a dam against which the general stream will break, and flow into a new bed.

It is, therefore, necessary that the National Socialist movement be recognized and established in the eyes of all as the champion of a definite political purpose. *Whatever Heaven may have in store for us, let men recognize us by our very visor!*

Once we ourselves recognize the crying need which must determine our conduct in foreign affairs, from this knowledge will flow the force of perseverance which we sometimes need when, beneath the drumfire of our hostile press hounds, one or another of us is seized with fear and there creeps upon him a faint desire to grant a concession at least in some field, and howl with the wolves, in order not to have everyone against him.

The Right of Emergency Defense

THE ARMISTICE of November, 1918, ushered in a policy which in all human probability was bound to lead gradually to total submission. Historical examples of a similar nature show that nations which lay down their arms without compelling reasons prefer in the ensuing period to accept the greatest humiliations and extortions rather than attempt to change their fate by a renewed appeal to force.

This is humanly understandable. A shrewd victor will, if possible, always present his demands to the vanquished in installments. And then, with a nation that has lost its character — and this is the case of every one which voluntarily submits — he can be sure that it will not regard one more of these individual oppressions as an adequate reason for taking up arms again. The more extortions are willingly accepted in this way, the more unjustified it strikes people finally to take up the defensive against a new, apparently isolated, though constantly recurring, oppression, especially when, all in all, so much more and greater misfortune has already been borne in patient silence.

The fall of Carthage is the most horrible picture of such a slow execution of a people through its own deserts.

That is why Clausewitz in his *Drei Bekenntnisse* [1] incomparably singles out this idea and nails it fast for all time, when he says:

[1] Karl von Clausewitz (1780–1831), the eminent military strategist. His chief work is *Vom Krieg*.

'That the stain of a cowardly submission can never be effaced; that this drop of poison in the blood of a people is passed on to posterity and will paralyze and undermine the strength of later generations'; that, on the other hand, 'even the loss of this freedom after a bloody and honorable struggle assures the rebirth of a people and is the seed of life from which some day a new tree will strike fast roots.'

Of course, a people that has lost all honor and character will not concern itself with such teachings. For no one who takes them to heart can sink so low; only he who forgets them, or no longer wants to know them, collapses. Therefore, we must not expect those who embody a spineless submission suddenly to look into their hearts and, on the basis of reason and all human experience, begin to act differently than before. On the contrary, it is these men in particular who will dismiss all such teachings until either the nation is definitely accustomed to its yoke of slavery or until better forces push to the surface, to wrest the power from the hands of the infamous spoilers. In the first case these people usually do not feel so badly, since not seldom they are appointed by the shrewd victors to the office of slave overseer, which these spineless natures usually wield more mercilessly over their people than any foreign beast put in by the enemy himself.

The development since 1918 shows us that in Germany the hope of winning the victor's favor by voluntary submission unfortunately determines the political opinions and the actions of the broad masses in the most catastrophic way. I attach special importance to emphasizing the *broad masses*, because I cannot bring myself to profess the belief that the commissions and omissions of our people's *leaders* are attributable to the same ruinous lunacy. As the leadership of our destinies has, since the end of the War, been quite openly furnished by Jews, we really cannot assume that faulty knowledge alone is the cause of our misfortune; we must, on the contrary, hold the conviction that conscious purpose is destroying our nation. And once we examine the apparent madness of our nation's leadership in the field of foreign affairs from this standpoint, it is revealed as the subtlest, ice-cold logic, in the service of the Jewish idea and struggle for world conquest.

And thus, it becomes understandable that the same time-span, which from 1806 to 1813 sufficed to imbue a totally collapsed Prussia with new vital energy and determination for struggle, today has not only elapsed unused, but, on the contrary, has led to an ever-greater weakening of our state.

Seven years after November, 1918, the Treaty of Locarno was signed.

The course of events was that indicated above: Once the disgraceful armistice had been signed, neither the energy nor the courage could be summoned suddenly to oppose resistance to our foes' repressive measures, which subsequently were repeated over and over. Our enemies were too shrewd to demand too much at once. They always limit their extortions to the amount which, in their opinion — and that of the German leadership — would at the moment be bearable enough so that an explosion of popular feeling need not be feared. But the more of these individual dictates had been signed, the less justified it seemed, because of a *single* additional extortion or exacted humiliation, to do the thing that had not been done because of so many others: to offer resistance. For this is the 'drop of poison' of which Clausewitz speaks: the spinelessness which once begun must increase more and more and which gradually becomes the foulest heritage, burdening every future decision. It can become a terrible lead weight, a weight which a nation is not likely to shake off, but which finally drags it down into the existence of a slave race.

Thus, in Germany edicts of disarmament alternated with edicts of enslavement, political emasculation with economic pillage, and finally created that moral spirit which can regard the Dawes Plan as a stroke of good fortune and the Treaty of Locarno as a success. Viewing all this from a higher vantage-point, we can speak of one single piece of good fortune in all this misery, which is that, though men can be befuddled, the heavens cannot be bribed. For their blessing remained absent: since then hardship and care have been the constant companions of our people, and our one faithful ally has been misery. Destiny made no exception in this case, but gave us what we deserved. Since we no longer know how to value honor, it teaches us at least to appreciate freedom in the matter of bread. By now people have learned to cry out for bread, but one of these days they will pray for freedom.

Bitter as was the collapse of our nation in the years after 1918, and obvious at that very time, every man who dared prophesy even then what later always materialized was violently and resolutely persecuted. Wretched and bad as the leaders of our nation were, they were equally arrogant, and especially when it came to ridding themselves of undesired, because unpleasant, prophets. We were treated to the spectacle (as we still are today!) of the greatest parliamentary thick-heads, regular saddlers and glove-makers — and not only by profession, which in itself means nothing — suddenly setting themselves on the pedestal of states-men, from which they could lecture down at plain ordinary mor-tals. It had and has nothing to do with the case that such a 'statesman' by the sixth month of his activity is shown up as the most incompetent windbag, the butt of everyone's ridicule and contempt, that he doesn't know which way to turn and has pro-vided unmistakable proof of his total incapacity! No, that makes no difference, on the contrary: the more lacking the parliamentary statesmen of this Republic are in real accomplishment, the more furiously they persecute those who expect accomplishments from them, who have the audacity to point out the failure of their pre-vious activity and predict the failure of their future moves. But if once you finally pin down one of these parliamentary honora-bles, and this political showman really cannot deny the collapse of his whole activity and its results any longer, they find thou-sands and thousands of grounds for excusing their lack of success, and there is only one that they will not admit, namely, that they themselves are the main cause of all evil.

*   *   *

By the winter of 1922–23, at the latest, it should have been generally understood that even after the conclusion of peace France was still endeavoring with iron logic to achieve the war aim she had originally had in mind. For no one will be likely to believe that France poured out the blood of her people — never too rich to begin with — for four and a half years in the most decisive struggle of her history, only to have the damage previously done made good by subsequent reparations. Even Alsace-Lorraine in itself would not explain the energy with which the French carried on the War, if it had not been a part of French foreign policy's really great political program for the future. And this goal is: the dissolution of Germany into a hodge-podge of little states. That is what chauvinistic France fought for, though at the same time in reality it sold its people as mercenaries to the international world Jew.

This French war aim would have been attainable by the War alone if, as Paris had first hoped, the struggle had taken place on German soil. Suppose that the bloody battles of the World War had been fought, not on the Somme, in Flanders, in Artois, before Warsaw, Nijni-Novgorod, Kovno, Riga, and all the other places, but in Germany, on the Ruhr and the Main, on the Elbe, at Hanover, Leipzig, Nuremberg, etc., and you will have to agree that this would have offered a possibility of breaking up Germany. It is very questionable whether our young federative state could for four and a half years have survived the same test of strain as rigidly centralized France, oriented solely toward her uncontested center in Paris. The fact that this gigantic struggle of nations occurred outside the borders of our fatherland was not only to the immortal credit of the old army, it was also the greatest good fortune for the German future. It is my firm and heartfelt conviction, and sometimes almost a source of anguish to me, that otherwise there would long since have been no German Reich, but only 'German states.' And this is the sole reason why the blood of our fallen friends and brothers has at least not flowed entirely in vain.

Thus everything turned out differently! True, Germany collapsed like a flash in November, 1918. But when the catastrophe occurred in the homeland, our field armies were still deep in enemy territory. The first concern of France at that time was not the dissolution of Germany, but: How shall we get the German armies out of France and Belgium as quickly as possible? And so the first task of the heads of state in Paris for concluding the World War was to disarm the German armies and if possible drive them back to Germany at once; and only after that could they devote themselves to the fulfillment of their real and original war aim. In this respect, to be sure, France was already paralyzed. For England the War had really been victoriously concluded with the annihilation of Germany as a colonial and commercial power and her reduction to the rank of a second-class state. Not only did the English possess no interest in the total extermination of the German state; they even had every reason to desire a rival against France in Europe for the future. Hence the French political leaders had to continue with determined peacetime labor what the War had begun, and Clemenceau's utterance, that for him the peace was only the continuation of the War, took on an increased significance.

Persistently, on every conceivable occasion, they had to shatter the structure of the Reich. By the imposition of one disarmament note after another, on the one hand, and by the economic extortion thus made possible, on the other hand, Paris hoped slowly to disjoint the Reich structure. The more rapidly national honor withered away in Germany, the sooner could economic pressure and unending poverty lead to destructive political effects. Such a policy of political repression and economic plunder, carried on for ten or twenty years, must gradually ruin even the best state structure and under certain circumstances dissolve it. And thereby the French war aim would finally be achieved.

By the winter of 1922–23 this must long since have been recognized as the French intent. Only two possibilities remained: We might hope gradually to blunt the French will against the tenacity of the German nation, or at long last to do what would have to be done in the end anyway, to pull the helm of the Reich ship about on some particularly crass occasion, and ram the enemy. This, to be sure, meant a life-and-death struggle, and there existed a prospect of life only if previously we succeeded in isolating France to such a degree that this second war would not again constitute a struggle of Germany against the world, but a defense of Germany against a France which was constantly disturbing the world and its peace.

I emphasize the fact, and I am firmly convinced of it, that this second eventuality must and will some day occur, whatever happens. I never believe that France's intentions toward us could ever change, for in the last analysis they are merely in line with the self-preservation of the French nation. If I were a Frenchman, and if the greatness of France were as dear to me as that of Germany is sacred, I could not and would not act any differently from Clemenceau. The French nation, slowly dying out, not only with regard to population, but particularly with regard to its best racial elements, can in the long run retain its position in the world only if Germany is shattered. French policy may pursue a thousand détours; somewhere in the end there will be this goal, the fulfillment of ultimate desires and deepest longing. And it is false to believe that a purely *passive* will, desiring only to preserve itself, can for any length of time resist a will that is no less powerful, but proceeds *actively. As long as the eternal conflict between Germany and France is carried on only in the form of a German defense against French aggression, it will never be decided, but from year to year, from century to century, Germany will lose one position after another.* Follow the movements of the German language frontier beginning with the twelfth century until today, and you will hardly be able to count on the success of an attitude and a development which has done us so much damage up till now.

Only when this is fully understood in Germany, so that the vital will of the German nation is no longer allowed to languish in purely passive defense, but is pulled together for a final active reckoning with France and thrown into a last decisive struggle with the greatest ultimate aims on the German side — only then will we be able to end the eternal and essentially so fruitless, struggle between ourselves and France; presupposing, of course, that Germany actually regards the destruction of France as only a means which will afterward enable her finally to give our people the expansion made possible elsewhere. Today we count eighty million Germans in Europe! This foreign policy will be acknowledged as correct only if, after scarcely a hundred years, there are two hundred and fifty million Germans on this continent, and not living penned in as factory coolies for the rest of the world, but: as peasants and workers, who guarantee each other's livelihood by their labor.

In December, 1922, the situation between Germany and France again seemed menacingly exacerbated. France was contemplating immense new extortions, and needed pledges for them. The economic pillage had to be preceded by a political pressure and it seemed to the French that only a violent blow at the nerve center of our entire German life would enable them to subject our 'recalcitrant' people to a sharper yoke. With the *occupation of the Ruhr*, the French hoped not only to break the moral backbone of Germany once and for all, but to put us into an embarrassing economic situation in which, whether we liked it or not, we would have to assume every obligation, even the heaviest.

It was a question of bending and breaking. Germany bent at the very outset, and ended up by breaking completely later.

With the occupation of the Ruhr, Fate once again held out a hand to help the German people rise again. For what at the first moment could not but seem a great misfortune embraced on closer inspection an infinitely promising opportunity to terminate all German misery.

From the standpoint of foreign relations, the occupation of the Ruhr for the first time really alienated England basically from France, and not only in the circles of British diplomacy which had concluded, examined, and maintained the French alliance as such only with the sober eye of cold calculators, but also in the broadest circles of the English people. The English economy in particular viewed with ill-concealed displeasure this new and incredible strengthening of French continental power. For not only that France, from the purely politico-military point of view, now assumed a position in Europe such as previously not even Germany had possessed, but, economically as well, she now obtained economic foundations which almost combined a position of economic monopoly with her capacity for political competition. The largest iron mines and coal fields in Europe were thus united in the hands of a nation which, in sharp contrast to Germany, had always defended its vital interests with equal determination and activism, and which in the Great War had freshly reminded the whole world of its military reliability. With the occupation of the Ruhr coal fields by France, England's entire gain through the War was wrested from her hands, and the victor was no longer British diplomacy so industrious and alert, but Marshal Foch and the France he represented.

In Italy, too, the mood against France, which, since the end of the War, had been by no means rosy to begin with, shifted to a veritable hatred. It was the great, historical moment in which the allies of former days could become the enemies of tomorrow. If things turned out differently and the allies did not, as in the second Balkan War, suddenly break into a sudden feud among themselves, this was attributable only to the circumstance that Germany simply had no Enver Pasha, but a Reich Chancellor Cuno.

Yet not only from the standpoint of foreign policy, but of domestic policy as well, the French assault on the Ruhr held great future potentialities for Germany. A considerable part of our people which, thanks to the incessant influence of our lying press, still regarded France as the champion of progress and liberalism, was abruptly cured of this lunatic delusion. Just as the year 1914 had dispelled the dreams of international solidarity between peoples from the heads of our German workers and led them suddenly back into the world of eternal struggle, throughout which one being feeds on another and the death of the weaker means the life of the stronger, the spring of 1923 did likewise.

When the Frenchman carried out his threats and finally, though at first cautiously and hesitantly, began to move into the lower German coal district, a great decisive hour of destiny had struck for Germany. If in this moment our people combined a change of heart with a shift in their previous attitude, the Ruhr could become a Napoleonic Moscow for France. There were only *two possibilities: Either we stood for this new offense and did nothing, or, directing the eyes of the German people to this land of glowing smelters and smoky furnaces, we inspired them with a glowing will to end this eternal disgrace and rather take upon themselves the terrors of the moment than bear an endless terror one moment longer.*

To have discovered a third way was the immortal distinction of Reich Chancellor Cuno, to have admired it and gone along, the still more glorious distinction of our German bourgeois parties.

Here I shall first examine the second course as briefly as possible.

With the occupation of the Ruhr, France had accomplished a conspicuous breach of the Versailles Treaty. In so doing, she had also put herself in conflict with a number of signatory powers, and especially with England and Italy. France could no longer hope for any support on the part of these states for her own selfish campaign of plunder: She herself, therefore, had to bring the adventure — and that is what it was at first — to some happy conclusion. For a national German government there could be but a single course, that which honor prescribed. It was certain that for the present France could not be opposed by active force of arms; but we had to realize clearly that any negotiations, unless backed by power, would be absurd and fruitless. Without the possibility of active resistance, it was absurd to adopt the standpoint: 'We shall enter into no negotiations'; but it was even more senseless to end by entering into negotiations after all, without having meanwhile equipped ourselves with power.

Not that we could have prevented the occupation of the Ruhr by *military measures*. Only a madman could have advised such a decision. But utilizing the impression made by this French action and while it was being carried out, what we absolutely should have done was, without regard for the Treaty of Versailles which France herself had torn up, to secure the military resources with which we could later have equipped our negotiators. For it was clear from the start that one day the question of this territory occupied by France would be settled at some conference table. But we had to be equally clear on the fact that even the best negotiators can achieve little success, as long as the ground on which they stand and the chair on which they sit is not the shield arm of their nation. A feeble little tailor cannot argue with athletes, and a defenseless negotiator has always suffered the sword of Brennus on the opposing side of the scale, unless he had his own to throw in as a counterweight. Or has it not been miserable to watch the comic-opera negotiations which since 1918 have always preceded the repeated dictates? This degrading spectacle presented to the whole world, first inviting us to the conference table, as though in mockery, then presenting us with decisions and programs prepared long before, which, to be sure, could be discussed, but which from the start could only be regarded as unalterable. It is true that our negotiators, in hardly a single case, rose above the most humble average, and for the most part justified only too well the insolent utterance of Lloyd George, who contemptuously remarked, à propos of former Reich Minister Simon, 'that the Germans didn't know how to choose men of intelligence as their leaders and representatives.' But even geniuses, in view of the enemy's determined will to power and the miserable defenselessness of our own people in every respect, would have achieved but little.

But anyone who in the spring of 1923 wanted to make France's occupation of the Ruhr an occasion for reviving our military implements of power had first to give the nation its spiritual weapons, strengthen its will power, and destroy the corrupters of this most precious national strength.

Just as in 1918 we paid with our blood for the fact that in 1914 and 1915 we did not proceed to trample the head of the Marxist serpent once and for all, we would have to pay most catastrophically if in the spring of 1923 we did not avail ourselves of the opportunity to halt the activity of the Marxist traitors and murderers of the nation for good.

Any idea of real resistance to France was utter nonsense if we did not declare war against those forces which five years before had broken German resistance on the battlefields from within. Only bourgeois minds can arrive at the incredible opinion that Marxism might now have changed, and that the scoundrelly leaders of 1918, who then coldly trampled two million dead underfoot, the better to climb into the various seats of government, now in 1923 were suddenly ready to render their tribute to the national conscience. An incredible and really insane idea, the hope that the traitors of former days would suddenly turn into fighters for a German freedom. It never entered their heads. *No more than a hyena abandons carrion does a Marxist abandon treason.* And don't annoy me, if you please, with the stupidest of all arguments, that, after all, so many workers bled for Germany. German workers, yes, but then they were no longer international Marxists. If in 1914 the German working class in their innermost convictions had still consisted of Marxists, the War would have been over in three weeks. Germany would have collapsed even before the first soldier set foot across the border. No, the fact that the German people was then still fighting proved that the Marxist delusion had not yet been able to gnaw its way into the bottommost depths. But in exact proportion as, in the course of the War, the German worker and the German soldier fell back into the hands of the Marxist leaders, in exactly that proportion he was lost to the fatherland. If at the beginning of the War and during the War twelve or fifteen thousand of these Hebrew corrupters of the people had been held under poison gas, as happened to hundreds of thousands of our very best German workers in the field, the sacrifice of millions at the front would not have been in vain. On the contrary: twelve thousand scoundrels eliminated in time might have saved the lives of a million real Germans, valuable for the future. But it just happened to be in the line of bourgeois 'statesmanship' to subject millions to a bloody end on the battlefield without batting an eyelash, but to regard ten or twelve thousand traitors, profiteers, usurers, and swindlers as a sacred national treasure and openly proclaim their inviolability. We never know which is greater in this bourgeois world, the imbecility, weakness, and cowardice, or their deep-dyed corruption. It is truly a class doomed by Fate, but unfortunately, however, it is dragging a whole nation with it into the abyss.

And in 1923 we faced exactly the same situation as in 1918. Regardless what type of resistance was decided on, the first requirement was always the elimination of the Marxist poison from our national body. And in my opinion, it was then the very first task of a truly national government to seek and find the forces which were resolved to declare a war of annihilation on Marxism, and then to give these forces a free road; it was their duty not to worship the idiocy of 'law and order' at a moment when the enemy without was administering the most annihilating blow to the fatherland and at home treason lurked on every street corner. No, at that time a really national government should have desired disorder and unrest, provided only that amid the confusion a basic reckoning with Marxism at last became possible and actually took place. If this were not done, any thought of resistance, regardless of what type, was pure madness.

Such a reckoning of real world-historical import, it must be admitted, does not follow the schedules of a privy councilor or some dried-up old minister, but the eternal laws of life on this earth, which are the struggle for this life and which remain struggle. It should have been borne in mind that the bloodiest civil wars have often given rise to a steeled and healthy people, while artificially cultivated states of peace have more than once produced a rottenness that stank to high Heaven. You do not alter the destinies of nations in kid gloves. And so, in the year 1923, the most brutal thrust was required to seize the vipers that were devouring our people. Only if this were successful did the preparation of active resistance have meaning.

At that time I often talked my throat hoarse, attempting to make it clear, at least to the so-called national circles, what was now at stake, and that, if we made the same blunders as in 1914 and the years that followed, the end would inevitably be the same as in 1918. Again and again, I begged them to give free rein to Fate, and to give our movement an opportunity for a reckoning with Marxism; but I preached to deaf ears. They all knew better, including the chief of the armed forces, until at length they faced the most wretched capitulation of all time.

Then I realized in my innermost soul that the German bourgeoisie was at the end of its mission and is destined for no further mission. Then I saw how all these parties continued to bicker with the Marxists only out of competitors' envy, without any serious desire to annihilate them; at heart they had all of them long since reconciled themselves to the destruction of the fatherland, and what moved them was only grave concern that they themselves should be able to partake in the funeral feast. That is all they were still 'fighting' for.

In this period — I openly admit — I conceived the profoundest admiration for the great man south of the Alps, who, full of ardent love for his people, made no pacts with the enemies of Italy, but strove for their annihilation by all ways and means. What will rank Mussolini among the great men of this earth is his determination not to share Italy with the Marxists, but to destroy internationalism and save the fatherland from it.

How miserable and dwarfish our German would-be statesmen seem by comparison, and how one gags with disgust when these nonentities, with boorish arrogance, dare to criticize this man who is a thousand times greater than they; and how painful it is to think that this is happening in a land which barely half a century ago could call a Bismarck its leader.

In view of this attitude on the part of the bourgeoisie and the policy of leaving the Marxists untouched, the fate of any active resistance in 1923 was decided in advance. To fight France with the deadly enemy in our own ranks would have been sheer idiocy. What was done after that could at most be shadow-boxing, staged to satisfy the nationalistic element in Germany in some measure, or in reality to dupe the 'seething soul of the people.' If they had seriously believed in what they were doing, they would have had to recognize that the strength of a nation lies primarily, not in its weapons, but in its will, and that, before foreign enemies are conquered, the enemy within must be annihilated; otherwise God help us if victory does not reward our arms on the very first day. Once so much as the shadow of a defeat grazes a people that is not free of internal enemies, its force of resistance will break and the foe will be the final victor.

This could be predicted as early as February, 1923. Let no one mention the questionableness of a military success against France! For if the result of the German action in the face of the invasion of the Ruhr had only been the destruction of Marxism at home, by that fact alone success would have been on our side. A Germany saved from these mortal enemies of her existence and her future would possess forces which the whole world could no longer have stifled. *On the day when Marxism is smashed in Germany, her fetters will in truth be broken forever.* For never in our history have we been defeated by the strength of our foes, but always by our own vices and by the enemies in our own camp.

Since the leaders of the German state could not summon up the courage for such a heroic deed, logically they could only have chosen the first course, that of doing nothing at all and letting things slide.

But in the great hour Heaven sent the German people a great man, Herr von Cuno. He was not really a statesman or a politician by profession, and of course still less by birth; he was a kind of political hack, who was needed only for the performance of certain definite jobs; otherwise he was really more adept at business. A curse for Germany, because this businessman in politics regarded politics as an economic enterprise and acted accordingly.

'France has occupied the Ruhr; what is in the Ruhr? Coal. Therefore, France has occupied the Ruhr on account of the coal.' What was more natural for Herr Cuno than the idea of striking in order that the French should get no coal, whereupon, in the opinion of Herr Cuno, they would one day evacuate the Ruhr when the enterprise proved unprofitable. Such, more or less, was this 'eminent' 'national' 'statesman,' who in Stuttgart and elsewhere was allowed to address *his people,* and whom the people gaped at in blissful admiration.

But for a strike, of course, the Marxists were needed, for it was primarily the *workers* who would have to strike. Therefore, it was necessary to bring the worker (and in the brain of one of these bourgeois statesman he is always synonymous with the Marxist) into a united front with all the other Germans. The way these moldy political party cheeses glowed at the sound of such a brilliant slogan was something to behold! Not only a product of genius, it was national at the same time — there at last they had what at heart they had been seeking the whole while. The bridge to Marxism had been found, and the national swindler was enabled to put on a Teutonic face and mouth German phrases while holding out a friendly hand to the international traitor. And the traitor seized it with the utmost alacrity. For just as Cuno needed the Marxist leaders for his *'united front,'* the Marxist leaders were just as urgently in need of Cuno's money. So it was a help to both parties. Cuno obtained his united front, formed of national windbags and anti-national scoundrels, and the international swindlers received state funds to carry out the supreme mission of their struggle — that is, to destroy the national economy, and this time actually at the expense of the state. An immortal idea, to save the nation by buying a general strike; in any case a slogan in which even the most indifferent good-for-nothing could join with full enthusiasm.

It is generally known that a nation cannot be made free by prayers. But maybe one could be made free by sitting with folded arms, and that had to be historically tested. If at that time Herr Cuno, instead of proclaiming his subsidized general strike and setting it up as the foundation of the 'united front,' had only demanded two more hours of work from every German, the 'united front' swindle would have shown itself up on the third day. Peoples are not freed by doing nothing, but by sacrifices.

To be sure, this so-called passive resistance as such could not be maintained for long. For only a man totally ignorant of warfare could imagine that occupying armies can be frightened away by such ridiculous means. And that alone could have been the sense of an action the costs of which ran into billions and which materially helped to shatter the national currency to its very foundations.

Of course, the French could make themselves at home in the Ruhr with a certain sense of inner relief as soon as they saw the resisters employing such methods. They had in fact obtained from us the best directions for bringing a recalcitrant civilian population to reason when its conduct represents a serious menace to the occupation authorities. With what lightning speed, after all, we had routed the Belgian *franc-tireur* bands nine years previous and made the seriousness of the situation clear to the civilian population when the German armies ran the risk of incurring serious damage from their activity. As soon as the passive resistance in the Ruhr had grown really dangerous to the French, it would have been child's play for the troops of occupation to put a cruel end to the whole childish mischief in less than a week. For the ultimate question is always this: What do we do if the passive resistance ends by really getting on an adversary's nerves and he takes up the struggle against it with brutal strong-arm methods? Are we then resolved to offer further resistance? If so, we must for better or worse invite the gravest, bloodiest persecutions. But then we stand exactly where active resistance would put us — face to face with struggle. Hence any so-called passive resistance has an inner meaning only if it is backed by determination to continue it if necessary in open struggle or in undercover guerrilla warfare. In general, any such struggle will depend on a conviction that success is possible. As soon as a besieged fortress under heavy attack by the enemy is forced to abandon the last hope of relief, for all practical purposes it gives up the fight, especially when in such a case the defender is lured by the certainty of life rather than probable death. Rob the garrison of a surrounded fortress of faith in a possible liberation, and all the forces of defense will abruptly collapse.

Therefore, a passive resistance in the Ruhr, in view of the ultimate consequences it could and inevitably would produce in case it were actually successful, only had meaning if an active front were built up behind it. Then, it is true, there is no limit to what could have been drawn from our people. If every one of these Westphalians had known that the homeland was setting up an army of eighty or a hundred divisions, the Frenchmen would have found it thorny going. There are always more courageous men willing to sacrifice themselves for success than for something that is obviously futile.

It was a classical case which forced us National Socialists to take the sharpest position against a so-called national slogan. And so we did. In these months I was attacked no little by men whose whole national attitude was nothing but a mixture of stupidity and outward sham, all of whom joined in the shouting only because they were unable to resist the agreeable thrill of suddenly being able to put on national airs without any danger. I regarded this most lamentable of all united fronts as a most ridiculous phenomenon, and history has proved me right.

As soon as the unions had filled their treasuries with Cuno's funds, and the passive resistance was faced with the decision of passing from defense with folded arms to active attack, the Red hyenas immediately bolted from the national sheep herd and became again what they had always been. Quietly and ingloriously Herr Cuno retreated to his ships, and Germany was richer by one experience and poorer by one great hope.

Down to late midsummer many officers, and they were assuredly not the worst, had at heart not believed in such a disgraceful development. They had all hoped that, if not openly, in secret at least, preparations had been undertaken to make this insolent French assault a turning point in German history. Even in our ranks there were many who put their confidence at least in the Reichswehr. And this conviction was so alive that it decisively determined the actions and particularly the training of innumerable young people.

But when the disgraceful collapse occurred and the crushing, disgraceful capitulation followed, the sacrifice of billions of marks and thousands of young Germans — who had been stupid enough to take the promises of the Reich's leaders seriously — indignation flared into a blaze against such a betrayal of our unfortunate people. In millions of minds the conviction suddenly arose bright and clear that only a radical elimination of the whole ruling system could save Germany.

Never was the time riper, never did it cry out more imperiously for such a solution than in the moment when, on the one hand, naked treason shamelessly revealed itself, while, on the other hand, a people was economically delivered to slow starvation. Since the state itself trampled all laws of loyalty and faith underfoot, mocked the rights of its citizens, cheated millions of its truest sons of their sacrifices and robbed millions of others of their last penny, it had no further right to expect anything but hatred of its subjects. And in any event, this hatred against the spoilers of people and fatherland was pressing toward an explosion. In this place I can only point to the final sentence of my last speech in the great trial of spring, 1924:

'The judges of this state may go right ahead and convict us for our actions at that time, but History, acting as the goddess of a higher truth and a higher justice, will one day smilingly tear up this verdict, acquitting us of all guilt and blame.'

And then she will call all those before her judgment seat, who today, in possession of power, trample justice and law underfoot, who have led our people into misery and ruin and amid the misfortune of the fatherland have valued their own ego above the life of the community.

In this place I shall not continue with an account of those events which led to and brought about the 8th of November, 1923. I shall not do so because in so doing I see no promise for the future, and because above all it is useless to reopen wounds that seem scarcely healed; moreover, because it is useless to speak of guilt regarding men who in the bottom of their hearts, perhaps, were all devoted to their nation with equal love, and who only missed or failed to understand the common road.

In view of the great common misfortune of our fatherland, I

today no longer wish to wound and thus perhaps alienate those who one day in the future will have to form the great united front of those who are really true Germans at heart against the common front of the enemies of our people. For I know that some day the time will come when even those who then faced us with hostility, will think with veneration of those who traveled the bitter road of death for their German people.

I wish at the end of the second volume to remind the supporters and champions of our doctrine of those eighteen [1] heroes, to whom I have dedicated the first volume of my work, those heroes who sacrificed themselves for us all with the clearest consciousness. They must forever recall the wavering and the weak to the fulfillment of his duty, a duty which they themselves in the best faith carried to its final consequence. And among them I want also to count that man, one of the best, who devoted his life to the awakening of his, our people, in his writings and his thoughts and finally in his deeds:

Dietrich Eckart [2]

[1] Second edition reduces the number of heroes to sixteen. These are the men who fell in the Munich Beer Hall *Putsch* on November 11, 1923.

[2] Dietrich Eckart was the spiritual founder of the National Socialist Party. He was the type of homespun political philosopher often found in the Schwabing quarter of Munich. He was an habitué of the Brennessel Cabaret, where Heiden quotes him as saying in 1919: 'We need a fellow at the head who can stand the sound of a machine gun. The rabble need to get fear into their pants. We can't use an officer, because the people don't respect them any more. The best would be a worker who knows how to talk.... He doesn't need much brains, politics is the stupidest business in the world, and every marketwoman in Munich knows more than the people in Weimar. I'd rather have a vain monkey who can give the Reds a juicy answer, and doesn't run away when people begin swinging table legs, than a dozen learned professors. He must be a bachelor, then we'll get the women.'
Eckart died from overdrinking in 1923.

Conclusion

ON NOVEMBER 9, 1923, in the fourth year of its existence, the National Socialist German Workers' Party was dissolved and prohibited in the whole Reich territory. Today in November, 1926, it stands again free before us, stronger and inwardly firmer than ever before.

All the persecutions of the movement and its individual leaders, all vilifications and slanders, were powerless to harm it. The correctness of its ideas, the purity of its will, its supporters' spirit of self-sacrifice, have caused it to issue from all repressions stronger than ever.

If, in the world of our present parliamentary corruption, it becomes more and more aware of the profoundest essence of its struggle, feels itself to be the purest embodiment of the value of race and personality and conducts itself accordingly, it will with almost mathematical certainty some day emerge victorious from its struggle. Just as Germany must inevitably win her rightful position on this earth if she is led and organized according to the same principles.

A state which in this age of racial poisoning dedicates itself to the care of its best racial elements must some day become lord of the earth.

May the adherents of our movement never forget this if ever the magnitude of the sacrifices should beguile them to an anxious comparison with the possible results.

Index